# Innovative Processing and Synthesis of Ceramics, Glasses, and Composites II

Volume 94

# Innovative Processing and Synthesis of Ceramics, Glasses, and Composites II

*Edited by*
**Narottam P. Bansal**
National Aeronautics and Space Administration
Lewis Research Center

**J.P. Singh**
Argonne National Laboratory

*Published by*
The American Ceramic Society
735 Ceramic Place
Westerville, Ohio 43081

*Proceedings of the International Symposium on Innovative Processing and Synthesis of Ceramics, Glasses, and Composites, held at the 100th Annual Meeting of The American Ceramic Society in Cincinnati, Ohio, May 3–6, 1998.*

Copyright 1999, The American Ceramic Society. All rights reserved.

No part of this book may be reproduced, stored in a retrieval system, or transmitted in any form or by any means, electronic, mechanical, photocopying, microfilming, recording, or otherwise, without written permission from the publisher.

Permission to photocopy for personal or internal use beyond the limits of Sections 107 and 108 of the U.S. Copyright Law is granted by the American Ceramic Society, provided that the base fee of US$5.00 per copy, plus US$.50 per page, is paid directly to the Copyright Clearance Center, 222 Rosewood Dr., Danvers, MA 01923, USA. The fee code for users of the Transactional Reporting Service for *Ceramic Transactions Volume 94* is 1-57498-060-2/98 $5.00+$.50. This consent does not extend to other kinds of copying, such as copying for general distribution, for advertising or promotional purposes, or for creating new collective works. Requests for special photocopying permission and reprint requests should be directed to the Director of Publications, The American Ceramic Society, 735 Ceramic Place, Westerville OH 43081, USA.

COVER PHOTO: "Scanning electron micrographs of ZnO particles prepared by the hydrothermal process at $[Zn^{2+}]=0.1$ M and $[MEA]=0.5$ M," is courtesy of Chung-Hsin Lu, Yuan Cheng Lai, and Chi-Hsien Yeh, and appears as figure 2(a) in their paper "Hydrothermal Synthesis of Zinc-Oxide Particles," which begins on page 387.

Library of Congress Cataloging-in-Publication Data
A CIP record for this book is available from the Library of Congress.

For information on ordering titles published by The American Ceramic Society, or to request a publications catalog, please call 614-794-5890.

Printed in the United States of America.

1 2 3 4–01 00 99 98

ISSN 1042-1122
ISBN 1-57498-060-2

# Contents

Preface ................................................................. xi

## Combustion Synthesis

Synthesis of AlN-SiC Solid Solution by Combustion Nitridation ....... 2
M. Ohyanagi, N. Balandina, K. Shirai, M. Koizumi, and Z.A. Munir

Chemically Assisted Combustion Synthesis of Silicon Carbide from
Elemental Powders ...................................................... 13
J.A. Puszynski and S. Miao

Combustion Synthesis of Cd-In-Ga-O Powder for Novel Thin Film
Growth Substrates ...................................................... 23
S.-C. Lin, R. Wilkins, M. Nersesyan, and D. Luss

## Microwave Processing

Step Sintering of Microwave Heating and Microwave Plasma
Heating for Ceramics ................................................... 37
J. Zhang, L. Cao, Y. Yang, Y. Diao, and X. Shen

Microwave-Assisted Sintering of $ZrO_2$-8 mol% $Y_2O_3$: Enhancement
of Densification ....................................................... 45
S.M. Rathbone, R. Freer, F.R. Sale, and F.C.R. Wroe

Pressureless Microwave Sintering of Functional Gradient Materials .... 57
R. Borchert and M. Willert-Porada

The Effect of Casket Geometry on Microwave Heating and Its
Modeling ............................................................... 69
M.L. Traub, K.-Y. Lee, and E.D. Case

## Spark Plasma Processing

Seeding With $\alpha$-Alumina for Transformation and Densification of
Boehmite-Derived $\gamma$- and $\delta$-Alumina by Spark Plasma Sintering ....... 83
S.D. De la Torre, A. Kakitsuji, H. Miyamoto, K. Miyamoto, H. Balmori-R,
J. Zárate-M, and M.E. Contreras

Spark Plasma Reaction-Sintering of Mullite-ZrO$_2$ Composites ........ 91
  E. Rocha-Rangel, S.D. De la Torre, H. Miyamoto, M. Umemoto, K. Tsuchiya,
  J.G. Cabañas-Moreno, and H. Balmori-Ramírez

Mechanical Properties of ZrO$_2$(Y$_2$O$_3$)/20 mol% Al$_2$O$_3$ Composites
Prepared by Spark Plasma Sintering ............................... 97
  J.S. Hong, S.D. De la Torre, K. Miyamoto, H. Miyamoto, and L. Gao

## Reaction Forming/Bonding

A Range of SiAlONs by Carbothermal Reduction and Nitridation .... 107
  G.V. White, T.C. Ekström, G.C. Barris, and I.W.M. Brown

Carbothermal Synthesis of α-Si$_3$N$_4$ Powders Using Carbon Coated
Silica Precursors ................................................ 119
  S. Kaza and R. Koc

Near Net-Shaped Magnesium Aluminate Spinel by the Oxidation
of Solid Magnesium-Bearing Precursors ........................... 129
  P. Kumar and K.H. Sandhage

The Displacive Compensation of Porosity (DCP) Method for
Fabricating Dense Oxide/Metal Composites at Modest
Temperatures with Small Dimensional Changes.................... 141
  K.A. Rogers, P. Kumar, R. Citak, and K.H. Sandhage

Innovative Processing for Nonoxides by Adding Small Amounts of
Elements ........................................................ 153
  K. Komeya and T. Meguro

## Sol-Gel Synthesis

Application of Sol-Gel Concepts to Synthesis of Nonoxide
Ceramics ........................................................ 163
  P.N. Kumta, J.Y. Kim, and M.A. Sriram

Composite Alumina Sol-Gel Ceramics............................. 185
  T. Troczynski and Q. Yang

Aqueous Sol-Gel Method for the Synthesis of Nanosized Ceramic
Powders......................................................... 195
  P. Pramanik and N.N. Ghosh

Innovative Processes for Ceramic Synthesis Using Liquid Metal
Carboxylates .................................................... 205
  E.H. Walker, Jr. and A.W. Apblett

Lanthanumhexaluminate as Interphase Material in Oxide-Fiber-
Reinforced Oxide Matrix Composites ............................. 215
  B. Saruhan, L. Mayer, and H. Schneider

Introduction of Sintering Aid to Silicon Nitride Systems via
Colloidal Processing . . . . . . . . . . . . . . . . . . . . . . . . . . . . . . . . . . . . . . . . . 227
   A.C. Orlando and R.A. McCauley

## Polymer Processing

Synthesis of SiC-Si$_3$N$_4$ Composites from a Polymeric Precursor . . . . . 241
   X. Bao, M.J. Edirisinghe, G.F. Fernando, and M.J. Folkes

Effect of Polymer Architecture on the Formation of Si-O-C Glasses. . . 251
   S. Dire, M. Oliver, and G.D. Sorarù

Polymer Derived Ceramic Hard Materials by Addition of Tungsten . . . 263
   K.-T. Kang, D.-J. Kim, A. Kaindl, and P. Greil

## Shock Synthesis

Shock Synthesis of MoSi$_2$-SiC$_p$/SiC$_w$ Composites from Mechanical
Alloying Pretreated Precursors. . . . . . . . . . . . . . . . . . . . . . . . . . . . . . . . . 273
   T. Aizawa and N.N. Thadhani

## Mechanical Alloying

Mechanochemical Synthesis of Refined Ag- and Zn-Composite
Powders Starting from Oxides . . . . . . . . . . . . . . . . . . . . . . . . . . . . . . . . 287
   S.D. De la Torre, K.N. Ishihara, P.H. Shingu, D. Rios-Jara, and H. Miyamoto

Ceramic Oxide (MeO$_2$) Solid Solutions Obtained by Mechanical
Alloying . . . . . . . . . . . . . . . . . . . . . . . . . . . . . . . . . . . . . . . . . . . . . . . . . . 295
   F. Bondioli, M. Romagnoli, L. Barbieri, and T. Manfredini

## Films/Coatings

Formation of Self-Assembled Monolayers and Ceramic Films on
Semiconductor and Oxide Substrates . . . . . . . . . . . . . . . . . . . . . . . . . . 307
   U. Sampathkumaran, M.R. De Guire, A.H. Heuer, T. Niesen, J. Bill, and F. Aldinger

Synthesis of β-SiC Conversion Coating Layer by Chemical Vapor
Reaction Process. . . . . . . . . . . . . . . . . . . . . . . . . . . . . . . . . . . . . . . . . . . 319
   Y.-H. Yun and S.-C. Choi

Adhesion of Copper Nitride Film to Silicon Oxide Substrate. . . . . . . . 329
   K.H. Kim, S.O. Chwa, H.C. Park, and D.W. Shin

Synthesis and Characterization of SiO$_2$-Based Porous Glass
Membranes for CO$_2$ Separation . . . . . . . . . . . . . . . . . . . . . . . . . . . . . . 337
   J.-J. Park, C. Kawai, S. Nakahata, M. Yamagiwa, T. Nishioka, H. Takeuchi,
   and A. Yamakawa

Mechanical Properties of Plasma-Sprayed $Mo_5Si_3$-MoB-$MoSi_2$ System ... 347
S.C. Okumus, O. Unal, M.J. Kramer, and M. Akinc

Effect of Si-O-C Coatings on the Strength of Saphicon™ and Nicalon™ Fibers ... 361
J.R. Hellmann, M.P. Petervary, J.M. Priest, and C.G. Pantano

## Electronic Ceramics

Synthesis, Microstructure and Magnetic Properties of Sintered Mn-Zn Ferrites from Hydrothermal Powders ... 373
R. Lucke, E. Schlegel, and R. Strienitz

Hydrothermal Synthesis of Zinc Oxide Particles ... 387
C.-H. Lu, Y.C. Lai, and C.-H. Yeh

Synthesis of Potassium Niobate Ceramic Powder under Hydrothermal Conditions ... 397
C.-H. Lu and S.-Y. Lo

Processing of Pure and Mn, Ni, and Zn Doped Ferrite Particles in Microemulsions ... 407
D.O. Yener and H. Giesche

Grain Growth Inhibition of Hard Ferrites through Sol-Gel Particulate Coating ... 419
J.W. Lee, Y.S. Cho, and V.R.W. Amarakoon

Preparation and Characterization of Lithium-Titanium Oxide by Alcohol Burning out Process ... 427
C.-F. Kao and C.-L. Jeng

## Nanotechnology

Strong Machinable Nanocomposite Ceramics ... 443
T. Kusunose, Y.H. Choa, T. Sekino, and K. Niihara

New Mechanism and Kinetics of Nanoamorphous Metals Synthesis Process ... 455
R.T. Malkhasyan and S.L. Grigoryan

## Fibers/Whiskers

Polycrystalline Ceramic Fibers by Way of Conventional Powder Processing ... 465
M. Wegmann, B. Gut, and K. Berroth

Synthesis of $Ta_{0.5}Ti_{0.5}C$ and $Ta_{0.33}Ti_{0.33}Nb_{0.33}C$ Whiskers ... 473
M. Johnsson, M. Carlsson, N. Ahlén, and M. Nygren

## Crystal Growth

A New Approach to Growth of Bulk ZnO Crystals by Melt Solidification. . . . . . . . . . . . . . . . . . . . . . . . . . . . . . . . . . . . . . . . . 483
   J.E. Nause, G. Agarwal, and D.N. Hill

## Porous Metals

Preparation of Porous Metals from Ceramic Precursors. . . . . . . . . . . . 495
   S.M. Landin and D.W. Readey

## Joining

Joining of Diamond Thin Film to Optical and IR Materials. . . . . . . . . . 509
   J.G. Lee, K.Y. Lee, and E.D. Case

## Laminated Object Manufacturing

Laminated Object Manufacturing Using Ceramic Paper Products. . . . . 523
   B.J. Kellett and W. Guo

## Kinetics and Mechanism

Neutron Diffraction Studies of the Partial Reduction of $NiAl_2O_4$: Phase and Strain Evolution . . . . . . . . . . . . . . . . . . . . . . . . . . . . . . . . . 537
   E. Üstündag, R.H. Woodman, J.C. Hanan, B. Clausen, T. Hartmann, and M.A.M. Bourke

Surface Area and Oxidation Effects on Nitridation Kinetics of Silicon Powder Compacts. . . . . . . . . . . . . . . . . . . . . . . . . . . . . . . . . . 549
   R.T. Bhatt and A.R. Palczer

Supercritical Debinding of Ceramics. . . . . . . . . . . . . . . . . . . . . . . . . . . 561
   T. Chartier, E. Delhomme, J.-F. Baumard, P. Marteau, and R. Tufeu

Index. . . . . . . . . . . . . . . . . . . . . . . . . . . . . . . . . . . . . . . . . . . . . . . . . . . 571

# Preface

An international symposium, "Innovative Processing and Synthesis of Ceramics, Glasses, and Composites" was held during the centennial meeting of The American Ceramic Society in Cincinnati, Ohio, May 3–6, 1998. The objective of this symposium was to provide an international forum for scientists, engineers, and technologists to exchange ideas, information, and technology on advanced methods and approaches for processing and synthesis of ceramics, glasses, and composites. A total of 134 papers, including six invited talks, were presented in the form of oral and poster presentations indicating high interest in the scientifically and technologically important field of ceramic processing. Authors from 21 countries (Armenia, Austria, Brazil, Canada, France, Germany, India, Israel, Italy, Japan, Korea, Mexico, New Zealand, People's Republic of China, Russia, Spain, Sweden, Switzerland, Taiwan, United Kingdom, and the United States) participated. The speakers represented universities, industry, and government research laboratories.

These proceedings contain 50 contributions on various aspects of synthesis and processing of ceramics, glasses, and composites that were discussed at the symposium. Latest developments in combustion and shock synthesis, reaction forming/bonding, mechanical alloying, sol-gel, polymer, microwave, and spark plasma processing are described. Papers describing preparation of oxide and nonoxide ceramics in the form of powders, thin films, coatings, fibers, whiskers, porous metals, crystals, laminates, and composites, etc., are included in this volume. Some aspects of joining and laminated object manufacturing are also described. Each manuscript was peer reviewed using The American Ceramic Society review process.

The editors wish to extend their gratitude and appreciation to all the authors for their cooperation and contributions, to the session chairs for their time and efforts in keeping the sessions on schedule, and to all the reviewers for their useful comments and suggestions. Financial support of The American Ceramic Society is gratefully acknowledged. Thanks are due to the staff of the meetings and publications departments of The American Ceramic Society for their invaluable assistance. Finally, we are grateful to Sarah Godby of The American Ceramic Society for her efforts in coordinating the on-site review of the manuscripts.

We hope that this volume will serve as a valuable reference for the researchers as well as the technologists interested in innovative approaches for synthesis and processing of ceramics, glasses, and composites.

Narottam P. Bansal
J.P. Singh

# Combustion Synthesis

# SYNTHESIS OF ALN-SIC SOLID SOLUTION BY COMBUSTION NITRIDATION

Manshi Ohyanagi, Nadejda Balandina, Kenshiro Shirai, Mitsue Koizumi
High-tech Research Center, Ryukoku University
Seta, Ohtsu 520-2194, JAPAN

Zuhair A. Munir
Department of Chemical Engineering and Materials Science
University of California, Davis, CA 95616 USA

## ABSTRACT

Solid solutions of AlN-SiC were synthesized by combustion nitridation process. The reactants consisted of Al, 3C-SiC and gaseous nitrogen at pressures in the range of 0.1-8.0 MPa. The reactant powders were placed in a porous graphite crucible and ignited by an ignition pellet of Ti and C placed on top of the powders. The reaction was found to be dependent on the reactant packing density. The maximum combustion temperature decreased from about 2500 to 2180 °C as the green density increased from 0.75 to 1.16 g/cm$^3$ under a nitrogen pressure of 4.0 MPa. The wave propagation was extinguished at a density higher than 1.40g/cm$^3$. To investigate the formation mechanism of the solid solution, the layer close to the wave front was characterized by X-ray diffraction and SEM. In this study, use of 3C-SiC in whisker form was also investigated.

## INTRODUCTION

Over the past decade SiC-AlN ceramics have attracted considerable interest as promising materials, primarily because of the improvement of the fracture toughness of silicon carbide resulting from the addition of aluminum nitride. Figure 1 shows the tentative phase diagram of AlN-SiC system [1]. The presence of aluminum and nitrogen stabilizes the 2H-structure of SiC which has nearly the same parameters as the 2H structure of AlN. An extensive solid solubility exists between these phases in the compositional range of 25 to 100 mol% [2-6]. The solid solutions undergo spinodal decomposition below about 1900°C, leading to the formation of modulated structures [7] with improved fracture toughness [8].

To the extent authorized under the laws of the United States of America, all copyright interests in this publication are the property of The American Ceramic Society. Any duplication, reproduction, or republication of this publication or any part thereof, without the express written consent of The American Ceramic Society or fee paid to the Copyright Clearance Center, is prohibited.

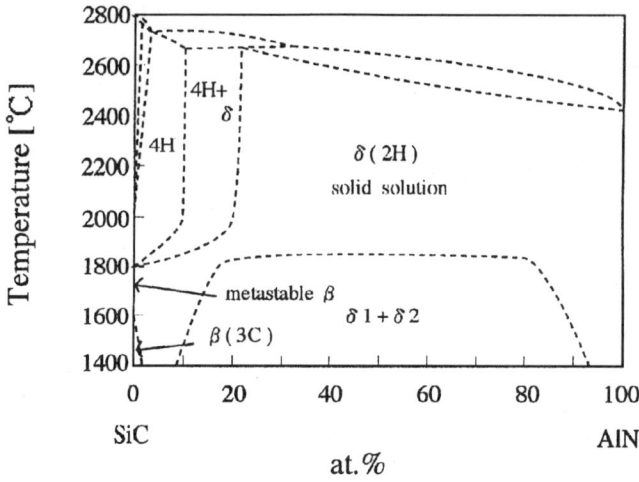

Figure 1. Phase diagram of AlN-SiC system.

The synthesis of the solid solutions has been attemped by different techniques. In the first report by Ervin [2] AlN-SiC solid solutions were obtained by the vapor deposition of aluminum nitride onto silicon carbide porous bodies. Subsequently, the solid solutions were prepared by hot-pressing of mixtures of AlN-SiC [3, 4] or $Al_4C_3$-$Si_3N_4$ [5]. Detais of these processes are described by Zangvil and Ruh [6].

The typical hot-pressing parameters [3-6] include the annealing at temperatures in the range 2100-2300°C for times of the order of hours under an atmosphere of nitrogen or argon at a pressure of several tens MPa. The formation of the solid solution is governed by interdiffusion between AlN and SiC, a process recognized to be very slow [9].

A new effective approach to produce AlN-SiC solid solution by combustion synthesis was recently employed [10, 11]. In this new approach, the formation of solid solutions is complete within seconds, making it an attractive energy-saving process. There are several possible routes to synthesize AlN-SiC solid solution through a combustion reaction. The one used in the cited work was based on the following reaction:

$$Si_3N_4 + 4Al + 3C = 3SiC\text{-}4AlN(ss), \qquad (1)$$

where (*ss*) refers to the solid solution. However, it is difficult to initiate the reaction in Eq. (1) without a form of activation. In the cited work, an electrical field was applied to initiate and sustain the reaction [11].

In the present study, the combustion synthesis of SiC and Al powders was performed according to :

$$x(Al + 0.5N_2) + (1-x)SiC = xAlN-(1-x)SiC_{ss}. \qquad (2)$$

The reaction of Eq. (2) can be easily ignited and the resulting combustion wave propagated through the sample leading to the formation of the solid solutions within seconds.

Thus it is interesting to understand the mechanism of this combustion reaction and how it provides such high rates of transformation compared to conventional processing methods. This is the broad objective of the research described in this paper.

## EXPERIMENTAL

Commercially available powders of aluminum and silicon carbide were dry-mixed for one hour in proportions that gave a product with a molar ratio of AlN/SiC =6/4. The 3C-SiC powder (Kojundo Chemical Lab. Co. Ltd, Japan) was 99% pure with a particle size in the range of 2-3 μm. The Al powder (Kojundo Chemical Lab. Co. Ltd, Japan) was 99.9% purity and had an average particle size of 20 μm. The green mixture was poured into a porous graphite crucible (diameter: 15 mm, height: 30mm). In order to understand the morphology of the transformation during combustion, whiskers of 3C-SiC were used in some experiments. The whiskers were 98 % pure and were about 0.3-1.4 μm in diameter and 5-30 μm in length. When whiskers were used, the mixture was prepared by wet-milling in ethanol in a $Si_3N_4$ jar with $Si_3N_4$ balls for 12 hours. After milling the suspension was dried and the mixture was sifted through a sieve (10mesh).

The combustion experiments were performed in the reaction chamber (Ueno Metal Inc., Japan) under a nitrogen atmosphere. The nitrogen gas pressure was varied from 0.1 to 8.0MPa. To initiate the reaction between Al, gaseous nitrogen and SiC, the ignition pellet of Ti-C mixture (diameter: 16 mm, height: 6.0-8.0 mm; molar ratio: 1/1) was placed on top of the crucible. A current of ca. 70 A (at 40V) was passed through a carbon ribbon in contact with the pellet, causing it to ignite. This in turn ignited the green mixture of Al-SiC under the set nitrogen gas pressure. If a combustion reaction was initiated, it was completed within several seconds.

The temperature was measured by means of dual-wavelength optical pyrometer (Model IR-AQ, Chino Co. Ltd.) focused on a hole in the middle part of the crucible (diameter of the hole was about 5 mm). To measure the wave propagation velocities, two holes at 10 mm between centers were made in the middle part of the crucible. The combustion process was video-recorded (Sony CCD-V800 with a Kenko ND400 filter) and the velocity was determined from the optical images.

The products were analyzed by X-ray powder diffraction, using CuKα radiation (RINT-2500, Rigaku Co. Ltd.). For phase identification, the scanning rate was set at 2° of 2θ/min., and for peak curve comparison, the (110) peak pattern near 2θ = 60° was run in a mode of step-scanning under conditions of Δ2θ =0.02° and a collecting time of 2 seconds. The lattice constant was also measured by the same scanning method using five peaks each of internal standard Si and the sample. The morphology of products was determined by SEM (JSM-5410, JEOL).

## RESULTS AND DISCCUSION

Reactants consisting of Al and SiC in the reactant ratios of Al/SiC=7/3, 6/4, 5/5 burned spontaneously after an ignition under a nitrogen gas pressure (0.1-8.0MPa). Figure 2 shows the typical temperature profiles obtained during the combustion of reactants with a ratio of Al/SiC=6/4 under a nitrogen gas pressure of 1.0, 6.0 and 8.0 MPa. The maximum combustion temperature increased with an increase in the nitrogen gas pressure. The combustion wave velocities were about 1.0mm/sec. The temperature profile exhibited a double peak pattern. But, this became less prominent as with the gas pressure was increased. The observation of a double peak suggests that an "afterburn" reaction occurs. The combustion reaction is believed to be dominated by the local supply of nitrogen.

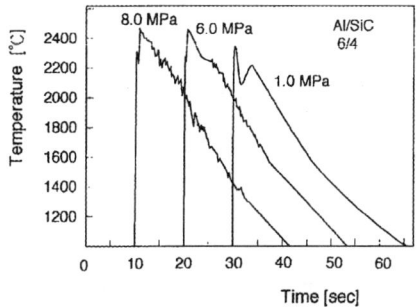

Figure 2. Temperature profiles of combustion reaction, Al/SiC=6/4, nitrogen gas pressure: 1.0 to 8.0MPa.

Figure 3 shows the effect of packing density of powders on maximum combustion temperature for samples with Al/SiC ratio of 7/3 and 6/4 combusted under a nitrogen pressure of 4.0 MPa. The figure shows that the temperature is higher for the system with higher Al content and that the temperature increases as the green density is decreased. The formation enthalpy of AlN by the reaction of Al with a gaseous nitrogen is -318.0kJ/mol at 298K which is enough to self-sustain the combustion reaction, while SiC works as a diluent in the reaction. Thus the increase in combustion temperature

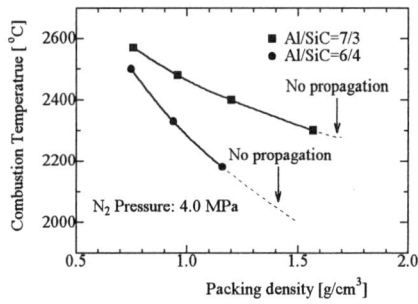

Figure 3. Effect of packing density of powders on maximum combustion temperature and the wave propagation.

is a reflection of the higher heat evolution when more aluminum is present. The increase of temperature with packing density relates to the local availability of nitrogen. As the packing density decreases, the availability of nitrogen is enhanced. The opposite trend in the packing density has an obvious limit, as shown in Figure 3. A value is reached when the porosity of the reactants is insufficient to allow for the permeation the nitrogen to allow for the continuation of the reaction and thus the waves become extinguished.

Figure 4 shows X-ray diffraction patterns (XRDP) of the products from the reaction system of Al/SiC=6/4 under nitrogen gas pressure ranging from 0.1 to 8.0 MPa. Each pattern show the presence of the 2H AlN-SiC solid solution as the major phase, with a slight residue of Si. The lattice constant of the solid solution phase is seen not to be significantly affected by the nitrogen pressure during synthesis. For example, for samples synthesized with pressures of 1.0, 4.0 and 8.0MPa, the respective lattice constants were calculated as : a=3.103, 3.102, 3.102A, c=5.017, 5.012, 5.013A. The lattice constants of the monolithic compounds AlN and 2H-SiC are, respectively, a=3.111, c= 4.979A (JCPDS 25-1133) and

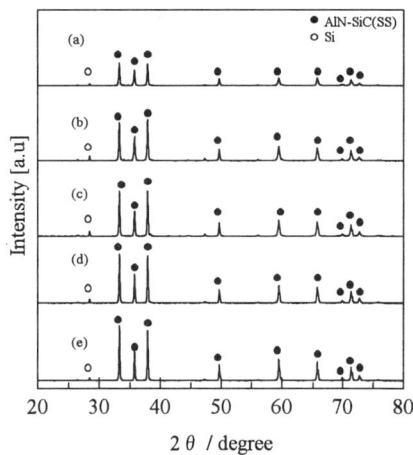

Figure 4. X-ray diffraction patterns of AlN-SiC solid solution synthesized by combustion synthesis, reactant ratio: Al/SiC=6/4, N2= (a) 0.1, (b) 1.0, (c) 4.0, (d) 6.0, (e) 8.0MPa.

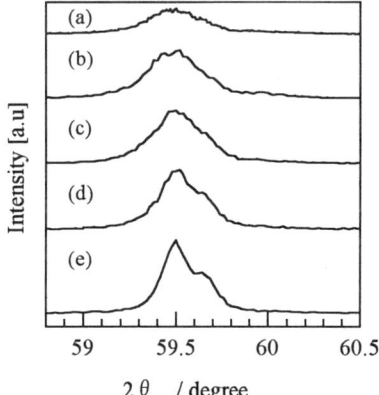

Figure 5. X-ray diffraction patterns of AlN-SiC solid solution synthesized by combustion synthesis, reactant ratio: Al/SiC=6/4, N2= (a) 0.1, (b) 1.0, (c) 4.0, (d) 6.0, (e) 8.0MPa, (110) lines in Figure 4.

a=3.081, c=5.048A in 2H-SiC (JCPDS 29-1126). The concentrations of AlN and SiC in the product must, therefore, be the same regardless of nitrogen pressure.

The hexagonal structures of 2H-AlN and -SiC are very similar with only a small difference in their lattice parameters as indicated above. The biggest difference in 2θ of the diffraction lines, from the (110) planes, is 0.64°. The (110) diffraction lines of commercial α-SiC and AlN can be easily resolved into two peaks. However, in the AlN-SiC solid solution, the full width at half maximum (FWHM) of the (110) peak is smaller than 0.64°. Figure 5 shows X-ray diffraction patterns of the (110) planes of AlN-SiC solid solutions synthesized with a reactant ratio of Al/SiC=6/4 under nitrogen pressures corresponding to those shown in Figure 3. Both of the diffraction peaks due to X-rays of $K\alpha_1$ and $K\alpha_2$ can be also observed clearly in the product synthesized at the higher gas pressure. These results show that the products synthesised by the present method are AlN-SiC solid solutions. Figure 6 shows the effect of nitrogen gas pressure on the FWHM of the diffraction line of (110) plane in the solid solutions. This value decreases by nearly a factor of two as the nitrogen pressure is increased from 0.1 to 8.0 MPa. At the highest pressure, the FWHM is approximately 0.15.

To attempt to understand the mechanism of the solid solution formation during the combustion reaction between Al, SiC and gaseous nitrogen, we performed morphological observation of the products and reagents by SEM, using 3C-SiC

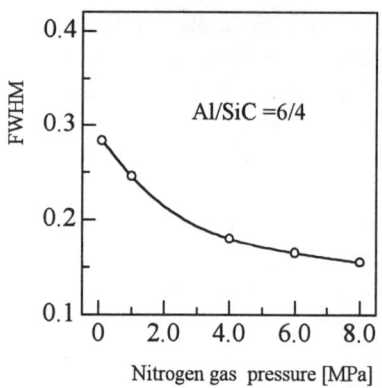

Figure 6. Effect of nitrogen gas pessure on FWHM of the diffraction line of (110) plane in AlN-SiC solid solution synthesized by combustion nitridation.

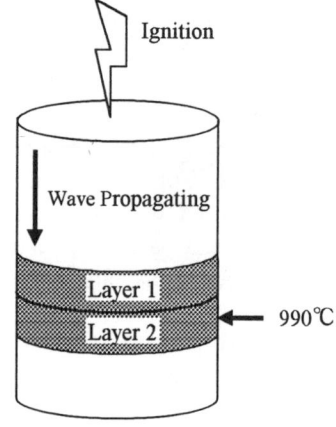

Figure 7. Scheme of extinct of wave propagation in the interface between layers 1 and 2.

Figure 8. SEM photographs of reactant mixture of Al and SiC whisker (a) and the magnification (b).

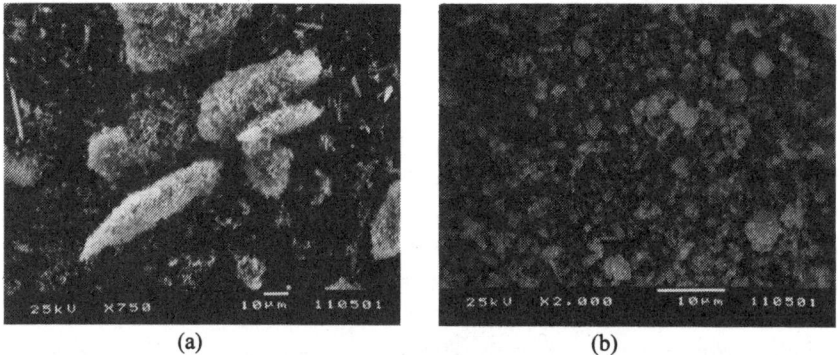

Figure 9 SEM photographs of materials in layer 2: preheat zone just below the wave front (a) and layer 1: combustion zone just behind the extinguished wave (b) under nitrogen pressure of 0.1MPa.

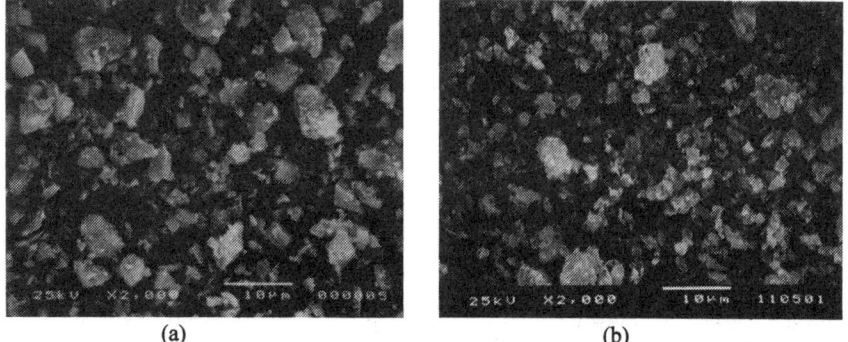

Figure 10. SEM photographs of AlN-SiC solid solution by combustion nitridation of Al-SiC=6/4 (a) and Al-SiC whisker=6/4 (b) under nitrogen gas pressure of 6.0MPa.

Innovative Processing/Synthesis: Ceramics, Glasses, Composites II

whiskers as reactants. The green powder mixture was slightly pressed (200 MPa) and ignited under a nitrogen gas pressure of 0.1MPa to suppress the rigorous reaction. The combustion wave became extinguished in the interface between two visually distinguishable layers (layers 1 and 2), as shown schematically in Figure 7. At a point close to the interface, the temperature was about 1000 °C. This sample was carefully divided layer by layer and analyzed by SEM and XRD method. Figure 8 shows the SEM photographs of reactant mixture of Al, SiC whiskers. Figure 9 shows the micrographs of materials in layer 2 (part a) and layer 1 (part b). The particle size of the starting Al is about 20μm and the SiC whisker is about 0.3-1.4 μm in diameter and 5-30μm in length. However, by wet milling in ethanol (in a $Si_3N_4$ jar with the $Si_3N_4$ balls) for 12 hours, the material appears to have agglomerated considerably forming particles of 50 to 100μm in dimension, as shown in Figure 8 (a). The shape of SiC whiskers was not altered even after this long milling, as shown in Figure 8 (b). Figure 9 (a) corresponds to layer 2, the preheat zone (just below the wave front) which had a peak temperature of about 990 °C. The particles containing the SiC whiskers seem to have agglomerated by the solidified molten liquid. On the other hand, Figure 9 (b) is the micrograph of layer 1 of the combustion zone (just behind the extinguished wave). The particle size of material here is about 1.0 to 3.0μm which is quite smaller than the reactant secondary particles and also the raw Al powder. In this region, no SiC whiskers could be observed. In case of the combustion nitridation under a nitrogen gas pressure of 6.0 MPa, the products from both of the reactants of Al/SiC powders and Al/SiC whiskers were AlN-SiC solid solutions with a particle size of about 5.0 to 10μm, as shown in Figure 10.

The materials of layers 1 and 2 were identified by X-ray diffraction patterns. The XRDP of layer 2, shown in Figure 11 (b), indicates that the material is the reactant mixture of Al and 3C-SiC. Figure 11 (a) corresponds to the XRDP of layer 1 which shows that the region contains AlN, 3C-SiC, unreacted Al and free Si. The 3C-SiC in the layer 1 is also not whisker form, as shown in Figure 9 (b). The existence of free Si and the absence 3C-SiC in whisker form suggest that the whiskers had dissolved in molten Al to from Al-Si-C liquid at about 2150 °C, as estimated from

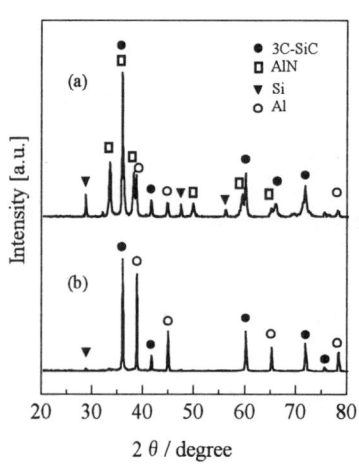

Figure 11. X-ray diffraction patterns of materials in layer 1 (a) and layer 2 (b) in Figure 7.

the phase diagram [12]. The subsequent precipitation of SiC from the liquid does not maintain the whisker form. The rise in temperature up to about 2000°C is a consequence of the formation of AlN when Al is partially nitrided, as shown in Figure 11(a).

If the formation of AlN in the reaction of Eq. (2) occurs completely before any other reactions, the solid solution is not likely to form during the short time of the whole process. If it did form, then a considerably long time is required to form the solid solution, as is experienced in the conventional methods. However, in this reaction, we expect that at first that heat generation comes from the partial nitridation of Al. As the temperature increases, the remaining unreacted molten Al can dissolve SiC to form an Al-Si-C liquid at about 2150 °C and the previously formed AlN can also dissolve into this liquid to form an Al-Si-C liquid containing a nitrogen gas inside the solution at about 2400°C as suggested by the Al-Si-C[12] and Al-Si-N[13] phase diagrams. Finally, we propose, that a transient liquid solution forms at the temperature close to maximum combustion temperature which, upon cooling, results in the formation of the AlN-SiC solid solution.

CONCLUSION

Solid solutions of AlN-SiC were synthesized by combustion nitridation of Al-SiC under the nitrogen gas pressures in the range of 0.1 to 8.0 MPa. The degree of homogeneity of the solid solution increases with nitrogen pressure, as concluded from measurements of FWHM of diffraction profile of the (110) plane. Using pressed reactant containing 3C-SiC whiskers, wave propagation is limited and becomes extinguished in middle of the compact. Analyses of the regions close to the combustion wave front were made by SEM and XRDP. From these, a mechanism of the formation of the solid solution was postlated, involving a solution-precipitation process.

ACKNOWLEDGMENTS

We are grateful for the support given to M.O. by the Japanese Ministry of Education and High-Tech Research Center, and to Z.A.M. by the Army Office of Research. This work is part of a Collaborative Agreement between Ryukoku University in Japan and the University of California (Davis).

REFERENCES
1. A. Zangvil and R.Ruh, "Phase Relationship in the Silicon Carbide-Aluminum Nitride System," J. Amer. Ceram. Soc., 71, 884-890 (1988).
2. G. Ervin, "Silicon Carbide-Aluminum Nitride Refractory Composite", U.S. Pat. No 3,492,153, North American Rockwell Corp., Jan 27, 1970.

3. W.Rafaniello, K.Cho and A.V. Virkar, " Fabrication and Characterization of SiC-AlN Alloys," *J. Mat. Sci.* **16**, 3479-3488 (1981).
4. R. Ruh and A. Zangvil, "Composition and Properties of Hot-Pressed SiC-AlN Solid Solutions," *J. Amer. Ceram. Soc.*, **65**, [5], 260-265 (1982).
5. P.C.Yates, "Dense, Submicron Grain AlN-SiC Bodies". U.S. Pat. No 3,649, 310, March 1972.
6. A. Zangvil, R. Ruh, *Proc. of "Silicon Carbide 1987" Symposium held in Columbus, OH, USA. Ceramic Trans. SiC-87* **2**, 63 (1989).
7. S.-Y. Kuo, A. Virkar, W. Rafaniello, "Modulated Structure in SiC-AlN Ceramics," *J. Amer. Ceram. Soc.*,**70**, [6], C-125-C-128 (1987).
8. M. Miura, T. Yoyo, S. Hirano, " Phase Separation and Toughening of SiC-AlN Solid-Solution Ceramics," *J. Mater. Sci.*, **28**, 3859-3865 (1993).
9. Q. Tian and A. Virkar, "Interdiffusion in SiC-AlN and AlN-Al$_2$OC Systems,"*J. Amer. Ceram. Soc.,* **79**, [8], 2168 (1996)
10. H. Xue and Z. Munir, " The Synthesis of Composites and Solid Solutions by Field-Activated Combustion," *Scripta Metall. Mater.*, **35**, 979-982 (1996).
11. H. Xue and Z. Munir, "Synthesis of AlN-SiC Composites and solid solutions by Field-Activated self-Propagating Combustion,"*J. Euro. Ceram. Soc.*, **17**, 1787-1792 (1997)
12. *Ternary Alloys*, Vol.3, ed. by G. Petzow and G. Effenberg (1988)
13. *Ternary Alloys*, Vol.7, ed. by G. Petzow and G. Effenberg (1988)

# CHEMICALLY-ASSISTED COMBUSTION SYNTHESIS OF SILICON CARBIDE FROM ELEMENTAL POWDERS

Jan A. Puszynski and Shuxia Miao
Chemistry and Chemical Engineering Department
South Dakota School of Mines and Technology
Rapid City, SD 57701 USA

The feasibility of synthesizing submicron silicon carbide from elemental silicon and carbon powders by chemically-assisted self-propagating high-temperature synthesis has been demonstrated. The effect of carbon powder properties, initial concentration of potassium chlorate as an reactive additive, gas atmosphere, and operating pressure on combustion front characteristics, product morphology and its purity as well as phase composition has been studied in detail. It has been found that combustion synthesis of silicon carbide can be accomplished both in nitrogen or argon atmosphere. It has been shown that a formation of silicon carbide at elevated nitrogen pressures is possible when carbon powder with a very high specific surface area is used. Average particle size of silicon carbide synthesized from high purity silicon and carbon powders in a presence of potassium chlorate at elevated nitrogen and argon pressures (P>0.3 MPa) is approximately 200 nm.

## INTRODUCTION

There exist many techniques of synthesizing silicon carbide powders, whiskers, and platelets [1-3]. Normally, silicon carbide is synthesized by a carbothermal reduction of silica. This process is strongly endothermic and requires a significant amount of energy to convert reactants to silicon carbide product. There have been developed numerous methods of synthesizing silicon carbide since the Acheson process was discovered, however most of them are not very energy efficient. The excellent overview of these techniques is presented by Weimer et al. [4].

A new class of processes, with an enormous technological flexibility and capability, so called combustion synthesis or self-propagating high-temperature synthesis (SHS) was developed and successfully applied in the last three decades. Many exothermic non-catalytic gasless reactions liberate enough heat so they can be self-sustained after being locally ignited. Over five hundred ceramic and composite powders have been synthesized using this technique [5-7].

The combustion synthesis technique is advantageous to conventional synthesis methods because of very low energy requirements, simplicity, and versatility of a combustion reactor. In addition, a product formed during the combustion usually has a higher purity than starting reactants. However, this technique can be only applied to exothermic reacting systems which generate a sufficient amount energy to be self-sustaining (e.g., Ti + C, Al + $Fe_2O_3$, Ti-B, Si-$N_2$, Ni-Al, and many others).

The reaction between Si and C is weakly exothermic and is not self-sustaining without a significant preheating of reactants (>1000°C). Therefore, economical advantage of SHS method in this case is in question. Munir and his co-workers has modified the "traditional" SHS process by applying an external electrical field to the reactant mixture [8]. They demonstrated that many weakly exothermic reacting systems can be self-sustaining in the presence of such external energy field. They showed that SiC can be directly synthesized from elemental powders without any additional preheating [9].

Recently, Liebig and Puszynski demonstrated that the addition of small quantities of volatile and nonvolatile additives may considerably affect the mechanism of formation of various combustion synthesized products [10]. They have shown that phase composition and particle size of silicon nitride can be controlled by an addition of ammonium fluoride or ammonium chloride. The addition of those compounds leads to a formation of gaseous intermediates which alter the combustion process and the formation of the desired silicon nitride phase. The same concept of changing an overall mechanism of a combustion reaction in Si-C reacting system has been a motivation for this research.

There exist chemicals which may react exothermically with carbon and silicon or decompose with a generation of gaseous or condensed by-products. By-products can be usually easily separated using standard chemical engineering techniques. Example of such chemicals are: ammonium perchlorate, potassium chlorate or perchlorate, Teflon, and many others. Therefore, when such reactants are added into Si-C system self-sustaining process may be expected.

The main objective of this research has been to study the effect of carbon source, $KClO_3$ as a reactive additive, gas atmosphere, and operating pressure on combustion front characteristics, product morphology, and phase composition of SiC synthesized from elemental silicon and carbon powders.

EXPERIMENTAL

Reactant powders consisting of equimolar mixture of silicon and carbon and 0-15 wt% $KClO_3$ were dry mixed for 8 hours. The mixed reactants were placed in a graphite container in a form of a loose powder (Figure 1). Prior to an ignition, the reactants were subjected to vacuum-inert gas purge operation in order to reduce oxygen and water vapor content. After three consecutive purge operations, the reactor was filled with nitrogen or argon up to a desired pressure. The reactant mixture was ignited by a chemical igniter consisting of Ti and carbon mixture in order to make the ignition of silicon-carbon reacting system uniformly and suddenly. The chemical igniter was initiated by a resistively heated molybdenum wire. Combustion syntheses were carried out in both nitrogen or argon atmosphere. Nitrogen and argon with a minimum purity of 99.9% were used in all experiments. Silicon powder from Hermann C Starck, Inc. has minimum purity of 99.99% (metal base), oxygen content less than 1 wt%, carbon content less than 0.15 wt%, and average particle size of 3.0 µm. Two different carbon sources were tried in this research. Raven 7000 powder from Columbian Chemicals Company has average particle size of 11 nm, specific surface area (NSA) of 575 $m^2/g$, and very high surface oxygen content, 8.5 wt%. Another high purity carbon powder has been provided by Chevron Chemical Company. This powder has an average particle size of 42 nm, specific surface area (BET) of 80 $m^2/g$, and moisture content less than 0.02 wt%. Other major impurities consists of Al<4ppm, Ca<4ppm, Fe<7ppm, Si<10ppm, Cr<1ppm, and Cu<0.5ppm. Products from combustion synthesis were examined by JSM - 840 Scanning Electron Microscope. Phase identification was done using PHILIPS X-ray diffractometer.

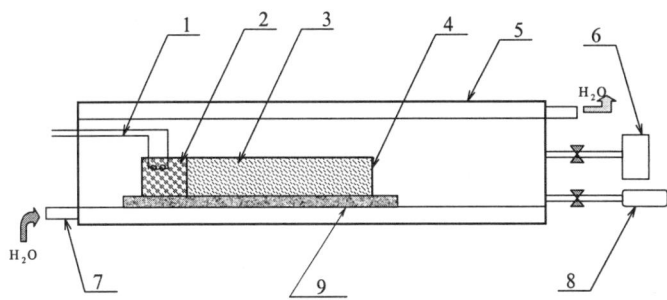

Figure 1. Schematic of the reactor. 1) resistively heated molybdenum filament; 2) igniter mixture (Ti+C); 3) reactants; 4) graphite boat; 5) reactor; 6) gas tank; 7) cooling water; 8) vacuum system; 9) refractory support.

## RESULTS AND DISCUSSION

It has been demonstrated that a reaction between silicon and carbon (Raven7000) powders is self-sustaining at nitrogen pressures greater than 3 MPa without any reactive additives. Similar experiments conducted in an argon atmosphere have shown that this reacting system is not self-sustaining. The analysis of combustion synthesized silicon carbide at elevated nitrogen pressures has shown that in addition to a major silicon carbide phase, silicon oxynitride phase was formed (see Figure 2).

Corresponding SEM photograph of the combustion synthesized powder is shown in Figure 3. It can be clearly seen that in this case the silicon carbide powder with a relatively large average particle size was obtained ($d_{av} > 2$ μm).

Due to the fact that Raven7000 carbon powder has a very high specific surface area and oxygen content, it was expected that additional heat generation from oxidation reaction and interaction of nitrogen with a silicon reactant may significantly contribute to the combustion process. The formation of silicon nitride and silicon carbide composites with a SiC content up to 20 wt% was reported previously by Agrafiotis et al. [11]. Their results have indicated that high pressure nitridation of silicon is responsible for a self-sustaining character of the overall combustion process. The thermodynamic calculations performed using HSC Chemistry software from Outokumpu Company has confirmed experimental results, regarding a formation of silicon nitride and silicon oxynitride products (see Figure 4).

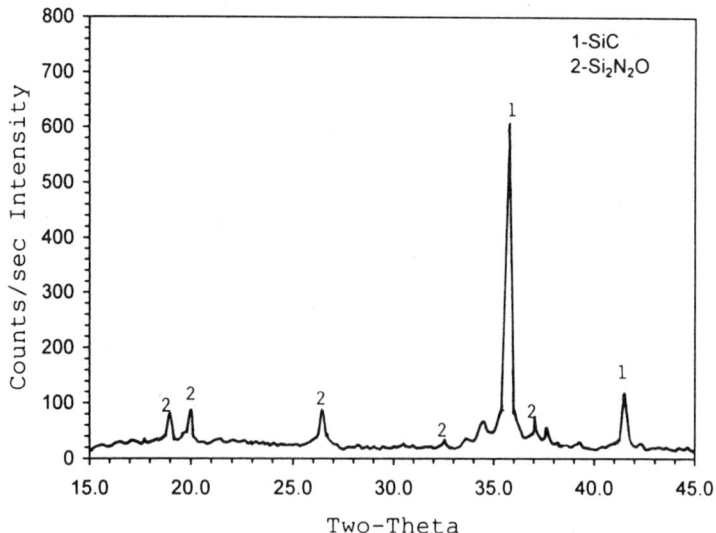

Figure 2. X-Ray diffraction of combustion synthesized silicon carbide from Raven7000 carbon source at 3 MPa nitrogen pressure.

Figure 3. SEM photograph of combustion synthesized silicon carbide from Raven7000 carbon source at 3 MPa nitrogen pressure.

The next series of experiments was focused on a formation of silicon carbide using a very high purity carbon powder with a low oxygen surface content. The experiments performed in nitrogen atmosphere have shown that the reaction between silicon and carbon (Chevron Chemical Company) is not self-sustaining. We have varied nitrogen pressure and initial powder density and we have not found any evidence of self-sustainability of the reaction between silicon and carbon. These results have led us to the conclusion that both high carbon surface area and the initial surface oxygen content may play an important role in the overall combustion process.

In order to explore the effect of other reactive additives on an ignition and combustion characteristics in silicon – carbon system, we have decided to examine the effect of potassium chlorate and perchlorate, ammonium perchlorate, and Teflon. In this paper we present the results obtained with potassium chlorate only. The effect of other additives listed above will be presented elsewhere.

The experiments with $KClO_3$ as a reactive additive were performed in a range of its concentration up to 15 wt%. It has been found that the silicon – carbon system is self-sustaining with a minimum content of $KClO_3$ to be 5 wt%. This reacting system is self-sustaining in both nitrogen or argon atmosphere. However, the minimum pressure of 0.3 MPa must be maintained in the reacting system. Higher nitrogen or argon pressures apparently reduce a fraction of $KClO_3$ vaporized prior to its decomposition. This new approach clearly indicate that the mechanism of

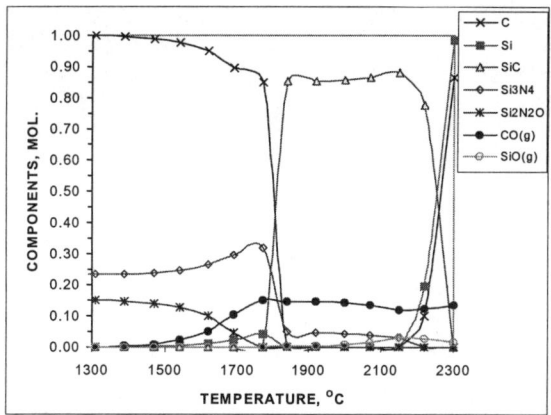

Figure 4. Equilibrium concentration profiles of reactants and products in the system consisting of 1 mole Si, 1 mole C and 0.075 mole $O_2$ in $N_2$ atmosphere(P=3MPa).

the reaction between silicon and carbon is different from the previous case when silicon reacted with high surface carbon powder in nitrogen atmosphere. In this case, the synthesis process is self-sustaining in pure argon atmosphere as well. The main exothermic reactions that may take place in the system are:

$$Si + C = SiC \qquad \Delta H_{298} = -69.1 \text{ kJ/mole} \qquad (1)$$

$$KClO_3 = KCl + 1.5\ O_2 \qquad \Delta H_{298} = -39.0 \text{ kJ/mole} \qquad (2)$$

$$KClO_3 + 3\ C = KCl + 3\ CO \qquad \Delta H_{298} = -370.6 \text{ kJ/mole} \qquad (3)$$

The second reaction, in addition to heat generation, results in a formation of oxygen which might promote a gas transport of both silicon and carbon. Studies of this complex reaction mechanism are currently underway. It has been found that a pure silicon carbide powder with very small average particle size is formed using this technique. X-Ray diffraction and SEM photograph of silicon carbide formed during chemically-assisted combustion synthesis and subsequently leached to remove KCl are shown in Figures 5 and 6. The equilibrium concentration profiles of products other than the main phase, SiC, are shown in Figure 7.

X-ray diffraction shown in Figure 5 indicates that the ultrafine silicon carbide product after leaching is mainly composed of β-SiC phase. A small quantity of α-SiC phase has been detected which implies that the reaction temperature is very high. The dynamic temperature profiles measured during a combustion process will be presented elsewhere. No crystalline silicon oxide has been detected in combustion synthesized products.

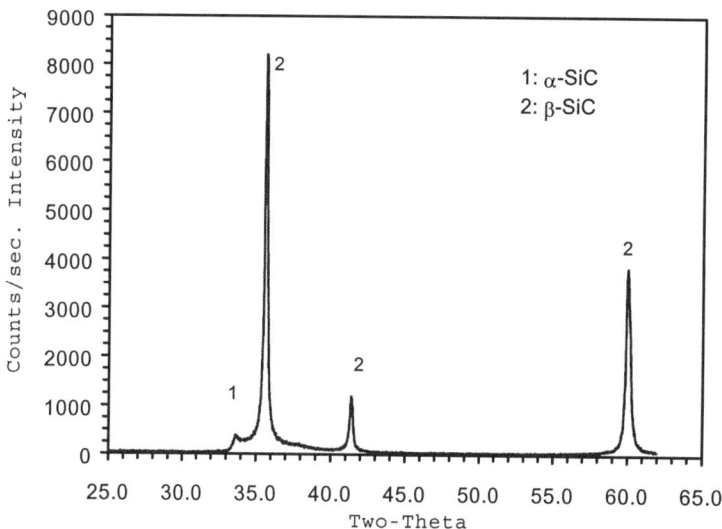

Figure 5. X-Ray diffraction of combustion synthesized silicon carbide in a presence of potassium chlorate (5 wt%).

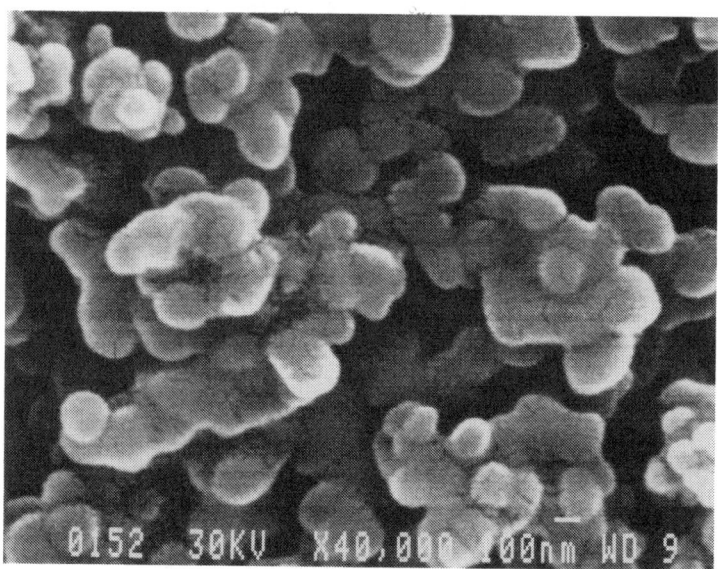

Figure 6. SEM photograph of combustion synthesized silicon carbide in a presence of potassium chlorate (5 wt%).

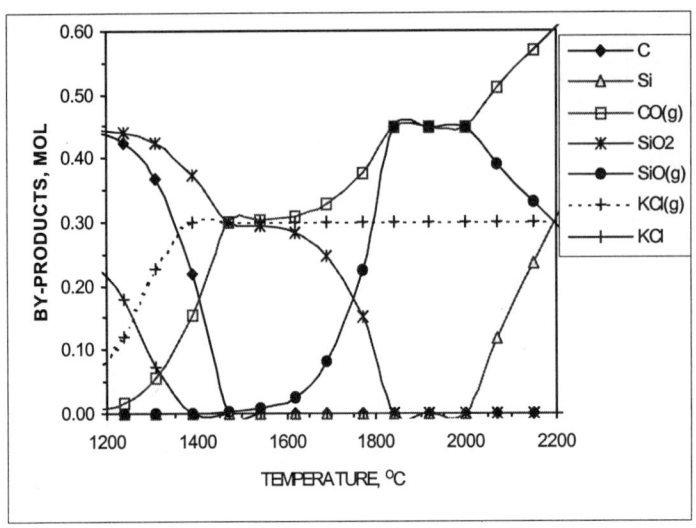

Figure 7. Equilibrium composition profiles in the reaction system consisting of 10 moles Si, 10 moles C and 0.30 mole KClO$_3$ in Ar atmosphere (P = 3 MPa). The main SiC phase is not shown.

CONCLUSIONS
- The reacting system consisting of high purity Si and C powders can be self-sustaining when KClO$_3$ is used as a chemical additive.
- Ultrafine silicon carbide powder with an average particle size of 200nm can be synthesized in a combustion regime in a presence of potassium chlorate. The properties and application of this ultrafine SiC powder is under investigating.
- Direct reaction between silicon and carbon powders without other additives might be self-sustaining at elevated nitrogen pressures (P$_{nitrogen}$≈ 3 MPa) when carbon with a very high surface area is used. It is expected that oxygen adsorbed on the carbon surface might play a significant role. However, further research is needed to find a definitive answer.

ACKNOWLEDGEMENT
The authors gratefully acknowledge the financial support from the National Science Foundation (Grant No. CTS-9700503). The authors would also like to acknowledge Mr. Alan Degraw from Novel Technologies, Inc. for his valuable suggestions and reactant powders.

REFERENCES
1. V. Kevorkijan, M. Komac and D. Kolar, J. Mater. Sci., 27 [10] 2705 (1992).
2. F. K. van Dijen and R. Metselaar J. Europ. Ceram. Soc., 7 177-84 (1991).
3. R. L. Beatty and F. H. Wyman, "Continuous Silicon Carbide Whisker Production", U. S. Pat. No. 4 637 924, 1987.
4. Alan W. Weimer, "Carbide, Nitride and Boride Materials Synthesis and Processing", Chapman & Hall, New York, 1997.
5. A. G. Merzhanov, Int. J. Self-propag. High-temp. Synth., 6 [2] 119-64 (1997).
6. Z. A. Munir and U. Anselmi-Tamburini, Mater. Sci. Rep., 3 277-365 (1989).
7. J. W. McCauley, Ceram. Eng. Sci. Proc., 119 [10] 1137-81 (1990).
8. Z. A. Munir, Int. J. Self-propag. High-temp. Synth., 6 [2] 165-86 (1997).
9. A. Feng and Z. A. Munir, Metall. & Mater. Trans. B, 26 [3] 587 (1995).
10. B. Liebig and J. A. Puszynski, Int. J. Self-propag. High-temp. Synth., 7 [1] (1997).
11. C. Agrafiotis, J. Lis, J. A. Puszynski and V. Hlavacek, Amer. Ceram. Soc. J. 73 [11] 3514 (1990).

COMBUSTION SYNTHESIS OF Cd-In-Ga-O POWDER FOR NOVEL THIN FILM GROWTH SUBSTRATE

Sy-Chyi Lin
NASA Center for Applied Radiation Research
Prairie View A&M University
Prairie View, TX 77446-4209

Richard Wilkins
NASA Center for Applied Radiation Research
Prairie View A&M University
Prairie View, TX 77446-4209

Mikael Nersesyan
Institute of Structural Macrokinetics
Russian Academy of Science
Moscow 142432, Russia

Dan Luss
Chemical Engineering Department
University of Houston
Houston, TX 77204-4792

ABSTRACT
Self-propagating High-temperature Synthesis (SHS) method was used for producing $CdIn_{1.673}Ga_{0.327}O_4$ complex oxide, a potential substrate for growing GaN thin film. The SHS process was conducted using two initial mixtures, Cd-$In_2O_3$-$Ga_2O_3$, and Cd-In-$Ga_2O_3$ reaching combustion temperatures 837 °C and 1,400 °C, respectively. The combination of these mixtures enabled us to carry out the synthesis below the sublimation temperature of the desired product ($\approx 1,100$ °C) and to increase its yield. Post-treatment increased the homogeneity of the product.

INTRODUCTION
Gallium nitride and associated III-V compounds are the most promising semiconductors not only for having energy gaps covering a wavelength from red ($E_g(InN) \cong 2.0$ eV) to UV ($E_g(AlN) \cong 6.0$ eV) but also for their high melting temperatures and their stability at high temperatures. Thin films of these materials are used for a variety of opto-electronic applications, and have potential utilization as high temperature and radiation resistant electronics[1-3]. However, the lack of an economical lattice matched substrate for III-V semiconductor thin films remains a

---

To the extent authorized under the laws of the United States of America, all copyright interests in this publication are the property of The American Ceramic Society. Any duplication, reproduction, or republication of this publication or any part thereof, without the express written consent of The American Ceramic Society or fee paid to the Copyright Clearance Center, is prohibited.

recognized problem in the further development of III-V semiconductor technology [4]. A good substrate material must have a lattice parameter and thermal property compatible with film. Thus, it will reduce the amount of defects in the crystalline material and prevent the formation of wurtzite expitaxial growth.

$CdIn_{2-\delta}Ga_{\delta}O_4$, having lattice constants ranged from 8.39Å ($\delta=2$) to 9.17Å ($\delta=0$) may be good substrates for growing various nitride thin films. For example, $CdIn_{1.673}Ga_{0.327}O_4$, having a lattice constant of 9.04Å (2x cubic GaN), and $CdIn_{0.952}Ga_{1.048}O_4$, having a lattice constant of 8.76Å (2x AlN), are substrates for GaN and AlN thin film growths, respectively. The most important commercial application for the III-V semiconducting materials is GaN thin films which can be used to produce blue to UV laser diodes in information storage systems. Operations at these wavelengths should increase the storage capacity by a factor of 10 over current technologies.

The most common substrate used for GaN thin film growth is sapphire (((0001) $Al_2O_3$), which has a 16% lattice mismatch. When such a large mismatch exists between the substrate and the film, a high dislocation density ($10^8$- $10^{10}$/cm$^2$) in the thin film is observed. Besides the dislocations, other defects, such as stacking mismatch boundaries, inversion domain boundary, and double positioning boundaries have also been found[5-6]. Silicon carbide, which has a smaller mismatch (3%), is a better substrate, but is very expensive. $LiGaO_2$ is being investigated for growing hexagonal GaN[7-9]. However, it still has a lattice mismatch of 0.9% and a long calcination is needed. Thus, fabrication of large quantities of lattice matched substrates at a low cost is necessary for the laser application of nitride thin films.

In-Ga-Cd-O ceramic materials are being investigated for possible GaN substrates. For example, theoretical calculations predict that $CdIn_{1.673}Ga_{0.327}O_4$ should have a spinel structure and a lattice constant of 9.04Å (2× cubic GaN) [10] and should be an ideal substrate for growing GaN thin film. Nitrate salts of cadmium, indium, and gallium were used to prepare $CdGa_{0.327}In_{1.673}O_4$ powder [10]. However, this solid state reaction is time consuming and not economical. To circumvent these problems, we used Self-propagating High-temperature Synthesis (SHS) to prepare $CdGa_{0.327}In_{1.673}O_4$ powder.

In an SHS process, solid powders are intimately mixed and pressed to form a dense pellet, which is then ignited by externally heating one end. The released heat causes a combustion front to propagate through the pellet, forming the desired product. The simplicity of the process minimizes equipment needs and makes the process economical. SHS has been successful used to produce complex oxides[11-15].

We reported preliminary results of combusting a mixture of Cd, $In_2O_3$, and $Ga_2O_3$ to form $CdIn_{1.673}Ga_{0.327}O_4$ via the following reaction[16]:

$$Cd + 0.8365\ In_2O_3 + 0.1635\ Ga_2O_3 + 0.5\ O_2 \rightarrow CdIn_{1.673}Ga_{0.327}O_4$$
$$\Delta H = -61.70\ kCal/mol. \qquad (1)$$

In this reaction system, cadmium metal was the fuel with indium oxide and gallium oxide acting as fillers. The oxidation of cadmium metal by the ambient oxygen provides the heat necessary to sustain the combustion front movement. Because of the lack of a solid oxidizer in this reaction and the relatively low heat released during the combustion, it is very difficult completely convert all the reactants during the combustion.

In order to increase the combustion temperature during the reaction, we decided to increase the fuel ratio and studied the following reaction:

$$Cd + 1.673\ In + 0.1635\ Ga_2O_3 + 1.7548\ O_2 \rightarrow CdIn_{1.673}Ga_{0.327}O_4$$
$$\Delta H = -431.88\ kCal/mol. \qquad (2)$$

In this reaction system, both cadmium metal and indium metal were the fuels, with gallium oxide acting as a filler. The rate of heat released by this reaction is much higher than that from reaction (1). We report here a study of the combustion synthesis of $CdIn_{1.673}Ga_{0.327}O_4$ using the reactant mixtures of Cd-$In_2O_3$-$Ga_2O_3$ and Cd-$In_2O_3$-$Ga_2O_3$ and the homogeneity of the reaction product.

EXPERIMENTAL

Stoichiometric proportions (Cd: In: Ga= 1: 0.673: 0.327) of cadmium (99.5% pure, average particle size = 20 µm, Cerac Inc., WI), indium (99.99% pure, particle size ≤ 325 mesh, Cerac Inc., WI), indium oxide (99.99% pure, average particle size = 10 µm, Cerac Inc., WI), and gallium oxide (99.999 pure, average particle size = 5 µm, Cerac Inc., WI) were mechanically mixed overnight in a ball mill (U.S. Stoneware, Mahwah, NJ). Loose reactant powders were used in most experiments. Some experiments were conducted with reactant powders that were pressed into cylindrical pellets of various sizes (5 - 25 mm in diameter, 2 - 5 cm in length) and densities (25 - 50% of the theoretical density, 7.77 g/cm$^3$) to study the impact of oxygen diffusion. The ignitor was a Chromel A wire, 0.6 mm in diameter (Hoskins Mfg. Co., Hamburg, MI), which could be heated to 1,100 °C. A variable autotransformer (Staco Inc., Dayton, OH) controlled the ignition current.

Figure 1 shows a schematic of the combustion reactor. Loose powder or a reactant pellet was placed inside a quartz tube (i.d. = 10 cm). A thin layer of booster containing cadmium, indium oxide, gallium oxide, and a small amount of glucose ($HOCH_2CH(CHOH)_4O$) (less than 5%) was placed next to the sample to assist in the ignition. The sample and the tube were kept at room temperature. Oxygen at the rate of one to five liter/min. (Extra Dry Grade, Linde Co.) was fed into the tube. After ignition, the ignitor was removed and the combustion wave propagated by itself. The combusted pellet had a dark yellow color. Some combusted powders were further calcined/sintered in an oxygen atmosphere to improve product quality.

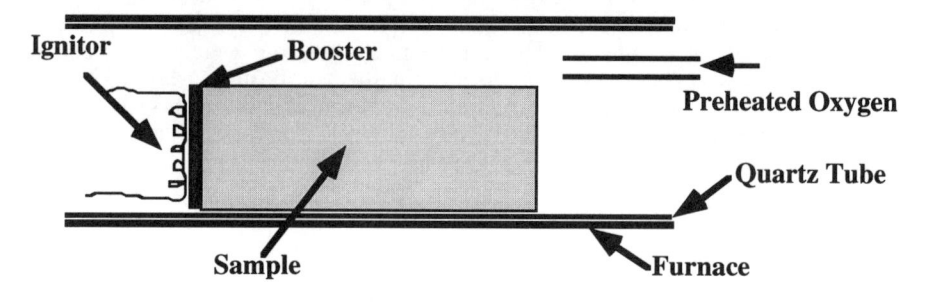

Figure 1. Schematic of the SHS reactor.

Thermal gravimetric/differential thermal analysis (TG/DTA) of the precursor mixture and the individual reactants were conducted in an oxygen environment using a Seiko TG/DTA 320 with an oxygen flow rate of 20 ml/min. and a heating rate of 80 °C/min. to simulate the rapid temperature rise during the combustion process. Alumina powder was placed on the bottom of the platinum sample cup to prevent reactions between the platinum cup and the samples.

B-type thermocouples, 0.003-inch diameter (CHAL-003, Omega Eng. Inc., Stamford, CT), inserted into a two-hole ceramic tubular insulator (TRM 005132 with two 0.005-inch i.d. holes, made by Omega Eng. Inc., Stamford, CT), were embedded at the center of the precursor powder to measure their temperatures during the combustion process. Their temperatures were recorded once per second.

Samples were analyzed using an X-ray Diffractometer (XRD) (Siemens D-5000 Diffractometer) to identify different phases in the precursors and products. The local product stoichiometry was determined by electron microprobe analysis (EMPA) using a JEOL JXA-8600 Superprobe.

EXPERIMENTAL RESULTS AND DISCUSSION

Loose pellets with a low bulk density (up to 50 % of the theoretical density) were prepared to minimize the resistance of oxygen diffusion into the pellet. A booster was used to increase the initial reaction rate and ignite the whole pellet cross section quickly. The heat released by the burning glucose helped to rapidly ignite the precursor.

As reported previously[16], the rate of heat released by the $CdIn_{1.673}Ga_{0.327}O_4$ formation via reaction (1) was relatively low compared to standard SHS processes when Cd, $In_2O_3$, and $Ga_2O_3$ were used in the precursor powder. When pellets of high densities (higher than 30% of the theoretical density) or large diameters (larger than 1 cm) were used, the combustion front velocity ranged from 1-3 mm/sec. However, the combusted pellets had a dark yellow color on the outside layer of the pellet and a light yellow color in the interior region, revealing that the reaction occurred only in the outer region and not in the interior region. This is caused by the difficulty of oxygen gas diffusion and reaction into the interior during the short time that it is at a high temperature.

In order to increase the combustion temperature, the rate of heat released, and the reaction time, part of $In_2O_3$ was replaced by In metal. The fuel ratio could be increased from 30 wt% to 91 wt% when all $In_2O_3$ was replaced by In metal in the precursor powders. The heat released by the $CdIn_{1.673}Ga_{0.327}O_4$ formation via reaction (2) was very high and produced a combustion front movement which was faster than 100 mm/sec. When reaction (2) was conducted, the combusted pellets had a dark brown color on the outside layer of the pellet and a dark yellow color in the interior region. Melting was also observed in the product pellets and $In_2O_3$ was found coated on the inner wall of the quartz tube after the combustion. These observations indicated that the reaction occurred vigorously in the outer region, the high temperature caused melting of the product, and the $In_2O_3$ formed in the reaction was vaporized. However, insufficient oxygen from the surroundings deterred the completion of the reaction (2) at the inner region of the pellet.

Based on experiments with reactions (1) and (2), we decided to test new reaction systems that contained cadmium metal, gallium oxide, and different ratios of indium metal and indium oxide in the precursor.

TG/DTA experiments were carried out to study the reactions among the reactant powders. TGA results allowed us to identify oxidation or decomposition processes by the weight gain or loss. DTA results enabled us to identify changes of phases or crystal structures by the change of temperature difference between the samples and the reference. In order to simulate the fast temperature rise during the combustion, a heating rate of 80 °C/min. was used. The TG/DTA results from the fast heating rate were different from those obtained at a lower heating rate (15 °C/min.) that we reported previously[16]. Figure 2 shows the result of the TG/DTA of cadmium metal. Cadmium metal starts to melt at 321 °C, leading to a significant

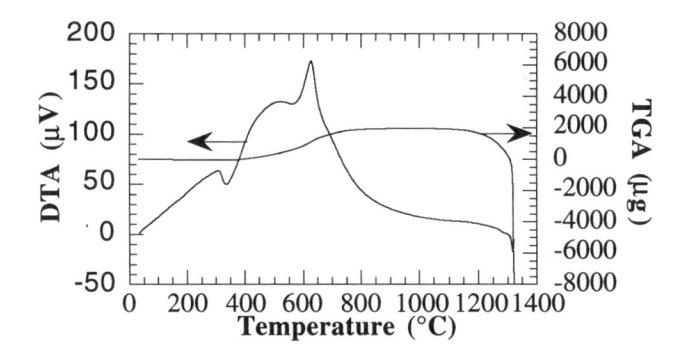

Figure 2. TG/DTA of cadmium metal.

dip in the DTA curve. After melting, cadmium is oxidized by the ambient oxygen leading to a broad DTA peak and a weight gain in the TGA curve between 342 °C and 927 °C. The metal is slowly oxidized at first and then ignites at 627 °C as indicated by the sharp peak in the DTA curve. The cadmium oxide sublimates at high temperatures which is shown by the weight loss above 1,011 °C.

Figure 3 shows the TG/DTA of indium metal. It starts to melt at 155 °C, leading to a significant dip in the DTA curve. After melting, indium is not oxidized by the ambient oxygen until the temperature reaches 280 °C. The oxidation process is shown by the broad DTA peak and the weight gain in the TGA curve between 280 °C and 1060 °C. Two extensively exothermic peaks are present at 506 °C and 818 °C. These two peaks indicate that the oxidation of indium metal is a two-step process.

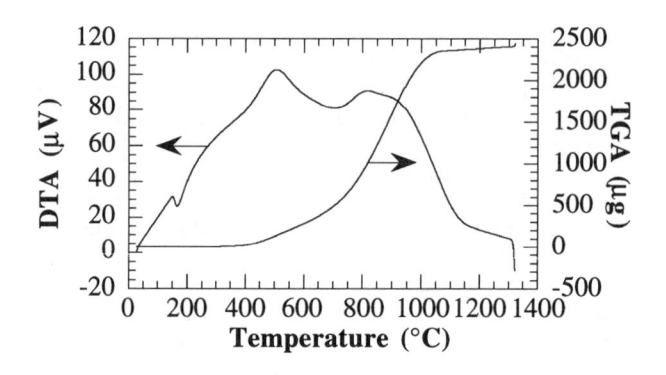

Figure 3. TG/DTA of indium metal.

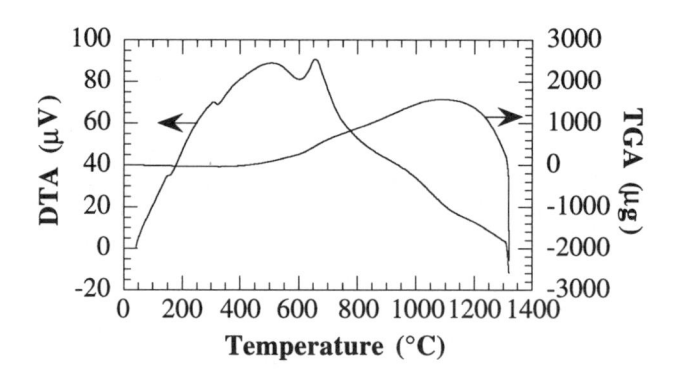

Figure 4. TG/DTA of Cd-In-Ga$_2$O$_3$ mixtures.

Figure 4 shows the TG/DTA of Cd-In-Ga$_2$O$_3$ reactant mixtures. The indium metal melts at 155 °C which is its normal melting temperature. However, cadmium metal melts at 315 °C which is below its normal melting point (321°C). The eutectic behavior helps the combustion front to propagate because the cadmium melt spreads on the outer surface of the gallium oxide powders, increasing the contact area and the reaction rate. The oxidation of cadmium and indium metals occurs between 350 °C and 1,090 °C as shown by the weight gain in this temperature range. Two combustion peaks are observed at 506 °C and 656 °C. The oxidized product sublimates above 1,100 °C.

Figure 5 shows the TG/DTA of Cd-In$_2$O$_3$-Ga$_2$O$_3$ reactant mixtures. Cadmium metal melts at 314 °C which is also below its normal melting temperature (321°C). This eutectic behavior also helps the combustion reaction to propagate because the cadmium melt spreads on the outer surface of the indium oxide and gallium oxide powders, increasing the contact area and the reaction rate. The oxidation of cadmium occurs between 410 °C and 880 °C which is shown by the weight gain in this temperature range. A combustion peak is observed at 379 °C. The oxidized product sublimates at temperatures above 938 °C.

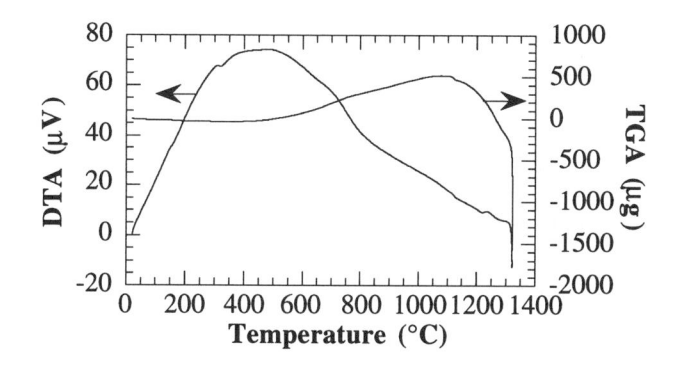

Figure 5. TG/DTA of Cd-In$_2$O$_3$-Ga$_2$O$_3$ mixtures.

From the above TG/DTA experiments, we conclude that eutectic melting happens in both Cd-In-Ga$_2$O$_3$ and Cd-In$_2$O$_3$-Ga$_2$O$_3$ reaction systems. The oxidation of Cd-In$_2$O$_3$-Ga$_2$O$_3$ occurs in a smaller temperature range than Cd-In-Ga$_2$O$_3$. However, the product in the Cd-In$_2$O$_3$-Ga$_2$O$_3$ system starts to sublimate at a lower temperature. In order to have a high reaction rate, a system giving a higher combustion temperature (Cd-In-Ga$_2$O$_3$) and a small oxidation temperature range (Cd-In$_2$O$_3$-Ga$_2$O$_3$) is desired. To prevent the sublimation reaction, a higher sublimation temperature (Cd-In-Ga$_2$O$_3$) is preferred. Considering all these factors, we decided to test the behavior of mixtures of these two reaction systems (Cd-In-Ga$_2$O$_3$/Cd-In$_2$O$_3$-Ga$_2$O$_3$.

To better understand the temperature change during the reaction, we recorded the temperature histories inside the pellets when they were combusted at room temperature by inserting type B thermocouples into the center of the pellets. Figure 6(a) shows the temperature history of the Cd-In-Ga$_2$O$_3$ system. The temperature rose from room temperature to 1,420 °C in 10 sec. and then slowly cooled down to room temperature. Figure 6(b) shows the temperature history of the Cd-In$_2$O$_3$-Ga$_2$O$_3$ system. The temperature rose from room temperature to 837 °C in 5 sec. and then slowly cooled down to room temperature.

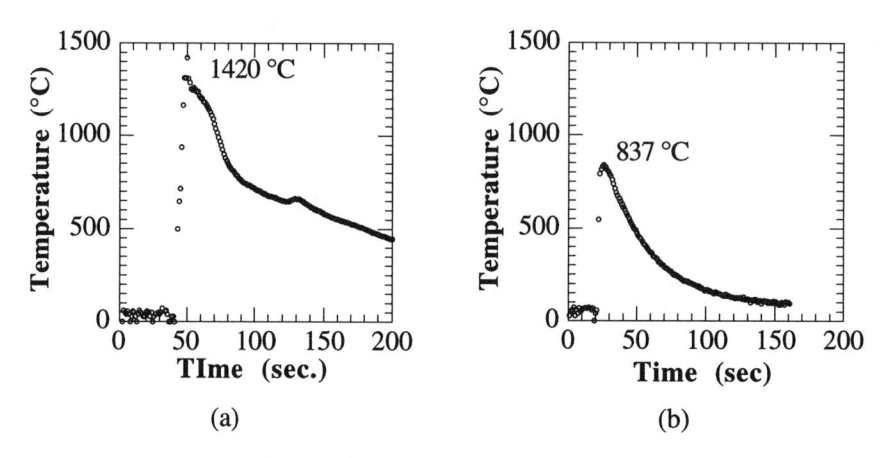

Figure 6. Temperature histories of (a) Cd-In-Ga$_2$O$_3$ and (b) Cd-In$_2$O$_3$-Ga$_2$O$_3$.

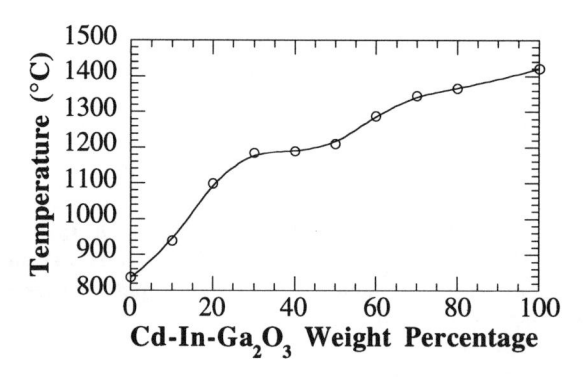

Figure 7. Combustion temperature at different compositions.

In order to find out a reaction system that has the combustion temperature just below the sublimation temperature, mixtures of different ratios of Cd-In-Ga$_2$O$_3$ to Cd-In$_2$O$_3$-Ga$_2$O$_3$ were prepared for temperature measurement. The measured combustion temperatures are shown in Figure 7. Figure 7 indicates that if we want to control the combustion temperature just below the sublimation temperature of the product (1,100 °C), we need to use mixtures that have less than

20 wt% of Cd-In-Ga$_2$O$_3$. Otherwise, the combustion temperature will exceed the sublimation temperature and decompose the product.

We used a mixture that contained 1/9 mixtures of Cd-In-Ga$_2$O$_3$ /Cd-In$_2$O$_3$-Ga$_2$O$_3$. Beside the main product, the product after the combustion contained CdO, In$_2$O$_3$, and CdIn$_2$O$_4$. These impurities could be converted into the main product by a short time sintering. Figure 8 is the XRD pattern of the product which was sintered for 5 hours at 1,000 °C after the SHS process. From the XRD pattern we could determine that the main product had a tetragonal lattice structure and the (440) plane was shown at diffraction angle 2θ = 57.02°. Thus, we calculated the lattice constant of the main product to be 9.12 Å and has a 0.9% mismatch for GaN thin film (9.04Å for 2× cubic GaN).

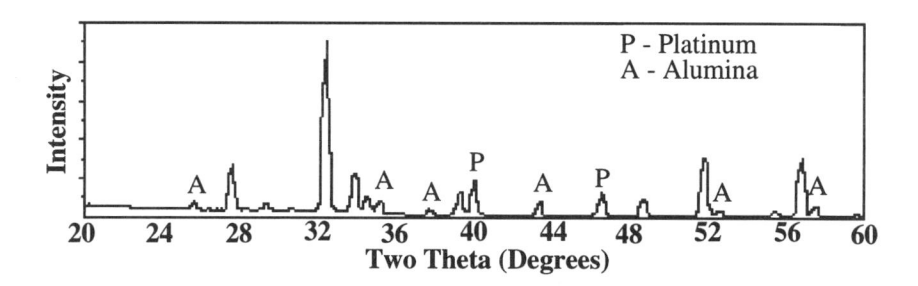

Figure 8. XRD patterns of an SHS sample starting from a mixture of Cd-In-Ga$_2$O$_3$ /Cd-In$_2$O$_3$-Ga$_2$O$_3$ = 1/9.

In order to determine the local compositions of powders after sintering were analyzed by electron microprobe analysis (EMPA) (Figure 9). Several phases were identified in the powders after the sintering. For example, CdIn$_{1.2}$Ga$_{0.8}$O$_x$, CdIn$_{1.6}$Ga$_{0.4}$O$_x$, CdIn$_{1.8}$Ga$_{0.2}$O$_x$, and CdIn$_{1.3}$Ga$_{0.6}$O$_x$ (not shown in Figure 9) were found in the samples. The instrument error of this measurement was estimated to be 10-15%. The heterogeneity of the samples indicated by EMPA may be due to the heterogeneity of the reactant powders and/or insufficient reaction time. Thus, using smaller reactant powders may improve the formation of the desired product (CdIn$_{1.673}$Ga$_{0.327}$O$_4$). The homogeneity of the product could be increased by increasing interaction time during the SHS (for example, using larger samples), and/or post-combustion treatment. Another way to increase the reaction temperature and time could be using higher gas pressure to suppress the sublimation and increase the homogeneity. An SHS process under high oxygen pressures is currently being studied to improve the product quality.

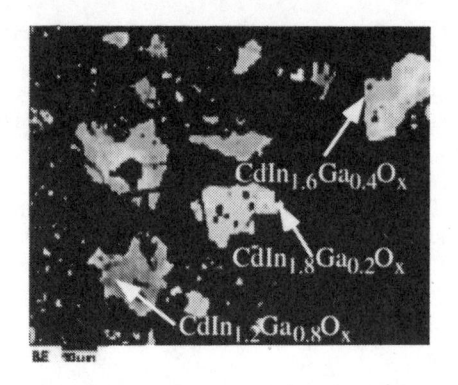

Figure 9. EMPA photographs of a sample after SHS.

CONCLUSIONS
SHS may be used to rapidly produce $CdIn_{1.673}Ga_{0.327}O_4$ Mixtures of two reaction systems ($Cd$-$In$-$Ga_2O_3$ and $Cd$-$In_2O_3$-$Ga_2O_3$) were tested. The results show that a mixture containing 10-20 wt% of $Cd$-$In$-$Ga_2O_3$ is suitable for the production of $CdIn_{1.673}Ga_{0.327}O_4$ by SHS. It generates a high combustion temperature and avoids sublimation during the combustion.

The main product after sintering the SHS product has a lattice constant of 9.12 Å which has a 0.9% mismatch for GaN thin film (9.04Å for 2× cubic GaN).

ACKNOWLEDGMENT
This work was supported by the NASA Center for Applied Radiation Research at Prairie View A&M University, Texas Center for Superconductivity at the University of Houston, and the NSF Materials Research Science and Engineering Center at the University of Houston. The authors wish to thank Q. Ming, A. Wagner, R. Scates, K. Ross, and J. Reifsnider for helpful assistance and discussions.

REFERENCES
[1] F. Fichter. *Z. Anorg. Chem.* **54** 322 (1907).
[2] J.V. Lirman and H.S. Zhdanov, *Acta Physicochim, USSR*, **6** 306 (1937).
[3] H.G. Grimmeiss and H. Koelmans, *Z. Naturf,* **14a** 264 (1959).
[4] G.M. Borsuk, "Recent Progress in GaN Devices", Semiconductor Devices, Krishan Lal, Ed., International Society for Optical Engineers, **2733** 154 (1996).
[5] D.J. Smith, D. Chandraskhar, B. Sverdlov, A. Botchkarev, A. Salvador, and H. Morkoc, Appl. Phys. Lett., **67**(13) 1830 (1995).
[6] L.T. Romano, J.E. Northrup, and M.A. O'Keefe, Appl. Phys. Lett., **69**(16) 2394, October (1996).
[7] T. Ishii, Y. Tazoh, and S. Miyazawa, "Polar Defects in $LiGaO_2$ Substrates Lattice Matched with GaN", Mat. Res. Soc. Symp. Proc., **468** 155 (1997).

[8] J.-H. Li, O.M. Kryliouk, P.H. Holloway, T.J. Anderson, and K.S. Jones, "Microstructures of GaN Films Grown on A LiGaO$_2$ New Substrate By Metalorganic Chemical Vapor Deposition", Mat. Res. Soc. Symp. Proc., **468** 167 (1997).

[9] J.F.H. Nicholls, H. Gallagher, B. Henderson, C. Trager-Cowan, P.G. Middleton, K.P. O'Donnell, T.S. Cheng, C.T. Foxon, and B.H.T. Chai, "Growth and Optical Properties of GaN Grown by MBE on Novel Lattice-Matched Oxide Substrates", Mat. Res. Soc. Symp. Proc., **395** 535 (1996).

[10] J.M. Reifsnider, D. Gotthold, K.S. Nanjundaswamy, J. Goodenough, and B. Streetman, Annual Public Review, Science and Technology Center, Pickle Research Center, University of Texas at Austin, Austin, TX, February (1996).

[11] A.G. Merzhanov, V.M. Shkiro and I.P. Borovinskaya, *USSR Inventor's Artificate*, No. 255221 (1967), Byull. Izobr. No. 10 (1971).

[12] A.G. Merzhanov, "Worldwide Evolution and Present Status of SHS as a Branch of Modern R&D (on the 30th Anniversary of SHS)", Int. J. S.H.S., **6**(2) 119 (1997).

[13] P.B. Avaicyan, M.D. Nersessyan, and A.G. Merzhanov, Amer. Cer. Soc. Bull., **75**(2) 50 (1996)

[14] S-C Lin, J. T. Richardson, and D. Luss, "YBa$_2$Cu$_3$O$_{6+x}$ Synthesis Using Vertical Self-propagating High-temperature Synthesis", Physica C, **233** 281 (1994).

[15] Lebrat, J-P and A. Varma, "Combustion Synthesis of the YBa$_2$Cu$_3$O$_{7-x}$ Superconductor", Physica C, **184** 220 (1991).

[16] S.-C. Lin, R. Willkins, and D. Luss, "Self-propagating High-temperature Synthesis of Cd-In-Ga-O Powder for GaN Thin Film Substrate", Ceramic Transactions, **86** 73 (1998).

# Microwave Processing

# STEP SINTERING OF MICROWAVE HEATING AND MICROWAVE PLASMA HEATING FOR CERAMICS

Jinsong Zhang, Lihua Cao, Yongjin Yang, Yunxiang Diao, and Xuexuan Shen

Institute of Metal Research, Chinese Academy of Sciences
72, Wenhua Road, Shenyang 110015 China

## Abstract

In this paper, according to this invention, we propose the step sintering of microwave heating and microwave plasma heating (SSMP). In this process, microwave heating and plasma heating are ingeniously combined in one chamber, and vacuum is unnecessary for microwave induced plasma. In the first step, the microwave energy directly heats ceramics to a given temperature, above which microwave will mainly ionize the gas and sustain a plasma. Then the plasma further heats ceramics to the final temperature to finish sintering. Maintaining the advantages of microwave heating and plasma heating, this step approach will overcome the limitation of both processes for processing ceramics. With this method, 50 $Al_2O_3$ nozzles were uniformly heated from room temperature to 1850 °C in 1.5h with no more than 15kWH energy consumption. The rate of finished product and final density reached 95% and 99% TD.. Industrial experiment shows that the nozzles sintered by SSMP have a life-span of three months and 0.85 rate of replacing coke with coal. These parameters are respectively 6 times and 1.5 times that of nozzles produced by conventional sintering.

## Introduction

Numerous studies have shown that microwave sintering of ceramics has some advantages over conventional sintering process, such as savings in energy and process time, volumetric and rapid heating, lower densification temperature, uniform microstructure of sintered ceramics, improved materials performance, production of new materials and microstructure etc.. Based on the form of microwave energy used, microwave sintering can be basically divided to microwave heating sintering, microwave plasma heating sintering and step sintering which combines microwave heating and microwave plasma heating.

Because microwave heating is strongly controlled by the capacity of material absorbing microwave, the number of ceramics that can be processed by microwave heating sintering is limited owing to the following reasons.

First, ceramics with low dielectric loss factors are difficult to heat from room temperature due to their minimal absorption of the incident microwave energy. For ceramics with permittivity and/or loss factors that increase rapidly with temperature, microwave heating easily causes uneven heating and thermal runaway, resulting in cracking ultimately. The hybrid heating technique proposed by Janney at el.[1] makes it possible to overcome the above difficulties in principle. With this approach, however, stable and uniform microwave heating of these types of ceramics is still hard to achieve because of poor match of dielectric loss factors between microwave susceptor and sample throughout the sintering process. Therefore this process is only applicable to limited species of ceramics. In order to solve this problem, we proposed an energy distributing technique[2] relying on the combination of dielectric multi-mode resonating and hybrid heating. With this method, it is possible to select a material as the insulator which is identical or similar to the sample in composition to reduce temperature gradient between the sample and the environment and guarantee rapid and uniform heating. Relying on this advantage, this approach therefore is capable of successfully sintering all ceramics without significant ionic and metallic conductivity when an extremely high sintering temperature is not required. But this characteristic also introduce a challenge as the efficiency of microwave energy utilization is not high and the processing cost is increased.

Second, during sintering of some semiconductor ceramics and ceramics with metallic or ionic conductivity, the coupling of microwave radiation of the sample decreases rapidly with increasing density, so these ceramics can not be heated efficiently to the target temperature.

Compared to microwave heating sintering, microwave plasma sintering[3-4] is not affected by the dielectric properties of the ceramics, and has shown an advantage of heating sample effectively to any required temperature. But some drawbacks restrict its application. First, a large area microwave plasma is very difficult to achieve in room temperature and atmospheric pressure, however low pressure plasma is prone to causing sample volatilization at elevated temperature; second, while sintering ceramic with large size and/or low thermal conductivity and high thermal expansion coefficient, rapid heating from room temperature easily leads to sample cracking due to great thermal gradient between interior and exterior of sample.

Step sintering process by combination of microwave heating and microwave plasma heating may be an ingenious approach to overcome the limitation of both processes. Samandi et al [5] conducted pioneer work in this field. They proposed microwave induced plasma assisted sintering. In this process, low pressure microwave plasma is used to heat ceramics with poor absorption of microwave energy from room temperature to a certain temperature at which they become absorbing. But other problems of microwave plasma sintering still remain due to the use of low pressure microwave plasma. The invention of microwave induced atmospheric pressure plasma during microwave heating [6] makes it possible to overcome the difficulties of MIP-assisted sintering. Basing on this invention, we propose a novel step sintering of microwave heating and microwave plasma (SSMP), which we think can be a universal method to sinter a wide variety of ceramics in precondition of using microwave energy efficiently. In the present paper its characteristics and application on processing high performance $Al_2O_3$ nozzles will be reported.

**EXPERIMENTAL**

Alumina nozzles 20mm in diameter and 70mm in height were formed by means of thermal die-casting, and are shown in fig.1. The composition of nozzles is 99.5%$Al_2O_3$+0.5% MgO. 50 such nozzles were loaded in a thermal insulator shown in fig.2, and were then sintered in MFM-863 II microwave sintering furnace with frequency of 2450 MHz and output power of

500W to 10 KW continuously adjustable.

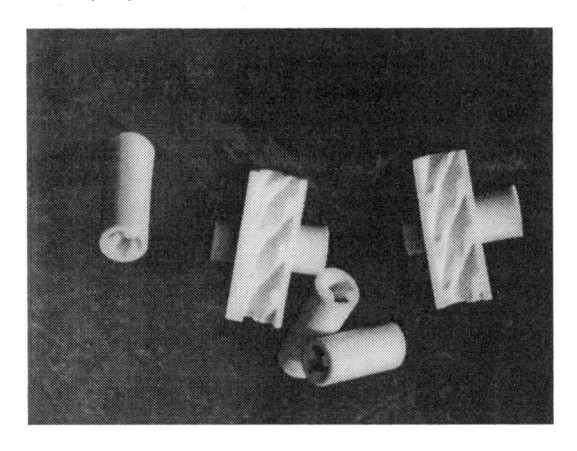

Fig.1 The photo of Al$_2$O$_3$ nozzles.

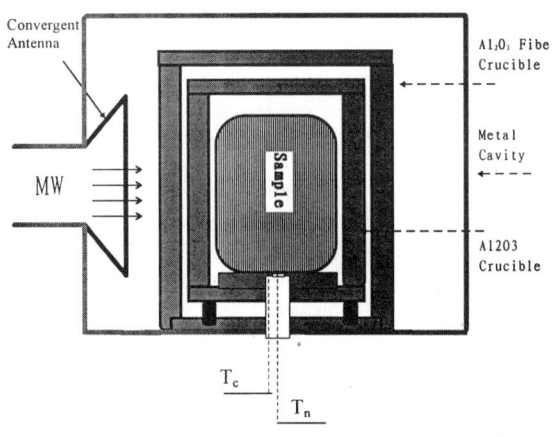

Fig.2 The scheme of structure of the microwave applicator used in the SSMP.

The outer part of the thermal insulator is made of Al$_2$O$_3$ fiber, and the inner part is a crucible made of small hollow Al$_2$O$_3$ balls with purity of 98%. Nozzles were heated from room temperature to 1850 ℃ in less than 80 min. and held for 10 minutes at 1850 ℃, then microwave power decreased to allow samples to cool to 1300 ℃. After that microwave power was turned out, and samples naturally cool down. The microstructure of sintered nozzles was examined by means of SEM. The life-span of sintered nozzles was tested in a blast furnace. Coal powders pass through the inner hole of nozzle at 1.5 ton per hour. The temperature at nozzle top is about 1000 ℃, and at nozzle tail is close to room temperature. It was taken out every 2 days to check if it was damaged.

## RESULTS AND DISCUSSION

Fig.3 shows the temperature and power curves during SSMP. According to fig.3, the step process can be divided into three phases. The first two phases correspond to microwave heating, and the later corresponds to microwave plasma heating.

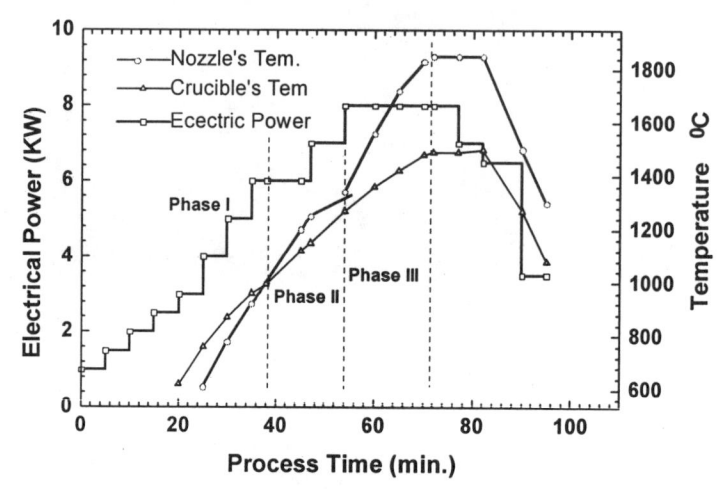

Fig3. The relationship of electric power and temperature of the nozzles and the inner crucible with process time during the SSMP.

The first phase is below 1000 ℃. In this phase, the temperature of crucible is always higher than that of nozzle, and the measured maximum temperature difference between the crucible and the nozzle is about 150 ℃. With temperature rising, the temperature difference rapidly decreases. This phenomenon indicates that samples are simultaneously heated by direct microwave heating and assisted heating of crucible, and the ratio of direct microwave heating to assisted heating increases with temperature rising. This heating mode is apparently more beneficial to rapidly and uniformly heating samples than previous hybrid heating due to the fact that the loss of alumina crucible is similar to that of the nozzle.

The second part is from 1000 to 1350 ℃. When temperature of samples reaches 1000 ℃, the temperature difference is inversed, and then the sample temperature is always higher than that of crucible. The temperature difference changes with nozzle's temperature, and reaches a maximum of about 100 ℃ when the nozzle temperature is 1250 ℃. When the nozzle temperature is close to 1350 ℃, the heating rate of nozzle rapidly decreases. It is obvious that the temperature inversion is a result of the marked increasing of dielectric loss of nozzle and exceeding the loss of the crucible; and the change of temperature difference should be caused by incessant redistribution of absorbed microwave energy between the samples and the crucible.

The third part begins from 1350 ℃. In this phase, heated air is ionized by microwave field to become microwave plasma in crucible. It is the normal pressure microwave plasma (NPMP) that rapidly heats nozzles to 1850 °C in less than 15 minutes. At the same time, however, crucible is heated only to 1500 °C. Therefore, microwave plasma mainly heats nozzles.

Alumina crucible plays an important role during this step process. First, at low temperature, as an insulator with feature of energy distribution, it is capable of assisting the heating of samples and raising ambient temperature of surrounding. When a small quantity of nozzles with poor microwave absorption are processed, it can effectively heat them from room temperature by using hybrid heating; when a great number of nozzles are heated, it absorbs enough microwave energy to raise the ambient temperature of the surrounding nozzles. Second, alumina crucible and the nozzles make up a dielectric multi-mode resonator, in which microwave energy is accumulated by means of dielectric multi-mode resonance. The enhanced microwave field within the resonator will benefit both direct microwave heating of the sample and the induced microwave plasma as well. The more the samples are loaded, the more efficient it became, and the more uniform and rapid microwave heating will be. Finally, It confines the NPMP. Generally speaking, both high temperature gas and the enhanced electric field benefit the ionization of gaseous molecule and maintaining of plasma. Inside the crucible, the air temperature is well above that outside the crucible, and the electromagnetic field is also higher than that outside the crucible due to dielectric multi-mode resonance effect. Therefore NPMP should be generated first inside the crucible and then naturally confined within it.

The key problem in step process of microwave sintering and plasma sintering is to uniformly induce and confine the plasma around the sample all along. This requirement can be met through proper thermal insulation design and appropriate control of heating rate. Fig.4 shows a typical temperature curve corresponding to uneven plasma distribution. It can be seen from this figure that when plasma was generated, sample temperature first decreases, and then rises again after half a minute or so. This phenomenon implies that the plasma was not first generated around the sample.

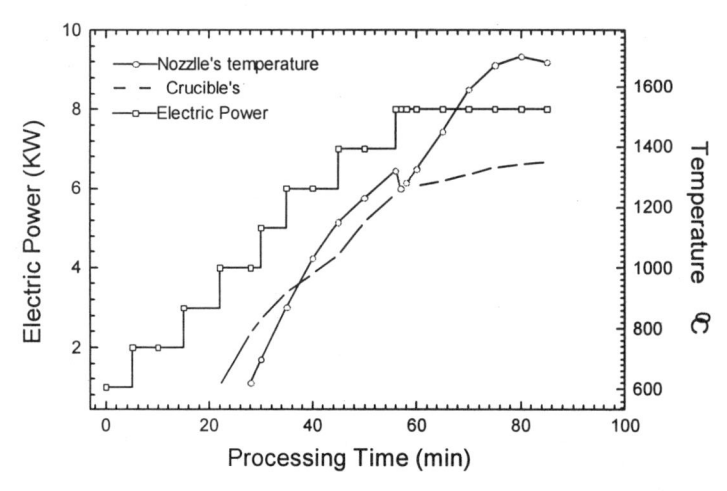

Fig4. The relationship of electric power and temperature of the nozzles and the inner crucible with process time corresponding to uneven ionization and distribution of plasma during the SSMP

With SSMP, 50 Al$_2$O$_3$ nozzles were uniformly heated from room temperature to 1850 °C in 1.5h with less than 15kWh energy consumption. The rate of finished product reached 95%. This result indirectly proves that the microwave plasma is uniformly distributed around all nozzles. The microstructure of nozzles sintered by this step process at 1600 °C and 1850 °C respectively is shown in fig.5(a,b). This figure shows that a uniform equiaxied crystalline structure with mean size of 2.8 μm is obtained in the nozzle sintered at 1600 °C, and that the nozzle sintered at 1850 °C has a relatively uniform lamellar structure with thickness of 8 to 10μm, which is beneficial to the wear-resistance and thermal shock-resistance of the nozzles. Industrial experiment carried out in a blast furnace shows that the nozzle sintered at 1850 °C by SSMP has a life-span of 3 month and a rate of replacing coke with coal of 0.85; while the life-span and replacing rate of the nozzle sintered at 1600 °C by the same process is 1.5 months and 0.7 respectively. These parameters are all better than those of nozzles processed by conventional sintering method, which are half a month for life-span and 0.6 for replacing ratio.

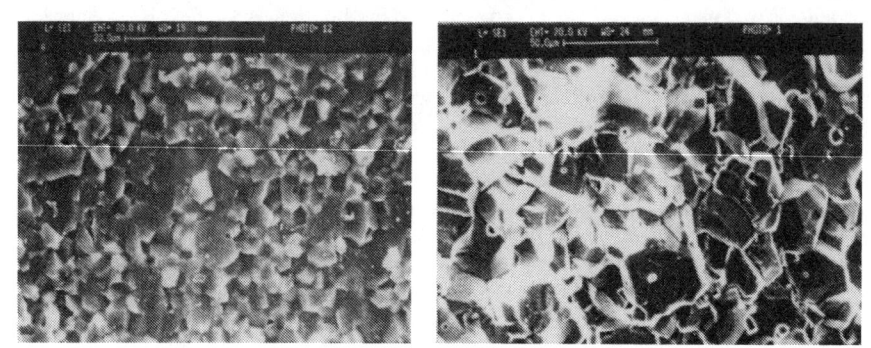

Fig.5, The SEM of microstructure of nozzles sintered by the SSMP at 1600 °C (fig.5(a)) and 1850 °C (fig.5(b))

## CONCLUSION

The experiment showed that microwave heating and plasma heating can be ingeniously combined in one chamber to form a novel step sintering process. Maintaining the advantages of microwave heating and plasma heating, this step approach will overcome the limitation of both processes, and has be successfully applied to the sintering and high temperature treatment of low-dielectric loss materials such as Al$_2$O$_3$. It also has the potential advantage to meet the requirement of high temperature sintering of most ceramics which are not available in microwave sintering or microwave plasma sintering alone because it does not rely on the dielectric property of materials at high temperature and vacuum is unnecessary to ionize and sustain plasma during the use of this approach. So this technique may promote the transfer of microwave sintering from laboratory-scale research to production-scale application. With this method, the high performance Al$_2$O$_3$ nozzles have been sintered with 95% rate of finished product and 99% relative density. Industrial experiment shows that the nozzles sintered by SSMP have a life-span of three months and 0.85 rate of replacing coke with coal, which are all better than that of nozzle produced by conventional sintering .

## REFERENCES

[1] M.A.Janney, C.J.Calhoun, and H.D.Kimrey, Ceram. Trans. 21, Amer. Ceram. Soc. (1991)311-318

[2] Jinsong zhang, Yongjin Yang, Lihua Cao, Yunxiang Diao, Chinese Patent, No. 96226611.6

[3] D. L. Johnson and R. R. Rizzio, Am. Ceram. Soc. Bull., 59, (4) pp467- 72,1980

[4] J. B. Salasman and S. P. Holdenfield, Mat. Res. Soc. Symop. Proc. Vol.124, 20-25, (1988)

[5] M. Samandi, M. Doroudian, Mat. Res.Soc.symp.,347,605-615

[6] Jinsong zhang, Yongjin Yang, Lihua Cao, Yunxiang Diao, Chinese Patent, No.94110112.6

# MICROWAVE-ASSISTED SINTERING OF $ZrO_2$-8 mol%$Y_2O_3$: ENHANCEMENT OF DENSIFICATION

Sharon M Rathbone, Robert Freer and Frank R Sale
Materials Science Centre
University of Manchester/UMIST
Grosvenor Street
Manchester M1 7HS, UK

F C Ruth Wroe
EA Technology Ltd
Capenhurst
Chester CH1 6ES, UK

## ABSTRACT

Zirconia powders containing 8 mol % $Y_2O_3$ were pressed into disc shaped pellets and sintered in a hybrid furnace developed by EA Technology Ltd. This allowed the simultaneous application of microwave and radiant energy. Samples were sintered by conventional and microwave-assisted heating at rates of 5, 10 and 15°C $min^{-1}$. Samples were characterised by optical and scanning electron microscopy and X-ray diffraction techniques. The fired samples reached densities up to 99% theoretical. Microwave-assisted heating resulted in densification at temperatures typically 50-100°C lower than conventionally heated samples. The highest densities were achieved by the use of microwave-assisted heating at fast heating rates, enabling savings in energy. There was no difference between the activation energies for microwave-assisted and conventional sintering (~305 kJ $mol^{-1}$), but microwave-assisted heating resulted in an increase of the pre-exponential factor by 130%.

## INTRODUCTION

The benefits of using microwaves to sinter ceramics are now well established for a variety of materials including alumina, zirconia and titanates [1-3]. The reduction in sintering time, sintering temperature or activation energy for sintering upon the use of microwaves are usually described in terms of a 'microwave effect'. Various models have been proposed to explain such effects including differences in ionic conductivity [3] and non-thermal processes, eg changes in lattice vibrations [4]. This study is concerned with the sintering of $ZrO_2$-8 mol % $Y_2O_3$, ie fully stabilized (FSZ) ceramics. A number of authors have explored direct microwave sintering of zirconia ceramics [2,3,5], but the present investigation employs 'microwave-assisted' technology (the use of both microwaves and radiant energy in the same sintering cycle) and compares the results with conventional sintering. The activation energies and pre-exponential factors for each type of sintering conditions are assessed.

## EXPERIMENTAL
### Starting Materials

The starting powder was Tosoh TZ-8Y FSZ (Tokyo, Japan). The nominal yttria content and impurities are (all wt %): $Y_2O_3$ 13.4; $Al_2O_3$ <0.005; $SiO_2$ < 0.005; $Fe_2O_3$ 0.003; $Na_2O$ 0.052. PEG binder was added to the powder (based on 1.5g PEG/250 ml deionised water). Disc-shaped samples were prepared by pressing 6g powder at 100 MPa in a 19mm steel die. Prior to sintering, the binder was burned out in a conventional Carbolite CSF1100 tube furnace. The green density of pressed specimens was approximately 48% theoretical.

### Sintering Experiments

The sintering experiments were carried out in a dedicated microwave-assisted furnace, invented and developed at EA Technology Ltd., UK [6]. It comprised a 2.45 GHz multimode microwave cavity combined with radiant heating elements [7]. This enabled conventional-only, or microwave-assisted sintering experiments to be performed in the same furnace. The construction of the 'microwave-assisted' furnace is shown in Figure 1.

Samples were housed in the upper part of an alumina fibreboard box within the body of the furnace. Two zirconia discs were located within a cavity in the fibreboard box, with a duplex type R thermocouple arranged centrally, 3mm from the samples (Figure 2). To avoid contamination by the fibreboard, the pellets were placed on FSZ powder. For microwave-assisted experiments, the level of microwave power was set to 300W and the thermal cycle maintained by varying the radiant input power.

Pairs of conventional-only and microwave-assisted sintering experiments were performed at temperatures in the range 1150-1500°C; heating rates were 5, 10 and 15°C $min^{-1}$. There was no hold at maximum temperature.

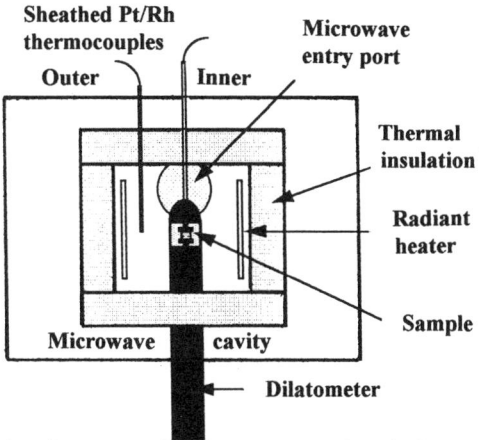

**Figure 1.** Schematic diagram of microwave-assisted furnace with (optional) dilatometer in position.

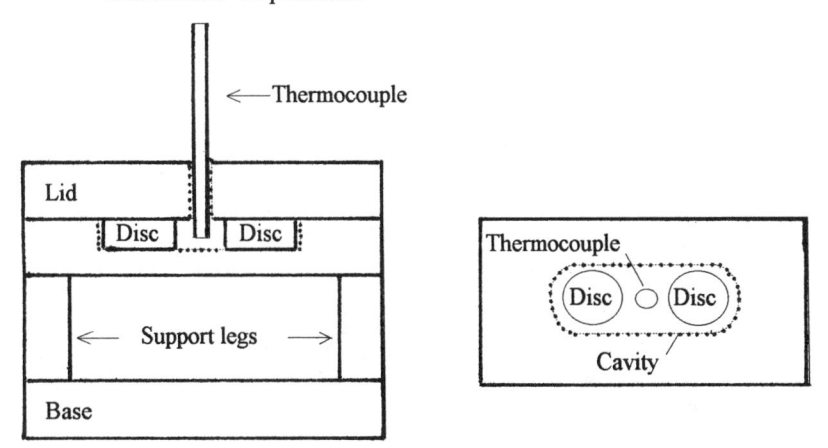

**Figure 2.** Front elevation and plan views of alumina fireboard box used to house samples in microwave-assisted furnace.

## Specimen Characterisation

Densities of fired discs were determined from mass and dimension measurements; the typical uncertainty in the fired densities was ±0.5%. Phase analysis of sintered specimens was performed using a Philips PW3710 X-ray diffractometer with monochromatic Cu $K_\alpha$ radiation. Spectra were compared with standard JCPDS data.

For microstructural analysis, specimens were ground (30μm diamond-impregnated wheel) and polished (down to 1μm diamond paste), thermally etched and examined by optical and scanning electron microscopy (Philips SEM 515). Grain sizes were determined using standard procedures [8].

## RESULTS AND DISCUSSION

X-ray diffraction confirmed that all the sintered samples (both conventional and microwave-assisted heating) had a cubic structure, as expected for fully stabilised zirconia.

Microstructure development in microwave-assisted and conventionally sintered zirconia will be reported elsewhere, but Figure 3 shows typical SEM micrographs. In general grain sizes are very small, < 0.5μm for specimens having densities less than 90% theoretical, but increase to 0.5-2.0μm for specimens having densities of 90-99% theoretical. Nightingale et al [5] reported grain sizes of approximately double this for zirconia-8mol % yttria sintered by conventional or microwave heating.

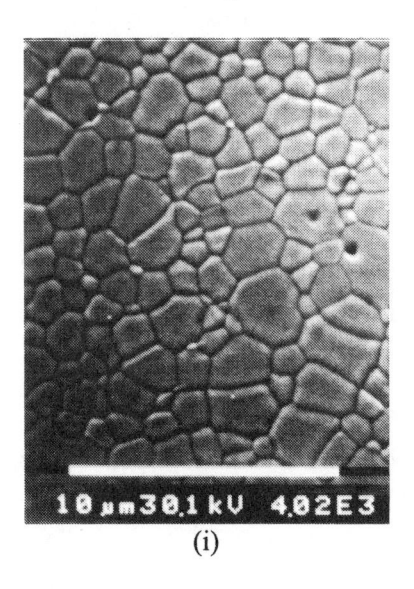

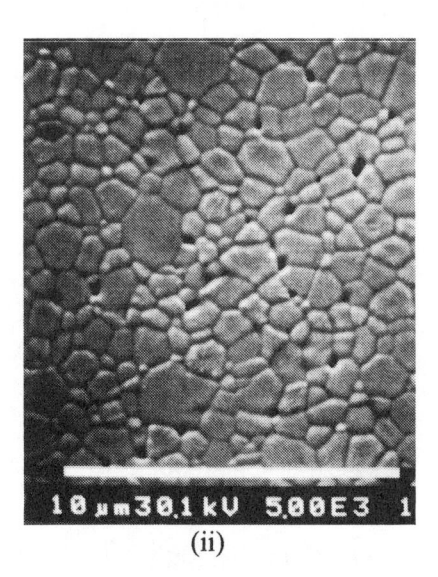

(i)  (ii)

**Figure 3.** SEM micrographs of Tosoh TZ-8Y samples fired at 10°C min$^{-1}$ :
(i) conventional sintering to 1350°C, (ii) microwave-assisted sintering to 1300°C.

Sintered densities depend critically on the heating rate, the form of power and final temperature; they varied from 54% to 99% theoretical (Figure 4).

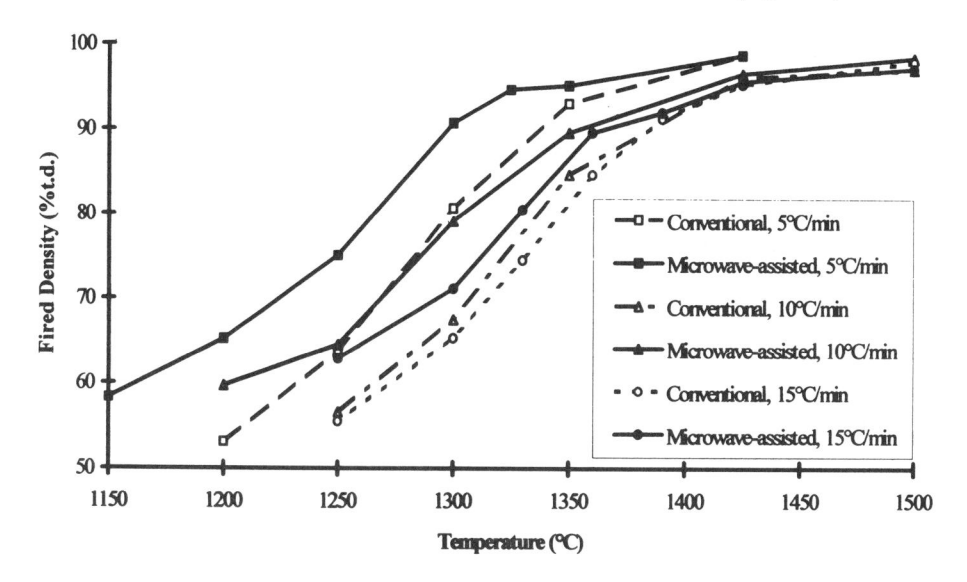

**Figure 4**. Densities of Tosoh TZ-8Y ceramics after conventional and microwave-assisted sintering at 5, 10 and 15°C min$^{-1}$.

Microwave-assisted enhancement of sintering of the FSZ is apparent, with microwave-assisted densification occurring at lower temperatures than for conventionally heated samples. The higher heating rates led to more rapid densification, but also a reduction in the enhancement of sintering (possibly because of a reduction in the electric field strength at higher temperatures). The microwave-assisted enhancement diminished at the higher densities (>95% theoretical), possibly because grain growth resulted in a reduction of the driving force for vacancy motion [4]. It is significant that enhancement diminished first in samples heated at 15°C min$^{-1}$ (once the density reached 90 % theoretical) although grain growth commenced first in such samples. The fact that samples can be heated rapidly by microwave-assisted techniques, without degrading density, means significant savings in terms of sintering time and energy consumption.

The activation energies for conventional and microwave-assisted sintering were determined from densification data (Figure 4) by numerical integration of the equation:

$$dp/dt=(1-\rho)A\exp(-Q/RT) \qquad (1)$$

where ρ is the density, A is the pre-exponential factor, Q is the activation energy, R is the gas constant and T is temperature. The values of the pre-exponential factor and activation energy (Table I) for both conventional and microwave-assisted sintering may be compared to literature data where there are claims of reduction in the activation energy [9], or the pre-exponential factor [10] for diffusion in the presence of a high frequency electric field.

**Table I** Pre-exponential factors and activation energies for conventional and microwave-assisted sintering of Tosoh TZ-8Y samples.

| Heating Rate(°Cmin$^{-1}$) | Power | Pre-exponential factor, A (min$^{-1}$) | Activation energy Q (kJ mol$^{-1}$) |
|---|---|---|---|
| 5 | Conventional | 6.54 x 10$^8$ | 298 |
| 10 | Conventional | 8.22 x 10$^8$ | 305 |
| 15 | Conventional | 2.27 x 10$^9$ | 309 |
| 5 | Microwave-assisted | 8.75 x 10$^8$ | 297 |
| 10 | Microwave-assisted | 1.32 x 10$^9$ | 303 |
| 15 | Microwave-assisted | 6.50 x 10$^9$ | 321 |

The average activation energies for conventional and microwave-assisted sintering were 304±3 kJ mol$^{-1}$ and 307 ±10 kJ mol$^{-1}$ respectively. Young and Cutler [11] calculated the 'effective' activation energy (nQ) of yttria stabilised zirconia (YSZ) to be 125.6 kJ mol$^{-1}$, with the value of n equal to 1/3 for grain boundary diffusion or ½ for volume diffusion. The activation energy of YSZ was therefore determined to be 251 kJ mol$^{-1}$ or 377 kJ mol$^{-1}$ when volume or grain boundary kinetics dominate respectively [11]. The activation energy for sintering YSZ determined in the present study is intermediate between the values obtained by Young and Cutler, suggesting that both grain boundary and volume diffusion were involved in the sintering process.

From Table I there appears to be a small increase in activation energy with heating rate. It has been shown that increasing the heating rate results in the displacement of densification curves to higher temperatures [12]. Volume diffusion is associated with a higher activation than grain boundary diffusion and the former predominates at higher temperatures [13]. Sintering at higher heating rates and, hence, higher temperatures, therefore leads to an increase in the activation energy for sintering because volume diffusion predominates for a larger proportion of the sintering process.

The pre-exponential factors tended to increase with heating rate (Table I) but average values were $1.25 \times 10^9$ min$^{-1}$ and $2.9 \times 10^9$ min$^{-1}$ for conventional and microwave sintering respectively. Whilst there is no significant difference between the activation energies for sintering by the two types of heating, the microwave-assisted heating appears to have resulted in an increase in the pre-exponential factor by 130%. These trends are consistent with the results of Binner et al [10] and Katz et al [14], who found that the use of microwaves did not affect the activation energy for diffusion but resulted in an apparent increase in the pre-exponential factors by multiples of 3.3 [10] and 3.0 [14] respectively.

The difference in magnitude of the increase in pre-exponential factor between this study and results in the literature [10,14] may be due to differences in the experimental conditions. Both Binner et al [10] and Katz et al [14] used microwave-only heating rather than microwave-assisted heating. The level of microwave power used by Binner et al (>500W) was higher than that used in the present study (mainly 300W) and it has been noted [15] that increasing levels of microwave power result in increased enhancement of diffusion. In this study, the levels of microwave power were generally higher in experiments with a heating rate of 5°C min$^{-1}$ than in experiments with faster heating rates. However, from Table I it can be seen that for the microwave-assisted experiments the pre-exponential factors were enhanced by factors of 1.33, 1.6 and 2.86 over those for the conventionally heated samples when the heating rates were 5, 10 and 15°C min$^{-1}$ respectively. In view of the uncertainties associated with the pre-exponential values it is perhaps more appropriate to compare the *average* enhancement value of 2.32 with literature values. Whilst it is not known how much microwave power was employed by Katz et al [14] it is believed that increased microwave power input can lead to greater enhancement of the pre-exponential factor.

Binner et al [10] and Katz et al [14] determined activation energies from plots of the natural logarithm of the reaction rate constant [10] or diffusivity [14] against reciprocal temperature in the manner described by Wang and Raj[16]. However, Katz et al determined Q values for interdiffusivities from straight lines through pairs of data points. They compared their results, based on sintering by microwaves, with data from experiments by other workers using conventional power. It is unlikely that the extrinsic defects in the diffusion couples were identical in the two cases [14] which makes direct comparisons more difficult.

The different methods of calculation may have contributed to the observed differences in the enhancement of the pre-exponential factors between this and other studies [10,14]. To ascertain whether this was the case, the activation energies for conventional and microwave-assisted sintering in this study were recalculated using the method of Wang and Raj [16]. The temperatures at which various densities were reached, for each set of sintering conditions, were determined from densification curves (Figure 4). Then, from equation (1), the relationship between the rate of densification (dρ/dt) and activation energy (Q) can be expressed as [16]:

$$\ln(T d\rho/dt) = -(Q/RT) + \ln[f(\rho)] + \ln A - n \ln d \qquad (2)$$

where T, R and A have the usual meanings and $d$ is grain size. Data for conventional and microwave-assisted sintering are shown in Figures 5 and 6.

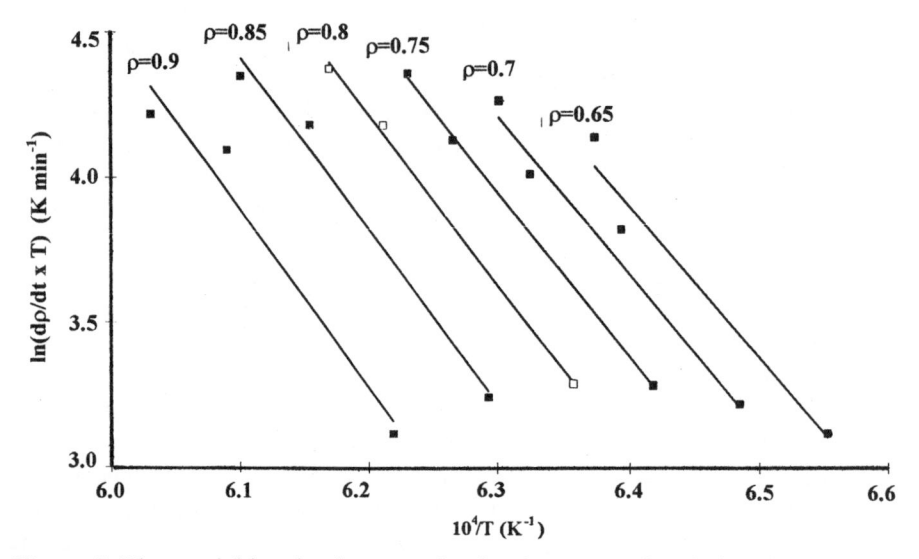

**Figure 5**. Plot to yield activation energies for the conventional sintering of Tosoh TZ-8Y samples, after the method of Wang and Raj [16].

Using the method of Wang and Raj [16] the activation energies for conventional and microwave-assisted sintering were determined (from Figures 5 and 6) to be 475±30 kJ mol$^{-1}$ and 353±12 kJ mol$^{-1}$ respectively. There is a significant difference (25%) between these activation energies. There is only 13% difference between the Q values for microwave-assisted sintering determined by the two methods of calculation, but the Q value for conventional sintering was 36% higher, when calculated by the method of Wang and Raj [16]. From the analysis of the data it became clear that any non-systematic behaviour in densification at one heating rate can significantly affect the Q value when the method of Wang and Raj is employed. At least four different heating rates would be required to significantly increase the accuracy of values determined by this method.

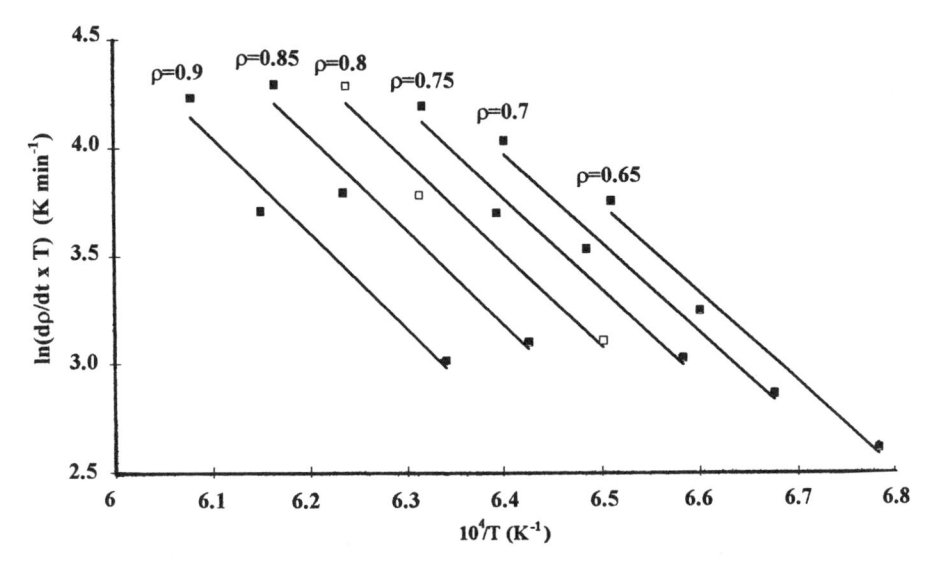

**Figure 6.** Plot to yield activation energies for microwave-assisted sintering of Tosoh TZ-8Y samples, after the method of Wang and Raj [16].

Although the calculated activation energies are sensitive to the method of analysis, there is comparatively little variation in the data for sintering in the presence of a microwave field. Despite the uncertainty in the magnitude of the differences in the pre-exponential factors [10,14], it is clear that there is enhancement when microwaves are applied. The pre-exponential factor is related to the material properties by an equation of the form [13]:

$$A = \gamma \lambda^2 \Gamma \qquad (3)$$

where $\gamma$ is a geometrical factor (including the number of nearest neighbours), $\lambda$ is the jump distance and $\Gamma$ is the jump frequency. Binner et al [10] noted that microwaves are more likely to affect the jump frequency $\Gamma$ than the geometrical factor $\gamma$ or the jump distance $\lambda$, since the latter two are controlled by the crystal structure. The jump frequency $\Gamma$ may be represented by [13]:

$$\Gamma = \nu \exp(-\Delta G^+/kT) \qquad (4)$$

where ν is the natural vibration frequency of the atoms, k is Boltzmann's constant, T the temperature and $\Delta G^+$ is the change in free energy. Since the latter is unaffected in many cases [10], it is possible that the microwave field affects the natural vibration frequency of the atoms. However, this is unlikely to account for the observed enhancement of sintering because typical resonant frequencies of ceramics are two or three orders of magnitude higher than microwave frequencies [17]. Equation (3) is based on the assumption of random walk motion for the atoms [13]. This assumption may not be valid for diffusion in the presence of a high frequency electric field since it has been suggested that, for positive and negative diffusing species having different diffusion coefficients, the drift of vacancies can be altered by space charge formation near surfaces and grain boundaries [4]. The pre-exponential factor may therefore increase because the high frequency electric field provides a driving force for vacancy motion in a particular direction.

## CONCLUSIONS

Microwave-assisted heating of yttria-stabilised zirconia ceramics enabled densification at temperatures 50-100°C lower than that for conventional sintering. The enhancement appears to depend on the strength of the electric field which may provide an additional 'non-thermal' driving force, perhaps affecting vacancy motion. The microwave-assisted enhancement diminished at higher densities (>95% theoretical). Samples sintered at higher heating rates using microwave-assisted techniques still achieved higher fired densities than samples produced by slower conventional heating, enabling significant time and energy savings. There was no significant difference between the activation energies for conventional and microwave-assisted heating (~305 kJ mol$^{-1}$) but microwave-assisted heating resulted in an apparent increase in the pre-exponential factor by 130%.

## ACKNOWLEDGEMENT
SMR was supported through the EPSRC, UMIST-EA Technology Ltd Postgraduate Training Partnership Scheme.

## REFERENCES
1. W.B. Harrison, M.R.B.Hanson and B.G.Koepke, "Microwave Processing and Sintering of PZT and PLZT Ceramics", *Mater. Res. Soc. Symp. Proc.* **124**, 279-286 (1988).
2. J.Samuels and J.R.Brandon, "Effect of Composition on the Enhanced Microwave Sintering of Alumina-Based Composites", *J. Mat. Sci.*, **27**, 3259-3265 (1992).

3. M.A.Janney, M.L.Jackson and H.D.Kimrey, Microwave Sintering of $ZrO_2$-12mol% $CeO_2$", in *Ceramic Transactions* 36, Microwaves: Theory and Applications in Materials Processing II, ed D E Clark, W R Tinga and J R Laia (Am. Ceram. Soc., Westerville Ohio, 1993) pp 101-108.
4. K.I.Rybakov and V.E.Semenov, *Phys. Rev. B*, **49**, 64-68 (1994).
5. S.A.Nightingale, H.H.Worner and D.P.Dunne, " Microstructural Development During Sintering of Yttria-Zirconia Ceramics", *J. Am. Ceram. Soc.*, **80**, 394-400 (1997).
6. F.C.R. Wroe, *EA Technology Report*: N2720 (Feb. 1993).
7. EA Technology Ltd., *Patent Application*: GB 2 281 016 A (Feb. 1995).
8. *DD ENV* 623-3:1993, Advanced Technical Ceramics, Monolithic Ceramics - General and Textural Properties, Part 3. Determination of Grain Size (1993) 3-9.
9. M.A.Janney and H.D.Kimrey, *Mat. Res. Soc. Symp. Proc.* **189**, 215-227 (1991)
10. J.G.P.Binner, N.A.Hassine and T.E. Cross, *J. Mat. Sci.*, **30**, 5389-5393 (1995).
11. W.S.Young and I.B.Cutler, *J. Am. Ceram.* Soc., **53**, 659-663 (1970).
12. S.M.Rathbone, *PhD Thesis*, University of Manchester (1996).
13. W.D. Kingery, H.K.Bowen and D.R. Uhlman, *Introduction to Ceramics*, 2$^{nd}$ ed. (Wiley-Interscience, New York, 1976).
14. J.D.Katz, R.D.Blake and W.H.Sutton, in *Ceramic Transactions*, vol 21, Microwaves: Theory and Applications in Materials Processing, eds D.E. Clark, F.D.Gac and W.H. Sutton (American Ceramic Society, Westerville, Ohio, 1991) pp 95-105.
15. R.Wroe and A.T.Rowley, *J. Mater. Sci.*, **31**, 2019-2026 (1996).
16. J.Wang and R.Raj, *J. Am. Ceram. Soc.*, **73**, 1172-1175 (1990).
17. V.M.Kenkre, in *Ceramic Transactions*, vol 21, Microwaves: Theory and Practice in Materials Processing, eds D.E Clark, F.D.Gac and W.H. Sutton, (American Ceramic Society, Westerville, Ohio 1991) pp 69-80.

PRESSURELESS MICROWAVE SINTERING OF FUNCTIONAL GRADIENT MATERIALS

R. Borchert *, M. Willert-Porada, University of Dortmund, Department of Chemical Engineering, Institute of Materials Sciences, P.O.Box 500 500, 44221 Dortmund, Germany

INTRODUCTION

Functional Gradient Materials (FGM) are composite materials with a continuous spatial change of properties. They can be described as a combination of macroscopic and microscopic composites, e.g., a composite layered material as shown in Fig. 1.

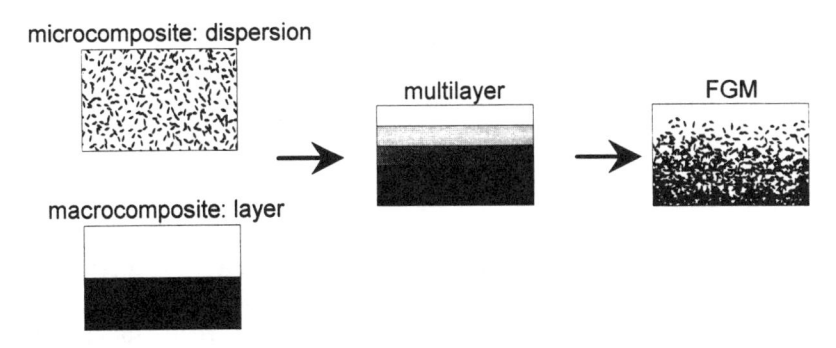

Fig.1: FGM are a combination of microcomposites and macrocomposites

FGM show several advantages as compared to conventional composites. Especially in metal-ceramic FGM the advantages of their components can be fully exploited. The possible applications of FGM are in energy conversion and generation [1]. This paper will concentrate on the most interesting material system for thermal barrier coatings, a FGM composed of a Ni based alloy and $ZrO_2$. The goal of the FGM-development for thermal barrier applications is to increase the operation temperature and the thermal shock resistance as well as the oxidation resistance [2].

An economic production of FGM can be achieved by pressureless sintering of powder metallugical parts with a compositional gradient. However, if one component of the FGM is thermally unstable under the conditions necessary to densify the other component, conventional pressureless sintering can not be applied. For some metal-ceramic systems, the sintering temperature necessary for densification of the ceramic phase is higher than the melting temperature of the metal phase. Upon shrinkage of the ceramic component due to sintering the molten metal is redistributed within the FGM and partially removed.

The application of a temperature gradient matching the compositional gradient is helpful, however, it is difficult to obtain a strong temperature gradient by conventional heating.

Because of the different heating behaviour of metallic and ceramic materials in a microwave field, microwave sintering can produce the desired temperature gradient in metal-ceramic FGM. Furthermore, the thermal gradient is preserved as steady state condition, therefore melting of the metal is prevented througout the sintering process.

Nevertheless, the shape and slope of a thermal gradient is very difficult to measure within an FGM sintered in a microwave field. Thermocouples cannot be used in microwave fields because of their "antenna-effect", the direct and strong absorption of microwave energy in the thermocouple. Pyrometric measurement is used to measure the temperature of the upper surface of the samples, but the spatial resolution is limited and the temperature profile is influenced by heat flow due to the openings in the sample insulation, which are necessary for pyrometric measurement. Therefore, a simulation program was developed to analyse the thermal gradient in metal-ceramic FGM upon microwave sintering.

EXPERIMENTAL

a) Sample preparation

The green parts were prepared by a powder metallurgical process. Powder fractions of the metallic and ceramic powders were homgenised by ball milling, dried and sieved. The composition changed in steps of 10 %, to give a full gradient from 100 % metal to 100 % ceramic within 11 layers, with a layer extension yielding an exponential gradient of composition as given by eq. (1). The influence of the exponent on the thermomechanical properties of the FGM can therefore be investigated and the compositional gradient is variable in a wide range [3].

$$c_{ceramic}(d) = c_0 \left(\frac{d}{H}\right)^x \qquad (1)$$

The stacked powder fractions were pressed in three different ways:
- The powders were filled in silicon dies, sealed into 0.1 mm polyethylene foil and pressed in an isostatic press with water as incompressible fluid.

- The powder fractions were layered in an adjustable die. A larger amount of powder is poured into the die and the surplus powder is removed with a blade. The stacked powder fractions are compacted by axial pressing.
- A combination of axial and isostatic pressing was applied in order to combine the more homogeneous pressure distribution of the isostatic pressing with the more accurate gradient formation of the axial pressing.

A scheme of the FGM-preparation is shown in Fig. 2.

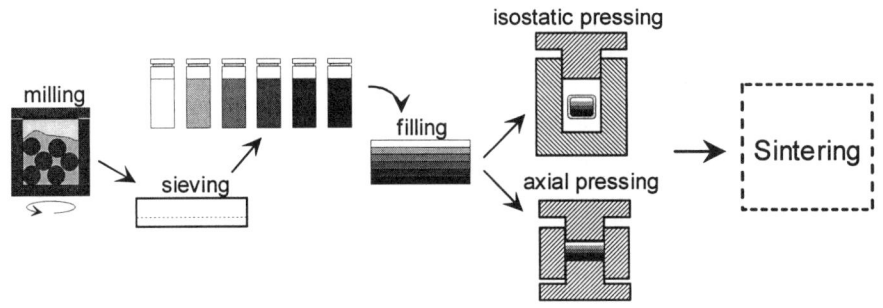

Fig.2: Green processing

Sintering was performed in a microwave oven with additional electric heating, as shown in Fig. 3 [4].

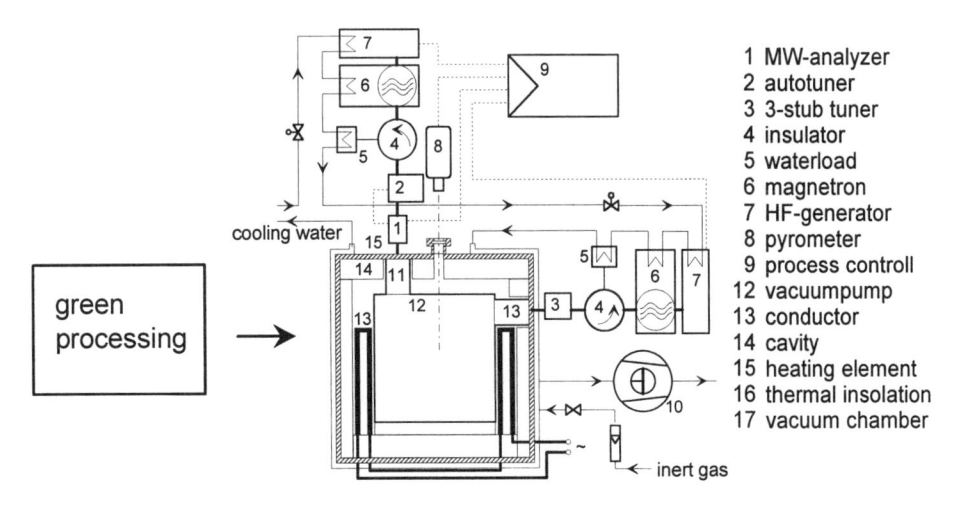

1 MW-analyzer
2 autotuner
3 3-stub tuner
4 insulator
5 waterload
6 magnetron
7 HF-generator
8 pyrometer
9 process controll
12 vacuumpump
13 conductor
14 cavity
15 heating element
16 thermal insulation
17 vacuum chamber

Fig.3: Microwave ofen for sintering metal-ceramic FGM

Electric heating at lower temperatures can prohibit the unwanted microwave effect of thermal runaway. At temperatures up to 800°C microwave absorption of some ceramic materials is highly temperature dependent, so that small regions of the material can be superheated. If the part is electroheated in the beginning, and microwave heating is added at T>800°C, thermal runaway can be avoided.

b) Microwave sintering

Microwave heating generates thermal gradients, if the material absorbs the radiation within the volume and the process is performed in a "cold" environment [5]. For homogeneous materials the temperature profile is symmetric with a hot center and a colder surface. For conventional heating the temperature profile is inverse to the microwave temperature profile, with hot surface and a colder center as shown in Fig. 4.

microwave heating
⇨ volume heating
⇨ heat flow from center to surface

conventional heating
⇨ surface heating
⇨ heat flow from surface to center

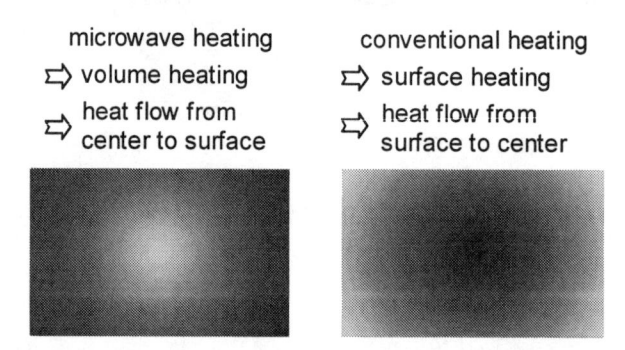

Fig.4: Temperature profiles for microwave heating and conventional heating

In metal-ceramic FGM at sintering temperature the absorption of microwaves is much higher within the ceramic phase than within the metal phase, therefore a non symetric heat flow and a non symetric thermal gradient is developed. The slope of the thermal gradient can be influenced by additional heat sinks and heat sources, as shown in Fig.5. Heat sources can be ceramic suszeptor materials and heat sinks can be developed by different thicknes of the thermal insulation. The side with thin insulation will be cooled more than the other side by radiation.

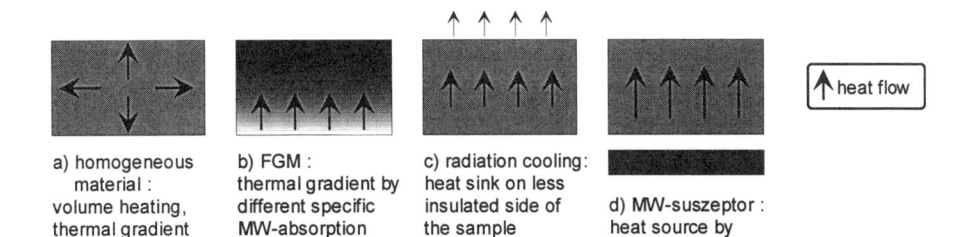

a) homogeneous material : volume heating, thermal gradient

b) FGM : thermal gradient by different specific MW-absorption

c) radiation cooling: heat sink on less insulated side of the sample

d) MW-suszeptor : heat source by good MW-absorber

↑ heat flow

Fig.5: Different ways to modify a thermal gradient

RESULTS AND DISCUSSION

As input for the simulation several material properties are used. Some material properties can be found in literature and other data had to be determined by experiments.

Material properties such as thermal conductivity and heat capacity necessary for the calculation were taken from literature [6]. The microwave absorption of metal ceramic composites, which is highly dependent on microstructure and thermal treatment, was estimated from experiments.

Metallic and ceramic powder mictures of different composition were pressed and prefired at different temperatures. The microwave absorption was determined from microwave heating experiments [5]. The resulting microwave absorption data as a function of composition and temperature are shown in Fig.6.

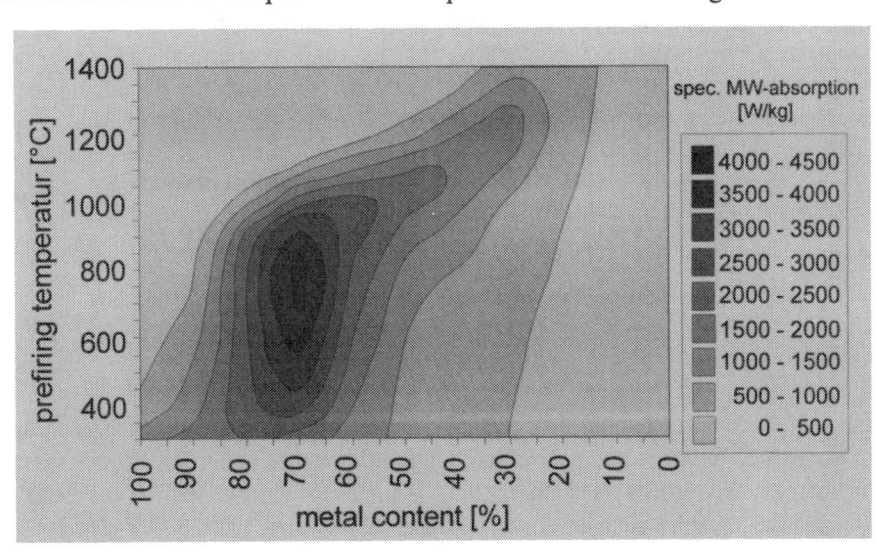

Fig.6: Specific microwave absorption of metal-ceramic dispersions

Innovative Processing/Synthesis: Ceramics, Glasses, Composites II

In a first stage, the simulation was performed by a 1-dimensional finite difference model [4]. The cylindrical sample is divided into a stack of cylindrical plates as shown in Fig.7.

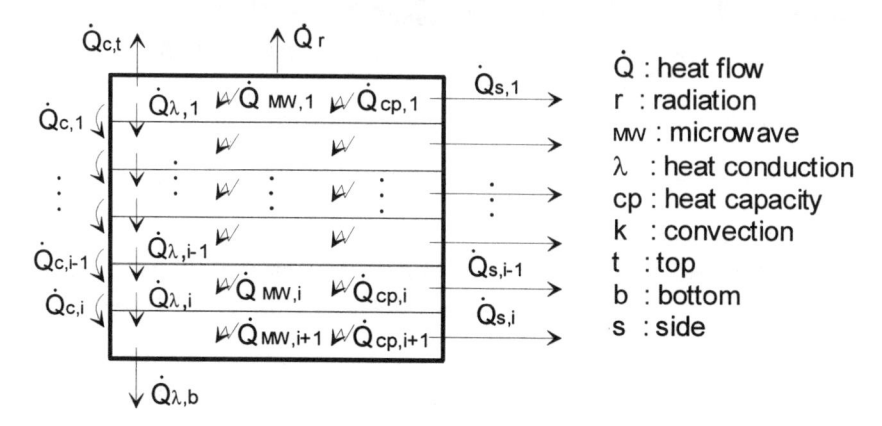

$\dot{Q}$ : heat flow
r : radiation
MW : microwave
$\lambda$ : heat conduction
cp : heat capacity
k : convection
t : top
b : bottom
s : side

Fig.7: 1-dimensional finite difference model for calculation of the thermal gradient within a microwave sintered metal-ceramic FGM

Based on an initial temperature profile, density, heat capacity and heat conductivity were calculated. From these material properties and the current temperature level the specific microwave dissipation and heat flows were calculated. The integrated heat flow and material properties yielded a new temperature profile, from which the next material properties can be calculated, as shown in the scheme in Fig.8.

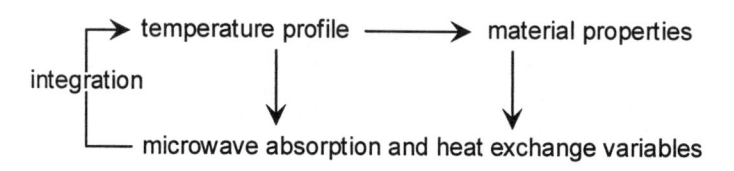

Fig.8: Flow chart of the calculation procedure

Using this one-dimensional model, temperature profiles could be calculated as a function of different variables. In Fig. 9a the influence of the gradient thickness on the temperature profile is shown, for a cylindrical sample with a diameter of 40 mm and a thickness of 8 mm, composed of Ni Cr 80 20 / 8Y-ZrO2. The thermal gradient of this system as a function of the exponent $x$ is shown in Fig. 9b.

To optimize the time increment for integration, a high starting value of 1 second is chosen which is surely not convergent. As long as the simulation is not convergent, temperature values begin to oscillate after some integration steps and the simulation is restarted with reduced time increment. By this way calculation could be performed on a normal Pentium PC in a few minutes.

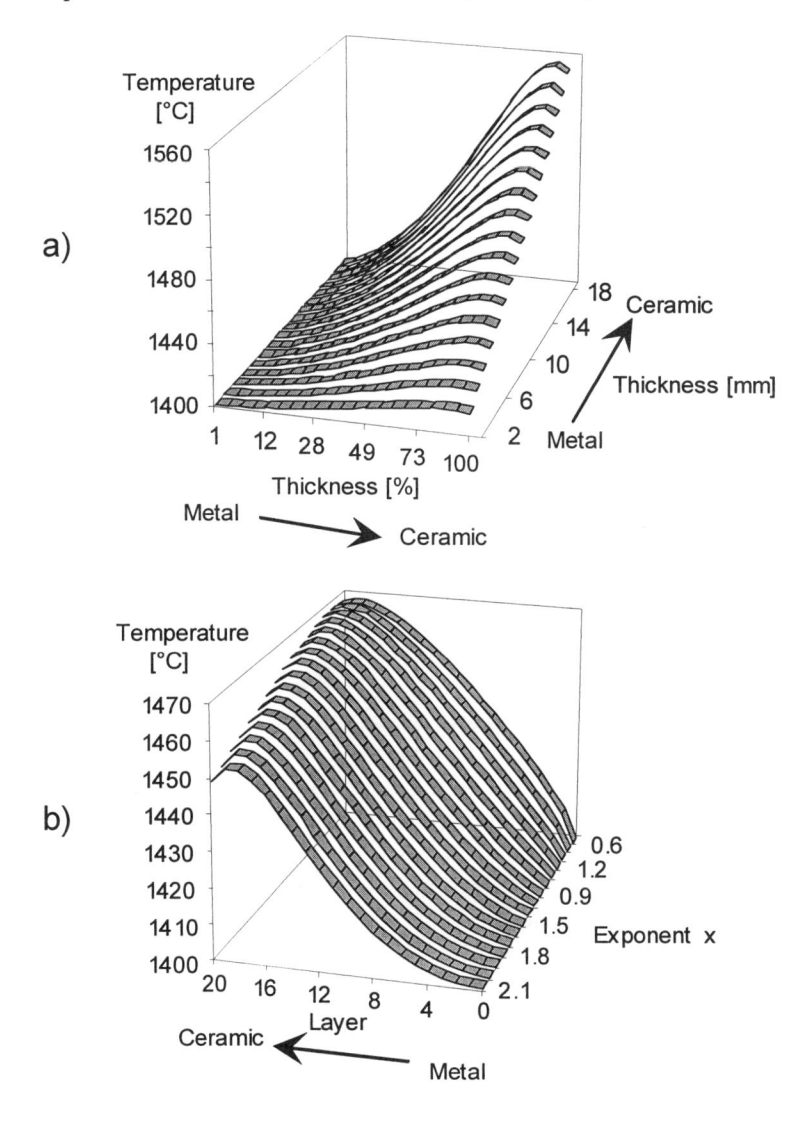

Fig.9: Results of the thermal gradient calculation for the 1-dimensional model

Innovative Processing/Synthesis: Ceramics, Glasses, Composites II

It is shown that the temperature gradient depends linearly on the thickness. For the given geometry an average temperature gradient of 9 K/mm is developed. The maximum of the temperature profile is located in the sample at a ceramic concentration of 90%. The maximal thermal gradient is about 10 K/s on the metallic side and the minimal thermal gradient is zero in the 90% ceramic layer.

If the exponent of the compositional gradient is varied, for an exponent of 1.1 the most linear temperature profile is developed. If the exponent is lower, the temperature in the center of the sample is higher, for higher exponents the middle temperature is lower.

A more detailed description of the temperature distribution in metal-ceramic FGM during microwave heating could be achieved by a 2-dimensional model. According to Fig. 10, the cylindric plates of the 1-dimensional model were subdivided into several rings, allowing for radial heat flow. From this model axial and radial temperature profiles could be calculated. Within this model, the influence of heat sinks or of additional heat sources can also be analysed.

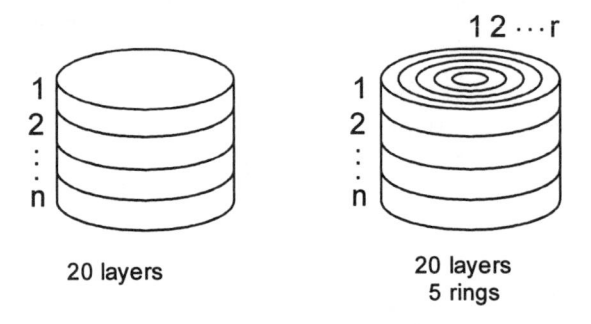

Fig.10: Conversion of the 1-dimensional model into a 2-dimensional model

The 2-dimensional simulation shows the temperature profile in a sample as a group of curves, therefore it is not possible to show the dependance of the temperature profile from a variable within a 2D plot.

The influence of the thickness of the gradient and of the compositional exponent are expected to be similar to the 1-dimensional model. A question which could be answered with a few simulations is the influence of the ambient temperature on the temperature profile within the sample investigated, as shown in Fig.11. If the hybrid heating system of the microwave oven (see Fig.3) is used, the temperature of the surrounding for the system sample-insulating casket can be set to different values.

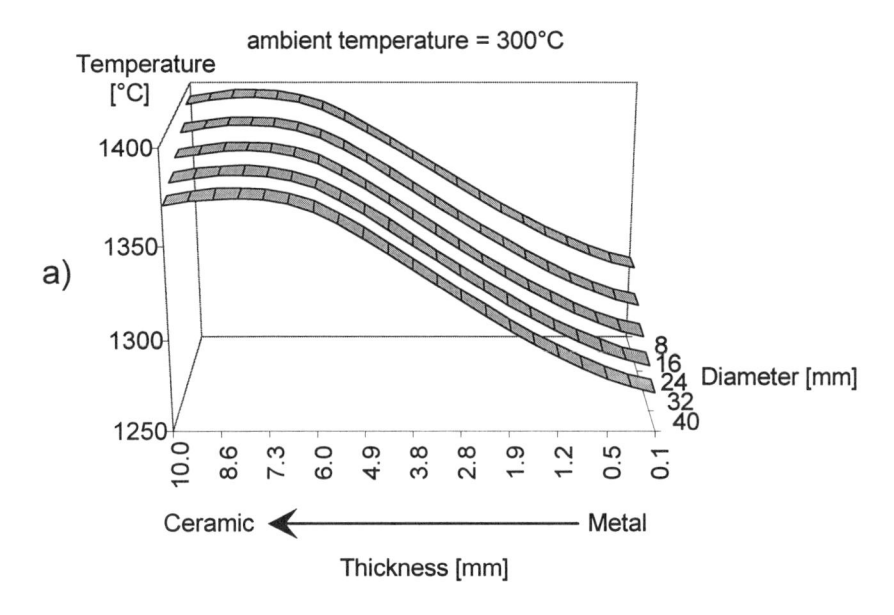

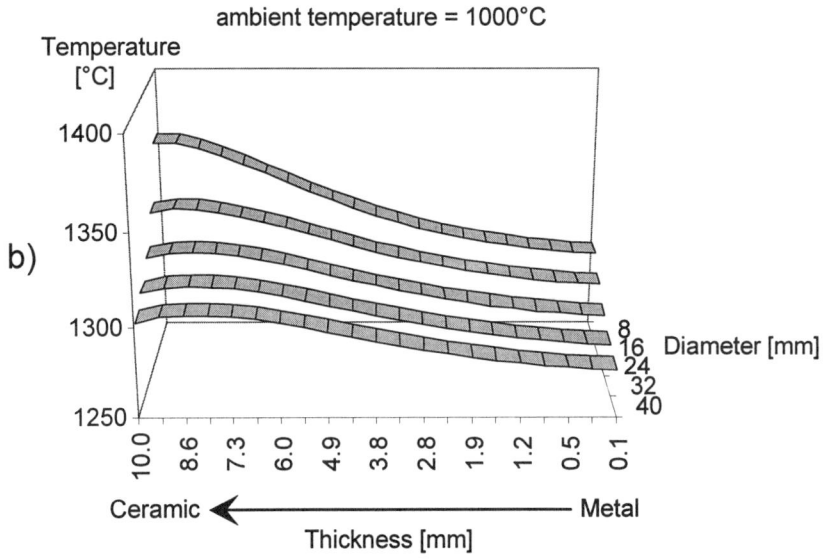

Fig.11: 2-dimensional temperature profiles for temperature of the surrounding of 300°C and 1000°C

The radiation cooling of the sample due to the pyrometric temperature measurement was taken into account within the 2D-model. It reaches a maximum in the center of the metal side and decreases with the radius of the FGM-cylinder. This effect is compensated by the radial heatflow, therefore in each cylindrical plate the temperature reaches its maximum in the center of the FGM.

For a temperature of the surrounding of 300°C the radial thermal gradient is only 20 K while the axial thermal gradient reaches 100 K througout the gradient. For a temperature of the surrounding of 1000°C the axial thermal gradient decreases to 60 K and the radial thermal gradient increases to 50 K.

CONCLUSION

Microwave sintering is a versatile method for pressureless sintering of metal-ceramic functional gradient materials. Due to different material properties FGM can develop temperature gradients of up to 100 K/cm. Therefore, the melting of the metal phase can be avoided.

If sintering is performed in a microwave oven with additional electric heating unwanted microwave effects can be avoided. Nevertheless, the thermal gradient within the sample is diminished.

Therefore, material systems that require a thermal gradient of 100 K must be sintered with pure microwave heating or the thermal gradient has to be increased by an additional heat source.

REFERENCES

[1] M.I. Mendelson, T.N. McKechnie, "Functionally Gradient Thermal Barrier Coatings : Design", "Ceramic Transactions, Vol. 34 - Functionally Gradient Materials, (Proceedings of the 2$^{nd}$ International Symposium on Functionally Gradient Materials)", Hrsg.: J.B. Holt, M. Koizumi, T. Hirai, Z.A. Munir, Westerville, OH, American Ceramic Society, ISBN 0-944904-64-5, 1993, 417-424

[2] M. Alaya, G. Grathwohl, J. Musil, "A Comparison of Thermal Cycling and Oxidation Behaviour of Graded and Duplex ZrO2-Thermal Barrier Coatings", in: "FGM 94, Proceedings of the 3rd International Symposium on Functionally Gradient Materials", Hrsg.: B. Ilschner, N. Cherradi, 405-411

[3] L.M. Zhang, R. Tu, R.Z. Yuan, "A General Rule of Design for The Optimum Compositional Gradation of Ceramic/Metal FGM", in: "FGM 94, Proceedings of the 3$^{rd}$ International Symposium on Functionally Gradient Materials", Hrsg.: B. Ilschner, N. Cherradi, Presses Polytechniques, Lausanne, Schweiz, ISBN 2-88074-290-0, 1994, 273-278

[4] "Verfahrensentwicklung zur Herstellung metallisch-keramischer Gradientenwerkstoffe durch Mikrowellensintern", Ph.D. Thesis, R. Borchert, University of Dortmund, Germany, 1998, 28-29

[5] M. Willert-Porada, T. Gerdes, R. Borchert, "Application of Microwave Processing to Preparation of Ceramic and Metal-Ceramic FGM", in: "FGM 94, Proceedings of the 3$^{rd}$ International Symposium on Functionally Gradient Materials", Hrsg.: B. Ilschner, N. Cherradi, Presses Polytechniques, Lausanne, Schweiz, ISBN 2-88074-290-0, 1994, 15-20

[6] Y.Touloukian, "Thermochemical Properties of High-Temperature Refractory Solid Materials", Hrsg.: Y.Touloukian, Collier Macmillan Ltd., London, 1967

# THE EFFECT OF CASKET GEOMETRY ON MICROWAVE HEATING AND ITS MODELING

Martin L. Traub, Ki-Yong Lee, and Eldon D. Case

Michigan State University
Materials Science and Mechanics Department
East Lansing, MI 48824

## ABSTRACT

During microwave heating, a casket composed of a microwave susceptor material such as zirconia or silicon carbide preheats low-dielectric-loss materials and provides thermal insulation. The casket geometry and microwave input power systematically affect the casket's internal steady-state temperature. At a fixed microwave input power of 600 W, the measured steady-state casket temperatures, $T_i$, range from about 1100°C to 1500°C as casket geometry changes. This paper discusses and compares two models for determining $T_i$ as a function of casket geometry and input power: (1) a simple, approximate relation based on an energy balance concept and (2) a Finite Element Method (FEM) analysis.

## 1. INTRODUCTION

In this paper, we shall first discuss a model for steady-state casket temperature in terms of a simple and approximate energy balance model [1,2]. Second, we discuss the experimental data obtained for refractory caskets composed of a hollow cylindrical section and two disc-shaped end plates. We then present a Finite Element Method (FEM) analysis for the steady-state casket temperature. Finally, we compare both models to the experimental data.

## 2. ENERGY BALANCE MODEL
### 2.1. Review of Previous Work

Two of the present authors (K.Y. Lee and E.D. Case) [1,2] used an energy balance concept to model the inner wall casket temperature, $T_i$, in terms of casket geometric variables and microwave power level. For a cylindrical casket composed of a hollow zirconia cylinder and disk-shaped end plates, $T_i$ is given by

---

To the extent authorized under the laws of the United States of America, all copyright interests in this publication are the property of The American Ceramic Society. Any duplication, reproduction, or republication of this publication or any part thereof, without the express written consent of The American Ceramic Society or fee paid to the Copyright Clearance Center, is prohibited.

$$T_i = \frac{P_C}{2\pi k} \cdot \frac{L_{SA}\left[\dfrac{1}{H} + b\ln(b/a)\right]}{bL_T L_{SA} + 2b^2\left[\dfrac{1}{H} + b\ln(b/a)\right]} + T_o \qquad (1)$$

where
- $P_C = P_I - P_R$
- $P_I$ = microwave input power
- $P_R$ = reflected microwave power [1,2]
- $k$ = thermal conductivity
- $T_i$ = temperature at inner wall of hollow cylinder
- $T_o$ = temperature at outer wall of hollow cylinder
- $H = h/k$
- $h$ = surface heat transfer coefficient for the cylinder
- $L_T = L_{SALI} + L_{ZYC}$ = total length of the casket, where
- $L_{ZYC}$ = length of zirconia cylinder
- $L_{SALI}$ = total thickness of the end plates
- $b$ = outer radius of the cylinder
- $a$ = inner radius of the cylinder.

In previous papers [1,2], we reported that the inner wall casket temperature, $T_i$, was a function of casket geometry. For example, for a fixed casket geometry, $T_i$, changed from about 1100°C to 1500°C at a fixed microwave input power level of 600 Watts. Equation 1 shows the systematic changes in $T_i$ in terms of geometric variables $b$, $b/a$, $L_T$, $L_{ZYC}$, and $P_C$. In order to least-squares fit equation 1 to the $T_i$, casket geometry, and microwave power data, equation 1 was recast [1,2] as equation 2, with the three free variables $C_1$, $C_2$, and $C_3$

$$T_i = \frac{P_C L_{SA}[C_1 + C_2 b\ln(b/a)]}{bL_T L_{SA} + 2b^2[C_1/C_2 + b\ln(b/a)]} + C_3 \qquad (2)$$

where
$$C_1 = \frac{1}{2\pi kH} = \frac{1}{2\pi h} \qquad (2a)$$

$$C_2 = \frac{1}{2\pi k} \qquad (2b)$$

$$\frac{C_1}{C_2} = \frac{1}{H} = \frac{k}{h} \qquad (2c)$$

$$C_3 = T_o. \qquad (2d)$$

Using equation 2, an $R^2$ value of 0.957 was obtained for a total of 132 sets of casket geometry/microwave power level data [2].

## 2.2. Simplifications and Approximations for the Energy Balance Model

In our previous two papers [1,2], we ignored the temperature dependencies of the following parameters: (1) thermal conductivity $k$, (2) surface heat transfer coefficient $h$, and (3) dielectric properties of the casket. Thus a least-squares fit to equation 2 yielded constant values of $k$ and $h$. For example, for temperatures between 400°C and 1650°C, vendor-specified (Zircar Products Inc.) thermal conductivity for zirconia cylinder (ZYC) and aluminosilicate board (SALI) varies from 0.09 to 0.23 W/(m·°C) and from 0.20 to 0.39 W/(m·°C), respectively [3]. Since temperature of the cylindrical casket would change in radial direction, $k$, $h$, and the dielectric properties of the casket would vary as functions of spatial coordinates.

In deriving equation 1, only heat flow from inner wall to outer wall of the casket was considered. However, at the casket outer wall, there should be thermal energy losses transported by radiation and convection. The loss due to radiation increases as a function of $T^4$ according to a well-known Stefan-Boltzmann law [4]. Although the casket is placed in the microwave cavity, the casket outer wall is still exposed to air during microwave heating. Thus for better modeling of the steady-state temperature of the casket, we should consider both radiation and convection losses in addition to heat flow through the casket wall. However, possible additive variables with the casket's geometric variables and microwave power obviously would make the current analytical modeling to be much more complicated and difficult to approach. One way to consider various thermal energy transport mechanisms and energy balances among the heat energy generated by dissipating microwave energy and the energy losses is to solve the problem by finite element analysis considering possible boundary conditions.

In this study, we employ the finite element analysis method (FEM) to describe the steady-state temperature of the casket and compare the relative differences between the measured values and the FEM-calculated values to the relative differences resulted by the current analytical modeling.

## 3. EXPERIMENTAL DATA FOR REFRACTORY CASKET HEATING

### 3.1. Casket Preparation

Caskets were constructed from commercial zirconia insulation cylinder (ZYC, Zircar Products Inc.) and aluminosilicate insulation board (SALI, Zircar Products Inc.). As-received ZYC cylinders and SALI boards were cut to construct 12 caskets which have total casket lengths, $L_T$, of 4, 5, 6, and 7 cm for a fixed thickness of SALI end plate, $0.5L_{SALI} = 1$ cm. Outer casket radius $b$ was either 5.08 cm or 3.81 cm, while inner zirconia cylinder radius $a$ was either 3.81 cm or 2.54 cm. In this study the measured $T_i$ data were grouped according to the total length of casket, $L_T$. The four individual groups of data involve caskets with $b/a = 1.33$, 1.50 and 2.00. Detailed procedures for casket construction are given elsewhere [2].

### 3.2. Microwave Heating of Caskets and Temperature Measurements

Using a single-mode microwave cavity (Model CMPR250, Wavemat, Plymouth, MI) operated at 2.45 GHz, individual caskets were heated by microwave input power level ranging from 200 Watts to 700 Watts [2]. A steady state of the inner wall casket temperature was reached within about 10 to 15 minutes after the microwave input power level was increased to a given power level. Using an optical pyrometer (Model 10, Accufiber Optical Fiber Thermometer, Luxtron C., Oregon), the steady-state $T_i$ and $T_o$ temperatures were measured through view ports in the microwave cavity. Detailed microwave heating and temperature measurements of caskets are given elsewhere [2].

## 4. FINITE-ELEMENT ANALYSIS (FEM)

A standard finite element program (ANSYS 5.3) allowed us to simulate the steady-state temperature within a refractory casket using a two-dimensional cylindrical heat conduction analysis with radiating boundaries and non-linear material properties, since the thermal conductivities and the emissivities for both zirconia and aluminosilicate are functions of temperature [3,5]. The analysis was done for 12 different caskets and eleven input powers for each casket, so a set of 132 temperature distributions were compared with the measured inner casket wall temperature $T_{i,Measured}$. The comparison of the FEM results and the experimental data tests the validity of the model developed for the casket.

### 4.1. Boundary Conditions for FEM Analysis

The choice of the boundary conditions is a difficult part of modeling the casket. Accurate boundary conditions allow reliable predictions of the inner wall temperature $T_i$, yet the boundary condition formulation should be simple enough to allow a fast calculation of the steady-state temperature. The casket consists of three surfaces with different boundary conditions: (1) bottom plate, outer surface, (2) inner casket surface and (3) outer casket wall.

*Bottom plate, outer surface:* The outer surface of the bottom plate is in contact with the microwave cavity. The cavity, which is largely brass, has a thermal conductivity at least two hundred times greater than the thermal conductivity of the casket's bottom plate ($k_{SALI} < 0.4$ W/(m·K), $k_{Brass} > 80$ W/(m·K) [3,6]). Since the cavity is water-cooled, the bottom surface of the lower casket end plate adopts the temperature of the cavity and stays constant.

*Inner casket surface:* Inside the casket, no significant amount of heat is transported by convection, since the hollow space has a relatively small connection to the outer area. The air inside the casket reaches a high temperature during the experiment, and since there is likely limited exchange between the hot air and the cooler outer air, convective heat flux will not occur.

Since the casket's inner walls reach a high temperature, the heat transportation by radiation can not be neglected. Nearly all of the emitted radiation is absorbed by the casket's inner walls, so the inside cavity wall is adiabatic. The form factors for the given elements (i.e. the portion of radiation emitted by surface-element $i$ and absorbed by surface-element $j$) are calculated by ANSYS 5.3 to estimate the radiation effects inside the casket.

*Outer casket wall:* A large volume of air cooled by the cavity walls surrounds the outer casket wall. Therefore, free convection at the outer casket wall is taken into consideration.

The outer surface also emits radiation, which is partially absorbed by the cavity wall, partially reflected and reabsorbed by the casket. In contrast to the inner casket surface, a simpler model was chosen to calculate the radiative heat transfer at the outer wall. The casket is completely enclosed by the cavity, and the casket can not absorb its emitted radiation directly because it is a convex body. Siegel and Howell provide an analytical solution for the share of power transported by radiation [4]. Figure 1 illustrates the mechanisms of heat transfer at the casket's surfaces.

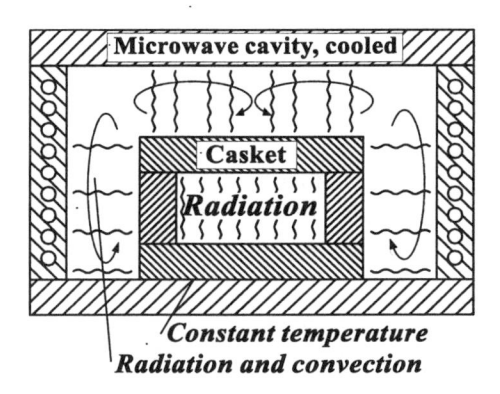

**Figure 1.** Schematic showing the radiation and convection boundary conditions for the microwave-heated refractory casket.

Since the radiative power dissipation is a function of the fourth power of the surface's temperature, the radiation heat transfer was calculated for different temperatures. The following scheme was used to linearize the radiation [7]:

$$\begin{aligned} P_{Rad} &= c_{Geo}\varepsilon(T)\sigma A \left(T_{Casket}^{4} - T_{Cavity}^{4}\right) \\ &= c_{Geo}\varepsilon(T)\sigma A \left(T_{Casket}^{2} + T_{Cavity}^{2}\right)\left(T_{Casket} + T_{Cavity}\right)\left(T_{Casket} - T_{Cavity}\right) \\ &= h_{Rad}(T) A \left(T_{Casket} - T_{Cavity}\right) \end{aligned} \quad (3)$$

where  $c_{Geo}$ = geometrical factor
$\varepsilon(T)$ = emissivity of the radiating surface
$A$ = surface area
$\sigma$ = Stefan-Boltzmann-constant
$T_{Casket}$ = outer casket wall temperature
$T_{Cavity}$ = inner cavity wall temperature
$h_{Rad}(T)$ = apparent heat transfer coefficient for radiation

$$h_{Rad}(T) = \varepsilon(T)\sigma \left(T_{Casket}^{2} - T_{Cavity}^{2}\right)\left(T_{Casket} - T_{Cavity}\right). \quad (4)$$

The apparent heat transfer coefficient is a function of both temperature and material, since $\varepsilon(T)$ is a material property. Superposing both the radiative and the convective heat transfer modes leads to equation (5):

$$P_{Surface} = P_{Rad} + P_{Conv} = (h_{Rad}(T) + h_{Conv}(T))A(T_{Casket} - T_{Cavity})$$
$$= h_{eff}(T) A (T_{Casket} - T_{Cavity}) \quad (5)$$

where $h_{Conv}(T)$ = heat transfer coefficient for convection
$h_{eff}(T)$ = effective heat transfer coefficient
$$= (h_{Rad}(T) + h_{Conv}(T)). \quad (6)$$

The casket's convective heat transfer coefficients $h_{Conv}$ were calculated using empirical relations for the free convection of vertical and horizontal surfaces [8]. The convective heat transfer coefficients are a function of temperature because the material properties of air (density, viscosity, thermal conductivity and effective heat capacity) are a function of temperature, too.

Since the casket geometry influences the radiation as well as the convection, the heat transfer coefficients also are a function of casket geometry. Both heat transfer phenomena are non-linear and neither phenomena can be neglected in the temperature range from room temperature to 1200 K. Using equation 4 and Touloukian's emissivity data [5], we calculated $h_{Rad}$ and $h_{eff}$ (equation 6, Figure 2).

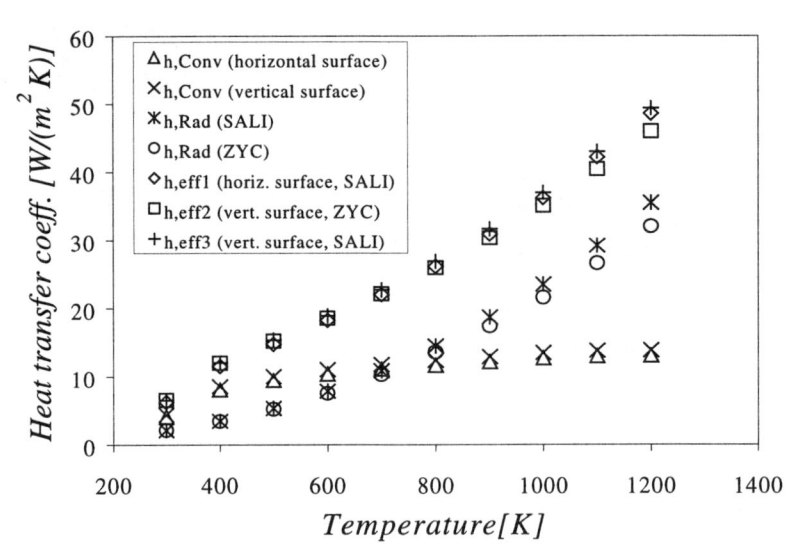

**Figure 2.** Heat transfer coefficients for convection, $h_{Conv}$, and radiation, $h_{Rad}$, as well as the effective heat transfer coefficient $h_{eff}$ (for casket 11).

The thermal conductivity of the casket's materials was interpolated and extrapolated from the vendor's $k_{SALI}$ and $k_{ZYC}$ data for steps of 100 K from 300 to 2600 K. The casket's material properties as well as the heat transfer coefficients $h_{eff1}$, $h_{eff2}$ and $h_{eff3}$ were imported to ANSYS and used to solve the non-linear heat conduction problem.

Although the finite-element analysis produced steady-state temperature estimates for the entire casket, only the solutions for one node were recorded in order to compare these values with the experimentally determined temperatures. The intention of the analysis was to test the assumed boundary conditions for the simulation.

### 4.2. Modeling and Discretization of the Casket

The analysis assumed that the microwave input power is homogeneously dissipated by the entire casket, so interactions between the casket and the electrical field pattern were neglected. The cylindrical symmetry of both the casket and the microwave cavity allows the casket heating problem (including boundary conditions) to be expressed in terms of two-dimensional cylindrical coordinates *(r, z)*. The simple geometry permitted us to use a regular and rectangular mesh.

The co-author of the paper measured the inner casket temperature for a fixed inside wall location 0.025 m above the lower surface of the bottom end plate (Figure 4). In the following analysis, the node of the model that has the same location as the measured point of the real casket was evaluated.

Increasing the mesh density increases the accuracy but decreases the calculation speed. Therefore, we first analyzed the convergence of the solution for different mesh densities (i.e. different number of elements). The estimation for casket 1 and $P_I = 600$ W (Figure 3) indicates the calculation converges as long as the number of elements exceeds 500, so we chose to use 1040 elements for the casket temperature simulations.

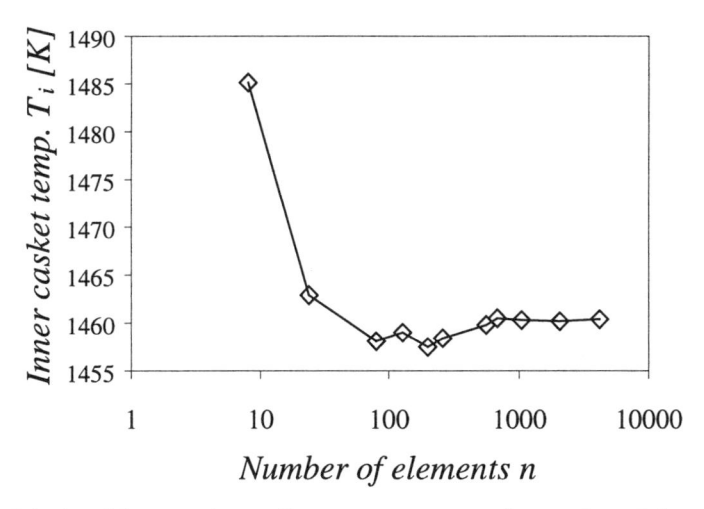

**Figure 3.** Calculated inner casket wall temperature versus the number of elements used in modeling the casket.

The SALI plates are represented by eight elements in the z-direction and in the r-direction, 20 elements for $r < a$ and eight elements for $r > a$. The hollow Zirconia cylinder is divided in 20 elements in r-direction and in eight elements in z-direction (Figure 4). Therefore, the model is built of 1040 Elements. If the casket geometry did not allow one to position a node at a point that corresponded exactly to the point at which the inner casket wall temperature was measured, then the nearest node to the measured point was chosen.

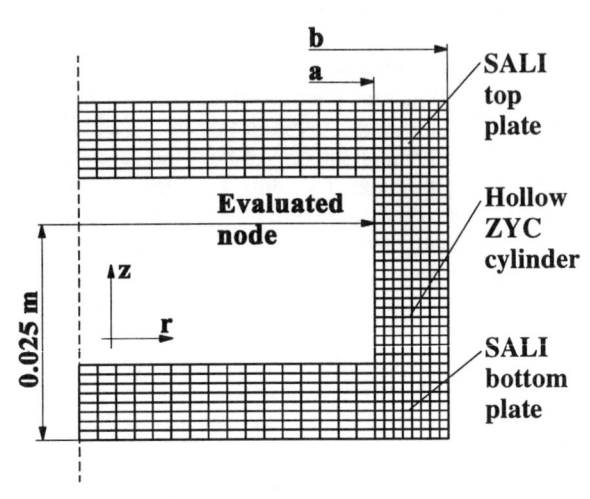

**Figure 4.** Meshed casket and evaluated node.

## 5. COMPARISON OF ENERGY BALANCE MODEL AND FEM ANALYSIS WITH EXPERIMENTAL DATA

Plots of the measured casket inner wall temperature, $T_i$, including 12 data points from reference [1] as a function of total volume of casket, $V_T$, or outer surface area of the casket, $S_{out}$, display no trend as a function of either $V_T$ (Figure 5a), $S_{out}$ or $V_T/S_{out}$ (Figure 5b). This indicates that $T_i$ can not be described by only a few geometric variables such as $V_T$ and $S_{out}$ in that caskets used for microwave processing of low-dielectrically-loss materials absorb microwave energy and dissipate it as heat on both inner wall surface and outer wall surface of the casket.

The 132 data sets of combined geometric variables ($b$, $b/a$, $L_{SALI}$, and $L_T$), the measured $T_i$ and the measured $P_C$ values were least-squares fit to equation 2, with a coefficient of determination $R^2$ of 0.957. The least-squares fit of the four data groups, including 33 data sets with a fixed $L_T$ of either 4, 5, 6, or 7 cm, gave $R^2 \geq 0.967$ [2].

The relative error $E_{rel}$ between the measured $T_i$ and the calculated $T_i$ (based on the FEM-analysis or parameters $C_1$, $C_2$, and $C_3$ determined by least-squares fitting the data in individual groups with a fixed $L_T$ to equation 2 [2]) can be calculated by

$$E_{rel} = \frac{T_{i,Calculated} - T_{i,Measured}}{T_{i,Measured}} \times 100 \ (\%) \qquad (7)$$

where  $T_{i,Calculated}$ = $T_i$ calculated by FEM-program or equation 2
  $T_{i,Measured}$ = measured $T_i$.

For the least-square fitting, plots of $E_{rel}$ versus $P_I$ show that for 8 of 12 caskets with either $b/a$ = 1.33 or 1.50, $E_{rel}$ is within ±5% over the entire input power range (for caskets with $b/a$ = 2.0, $E_{rel}$ is up to +15% for $P_I \leq 250W$). For the FEM-Analysis, 8 of 12 caskets with either $b/a$=1.50 or 2.00, $E_{rel}$ is within ±15% over the entire input power range (for caskets with b/a = 1.33, $E_{rel}$ is up to +25% for $P_I \leq 300$ W) (Figures 6a and 6b).

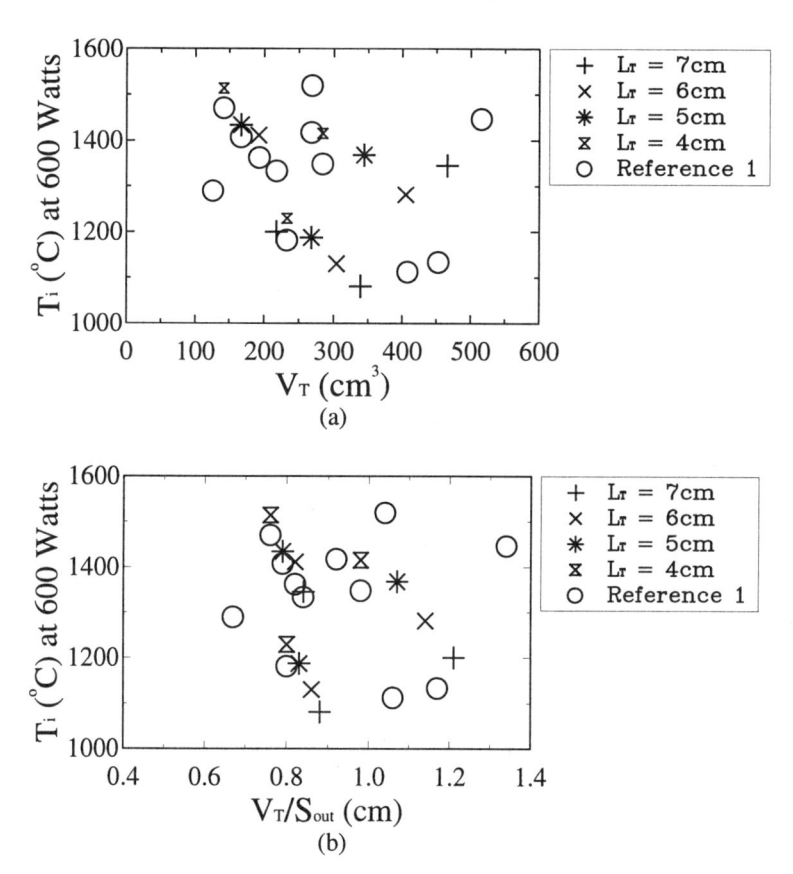

**Figure 5.** Measured $T_i$ at microwave input power of 600 Watts as a function of $V_T$ (a) and $V_T/S_{out}$ (b), where $V_T$ is total volume of casket and $S_{out}$ is outer surface area of the casket.

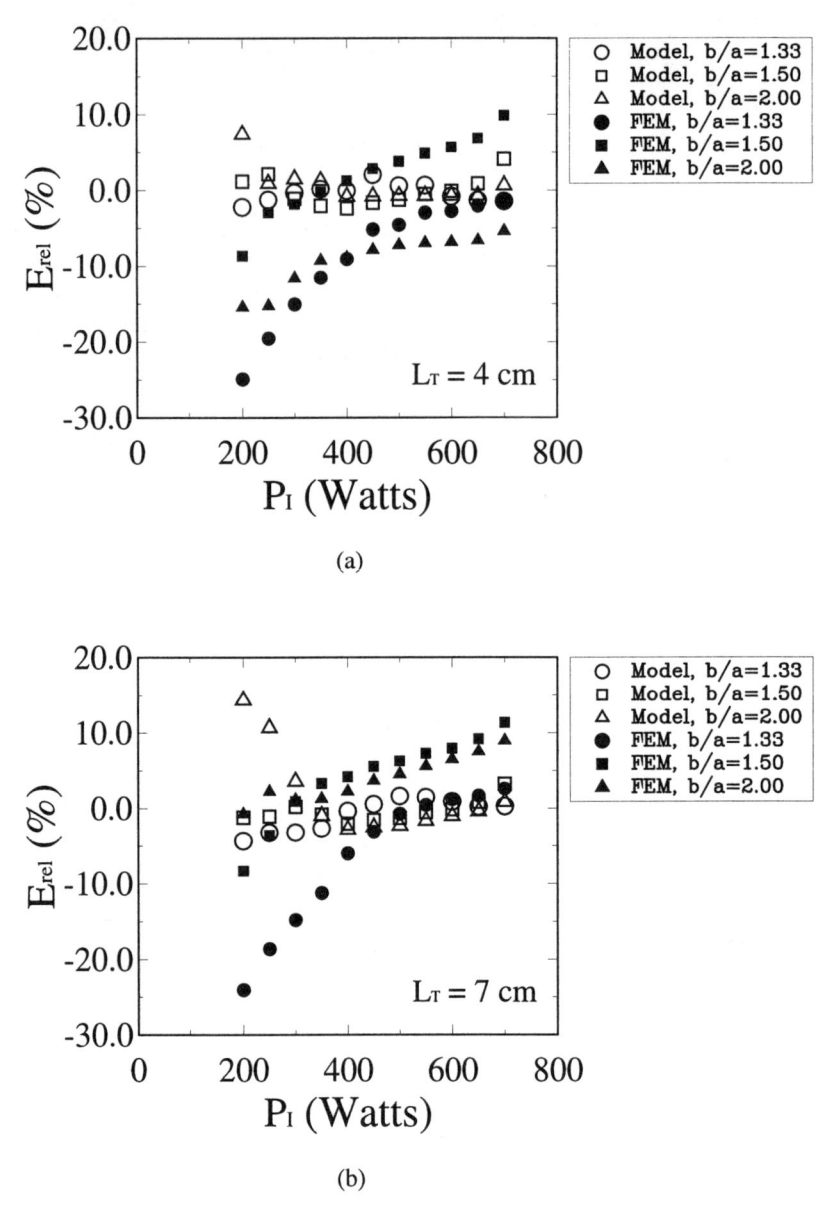

**Figure 6.** Relative error between calculated and measured $T_i$ for the analytical model (equation 2) and the FEM-analysis: (a) $L_T=0.04$ m, (b) $L_T=0.07$ m.

To estimate the quality of both the analytical modeling and the FEM-simulation, an average relative error $z$ was calculated for all 12 caskets as a function of input power $P_I$. The sum of the squares of the relative errors was divided by the number of caskets $n$:

$$z = \sqrt{\frac{1}{n}\sum_{m=1}^{n}\left(\frac{T_{i,Calculated,m} - T_{i,Measured,m}}{T_{i,Measured,m}}\right)^2} \cdot 100\% \qquad (8)$$

where  $n$ = number of caskets
$T_{i,Calculated,m}$ = by ANSYS 5.3 calculated inner temperature for casket $m$
$T_{i,Measured,m}$ = measured inner temperature for casket $m$.

At the lowest input power used in this study, $z$ attains a maximum value of about 16% for the FEM-Analysis, while a minimum value of $z$ of roughly 4% occurs for 450 W $\leq P \leq$ 500 W. For the analytical modeling, $z$ ranges from about 1% to 7% (Figure 7).

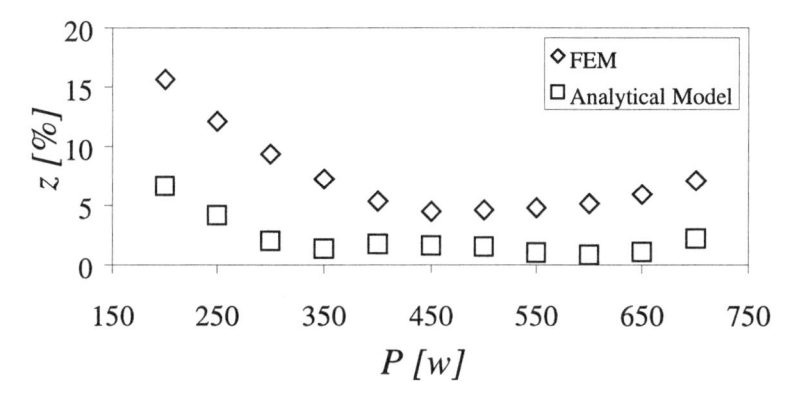

**Figure 7.** $z$ as a function of microwave input power for both analytical model and FEM analysis.

## 6. SUMMARY AND CONCLUSION

An analytical model, based on an energy balance concept developed previously [1,2], described well the experimentally-measured steady-state inner wall casket temperatures, $T_i$, as a function of casket geometry and microwave input power. For the caskets with $b/a$ = 1.33 and 1.50, the relative error in $T_i$ was within ±5% for the entire microwave input power range of 200 Watts to 700 Watts.

For low input power, the temperatures calculated by the FEM-program are too low and the relative error is high (Figures 6 and 7). A reason for the high error may be the neglect of the microwave field pattern inside the cavity, since we assumed homogeneous heat generation within the casket. We also ignored the temperature dependence of the casket material's dielectric properties although the electromagnetic field pattern changes as the casket temperature changes. In addition, the radiation outside the cavity differs from element to element as the form factors change, so strictly speaking the form factor should be calculated for each element. Furthermore, the reflected radiation was ignored.

Nevertheless, our FEM-model gave reasonable results, especially for input powers $P_i > 300$ W. The boundary conditions also will be used for a finite difference program, which will be developed later. The simulation of the boundary conditions will be enhanced by considering the reflected and reabsorbed radiation, which is currently ignored.

**REFERENCES**
1. K.Y. Lee, E.D. Case, and J. Asmussen, Jr., "The Steady-State Temperature as a Function of Casket Geometry for Microwave-Heated Refractory Caskets," Mat. Res. Innovat. 1[2]: 101-116 (1997).
2. K.Y. Lee, E.D. Case, "Steady-State Temperature of Microwave-Heated Refractories as a Function of Microwave Power and Refractory Geometry," submitted for publication, Mat. Sci. and Eng. A (1998).
3. Zircar Fibrous Ceramics Catalog, Zircar Products Inc, Florida, New York (1995).
4. R. Siegel, J.R. Howell, "Thermal Radiation Heat Transfer," $2^{nd}$ ed., Hemisphere Publishing Cooperation, Washington, pp. 25, 240-241 (1981).
5. Y.S. Touloukian, "Thermophysical Properties of High Temperature Solid Materials, Volume 4, Part I," Macmillian Company, New York, pp. 28, 589 (1967).
6. "Metal Handbook, Volume 2," The Materials Information Society, pp. 296-302 (1990).
7. P. Kohnke, "ANSYS Users' Manual for Revision 5.0 Part IV," Swanson Analysis System Inc., Houston, pp. 6-7 (1992).
8. "VDI-Waermeatlas," (Heat transfer handbook published by German Professional Association *VDI*) VDI-Verlag Duesseldorf (1984).

# Spark Plasma Processing

# SEEDING WITH α-ALUMINA FOR TRANSFORMATION AND DENSIFICATION OF BOEHMITE-DERIVED γ AND δ-ALUMINA BY SPARK PLASMA SINTERING

S. D. De la Torre*[a], A. Kakitsuji [a], H. Miyamoto [a], K. Miyamoto [a], H. Balmori-R [b], J. Zárate-M [c] and M.E. Contreras [c].

[a] *Tech. Research Institute of Osaka Pref.2-7-1 Ayumino, Izumi, Osaka 594-1157 Japan.*
[b] *IPN-ESIQIE, U.P.A.L.M.-Zacatenco A.P. 75-872,CP.07300, Mexico DF., Mexico.*
[c] *IIM-Universidad Michoacana. S.N.H., A.P.52B, CP.58000, Morelia, Mich., Mexico.*

## ABSTRACT

The spark plasma sintering (SPS) behavior of γ and δ-alumina seeded and unseeded with α-alumina particles has been investigated. Pseudo-boehmite (PB) precursor powder prepared from non-bauxitic materials (containing 98.90 wt% $Al_2O_3$) was mechano-chemically treated with $HNO_3$, diluted in water and spray dried. 3% of α-$Al_2O_3$ seeds were added before spray drying. The pseudoboehmite was calcined at 900°C for 1 h at a heating rate of 5°C/min turning into γ and δ-$Al_2O_3$. Although thermal analysis indicated that unseeded powder transforms into α-$Al_2O_3$ at 1227°C and seeded one at 1154°C, a hastened SPS processing considerably accelerates the α-phase precipitation. Abnormal α-$Al_2O_3$ grain growth took place during sintering, resulting in plate-like structures formation. EDAX-microprobe analysis revealed that the abnormal grains are composed exclusively of aluminium and oxygen. These plate-like grains nucleated from 1050°C and thickened from 1500°C. Their presence, however, limits the specimens strength.

## INTRODUCTION

On searching for alternating aluminium minerals for the production of alumina-based ceramics, non-bauxitic sources are potential candidates since they are more abundant on the earth crust than the traditional used bauxitic resources. With the constant progress on the development of more efficient chemical routes for the extraction and beneficiation of cost affordable materials and taking advantage of innovative densification techniques such as the spark plasma sintering (SPS) process [1], the preparation and study of new ceramic materials can further be explored.

---

<sub>To the extent authorized under the laws of the United States of America, all copyright interests in this publication are the property of The American Ceramic Society. Any duplication, reproduction, or republication of this publication or any part thereof, without the express written consent of The American Ceramic Society or fee paid to the Copyright Clearance Center, is prohibited.</sub>

**Table I**. Chemical analysis in wt% of as-received pseudoboehmite (PB) powder.

| $Al_2O_3$ | $SiO_2$ | CaO | $Fe_2O_3$ | ZnO | CuO | $SO_3$ | NiO | $K_2O$ |
|---|---|---|---|---|---|---|---|---|
| 98.9010 | 0.5560 | 0.4960 | 0.0259 | 0.00609 | 0.00533 | 0.00486 | 0.00324 | 0.00152 |
| Al | Si | Ca | Fe | Zn | Cu | S | Ni | K |
| 97.038 | 0.8890 | 0.8285 | 0.6800 | 0.1895 | 0.1640 | 0.0680 | 0.0979 | 0.0451 |

The authors have been working on the densification of a number of structural ceramic composites [2-4], from which the efficiency of the SPS has been recognized. SPS is a pressurized hot pressing-like sintering process in which pulsed electric current discharges are lead into the voids formed between adjacent particles of green bodies. General views of its functionality have been given elsewhere [1-2]. The merit of using this densification technique is based on the fact that superfast sintering operations of up to 2000°C can be executed within few minutes (<10min) with relatively fast cooling rates (from 1600~200°C in 5min). Because of those operational features the metastability and the nanometric size of precursor powders can virtually be retained on cooling.

The present study is relevant both to the growth of the $\alpha$-$Al_2O_3$ starting from non-bauxitic boehmite-derived transtition alumina precursors and to attempt its nearly full compaction.

**EXPERIMENTAL**

The procedure used for the chemical synthesis of aluminium oxide and hydroxides has been reported elsewhere [5]. Briefly, the procedure consists of starting with alunite (($Al_2(SO_4)_3$·$K_2SO_4$·$4Al(OH)_3$) mineral, industrial grade aluminium sulfate $(Al_2(SO_4)_3$·17 $H_2O$), aluminium basic sulfite $(Al(OH)_{10}$·$SO_3)$ and sulfate $(Al(OH)_{10}$·$SO_4)$, sodium aluminate $(Na(Al(OH)_4))$ and aluminium metal. An adequate performance of the precipitation process of aluminium oxide and hydroxides free of anions is fundamental so that pseudo-$(AlO(OH))$ can be obtained from the aluminium basic salts [5]. The prefix pseudo stands for a material which has not been completely crystallized because its water content is still high. PB powder was prepared and supplied from the University of Guanajuato in Mexico, via the UG-process [6]. In Table I is reported these raw powder chemical composition, as obtained from X-ray fluorescence analysis. Controlled transformation and sintering of alumina precursors was attempted by seeding with $\alpha$-$Al_2O_3$ particles, according to the method suggested by Kumagai and Messing [7]. 130g of PB were mechanochemically treated with 20 ml of concentrated $HNO_3$. At this stage, the powder was both unseeded and seeded with 3% of $\alpha$-$Al_2O_3$ particles (1μm, Tamei TM-10, Japan). An aqueous suspension (20% solids) of these mixture was then spray-dried to obtain agglomerated precursor powder having a wide particle size distribution range convenient for SPS-processing. Phase estimation was performed by X-ray diffraction (XRD) using a Hägg-Gunier camera with Cu $K_{\alpha 1}$ rad. Phase composition was analyzed by special software attached to a Rigaku RINT2500VHF system. The phase transformation of transition-$Al_2O_3$ was also examined by heating 30mg samples in pure $Al_2O_3$ cups at a rate of 10 °C/min under argon gas flow (50 ml/min) in a thermogravimetric-differential thermal analysis (TG-DTA) Setaram system. Pure α-Al$_2$O$_3$ was used as the reference standard. Scanning electron microscopy (SEM) observations were carried out using a Philips XL-20 microscope supplied with EDAX system. To compact the powder into 4 x 20 mm ϕ cylinders, SPS operations were run by using a commercial SPS-1020 Dr.Sinter apparatus (Sumitomo Mining Co., Japan) operated in vacuum applying 40MPa and 2000Amp on the green compacts (~ 40% dense). Highest set heating rate was 525°C/min. On/Off pulse conditions of 12/2 were fixed for the compacting runs. Density of specimens was determined by the Archimedes method.

The spark plasma sintering behavior of three kinds of powder was compared; (a) pure pseudoboehmite PB as prepared by the UG-process, (b) same as (a) but PB-spray dried without seeds, and (c) same as (a) but PB-spray dried and charged with 3% of α-Al$_2$O$_3$ seeds. After (a)-(c) powders were calcined in air at 900°C for 1h at a heating rate of 5°C/min, PB lost ~ 30% weight during dehydration, transforming the material into γ and δ-alumina. The classification given to (a)-(c) starting powders and the weight % of the phases present in the as-received, as-calcined and SPS-treated materials are reported in Table II.

**Table II.** Classification and the phases present in SPS-treated powder.

| Sample code | Powder description | Treatment | PB | α | θ | δ | γ |
|---|---|---|---|---|---|---|---|
| PB | Starting pseudobohemite. As prepared by the UG-Process. | As-received | 100 | | | | |
| | | As-calcined | | | | 40 | 60 |
| | | SPS: 1065°C/2min | | 100 | | | |
| | | SPS: 1100°C/2min | | 100 | | | |
| | | SPS: higher temps. | | 100 | | | |
| SD-PB | Spray-dried PB. | As-received | 100 | | | | |
| | | As-calcined | | | | 40 | 60 |
| | | SPS: 1050°C/2min | | 96 | 4 | | |
| | | SPS: 1190°C/2min | | 100 | | | |
| | | SPS: higher temps. | | 100 | | | |
| SD-PB3% | Spray-dried PB and charged with 3% α-Al$_2$O$_3$ seeds. | As-received | 97 | 3 | | | |
| | | As-calcined | | 27 | 20 | 40 | 13 |
| | | SPS: 1000°C/2min | | 40 | 60 | | |
| | | SPS: 1080°C/2min | | 100 | | | |
| | | SPS: 1140°C/2min | | 100 | | | |
| | | SPS: higher temps. | | 100 | | | |

## RESULTS AND DISCUSSION

### Phase Transformation

From Table II it can be seen that there is no substantial effect on the transformation rate occurred between the studied powders after the SPS-operations. There is, however, a significant temperature difference of about 130°C when the temperature required for the precipitation of the equilibrium $\alpha$-$Al_2O_3$ phase is compared either to the typically observed value of 1200°C [8] or to the temperatures observed from DTA analysis. The presence of an exothermic peak indicated that unseeded powder transforms into $\alpha$-$Al_2O_3$ at 1227°C whereas seeded one does at 1154°C. From TG-DTA analysis results, an endothermic peak was also observed at 410°C from each trace recorded from the SD-PB and SD-PB3% powders. Around this temperature both powders lost 19.03 and 26.36% weight, respectively. The transition aluminas obtained by calcining PB powder transformed into crystalline $\alpha$-$Al_2O_3$ as indicated in Table II. In the course of sintering, abnormally grown grains were detected. Fig.1 shows a typical SEM view of a fractured surface where these plate-like structures are in full growth. They eventually precipitated and developed in all studied powders. When analyzing SPS-treated specimens by XRD their patterns exclusively diffracted the Bragg peaks of $\alpha$-$Al_2O_3$ with no real evidence of any other compound present nor an amorphous background. To decipher the nature of such structures EDAX analyses were performed in the triangles marked in Fig.1. The results confirmed the sole presence of aluminium and oxygen elements therein.

**Fig.1** Microstructure of sintered samples. Plate-like structures in full growth.

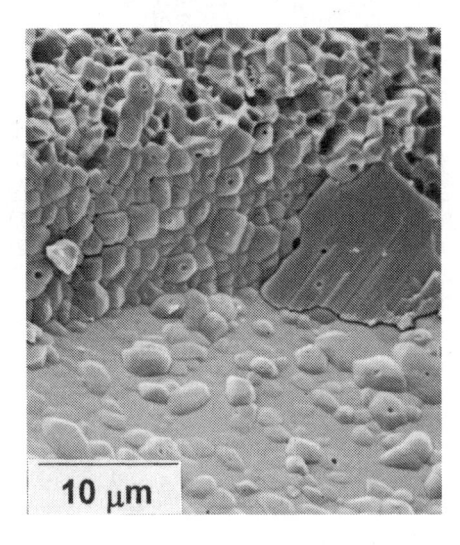

**Fig.2** Formed cavity of a plate-like structure. Sub-micron size pores are visible.

*Abnormal α-Al$_2$O$_3$ Structure Formation*

The gradual formation of fine irregular spots was observed on the microstructure of all SPS-treated specimens. Generally, it first occurred at around 1050°C. The size of these spots increased and they became more evident as the temperature rose. Its morphology adopted plate-like shape only after reaching 1500°C. Those spots are in fact nucleation sites where an abnormal α-Al$_2$O$_3$ grain structure develops from the originally transition alumina matrix. The fact that those grains developed looking translucent indicates that they densified faster than the matrix. These alumina grains developed faster when α-Al$_2$O$_3$ seeds were present (i.e., in SD-PB3%) while its slowest formation took place in PB powder. The volume and number of such structures increased as the temperature was higher. The average length of plates formed in a SD-PB specimen, SPS treated at 1630°C/4min was 80µm and 20~30µm width. Parallel to the abnormal grown grains fully dense and fine grain microstructure regions were obtained. A common feature observed in all treated specimens is that at the interface formed between these regions and the plate-like grains, there were sub-micron size residual pores as shown in Fig.2. In this photo one plate-like grain was pulled away so that only one small broken piece of it remained at the corner. Its finger print reveals the presence of highly compacted 1µm big grains having a round morphology that were in contact with the plate-like grains.

*Densification by SPS*

The implications of spray-drying and seeding the PB precursor powder on SPS-processing can be deduced from Figs.3 and 4. Fig.3. shows the apparent density attained for the three kinds of powders as a function of the sintering temperature. Even at low temperatures, i.e., from 1000 to 1350°C, SD-PB3% powder is better densified, followed by SD-PB. Evidently, by agglomerating and spherical-shaping the precursor powder one can bring the material particles in closer contact as to make better use of the pulsed electric energy generated by the SPS system. It is worth saying that spray dried powder had a particle size distribution ranging from 0.1 to 10 µm and that the biggest particles were not exactly spherical. Nevertheless, α-seeding further assisted in densifying these powders. In spite of the sudden increase observed in the apparent density of all powders after reaching 1400°C, the remaining porosity could not be any longer decreased since the intergranular pores formed in the plate-like grains could not be eliminated. Fig.4 shows a semi-logarithmic graph of the apparent porosity against the sintering temperature for the studied material. The effectiveness of seeding PB powder for developing higher density specimens is demonstrated in this figure.

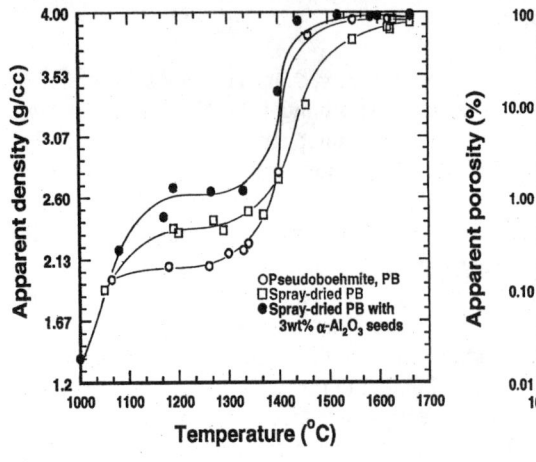

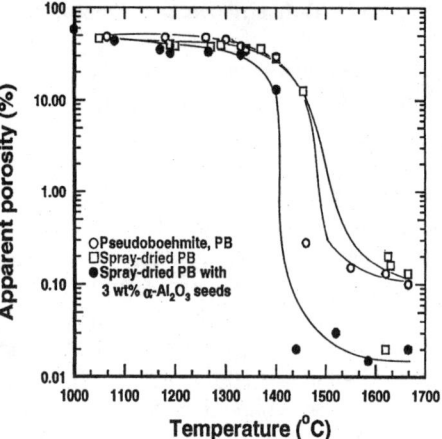

Fig.3 Apparent density of studied PB powders as a function of the SPS-temperature.

Fig.4 Apparent porosity of PB powders as a function of the SPS-temperature.

Preliminary attempts to evaluate the mechanical properties of sintered specimens were complex since the plate-like grains made it difficult, limiting the final products strength. Therefore, Vickers hardness Hv could be evaluated just on areas where no rods were present and achieved values as low as 170 MPa. Following the indentation marks propagated throughout the material surface (where no rods were present), and using Niihara's equation [9], a maximum toughness value of 3.3 MPa m$^{1/2}$ was obtained from the SD-PB samples SPS-treated at 1620°C/4min.

CONCLUSIONS

A hastened spark plasma sintering (SPS) processing accelerates the α-Al$_2$O$_3$ phase precipitation from transition δ and γ-aluminas, which in turn were synthesized from non-bauxitic sources. Sintering pseudoboehmite (PB) precursor powder from 1500°C in less than 5 min resulted in specimens disclosing nearly 0.15% porosity. Both spray drying and α-Al$_2$O$_3$ seeding such precursor powder further enhanced its densification, leaving 0.015% of pores. Full density specimens are not obtained since abnormally grown grains having plate-like structure nucleate and develop at high temperatures. XRD and EDAX-microprobe analysis revealed the sole existence of aluminium and oxygen constituting the plates-structure. Existence of these plate-like grains limits the sintered specimens strength.

## ACKNOWLEDGMENTS

This work has been supported by NEDO, under the Industrial Technology Fellowship Program, Japan. S.D.T. thanks to CIMAV of Chihuahua, Mexico for giving him the opportunity to temporary stay and work at TRI-Osaka. The support conferred to H.B.R. from S.N.I. and D.E.P.I.–I.P.N. of Mexico is also appreciated.

## REFERENCES

1. M.Tokita, "Trends in Advanced SPS Spark Plasma Sintering Systems and Technology, " *Journal of the Society of Powder Tech. Japan.*, 30 [11] 790-804 (1993) (in Japanese).
2. S.D.De la Torre, H.Miyamoto, K.Miyamoto, J.Hong, L.Gao, L.Tinoco, E.Rocha-R and H.Balmori-R., "Spark Plasma Sintering of Nano-Composite Ceramics, " pp. 892-897 in Proc. of the $6^{th}$ *International Symposium on Ceramic Materials & Components for Engines*, Edited by K.Niihara, S.Kanzaki, K.Komeya, S.Hirano and K.Morinaga. The Japan Keirin Association, 1997.
3. S.D.De la Torre, H.Miyamoto, K.Miyamoto, L.Gao, H.Balmori-R and D.Rios-Jara, "Phase Transformation of Transition-Alumina Upon Hastened Sintering, " in *Proc. of the $2^{nd}$ International Symposium on the Science of Engineering Ceramics (EnCera' 98)*. Osaka, September 6-9, 1998.
4. J.S.Hong, S.D.De la Torre, K.Miyamoto, H.Miyamoto and L.Gao, "Mechanical Properties of $ZrO_2(Y_2O_3)/20mol\%Al_2O_3$ Composites Prepared by Spark Plasma Sintering, " this book issue.
5. H.Juárez-M., J.M.Martínez-R., J.M.Ruvalcaba-L., O.A.Vargas-P. and J.Serrato-R. "Aluminum Oxide and Hydroxides from Non-Bauxitic Sources. " *American Ceramic Society Bulletin.* 76 [6] 55-59 (1997).
6. W.X.López "Three Methods to Produce Alumina From Alunite," *Light Metals* 2, 49-50 (1997).
7. M.Kumagai and G.I.Messing "Sintering of Boehmite Sol-Gel by α-Alumina Seeding," *Journal of American Ceramic Society.* 68 [9] 500-505 (1985).
8. K.Wefers and C.Misra, "Oxides and Hydroxides of Aluminum, " Alcoa Technical Report No.19, Revised, Alcoa Laboratories, page 50 (1987).
9. K.Niihara, R.Morena and D.P.Hasselman. "Evaluation of $K_{IC}$ of Brittle Solids by the Indentation Method With Low Crack-To-Indent Ratios." *J. Mat. Sci. Letters.*, 1 13-16 (1982).

# SPARK PLASMA REACTION-SINTERING OF MULLITE-ZrO₂ COMPOSITES

E. Rocha-Rangel[a], S.D. de la Torre[b], H. Miyamoto[b], M. Umemoto[c], K. Tsuchiya[c], J.G. Cabañas-Moreno[a], H. Balmori-Ramírez[a]

(a) Dept. of Metallurgical Engineering, ESIQIE-IPN, A.P.75-872, Mexico, D.F., 07300 Mexico; (b) Technology Research Institute of Osaka Prefecture.2-7-1 Ayumino, Izumi, Osaka 590-02, Japan; (c) Dept. of Production Systems Eng. Toyohashi University of Tech. Toyohashi, Aichi 441, Japan.

## ABSTRACT

Fully-dense mullite-$ZrO_2$ composites were produced by Spark Plasma Sintering (SPS) and in-situ reaction of $ZrSiO_4$ and $Al_2O_3$. A mixture of $ZrSiO_4$, $Al_2O_3$ and Al was milled in an attritor for 6 or 12 h and subsequently oxidized at 1100°C. During the fast densification processing, the heat generated from on-off DC-current discharges activates the reactants as to induce Spark Plasma Reaction-Sintering (SPRS) to obtain $ZrO_2$-toughened mullite. The phase transformations, the microstructure evolution and the mechanical properties are discussed as a function of the sintering temperature.

## INTRODUCTION

Spark Plasma Sintering (SPS) is a relatively new densification technique capable of sintering metals and/or ceramic powders at temperatures lower than the conventional sintering techniques [1]. This is due to the effective utilization of the phenomenon of microscopic electric discharging between the particles to be compacted, which are treated simultaneously under uniaxial pressure. The SPS densification mechanism has been recently reported [2] and discussed [3]. The mechanical properties of composite materials like Nb-Al [4], $ZrO_2/Al_2O_3$ whiskers [5] and SiC-based composite ceramics [6], treated by SPS have recently been studied.

In this study the SPS technique was used to energize selective reactants as to induce Spark Plasma Reaction-Sintering (SPRS) according to reaction (1).

$$3Al_2O_3 + 2ZrSiO_4 \rightarrow 3Al_2O_3 \bullet 2SiO_2 + 2ZrO_2 \qquad (1)$$

To the extent authorized under the laws of the United States of America, all copyright interests in this publication are the property of The American Ceramic Society. Any duplication, reproduction, or republication of this publication or any part thereof, without the express written consent of The American Ceramic Society or fee paid to the Copyright Clearance Center, is prohibited.

## EXPERIMENTAL PROCEDURE

50 g of a mixture composed of 64% $ZrSiO_4$ (Kreutz, Germany, particle size of 1 µm), 18.5% Al (Analytical de Mexico, 5 µm) and 17.5% $Al_2O_3$ (TM-10 Taimei, Japan, 0.2 µm) was used. Reagents were stoichiometrically mixed and placed into an attritor mill charged with 2 kg of zirconia balls. Wet milling was carried out using 125 ml of isopropyl alcohol for 6 or 12 h. Milled powder was dried at 75°C for 24 h and configured into specimens of 25 mm diameter x 3 mm thickness by cold isostatic pressing (CIP) at 400 MPa. To ensure oxidation of Al, specimens were heated at 1100°C at a rate of 1°C/min. Before SPS treatment, the specimens were vigorously ground in a planetary mill with YTZ balls for 35 min. The powder was then stacked into a graphite die set and sintered using a spark plasma sintering system (Dr. Sinter-SPS 1020, Sumitomo Coal Mining Co., Japan). Samples were heated to the required temperature at 500°C/min while uniaxially pressing the powder at 40 MPa. 3 min after the set temperature was reached, the load was removed and the specimens were cooled down inside the die set. Density and porosity of densified specimens were measured by the Archimedes method. Phase evolution and microstructure development were characterized by X-ray diffraction (XRD) and scanning electron microscopy (SEM). Fracture toughness was determined by indentation fracture [7] at 500 N. Young's modulus was estimated by a resonance technique. Vickers hardness was measured at 500 N.

## RESULTS AND DISCUSSION

### Microstructure

The XRD patterns of the powders milled for 6 h (R6) and 12 h (R12) before spark plasma-sintering are presented in Figure 1. There is no significant difference between the two XRD patterns, except for lower diffraction peak intensities of $Al_2O_3$ and $ZrSiO_4$ for powder milled for 12 h. Two extra peaks can be observed at 2θ 37.5 and 67.2°, that have been indexed as $\gamma$-$Al_2O_3$. This phase is formed from the oxidation of Al during milling and drying. The specific surface area increased from 2.9 $m^2$/g in the original powder to 16.46 and 21.71 $m^2$/g for powder R6 and R12 respectively. After oxidizing at 1100°C, the weight of samples R6 and R12 increased by 8.34 and 6.70 % respectively, but the diffraction peaks looked similar to those shown in Figure 1.

The XRD patterns of R12 specimen sintered at different temperatures are presented in Figure 2. As the sintering temperature increases the $ZrSiO_4$ and $\alpha$-$Al_2O_3$ peaks shorten and the formation of reaction products such as mullite, monoclinic and tetragonal $ZrO_2$ become evident. Although reaction (1) was partially induced, the fast sintering time inhibited completion of this reaction. The XRD patterns of sample sintered at 1500°C display the presence of mullite and tetragonal and monoclinic $ZrO_2$, but it still shows strong $ZrSiO_4$ and weak $\alpha$-$Al_2O_3$ diffraction peaks. The XRD patterns of R6 specimens are not reported because in these samples the evolution of reaction (1) was similar to samples R12 but less completed.

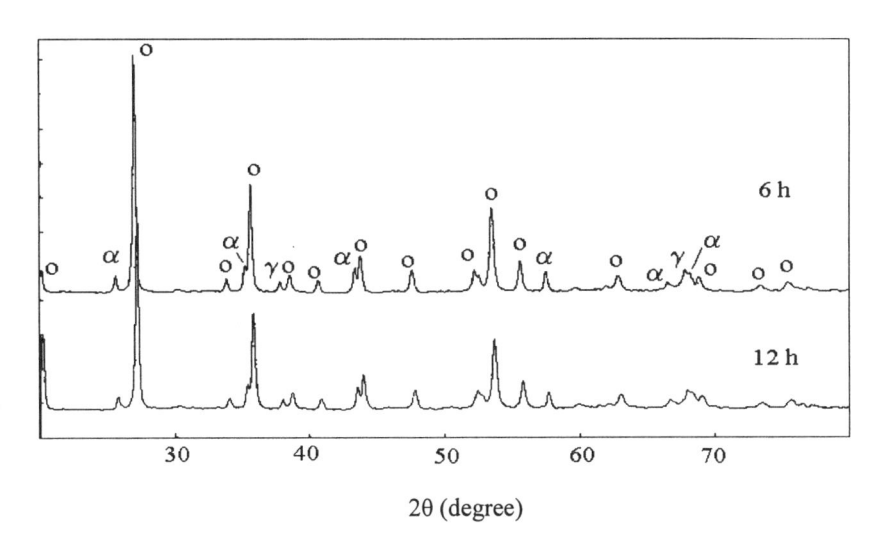

Figure 1.- XRD patterns of attrition milled powder.
o: ZrSiO$_4$, α:α-Al$_2$O$_3$, γ:γ-Al$_2$O$_3$

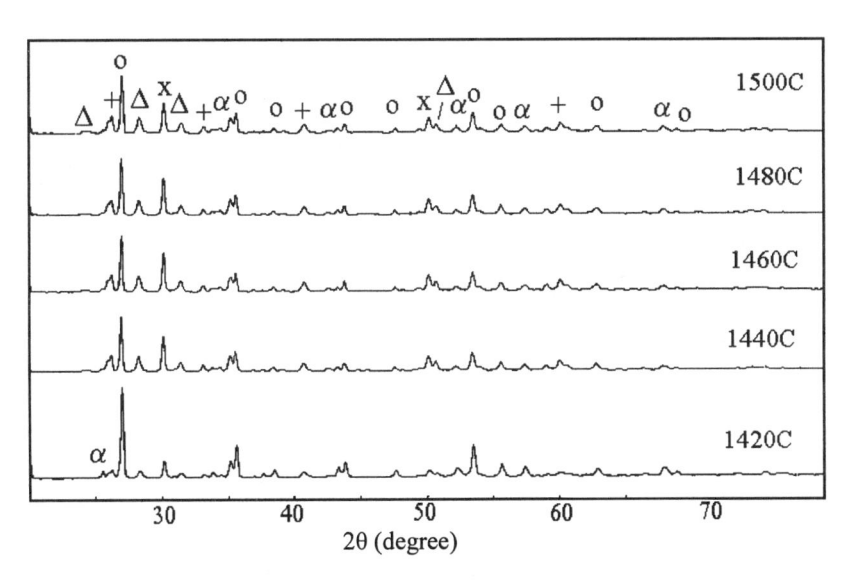

Figure 2.- XRD patterns of sample R12 sintered at different temperatures
+: mullite, Δ: monoclinic ZrO$_2$, x: tetragonal ZrO$_2$, o:ZrSiO$_4$, α: α-Al$_2$O$_3$.

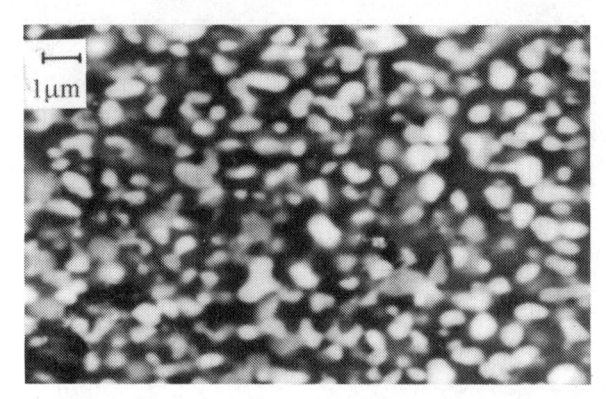

Figure 3.- Microstructure of sample R12 sintered by SPRS at 1500°C.

Microstructure characteristics of specimens after SPRS at 1500°C are presented in Fig. 3 and Table 1. The microstructure is characterized by a dispersion of inter and intragranular zirconia particles (white) in a mullite matrix (dark). $ZrSiO_4$ grains are disclosed as light gray. The microstructure is very homogeneous, fine and has negligible porosity.

**Density**

Figure 4 shows the effect of the milling time on the densification of mullite-$ZrO_2$ composites treated at 40 MPa and 3 min, as a function of the SPRS temperature. In both cases the relative density is > 90% of the theoretical, but it is higher in the R12 specimens. Increasing the sintering temperature resulted in denser sintered bodies. The spark plasma reaction-sintering technique leads to good densification of mullite-zirconia composites. This is probably due to factors such as self-heat generated due to the exothermic character of reaction (1), which is dissipated through the particles, but mainly to microscopic electric discharges taking place between particles, which in turn activates the particles surface to induce high speed mass transfer.

**Mechanical Properties**

Table 2 shows the results of estimating hardness, fracture toughness and Young's modulus of R6 and R12 specimens as a function of the SPRS temperature. In general, both kinds of specimens exhibited good mechanical properties. Lathabai et al [8] reported

Table 1. Microstructure characteristics of mullite-$ZrO_2$ sintered by SPRS at 1500°C.

| Sample | Average grain size (μm) Mullite | ZrO$_2$ | ZrO$_2$ Weight % | t-ZrO$_2$ (%) | Open Porosity (%) | Relative Density (%) |
|---|---|---|---|---|---|---|
| R6 | ~ 3 | ~ 1.5 | 38.93 | 28.37 | 2.9 | 95 |
| R12 | ~ 1 | ~0.5-0.7 | 34.13 | 55.14 | --- | 99.1 |

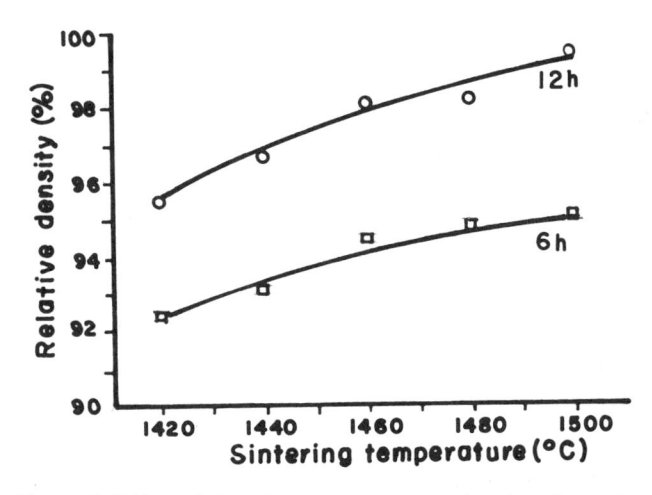

Figure 4. Effect of sintering temperature on density of samples milled for different times.

$K_{IC}$ values of 2.2 MPa•m$^{1/2}$ for a similar material obtained by the RBAO process. The higher $K_{IC}$ values measured in the present study are attributed to the good densification and fine grain size of the specimens that developed as a consequence of short sintering times. Although the amount of t-ZrO$_2$ in these composites suggests that stress-induced transformation could be responsible of toughening, mechanisms involving microcracking and other factors, possibly grain boundary strengthening produced by a metastable solid solution of zirconia in mullite[9] and crack deflection[10] can also contribute to the increase in toughness.

Table 2. Mechanical properties of SPRS-synthesized composites.

|  | Sample R6 | | | Sample R12 | | |
| --- | --- | --- | --- | --- | --- | --- |
| Sintering Temperature (°C) | HV (GPa) | E (GPa) | $K_{IC}$ (MPa•m$^{1/2}$) | HV (GPa) | E (GPa) | $K_{IC}$ (MPa•m$^{1/2}$) |
| 1420 | 1790 | 393 | 3.91 | 1865 | 398 | 3.97 |
| 1440 | 1432 | 318 | 4.55 | 1524 | 339 | 3.74 |
| 1460 | 1396 | 267 | 3.61 | 1530 | 359 | 3.97 |
| 1480 | 1394 | 225 | 3.85 | 1506 | 329 | 4.02 |
| 1500 | 1436 | 317 | 3.98 | 1515 | 335 | 4.09 |

**CONCLUSIONS.**

1. High-density mullite-zirconia composites can be produced by Spark Plasma Reaction-Sintering in 3 minutes.
2. Completion of the Spark Plasma Reaction-Sintering (SPRS) between high energy milled Al, Al$_2$O$_3$ and ZrSiO$_4$ precursors to produce mullite-ZrO$_2$ composites would

be developed either by prolonging the sintering time or increasing the sintering temperature.
3. The SPRS technique is a promising fabrication route for preparing high density composite ceramics.
4. The results obtained confirm that if a high density composite with fine microstructure is obtained, considerable toughening is possible. For this purpose, materials with small particle size both of the matrix and the second phase are needed.

**ACKNOWLEDGMENTS:**

ERR thanks the financial support of UAM-A, IPN, and AIEJ-Japan. HBR and JGCM are members of SNI and COFAA-IPN.

**REFERENCES**

1. M.Tokita, "Trends in Advanced SPS Spark Plasma Sintering System and Technology", J. Soc. Powder Tech. Japan, 30(1993)790-804.
2. M. Tokita, "Mechanism of Spark Plasma Sintering and Its Application for Ceramics", New Ceramics, 10(1997)43 (in Japanese).
3. S.D. De la Torre, H.Miyamoto, K.Miyamoto, J.Hong, L.Gao, L.Tinoco-D. E.Rocha and H.Balmori, "Spark Plasma Sintering of Nano-Composite Ceramics", pp. 892-897, Proc. 6$^{th}$ Int. Symp. on Ceramic Materials & Components for Engines, Arita, Japan (1997).
4. T.Nagae, M.Nose, M.Yokota and S.Saji, "Mechanical Alloying of Nb-Al Powder and Spark Plasma Sintering of MA Powder", Proc. Int. Symp. on Designing, Processing and Properties of Advanced Engineering Materials, Toyohashi, Japan, 1997 (in press).
5. J.S. Hong, S.D. de la Torre, H. Miyamoto and L.Gao, "Densification Behavior and Mechanical Properties of $ZrO_2$(3Y)/20% mol Composites Densified by SPS", Submitted to J. Am. Ceram. Soc.
6. N. Tamari, T. Tanaka, K. Tanaka, I. Kondoh, M. Kawahara and M. Tokita, "Effect of Spark Plasma Sintering on Densification and Mechanical Properties of Silicon Carbide", J. Ceram. Soc. of Japan, 103(1995)740-742.
7. K. Niihara, R. Morena and D.P.H. Hasselman, "Evaluation of KIC of Brittle Solids by The Indentation Method With Low Crack-to-Indent", J. Mater. Sci. Lett., 1(1982)13-16.
8. S. Lathabai, D.G. Hay, F. Wagner and N. Claussen, "Reaction-Bonded Mullite/Zirconia Composites", J. Am. Ceram. Soc., 79(1996)248-256.
9. J.S. Moya and M.I. Osendi, "Effect of $ZrO_2$(ss) in Mullite on the Sintering and Mechanical Properties of Mullite/$ZrO_2$ Composites" J Mater Sci Lett,2(1983)599-601.
10. G. Orange, G. Fantozzi, F. Cambier, C. Leblud, M.R. Anseau and A. Leriche, "High Temperature Mechanical Properties of Reaction-Sintered Mullite/Zirconia and Mullite/Alumina/Zirconia Composites", J. Mater. Sci., 20(1985)2533-2540.

# MECHANICAL PROPERTIES OF ZrO$_2$(Y$_2$O$_3$) / 20 mol% Al$_2$O$_3$ COMPOSITES PREPARED BY SPARK PLASMA SINTERING

J. S. Hong,[a,b] S. D. De la Torre,[*,a] K. Miyamoto,[a] H. Miyamoto,[a] and L. Gao[b]

[a] Tech. Research Institute of Osaka Pref., 2-7-1 Ayumino, Izumi, Osaka 594-1157, Japan.
[b] The State Key Laboratory of High Performance Ceramics and Superfine Microstructures, Shanghai Institute of Ceramics, Shanghai 200050, China.

## ABSTRACT

ZrO$_2$(1.5~3.0%Y$_2$O$_3$)/20mol%Al$_2$O$_3$ powders prepared by the coprecipitation method have been densified by spark plasma sintering (SPS). Nearly full dense specimens are obtained by processing them from 1400$^0$C. The effect of Y$_2$O$_3$-content on the fracture toughness, flexural strength and Vickers hardness of these materials has been investigated. The specimens with 2.0 mol% Y$_2$O$_3$ addition showed the highest fracture toughness values (11.4MPa m$^{1/2}$), while those with 2.5 mol% Y$_2$O$_3$ exhibited the largest flexural strength (1.25GPa) and Vickers hardness (Hv~1400 kg/mm$^2$) values. ZrO$_2$ phase transformation toughening is found the major toughening mechanism. The grain size difference of resulting particles limits the mechanical properties of these composites.

## INTRODUCTION

ZrO$_2$(Y$_2$O$_3$)/ Al$_2$O$_3$ composites are of importance in the field of structural materials because of their demonstrated high strength and high toughness [1-4]. Such composites are reported to reach strengths up to 2.4 Gpa with the addition of 28 vol% Al$_2$O$_3$ [5]. The mechanical properties improvement of ZrO$_2$-based materials is mainly attributed to a phase transformation toughening mechanism as the ZrO$_2$ phase is kept tetragonal on cooling [6], whereas microcrack toughening is claimed to be the main enhancement process as the ZrO$_2$ phase is monoclinic [7]. On the other hand, the addition of infiltrated Al$_2$O$_3$ to ZrO$_2$(Y), apparently produces additional toughening. This may result either from the improved transformability of ZrO$_2$, the modification of the grain boundary conditions, and/or the additional energy-absorbing effect originated from the second-phase dispersion

---

To the extent authorized under the laws of the United States of America, all copyright interests in this publication are the property of The American Ceramic Society. Any duplication, reproduction, or republication of this publication or any part thereof, without the express written consent of The American Ceramic Society or fee paid to the Copyright Clearance Center, is prohibited.

[8]. The addition of $Al_2O_3$ to $ZrO_2$, however, does not always produce positive toughening results. For instance, fine grained yttria-stabilized tetragonal zirconia polycrystals (Y-TZP) [9], with slight $Y_2O_3$ content showed a reduction of their toughness when $Al_2O_3$ was added [10]. This work is focused to understand the effect of $Y_2O_3$ content on the mechanical properties of $ZrO_2(Y_2O_3)/20$ mol% $Al_2O_3$ composites densified by spark plasma sintering (SPS). SPS is a powder densification technique by which the material to be compacted is uniaxially pressed and heated up via pulsed electric discharges. Green compacts are heated by the Joule effect. Advantages of this technique are the super fast heating rates available for ceramics and the feasibility of practising very short sintering runs (<10 min) to retain metastability and the fine grain size of precursors. An schematic of the SPS apparatus is shown in Fig.1. SPS principles and some features concerning sintering of nano-composite ceramics by this process have been reported elsewhere [11,12].

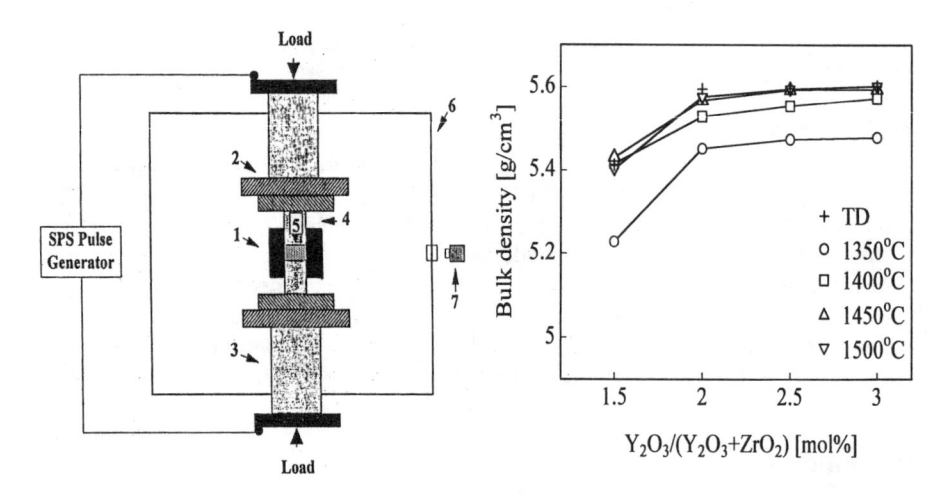

**Fig.1.** Schematic diagram of the SPS apparatus. Graphite die 1, plates 2, ram 3, punch 4, sample 5, vacuum chamber 6 and optical pyrometer 7.

**Fig.2** Bulk density of SPS-fabricated $ZrO_2(Y)/20$mol% $Al_2O_3$ composites as a function of the $Y_2O_3$ content in $ZrO_2$.

## EXPERIMENTAL PROCEDURE

$ZrO_2(Y_2O_3)/20$ mol% $Al_2O_3$ powders having different $Y_2O_3$ content were prepared by the coprecipitation method as described elsewhere [12]. $ZrOCl_2 \cdot 8H_2O$, $AlCl_3 \cdot 6H_2O$, $YCl_3 \cdot 6H_2O$ and $NH_4OH$ were used as starting materials. Powders were calcined at $700^0C$ for 2h. The SPS operations were carried out in vacuum with a Dr-Sinter 1020 apparatus (Sumitomo Coal Mining Co., Japan). Before each SPS run, powders were placed in a graphite dies set and cold pressed with a load of about 40 MPa. Along the SPS runs, samples were kept under the set load and heated in the furnace chamber up to the desired temperature at a heating rate of 200°C/min. Immediately after holding the specimen for 2 min at such temperature, the applied load was completely released and the specimen was cooled down under vacuum conditions. Specimens with final dimensions of 20 mm diameter and 5mm thickness were obtained in this way. Bulk density of the SPS processed samples was measured by the Achimedes' method. Test bars (3×4×15mm) were cut from each SPS sample with a diamond saw and then lapped with a diamond plate. The face of the bar to be subjected to tensile stress (during bending test) was polished with a SiC plate. Three-point bending tests were carried out with a span width of 10mm, setting a cross-head speed of 0.5 mm/min for strength measurements. Hardness was measured on the mirror-like polished face of the samples by the Vickers indentation technique, using a load of 196 N. Fracture toughness analysis was made by the indentation fracture (IF) technique using the results obtained with the 196 N Vickers load testing [13,14]. Quantification of the t- and m-$ZrO_2$ phases was performed from X-ray diffraction (XRD) data by measuring the relative intensities of the tetragonal (111), and the monoclinic (11$\bar{1}$) and (111) diffraction peaks [15,16].

The equations used for estimating the fraction of transformable $ZrO_2$ are given in [15,16] as follows:

m % = { $I_m$ (111) + $I_m$ (11$\bar{1}$) } / { $I_m$ (111) + $I_m$ (11$\bar{1}$) + $I_t$ (111) }
t % = 100 - m %

Where $I_m$ (111) is the relative intensity of monoclinic (111) Bragg-peaks, and $I_t$ (111) the equivalent of t-$ZrO_2$. The monoclinic fraction (m%) existing before inducing fracture is measured from the polished surface of sintered samples, whereas the m% generated after fracture is measured from the disrupted surface of strength-tested bars. The transformable amount is roughly approximated by subtracting m% after fracture from the m% before fracture. It should be noted, however, that our estimation of the transformable fraction of t-$ZrO_2$ only comprises the transformed quantity taking place locally at the surroundings (zone) of the fracture surface. Whereby reported data needs further analysis. That is, not all transformable t-$ZrO_2$ changes in one single fracture step because the stress conferred to each t-$ZrO_2$ grain is not necessarily the same.

## RESULTS AND DISCUSSION

According to XRD results, most of the tetragonal-$ZrO_2$ present in the $ZrO_2$(1.5 mol% $Y_2O_3$)/20 $Al_2O_3$ samples transformed into the monoclinic phase on cooling, whereas in the $ZrO_2$ ($\geq$ 1.5mol%$Y_2O_3$)-samples the tetragonal structure still was the major feature. Therefore, on calculating theoretical densities (TD); in the first case, it is assumed that all $ZrO_2$ is monoclinic (5.84 g/cm$^3$), meanwhile in the case of $ZrO_2$ ($\geq$1.5 mol% $Y_2O_3$)- samples the calculation is carried out considering it only as tetragonal (6.097 g/cm$^3$). In Fig.2 it is plotted the bulk density of samples sintered by SPS at different temperatures as a function of the $Y_2O_3$ content. It shows that all the samples can be densified to 99.9% of their TD by processing them at least at 1400°C. Fig.3 shows the fracture toughness ($K_{IC}$) of samples sintered at different temperatures, as a function of the $Y_2O_3$ content. The specimens having 2 mol% $Y_2O_3$ additions showed the highest $K_{IC}$ values. Literature [17] indicates that t- and m-$ZrO_2$ phases preferably enhance the fracture toughness by stress —

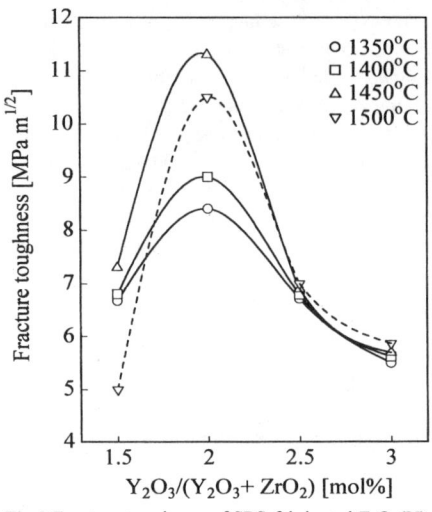

**Fig.3** Fracture toughness of SPS-fabricated $ZrO_2(Y)$ /20%$Al_2O_3$ composites as a function of the $Y_2O_3$ content in $ZrO_2$.

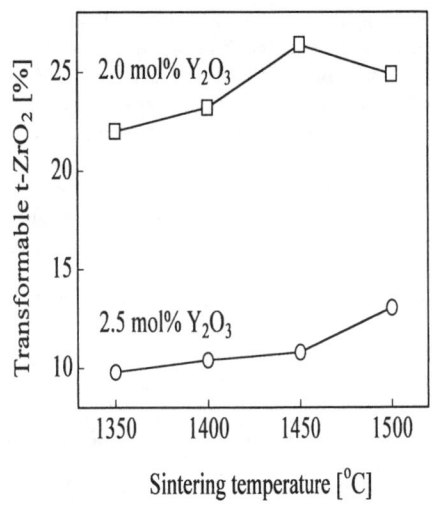

**Fig.4** The transformable t-$ZrO_2$ content in the $ZrO_2$(2.0 and -2.5 mol%$Y_2O_3$)/ 20%$Al_2O_3$ SPS-prepared composites as a function of temp.

induced transformation toughening and microcrack toughening mechanisms, respectively. Therefore, in spite of the fact that the material becomes microstructurally more stable the major reasons explaining why $K_{IC}$ decreased when the $Y_2O_3$ content is higher than 2.0 mol% are ascribed to a reduced phase transformation and to have the critical grain size (300 nm [18]) exceeded, as reported in Fig.6. On the other hand, XRD analysis revealed that in the 1.5 mol% $Y_2O_3$-samples a considerable volume fraction of t-$ZrO_2$ transformed into the monoclinic phase on cooling, and so micro-crack toughening developed in these samples can be thought to as the dominant toughening mechanism. Fig.4 shows the transformability of t-$ZrO_2$ estimated from the samples containing 2 and 2.5 mol% $Y_2O_3$ as a function of the sintering temperature. By contrast, the transformability of t-$ZrO_2$ in the 2 mol% $Y_2O_3$-case is, in average, two times larger whereby it is evident that this specimens experimented higher transformation and thus undergo larger toughening than in the latter case (see Fig.3). From Fig.4 it becomes clear that the materials toughening enhancement observed in this work mainly resulted via the transformation toughening effect. Note that in samples with 2 mol% $Y_2O_3$ addition, the fracture toughness increased as the sintering temperature rose, and reached its highest value at 1450°C. This trend follows the theory of dependence between the transformable t-$ZrO_2$ with the sintering temperature. By transformable $ZrO_2$ it is meant the fraction of t-$ZrO_2$ that can transform into the monoclinic phase. Such fraction of t-$ZrO_2$ provides an important contribution for toughening, via the transformation toughening mechanism. As above mentioned, however, most of the fraction of t-$ZrO_2$ can not transform at once into monoclinic during a given fracture course, even if the grain size were beyond its critical point. Therefore, although $ZrO_2$ is kept tetragonal after sintering, that toughening mechanism would not be considered as the only determinant factor.

The transformable $ZrO_2$ fraction is another important toughening factor influencing this mechanical property. Usually, when a given sample having large pct of the t-$ZrO_2$ phase has slightly precipitated a small amount of the monoclinic phase after being sintered, its balance t-$ZrO_2$ phase can easily be transformed into the monoclinic as a result of the applied stress, whereby the transformable fraction can reach a higher level. This actually happened in the case of samples containing 2 mol% $Y_2O_3$. Based on scanning electron micrograph (SEM) observations it appears that the morphology and particle size distribution of our precursor powders were not homogeneous and so, as it is established by Lee and Rainforth [20] unusual cubic grain sizes larger than those of t-$ZrO_2$ (0.2~2µm) can arose on most commercial TZPs containing 2~3mol%$Y_2O_3$. Fig.5 shows a typical SEM picture of the fracture surface of a 2.5mol%$Y_2O_3$ specimen SPS prepared at 1400°C/9 min. The large c-$ZrO_2$ particle (dark) is composed of at least three joined grains. Tetragonal $ZrO_2$ particles are 500 nm large. Fig.6 is another view of the same specimen of Fig.5 and is presented to illustrate in more detail the large grain size difference resulted in those specimens, the interface being a region where transgranular fracture preferentially took place upon induced fracture testing.

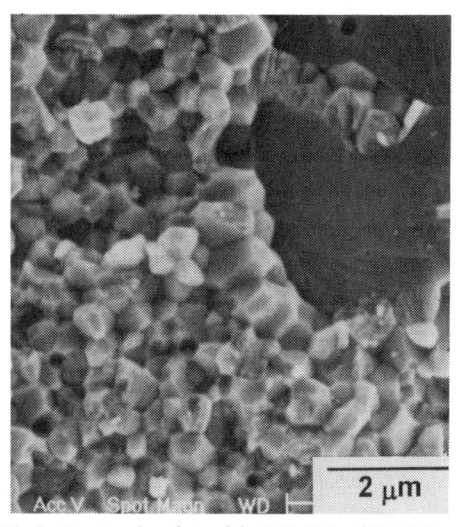

**Fig.5** Fractured surface of the $ZrO_2$(2.5mol%$Y_2O_3$) /20mol%$Al_2O_3$ sample SPS-processed at 1400°C/9m. tetragonal $ZrO_2$ particles are approx. 500nm large.

**Fig.6** Same specimen of Fig.5 disclosing ~2µm large particles of cubic-$ZrO_2$ embedded in a 500 nm t-$ZrO_2$ matrix.

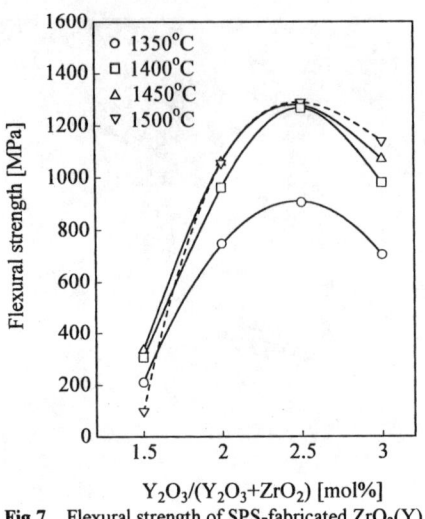

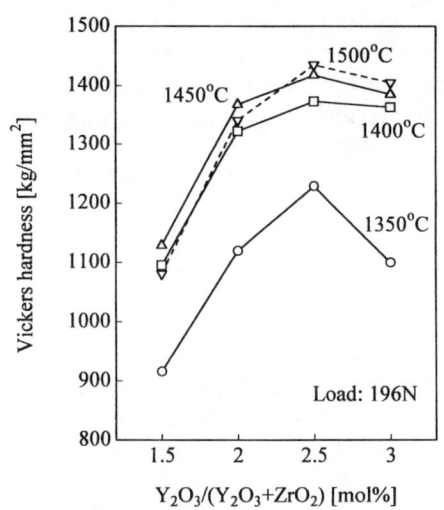

**Fig.7** Flexural strength of SPS-fabricated $ZrO_2(Y)$/20%$Al_2O_3$ composites as a function of the $Y_2O_3$ content in $ZrO_2$.

**Fig.8** Vickers hardness of SPS-fabricated $ZrO_2(Y)$/20%$Al_2O_3$ composites as a function of the $Y_2O_3$ content in $ZrO_2$.

Fig.7 shows the flexural strength ($\sigma$) of samples sintered at different temperatures as a function of the $Y_2O_3$ content. Samples with 2.5 mol% $Y_2O_3$ addition exhibited the largest $\sigma$ values. The lower strength values obtained from 1.5 mol% $Y_2O_3$-samples can be explained by the ample presence of m-$ZrO_2$. As above mentioned, in the field of $ZrO_2$-based materials, toughening may result from the micro-cracks originated upon m-$ZrO_2$ precipitation, but because of the intrinsic trade off existing between $K_{IC}$ and $\sigma$, the strength of the material can not be improved without a reduction on its toughness and vice versa. Fig.8 shows the Vickers hardness attained from samples sintered at different temperatures as a function of the $Y_2O_3$ content. Similar to the flexural strength data trend observed in this work materials their hardness reached maximum values at SPS temperatures > 1400°C, specially from the 2.5mol% $Y_2O_3$-samples.

## CONCLUSIONS

Nearly dense $ZrO_2$ (1.5 ~ 3.0 mol% $Y_2O_3$) / 20 mol% $Al_2O_3$ specimens can be prepared from 1400°C in 10 minutes by the SPS technique. The specimens with 2.0 and 2.5 mol% $Y_2O_3$ addition exhibited mechanical properties as good as those obtained from Y-TZP processed by the HIP technique [19]. The zirconia-phase transformation toughening is recognized as the major toughening mechanism. The inhomogeneity of precursor powders, however, generated specimens having traces of 2~3μm large cubic zirconia grains embedded in a 500 nm tetragonal zirconia matrix. These grain size difference limits the mechanical properties of the ceramic composites.

ACKNOWLEDGMENTS

The preparation and presentation of this research work has been sponsored by NEDO, as a part of the project "Nano-Composite Materials for Ceramic Ball-Bearing Applications" developed at TRI-Osaka.

REFERENCES

1. A. G. Evans, "Perspective on the Development of High-Toughness Ceramics," *J. Am. Ceram. Soc.*, **73** 187-206 (1990).
2. F. F. Lange, "Transformation Toughening, Part 4, Fabrication, Toughness and Strength of $Al_2O_3$-$ZrO_2$ Composites," *J. Am. Ceram. Soc.*, **17** 247-54 (1982).
3. D. W. Shin, K. K. Orr and H. Schubert, " Microstructure- Mechanical Property Relationships in Hot-Isostatically Pressed Alumina and Zirconia-Toughened Alumina, " *J. Am. Ceram. Soc.*, **73** 1181-88 (1990).
4. M. Kihara, T. Ogata, K. Nakamura and K. Kobayashi, "Effects of $Al_2O_3$ Additions on Mechanical Properties and Microstructure of Y-TZP," *J. Ceram. Soc. Jpn*, Int. Ed., **96** 635-42 (1988).
5. K. Tsukuma, K. Ueda and M. Shimada, "Strength and Fracture Toughness of Isostatically Hot-Pressed Composites of $Al_2O_3$ and $Y_2O_3$-Partially-Stabilized $ZrO_2$," *J. Am. Ceram. Soc.*, **68** C4-5 (1985).
6. R. McMeeking and A. G. Evans, "Mechanics of Transformation-Toughening in Brittle Materials," *J. Am. Ceram. Soc.*, **65** 242-24 (1982).
7. A. G. Evans and K. T. Faber, "Toughening of Ceramics By Circumferential Microscracking," *J. Am. Ceram. Soc.*, **64** 394-98 (1981).
8. S. J. Glass and D. J. Green, "Mechanical Properties of Infiltrated Alumina–Y-TZP Composites," *J. Am. Ceram. Soc.*, **79** 2227-236 (1996).
9. D.J. Green, R.H. Hannink and M.V. Swain, "Transformation Toughening of Ceramics", CRC Press, Boca Raton, FL., 1989.
10. T. Sato, H. Fujishiro, T. Endo and M. Shimada, "Thermal Stability and Mechanical Properties of Yttria-Doped Tetragonal Zirconia Polycrystals With Dispersed Alumina and Silicon Carbide Particles," *J. Mater. Sci.*, **22** 882-86 (1987).
11. S. D. De la Torre, H. Miyamoto, K. Miyamoto. J. S. Hong, L. Gao, L. Tinoco-D., E. Rocha-R. and H. Balmori-R., "Spark Plasma Sintering of Nano-Composite Ceramics," pp.892-897 in Proceedings of the 6$^{th}$ Int. Symposium on Ceramic Materials & Components for Engines. Eds. K.Niihara, S.Kanzaki, K.Komeya, S.Hirano and K.Morinaga. Technoplaza Co.,Ltd-The Japan Keirin Assoc., 1998.
12. J. S. Hong, S. D. De la Torre, H. Miyamoto, K. Miyamoto and L. Gao, "Densification Behavior and Mechanical Properties of $ZrO_2$ (3Y) / 20mol% $Al_2O_3$ Composites Prepared by Spark Plasma Sintering," Submitted to *J. Am. Ceram. Soc.*

13. A. G. Evans and E. A. Charles, "Fracture Toughness Determinations by Indentation," *J. Amer. Ceram. Soc.*, **59** 371-72 (1976).
14. K. Niihara, N. Nakahira and T. Hirai, "The Effect of Stoichiometry on Mechanical Properties of Boron Carbide," *J. Am. Ceram. Soc.*, **67** C13-14 (1984).
15. R.C.Garvie and P.S.Nicholson, "Phase Analysis in Zirconia Systems," *J. Am. Ceram. Soc.*, **55** 303-5 (1972).
16. H. Toraya, M. Yoshimura and S. Somiya, "Calibration Curve for Quantitative Analysis of the Monoclinic-Tetragonal $ZrO_2$ System By X-ray Diffraction," *J. Am. Ceram. Soc.*, **67** C119-121 (1984).
17. J.B.Wachtman "Mechanical Properties of Ceramics" John Wiley & Sons, Inc. 1996. Chapters 10-11 and 26.
18. R.Garvie "The Occurrence of Metastable Tetragonal Zirconia as a Crystallite Size Effect", Journal of Physical chemistry. **69** 4, 1238-43 (1965).
19. K.Tsukuma, Y.Kubota and T.Tsukidate, "Thermal and Mechanical Properties of $Y_2O_3$-Stabilized Tetragonal Zirconia Polycrystals", pp.382-90 in Science and Technology of Zirconia II. Advances in Ceramics. Eds. N.Claussen, M.Ruhle and A.Heuer. American Ceramic Soc. Vol.**12**, Westerville, OH, 1983.
20. W.E.Lee and W.M.Rainforth Eds."Ceramic Microstructures - Property Control by Processing" Chapman & Hall. First ed., 1994 pp.360-362.

# Reaction Forming/Bonding

# A RANGE OF SIALONS BY CARBOTHERMAL REDUCTION AND NITRIDATION

G.V.White, T.C. Ekström, G.C.Barris, & I.W.M.Brown.
The New Zealand Institute for Industrial Research and Development
PO Box 31310, Lower Hutt, New Zealand

## ABSTRACT

Sialon phases have been synthesized from starting mixtures of clay and fine silicon metal powder heated under flowing nitrogen at relatively low temperatures. The introduction of carbon or fine silicon carbide allows the preparation of alpha or beta sialons or mixtures of these phases. Composite bodies of sialon of varying compositions and properties can be prepared by a one step process which can be used to reaction bond silicon carbide. Sialon powders can also be produced.

A range of sialons were prepared from halloysite clay mixed with fine silicon powder and lampblack carbon. It was found that β-Sialon could be made phase pure (monophase by XRD) with a range of aluminum contents by varying the silicon content of the starting mix, and that alpha sialons could be stabilised by adding appropriate metal cations.

## INTRODUCTION

Silicon nitride based ceramics offer exceptional chemical and mechanical properties at high temperatures, with chemical resistance to molten metal being high on the list. From this group, sialon ceramics are of special interest for the following reasons:-

- They can be made from inexpensive raw materials e.g. clay, carbon, silicon, calcite.
- The use of clay greatly facilitates green forming.
- The use of silicon introduces the ability to reaction bond silicon carbide.
- Many properties can be controlled by adjusting the sialon composition.
- O'Sialon, which includes silicon oxynitride, is renowned for its resistance to oxidation.
- X-Phase sialon has been proposed to be resistant to molten iron and steel
- β-Sialon properties approach those of silicon nitride; β-Sialon is tougher and much easier to sinter than silicon nitride
- Low aluminum content β-Sialon is a precursor for the closely related, and very hard, α-Sialon.

## Review of β-Sialon and α-Sialon

The term SiAlON, or silicon aluminum oxynitride, encompasses a family of compounds or phases comprised of the elements: silicon, aluminum, oxygen and nitrogen, first reported in the early 1970s (1,2). Each phase is described by a composition range for which that particular structure is stable.

β-Sialon is stable over the composition range $Si_{6-z}Al_zO_zN_{8-z}$ where the z value, the aluminum content, can range from 0 to 4.2 at 1750°C. This includes silicon nitride as the z = 0 end member. β-Sialon has the same structure as silicon nitride (β-$Si_3N_4$), and can be regarded as a solid solution formed by substituting equal amounts of aluminum and oxygen for silicon and nitrogen respectively into the silicon nitride structure. The degree of substitution possible increases with temperature.

α-Sialon has a structure, derived from α-$Si_3N_4$, which is stabilized by a metal cation (M) such as Y, Li, Ca and some rare earth elements. In the formula $M_{m/v}Si_{12-(m+n)}Al_{m+n}O_nN_{16-n}$, m and n indicate the replacement of (m+n) (Si-N) bonds by m(Al-N) and n(Al-O) bonds and v represents the valence of the metal cation M. The range of composition which is stable varies with the stabilizing metal cation.

α-Sialon, when fully dense, is very hard. β-Sialon is less hard but readily assumes an elongate microstructure which imparts high fracture toughness. A composite of the two yields excellent mechanical strength and wear resistance (3,4).

## Reaction Mechanism for Carbothermal Reduction

One possible low-cost manufacturing route is the carbothermal reduction and nitridation of silicate or aluminosilicate minerals. This was first demonstrated using kaolinite clays in 1976 (5) and using halloysite in 1987 (6). New Zealand China Clays Ltd. halloysite clay is an excellent source mineral for carbothermal reduction and nitridation to produce β-Sialon powders, because it has a combination of high purity and fine particle size (6). The Al content of the β-Sialon will depend on the overall Al/Si ratio of the halloysite. Upon complete reaction, halloysite alone (Al/Si=1/1), will form β-Sialon z=3, $Si_3Al_3O_3N_5$. The use of this clay and the overall preparation route have been investigated, yielding β-Sialon powders with z=2.3-3.1 (6-9).

## The Effect of Additives on the Synthesis of β-Sialon

The effects of additives on the carbothermal reduction and nitridation of clay have been widely investigated, often in attempts to understand the effects of natural impurities in the clay and carbon raw materials (5,7,10-13). The formation of a silicon rich FeSi eutectic at 1209°C, has a marked effect and has been employed by a number of workers to promote the formation of β-Sialon but all found some evidence of additional phases in the product (5,7,10,11,14). There is also debate regarding the mechanism and formation of a liquid iron silicide phase is favored (5,10). Calcium promotes various phases: O'Sialon (12), 10% alumina (10), and also doubles the sialon yield (10).

The addition of yttria resulted in a high total crystalline phase content and formed alumina and yttrium aluminum garnet (YAG) in addition to promoting β-Sialon formation (10). The alumina and YAG would not degrade the properties of the ceramic product and should improve the strength. Yttria addition to carbothermal reduction and nitridation of clay has not been widely investigated as it is not a common clay impurity. Also, this route is promoted as low cost and yttria is relatively expensive.

The use of silicon in this study requires the effect of additives on silicon to be considered. There is considerable work reported on the formation of silicon nitride, and silicon nitride bonded materials prepared from silicon metal powder. Elimination of the inhibiting effect of the silica coating on the silicon surface is central to much work although the effect of additives on the properties of the ceramic product has also been widely considered (15). Impurities in the silicon have a large effect, particularly iron which is introduced by the use of steel ball mills to grind the silicon powder, but also aluminum which, with the iron, reduces the temperature of the FeSi eutectic (16). Iron catalyses the formation of atomic nitrogen (which promotes β-$Si_3N_4$), forms a liquid which provides rapid diffusion paths and volatile silicon, and aids devitrification of silica, supplying SiO and exposing underlying silicon (17). More recent work (18,19,20,23) has employed yttria and alumina to exploit the yttria-alumina-silica eutectic liquid phase formation at 1370°C to remove silica from the silicon surfaces and allow nitridation to proceed. Subsequently the silicon nitride ceramic can be annealed to crystallise YAG from the intergranular glass to improve the strength.

Reaction Mechanism for Carbothermal and Silicothermal Reduction
The addition of silicon to the clay/carbon carbothermal reaction mix offers a low cost technique for producing β-Sialon with lower z-values (≤2), especially z=0.5 as precursor for α-Sialon. In this study the influence of process parameters such as temperature and time on the reaction products has been investigated and the effect of yttria additions and increases in substituted aluminum (higher z-value) evaluated. The reaction proceeds through a series of steps; the overall reaction has been proposed (22,23) to be:-

*β-SiAlON z = 0.5, from NZCC halloysite clay, silicon, and carbon*

$$Al_2O_3 \cdot 2.4SiO_2 \cdot 2.2H_2O + 19.6Si + 15N_2 + 5.8C \Rightarrow$$
$$4Si_{5.5}Al_{0.5}O_{0.5}N_{7.5} + 5.8CO + 2.2H_2O$$

Additives to promote the reaction are particularly important for low z β-Sialon and α-Sialon, but they may also offer the possibility of lower firing temperature and shorter reaction time at the low temperature. Addition of yttria to the halloysite/silicon/carbon mixture has also been found to allow the synthesis of α-Sialon.

## METHODS AND MATERIALS

The starting materials were halloysite clay (NZ China Clays Ltd, $d_{50}$= 0.3μm, $Fe_2O_3$=0.25%, (6) ), carbon (Degussa Lampblack 101), silicon metal powder (Permascand $d_{50}$=4.6μm, Fe=0.08% KemaNord) and yttrium oxide ($Y_2O_3$ Fine, grade C, H.C. Starck Berlin). Halloysite clay, silicon, and carbon were combined to produce a series of reaction mixes calculated to yield β-Sialon powders with z values ranging from 0.25 to 2.0 (later referred to as nominal z values) in the β-Sialon formula $Si_{6-z}Al_zO_zN_{8-z}$. Yttria was added at 0.5 to 5 wt%, expressed as a percentage of the theoretical sialon yield. Experiments were performed to optimise the carbon content (21). An excess of 10% over the calculated stoichiometric amount gave optimum results and this was adopted as a standard.

The clay/silicon/yttria mixes were blended with carbon by ball-milling for 24 hours in isopropanol using HDPE bottles with $Si_3N_4$ milling media. They were dried in a rotary evaporator and the powder extruded into rods (diam.≈4mm, length≈12 mm).

The samples were heated under flowing nitrogen in alumina crucibles in a horizontal tube furnace (40 mm internal diameter) at 10°Cmin$^{-1}$ to holding temperatures ranging from 1300 to 1450°C and held at temperature for 2 to 96 hours. For temperatures >1350°C a soak of four hours at 1350°C was introduced to allow the silicon to undergo essential initial surface nitridation described under the next section

All fired products were examined by X-ray powder diffraction (XRD) using a Philips diffractometer with APD1700 software. The relative amounts of crystalline phases were obtained by comparing selected peak areas. For $α-Si_3N_4$ and β-Sialon the peak combinations (102) + (210) and (101) + (210) respectively, were used to establish the β/(β+α) Sialon ratio. To ascertain the relative phase development for less reacted samples of poor crystallinity, a simplified procedure compared the XRD peak heights of one major peak of each phase. The β-Sialon lattice parameters could be calculated and refined using 20 peaks. The β-Sialon phase ($Si_{6-z}Al_zO_zN_{8-z}$) z-value was obtained from the β-Sialon unit cell according to Ekström (13). When two or more β-Sialon phases with different z-values were present simultaneously, peaks overlapped and an additional technique was required. In this case the z-value of each β-phase was estimated from a single peak (101). This peak, which gives the best phase separation, was profile-fitted by computer (Philips APD 1700 software). The component peak positions were used to estimate the z-values of the β-Sialon phases present by comparison with standard sialon patterns, and the corresponding component peak areas were used to estimate the relative phase amounts as a percentage.

## RESULTS AND DISCUSSION

Reaction mixtures with a range of compositions (targeting five β-Sialon z values) were reacted at temperatures between 1300°C and 1450°C. The nominal z values, 0.25, 0.5, 0.7, 1.0, and 2.0, represented the expected compositions of the sialons when fully reacted.

At the lowest temperature used, 1300°C, the reaction was incomplete despite soaking times up to 96h. At this temperature unreacted mullite was detected by XRD together with broad, disordered peaks of other phases and a raised background indicating the presence of large amounts of amorphous material.

In Figure 1 the development of the phases at 1350°C after 96 hours is presented as a function of nominal z-value. The high z-value mixes form β-Sialon more readily. The reason may be the higher aluminum content or perhaps the large proportion of the clay because the clay introduces sufficient impurity ($Fe_2O_3$=0.25%) to promote the reaction. When a large proportion of the starting mix is silicon metal powder (low nominal z-value mixes), O'Sialon is formed with silicon carbide. Some residual unreacted cristobalite remains.

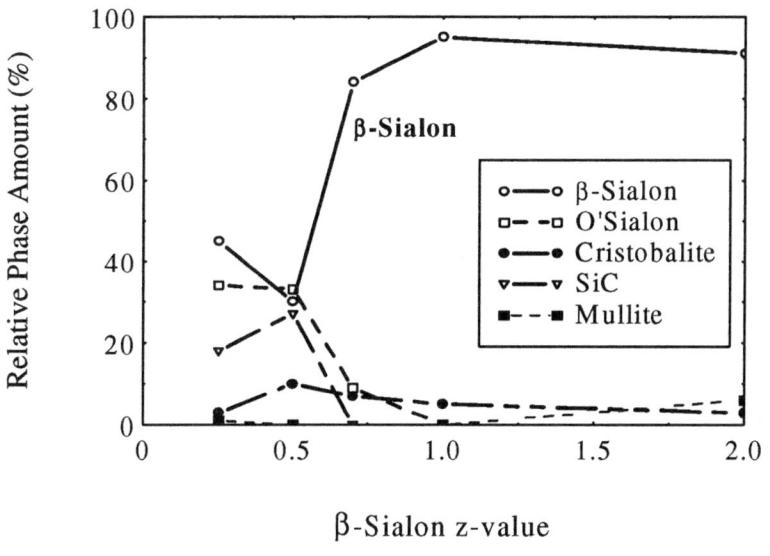

Figure 1. The extent of reaction for compositions z=0.25 to 2 β-Sialon mix after 96 hours at 1350°C.

The reaction products for a low nominal z-value mix (z=0.5) were measured at higher temperatures for times up to 96 hours. The results at 1400°C are presented in Figure 2. For this z-value, mixtures of β-Sialon with O'Sialon and silicon carbide were observed after 48 hours reaction at 1400°C and unreacted mullite was still detected after 22 hours. The z-value of β-Sialon was close to the nominal z=0.5 and the aluminum content of the O'Sialon corresponded to $Si_{1.8}Al_{0.2}O_{0.2}N_{1.8}$ (x=0.2), the highest possible Al substitution. Some of the silicon is taken up by the silicon carbide. It is clear from the figures that the reactions are time and temperature dependant. At 1400°C a fully reacted z=0.5 β-Sialon will be achieved after ≥96 hours but at 1450°C, 30 hours is sufficient (not illustrated).

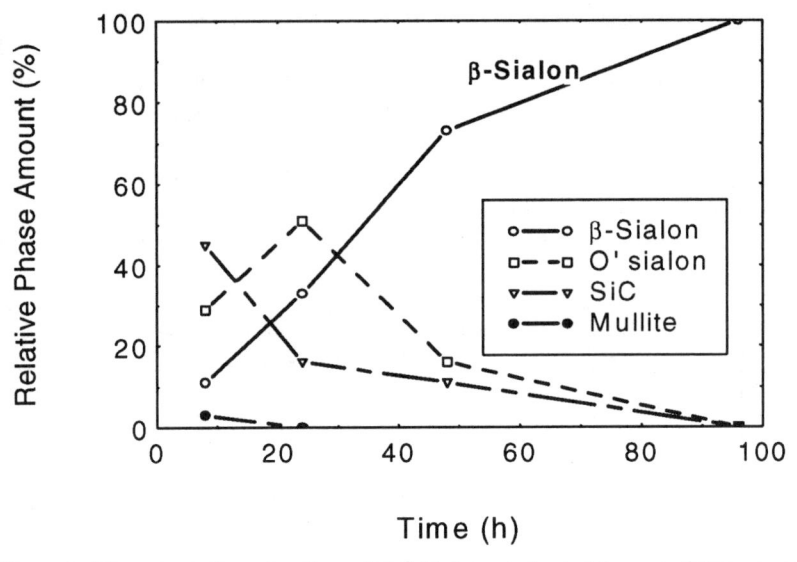

Figure 2. The extent of reaction for z=0.5 β-Sialon mix heated for up to 96 hours at 1400°C.

For samples of nominal z ≤ 0.7, the β-Sialon peaks in the 1400-1450°C reactions were sharp and observed z-values for the β-phase were in fair agreement with the nominal z-values. In the region with nominal z-values ≥1 broad peaks in the β-Sialon XRD patterns at low diffraction angles, which at higher angles tended towards two or three partly overlapping peaks, were better resolved at 1400°C and 1450°C. This indicated the presence of more than one β-Sialon with high-z phase present in small amounts besides the nominal β-phase. One of the additional β-phases was observed at a z-value around 1.5 and another had a higher but varying z-value around 2-2.8. The source or cause of these components may be the individual reactants. For example this hypothesis is in agreement with earlier findings that this halloysite clay forms β-Sialon powders with z=2.3-3.1 (5-9).

Yttria was added to exploit the eutectic liquid phase formation in the yttria-alumina-silica system at 1370°C which would be expected to speed the reaction at and above that temperature. The dramatic effect of the 0.5% yttria addition for the range of compositions (z values 0.25 to 2) is illustrated in Figure 3. With no additive the product was considerably under-reacted at z=0.5, even after 8 hours at 1400°C (Figure 2.). The addition of 0.5% yttria resulted in fully

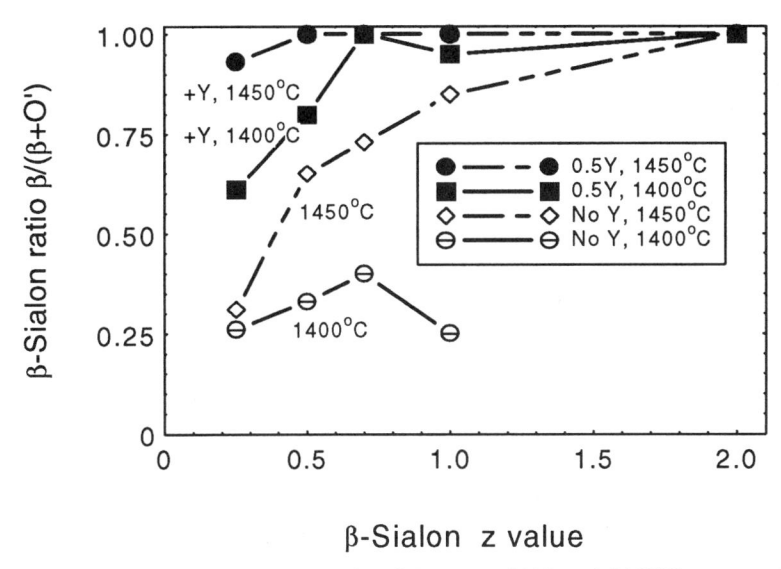

Figure 3. The extent of reaction after 8 hours at 1400 and 1450°C - comparison with no additive and 0.5% yttria

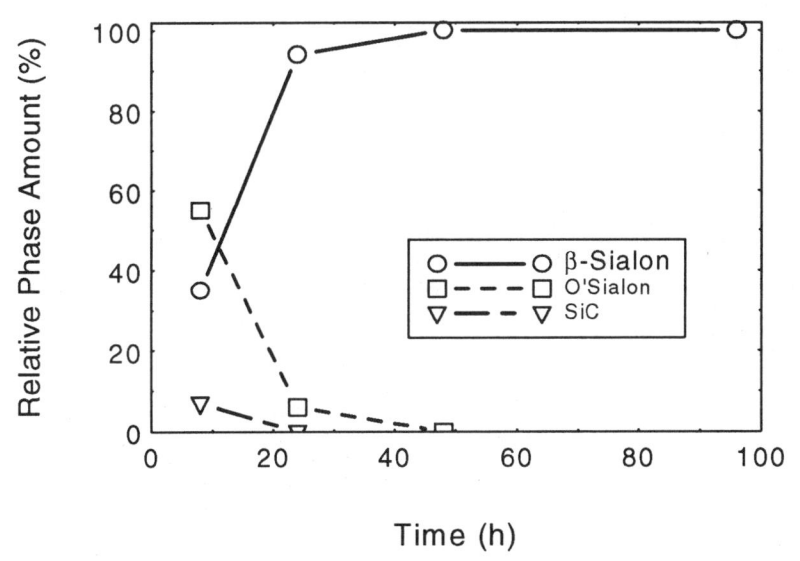

Figure 4. Phases with 0.5% yttria heated at 1400°C for up to 96 hours

Innovative Processing/Synthesis: Ceramics, Glasses, Composites II

reacted z=0.5 sialon which was one of the specific goals of the project. However at 1400°C low z compositions were still incompletely reacted. The effect of time at 1400°C for a z=0.5 mix with 0.5wt% yttria is shown in Figure 4. Here complete reaction took place after 48 hours (and 95% at 24h) whereas with no yttria (Figure 2.) 96 hours is required for complete reaction to β-Sialon. This time is reduced at 1450°C; with 0.5% yttria heating for more than 8 hours at 1450°C will lead to complete reaction (not illustrated).

A further benefit of the yttria addition was the complete elimination of fibrous, SiO gas derived "wool" on the surface of the sample bed. This was initially interpreted to indicate the complete elimination of gas phase reaction but some fibrous deposit was later found away from the sample, down the gas stream. Nevertheless the elimination of fibrous, and potentially carcinogenic, "wool" from the vicinity of the sample is another major benefit of yttria addition. In production the downstream byproduct could be collected on a cold finger. A detailed study (22) revealed the downstream product was in fact high in Si, sourced from SiO. Systematic discrepancies from the theoretical change in mass, also noted in other studies (10), could not be entirely explained by loss of volatiles. The formation in the sample of small amounts of SiC and AlN (mainly amorphous) was demonstrated by MAS NMR and shown to account for the discrepancies.

One advantage of the use of yttria to promote the sialon synthesis is that, following synthesis, it will be present in the sialon powder as a sintering aid. Although it has been shown that remarkably small proportions of yttria promote the synthesis, more will be required as a sintering aid. Also even more yttria will be required to move the composition into the mixed α/β sialon phase field of the Y-Si-Al-O-N system for the synthesis and densification of α-Sialon. The effect of increasing the yttria content in the z=0.5 synthesis mixture was tested.

β-Sialon mixes (nominal z=0.5) with 0-5% yttria content were reacted for 8 hours at 1350, 1400, and 1450°C (Figure 5). The β-Sialon content expressed as the β-Sialon ratio is sensitive to the temperature and the yttria content. There is an optimum yttria content of about 1%, at higher levels yttria definitely hinders the reaction. The optimum "window" or range of yttria content which gives the best reacted product is extended as the temperature increases. At 1450°C additions of 0.5 and 0.7% yttria give pure β-Sialon (by XRD) with z values close to z=0.5. Yttria addition of 2.4% gives almost pure β-Sialon with silicon carbide possibly present by XRD in addition to O'Sialon.

The reactions are time dependant. Longer times seem to overcome the reaction inhibiting effect of the high yttria concentration. These results may be viewed with the yttria-alumina-silica eutectic liquid phase formation at 1370°C and the melting point of silicon at ~1410°C in mind. After 1350°C for 48 hours the 1% yttria mix reacts completely, at 1400°C 24 hours is sufficient, and at 1450°C it is clear that the time is 8 hours or shorter; by far the larger shift is between 1350 and 1400°C. This suggest that the eutectic has the greater effect but it has been found that the nitridation of molten silicon (alloyed with impurities) starts well below 1400°C (see below Fig 6).

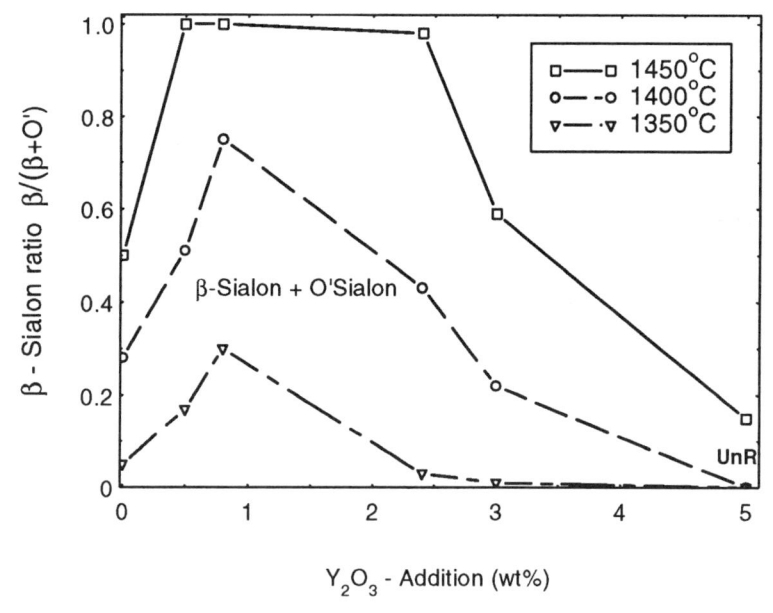

Figure 5. The development of z=0.5 β-Sialon with increasing yttria content after 8 hours at 1350, 1400, and 1450°C

When nitriding silicon, a low temperature soak is normally employed to allow pre-reaction of the metal surfaces. This is followed by a high temperature reaction. The pre-reaction facilitates the surface nitridation of silicon metal below its melting point, creating a nitride skeleton which prevents agglomeration of molten silicon and the consequent meltout.

A range of temperature schedules with holding times between 1250 and 1350°C were tried with the result that 1350°C for 4 hours was found to yield a better β-Sialon product. With no yttria added, a final soak at 1450°C for at least 8 hours was found to be required for complete reaction; longer for lower z values.

Thermogravimetric Analysis
Thermogravimetric experiments were performed to measure weight change and evolved gases while ramping the mixtures to 1450°C. They revealed that for clay/silicon/carbon mixes initial weight losses take place between 200 and 500°C; these are associated with water loss and dehydroxylation of the halloysite clay. Immediately following this a slow increase in weight

gradually accelerates with features (small inflections) at 1200 and 1300°C leading to an increasingly steep weight increase. This steep increase stops abruptly at 1402°C followed by a short plateau (no net wt change) and then a slow decline which accelerates at the start of the 1450°C soak but gradually levels off. For 0.5 to 5% yttria mixtures the reactions are initiated at the same temperature but the plateau at ~1300-1350°C is eliminated (not illustrated).

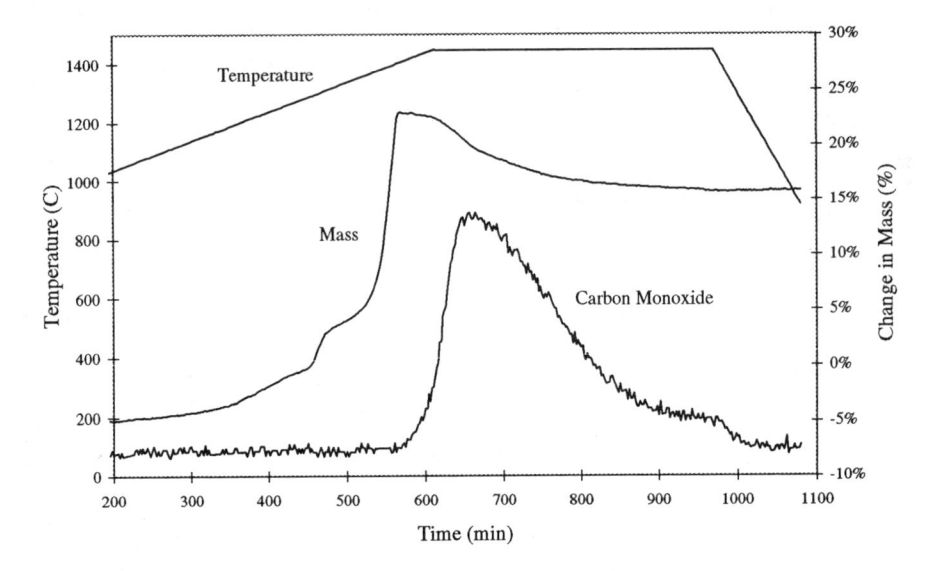

Figure 6. Change in mass and evolved carbon monoxide for β-Sialon z=0.5 mix heated to 1450°C at 1°/min. and held at 1450°C for 8 hours.

Monitoring the evolved gases by mass spectrometer showed that evolution of carbon monoxide and oxygen starts immediately after the completion of the steep weight increase at 1402°C. The steep weight increase represents the rapid nitridation of molten silicon (alloyed with impurities) to form O'Sialon. Complete reaction of all available silicon is the likely cause of the abrupt plateau. (It is possible that loss of weight from the carbon monoxide evolution is sufficiently large and rapid to cancel the final nitridation weight gain but the relative shapes of the curves support the former view.) The sudden production of carbon monoxide at this point is a consequence of the sudden availability of oxygen, possibly as an oxide. This phenomenon is considered to offer a new input to the proposed reaction path (23) and further work in this area should yield a greater understanding.

Synthesis of α-Sialon

α-Sialon is, if possible, even more interesting with the recent publication of a new tough and hard material (24). The clay/silicon/carbon synthesis route developed in the present study has been extended to make fully dense α-Sialon by HIPping partly reacted clay/silicon/carbon powder at 1800°C (25). In this study β and α-Sialon bonded bodies have also been made by pressureless sintering at 1450°C, achieving 20% apparent porosity after CIPping at 200MPa. It remains to refine this process and measure the physical properties of the product.

CONCLUSIONS

1. The combination of elemental silicon, clay, and carbon heated under nitrogen offers a new route for the preparation of β-Sialon $Si_{6-z}Al_zO_zN_{8-z}$, with low-z values, (z=0.25-0.70), yielding fully reacted β-Sialon at 1400°C.
2. The reactions of β-Sialon with $z \geq 1$ proceed well, but small additional amounts of higher aluminum content β-phases are present in the final product.
3. A pre- reaction step in the firing cycle gives the best results - for low-z sialons pre-reaction for 4 hours at 1350°C followed by 8 hours at 1450°C.
4. Small additions of yttria (0.5 to 5%) promoted the reaction. There was a clear optimum in the region 1 to 2% yttria with decreased yield apparent at 5 % additions.
5. Further work remains to be done on the reaction sequence and the production of commercial α-Sialon.

REFERENCES
1. Y.Oyama and O.Kamigaita, *Solid Solubility of Some Oxides in $Si_3N_4$*, Jpn. J. Appl. Phys., **10**, 1637 (1971).
2. K.H.Jack and W.I.Wilson, *Ceramics based on the Si-Al-O-N and related Systems*, Nature (London), Phys. Sci. **238**, 28-29 (1972).
3. T.Ekstrom and M.Nygren, *SiAlON Ceramics*, J. Am. Ceram. Soc. **75** 259-76 (1992).
4. J.W.T.van Rutten, R.A.Terpstra, J.C.T.van der Heijde, H.T.Hintzen and R.Metselaar, *Carbothermal Preparation and Characterisation of Ca-α-sialon*, J. Eur. Cer. Soc. **15** 599-604 (1995).
5. J.G. Lee and I.B. Cutler, *Sinterable Sialon Powder by Reaction of Clay with Carbon and Nitrogen*, Am. Ceram. Soc. Bull. **58**, 869-71 (1979).
6. D.S. Perera, *Conversion of Precipitated Silica from Geothermal Water to Silicon Nitride*, J. Aust. Ceram. Soc **23**, 11-20 (1987).
7. M.E. Bowden, K.J.D. MacKenzie and J.H. Johnston, *Reaction Sequence During the Carbothermal Synthesis of β'-Sialon from a New Zealand Halloysite*, Mater. Sci. Forum **34-36**, 599-603 (1988).
8. K.J.D.MacKenzie, R.H.Meinhold, G.V.White, C.M.Sheppard, and B.L.Sherriff, *Carbothermal Formation of βSialon from Kaolinite and Halloysite Studied by $^{29}Si$ and $^{27}Al$ Solid State MAS NMR*, J. Mater. Sci.. **29**, 2611-2619 (1994).
9. G.V.White, I.W.M.Brown, G.C.Barris, M.J.Ryan, W.R.Owers, C.M.Sheppard & J.M. Clarke, pp1043-1048 International Ceramic Monographs Ed. C.C.Sorrell and A.J.Ruys, Pub. Australasian Ceramic Society (1994).

10. I. Higgins and A. Hendry, *Production of β'-Sialon by Carbothermal Reduction of Kaolinite,* Proc. Brit. Ceram. Soc. **39**, 163-178 (1986).
11. E. Kokmeijer, C. Scholte, F. Blömer and R. Metselaar, *The influence of process parameters and starting composition on the carbothermal production of sialon,* J. Mater. Sci. **25**, 1261-67 (1990).
12. F.J. Narciso and F. Rodriguez-Reinoso, *Synthesis of β'-Sialon from clays: Effect of Starting Materials,* J. Mater. Chem. 1994. **4**[7], 1137-1141
13. T.C. Ekström, P -O. Käll, M. Nygren and P. -O. Olsson, *Single -Phase β-Sialon Ceramics by Glass-Encapsulated Hot Isistatic Pressing,* J. Mater. Sci., **24**, 1853-61 (1989).
14. A H. Mostaghaci, F.L. Riley, and J. -P. Torre, *The Milling and Densification Behaviour of β'Sialon Powders Prepared from Alumino-Silicate Minerals,* Int. J. High Technology Ceramics, **4** 51-71 (1988)
15. G.Ziegler, J.Heinrich, G.Wotting, *Relationships between processing, microstructure and properties of dense and reaction-bonded silicon nitride,* J. Mat Sci. **22**(7), 3041-3086 (1987).
16. A.J.Moulson, *Reaction-bonded silicon nitride: its formation and properties,* J. Mat Sci. **14**, 1017-1051 (1979).
17. R.G.Pigeon, A.Varma, A.E.Miller, *Some factors influencing the formation of reaction-bonded silicon nitride:* J. Mat Sci. **28**[7], 1919-1936 (1993).
18. T.C. Ekström, L.K.L.Falk, E.M.Knutsen-Wedel, *Pressureless-sintered $Si_3N_4$ -$ZrO_2$ composites with $Al_2O_3$ and $Y_2O_3$ additions,* J. Mater. Sci. Lett., **9**, 823-826 (1990).
19. G.V White, T.C. Ekström, I.W.M.Brown, G.C.Barris, & C.M.Sheppard. *Effects of Yttria on Carbothermal Reduction And Nitridation to form β'Sialons,* accepted for publication, Proceedings PACRIM 3 Cairns (1996).
20. K.J.D.MacKenzie, and R.H.Meinhold, *Additive pressureless sintering of carbothermal β'-sialon: an X-ray and solid state MAS NMR study,* J. Mater. Chem.. **6**(5), 821-831 (1996).
21. G.V White, T.C. Ekström, I.W.M.Brown, G.C.Barris, & C.M.Sheppard. *Synthesis Of Low Z β'Sialon Powder By Carbothermal Reduction And Nitridation Of Silica Enriched Halloysite Clay.* accepted for publication, Proceedings PACRIM 3 Cairns (1996).
22. T.Ekström, K.J.D.MacKenzie, G.V White, I.W.M.Brown, and G.C.Barris, *Volatile products formed by carboreduction and nitridation of clay mixtures with silica and elemental silicon,* J. Mater. Chem.. **6**[7], 1225-1230 (1996).
23. K.J.D.MacKenzie, T.C.Ekstrom, G.V.White. and J.S.Hartman. , *Carbothermal Synthesis of low-z β'-sialon from silica or elemental silicon in the presence and absence of $Y_2O_3$ an XRD and MAS NMR perspective,* J. Mater. Chem.. **7**[6]1057-1061 (1997).
24. I-Wei Chen and A. Rosenflanz, *A tough SiAlON ceramic based on α-$Si_3N_4$ with a whisker-like mio\crostructure,* Nature **389** 701-704 (1997).
25. T.C.Ekstrom, Z.-J.Shen, K.J.D.MacKenzie, I.W.M.Brown, and G.V.White. *α-Sialon Ceramics from a Clay Precursor by Carbothermal Reduction and Nitridation,* J. Mater. Chem.. J. Mater. Chem.. **8**[4], 977-983 (1998).

Acknowledgement: T.C.E. is grateful for the receipt of an IRL Senior Scientist Fellowship.

# CARBOTHERMAL SYNTHESIS OF α-SI$_3$N$_4$ POWDERS USING CARBON COATED SILICA PRECURSORS

**Swaroop Kaza and Rasit Koc**
Department of Mechanical Engineering and Energy Processes
Southern Illinois University at Carbondale
Carbondale, IL 62901.

## ABSTRACT

The synthesis of α-Si$_3$N$_4$ powders via carbothermal reduction and nitridation using novel carbon coated precursors was studied. The precursor starting materials contained starting silica powder coated with pyrolitic carbon obtained by the cracking of a hydrocarbon gas at 600°C. Two types of starting silica powders with various carbon contents were used. The resulting product powders were characterized and compared for the different starting materials and the various carbon contents. These were also compared with the powders produced by the conventional process of mixing the starting silica powder with carbon black. The studies show that the novel carbon coating process produces better final Si$_3$N$_4$ powders as compared to the conventional mixed carbon precursors.

## INTRODUCTION

Silicon nitride is a prominent ceramic material used in advanced engineering applications because of its excellent properties like high strength, low density, stability in oxidizing and reducing environments and low thermal expansion. It has especially found applications in reciprocating engine parts, automotive valves and gas turbine engines, but usage of silicon nitride in high temperature applications requires a highly dense final product. The density and final properties of the silicon nitride part depend on the sintering technique used and on the starting powder characteristics[1].

Various methods have been studied for the production of high quality silicon nitride powder. Carbothermal reduction has, however, received a lot of attention owing to its promise in producing large quantities of high quality silicon nitride powder economically.

## CARBOTHERMAL REDUCTION

Carbothermal synthesis of silicon nitride from silica has been studied for over a hundred years. It has, however, attracted a lot of attention in recent years because of its ability to produce high quality silicon nitride powder for advanced engineering applications. The overall reaction is

$$3SiO_2(s) + 6C(s) + 2N_2(g) = Si_3N_4(s) + 6CO(g) \quad \text{----------- (1)}$$

It is generally accepted that this overall reaction takes place through a number of intermediate steps. These intermediate reaction steps can be divided into two groups

SiO(g) generating:

$$SiO_2(s) + C(s) = SiO(g) + CO(g) \quad \text{----------- (2)}$$
$$SiO_2(s) + CO(g) = SiO(g) + CO_2(g) \quad \text{----------- (3)}$$
$$CO_2(g) + C(s) = 2CO(g) \quad \text{----------- (4)}$$

$Si_3N_4$ forming:

$$3SiO(g) + 3C(s) + 2N_2(g) = Si_3N_4(s) + 3CO(g) \quad \text{----------- (5)}$$
$$3SiO(g) + 3CO(g) + 2N_2(g) = Si_3N_4(s) + 3CO_2(g) \quad \text{----------- (6)}$$

Initially the generation of SiO is assumed to proceed via reaction (2). This is considered to take place readily owing to the abundance of $SiO_2$ and C and the relatively good contact between the solid reactants at the initial stage[2]. However, as the reaction proceeds, the reactants are consumed and there is a redistribution of the $SiO_2$ and C. This reduces the contact area and the generation of SiO becomes the rate limiting step[3]. SiO is also generated via reaction (3), but the flow of gas, which reduces the partial pressure of CO, generally considered good for the overall reaction, inhibits reaction (3). The SiO generated is converted to $Si_3N_4$ via reactions (5) & (6). The $Si_3N_4$ produced generally contains α-phase and is chemically pure except for some SiC that might be produced at higher temperatures.

Carbothermal reduction, for all its attractiveness, does have disadvantages, which have prevented its adoption for generation of high quality $Si_3N_4$ on a large scale. The initiation of the reaction is limited by the contact area of the reactants. The final powders contain a relatively high level of impurities, high agglomeration and un-equiaxed particles. Also, the reaction yield is low and the oxygen and carbon content in the final powders are unacceptably high. A new method[4] of precursor preparation for carbothemal reduction is presented in this paper. This novel method of carbon coating silica powders alleviates the various disadvantages of carbothermal synthesis and makes the process attractive for commercial production of high quality, low cost silicon nitride powders.

## EXPERIMENTAL PROCEDURE

In the present work, the novel synthesis method[4] described in US patent #5,324,494 was used to produce $Si_3N_4$ powders. Two different types of starting silica powders were used. Cabosil EH-5 fumed silica and Sylox 15 gel silica. The gel silica was coated with carbon at three different weight percentages: 28, 32 and 36. The fumed silica was coated with two different carbon weight percentages: 28 and 36. The coating was performed in a rotating coating reactor by cracking propylene gas at 600°C and 50 psi. Mixed precursors were also prepared by dry mixing the starting silica powders with carbon black in a Spex ball mill/mixer for two hours. The precursors were then seeded up to 10 wt % with $Si_3N_4$. The seed powder used was $\alpha$-$Si_3N_4$ (Hermann Stark, grade M11). The final precursors were then reacted in a tube furnace at 1500°C for 5 hours in a flowing nitrogen atmosphere with a gas flow rate of 1LPM, to form silicon nitride powder.

The starting material, the precursors and the reaction products were characterized using X-ray diffraction (XRD) with Cuk$\alpha$ radiation, BET surface area analysis and transmission (TEM) and scanning (SEM) electron microscopy. Free carbon percentages were determined by heating the reaction products in air at 750°C for 4 hours and measuring the weight loss. Oxygen and nitrogen content of the powders were analyzed by LECO.

## RESULTS AND DISCUSSION

The silica raw materials used had high surface area and were thus highly reactive. The fumed silica had a surface area of 310 $m^2/g$ and the gel silica had a surface area of 120 $m^2/g$. The gel silica contained about 10 wt. % of moisture in a combined form. However, this was eliminated during the coating process and did not influence of the final reaction.

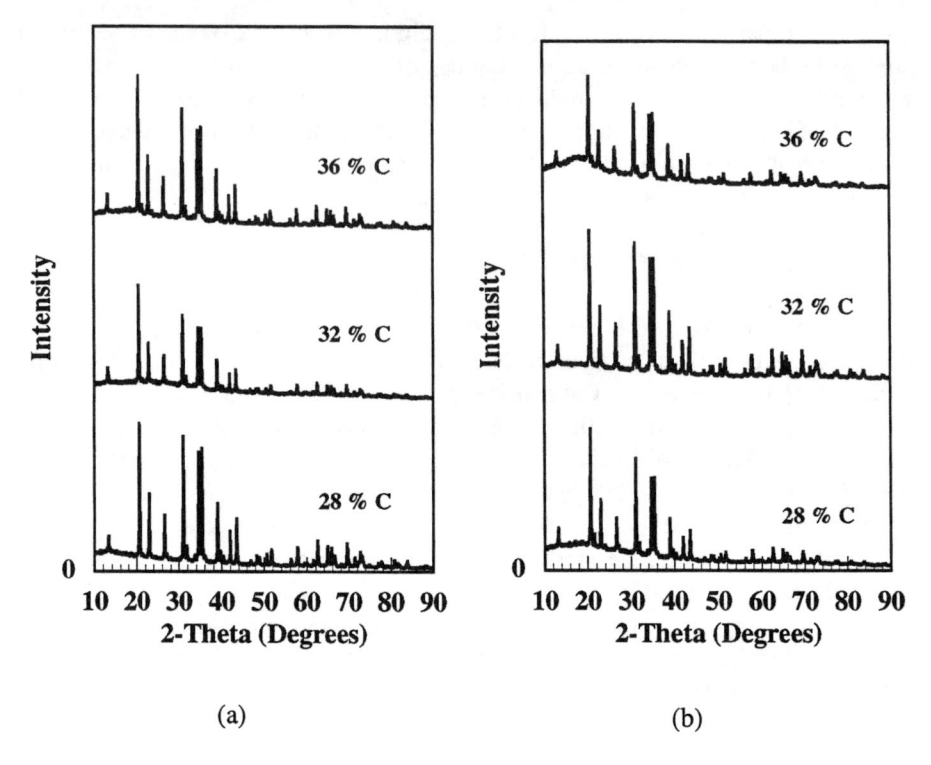

Fig 1. (a) XRD of $Si_3N_4$ produced from coated gel silica (b) XRD of $Si_3N_4$ produced from mixed gel silica

Fig. 1 shows the X-ray diffraction results of the $Si_3N_4$ powders from the gel silica raw material. Both the coated and mixed powders show pure $\alpha$-$Si_3N_4$. No secondary phase was detected. The X-ray diffraction was performed after the free carbon was removed.

The BET surface areas for the final silicon nitride powders from the various precursors, after carbon removal, are in the range 3 - 6 $m^2/g$. The surface area for the seed powder used was 9.6 $m^2/g$. Thus, the surface of the final silicon nitride powders produced seems to suggest that the silicon nitride particles during the reaction grow on the seed particles[5]. The $Si_3N_4$ from the mixed gel precursor shows a slightly higher surface area which is probably due to the presence of unreacted silica.

Table 1 shows the oxygen and nitrogen content in the $Si_3N_4$ from precursors after removal of free carbon. The oxygen content decreased with increase in initial carbon content. This is due to the fact that higher initial carbon content permits the reaction to proceed further towards completion thereby reducing the amount the unreacted $SiO_2$, which is primarily how oxygen is present in the final product. The mixed gel precursors show a higher oxygen content. The nitrogen content, however, is similar for all the produced powders and close to the theoretical amount of 39.9%. The measured oxygen content of the seed powder was 0.9%.

|  | CG28 | CG32 | CG36 | CF36 | MG28 | MG32 | MG36 | MF36 |
|---|---|---|---|---|---|---|---|---|
| **Oxygen %** | 1.84 | 1.77 | 1.70 | 1.70 | 2.35 | 2.11 | 2.02 | 1.46 |
| **Nitrogen %** | 38.7 | 38.7 | 38.9 | 37.6 | 38.4 | 38.4 | 38.7 | 38.3 |

**Table 1.** Oxygen and nitrogen contents for the various produced powders.
  CG28: Coated Gel Silica with 28 wt. % Carbon
  CG32: Coated Gel Silica with 32 wt. % Carbon
  CG36: Coated Gel Silica with 36 wt. % Carbon
  CF36: Coated Fumed Silica with 36 wt. % Carbon
  MG28: Mixed Gel Silica with 28 wt. % Carbon
  MG32: Mixed Gel Silica with 32 wt. % Carbon
  MG36: Mixed Gel Silica with 36 wt. % Carbon
  MF36: Mixed Fumed Silica with 36 wt. % Carbon

Fig. 2 shows the yield of $Si_3N_4$ from coated and mixed gel precursors with respect to the initial carbon content. The lower yield in mixed precursors is due to the higher amount of SiO loss during the reaction as compared to the coated precursors. In the coated precursors, the SiO generated is trapped by the carbon coating. The yield increases with increase in carbon content due to the fact that higher carbon content in the reaction mixture lets the reaction proceed further, preventing the loss of SiO by readily making carbon available for its conversion to $Si_3N_4$. But this increase is not very significant in the case of coated precursors because of the fact that the yield at low carbon content is already high.

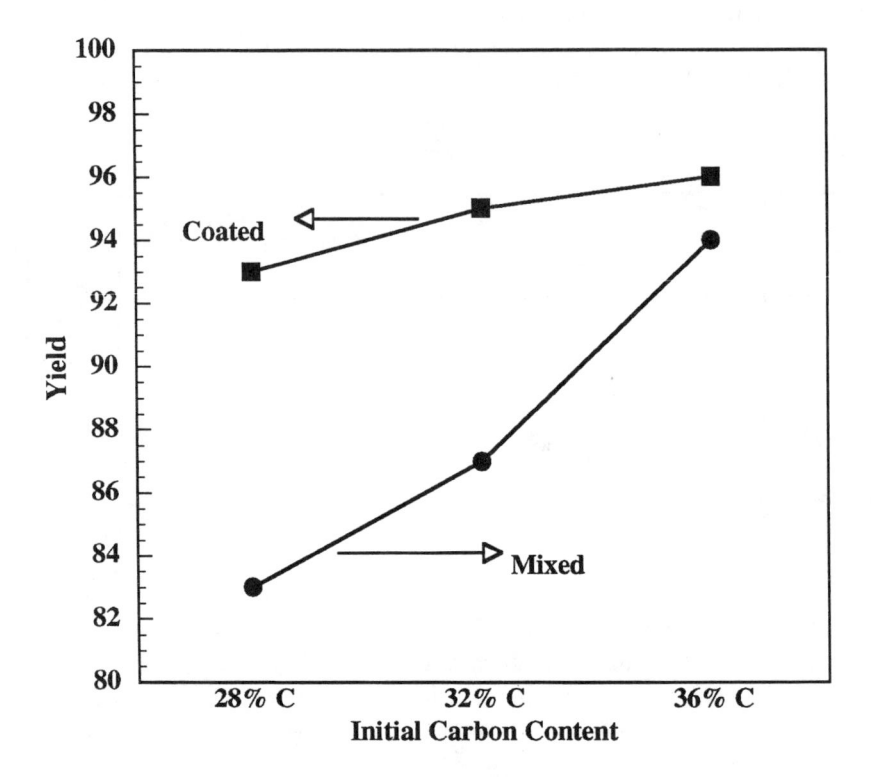

Fig 2. Yield data with respect to the initial carbon content for coated and mixed gel precursors

Fig. 3 shows the SEM pictures of $Si_3N_4$ from the coated precursors. The $Si_3N_4$ powders produced from the different starting materials with different carbon contents look similar except for the $Si_3N_4$ from fumed silica with 28 wt. % C, which seems to show a significant amount of whisker formation. These whiskers,

Fig 3: (a) $Si_3N_4$ from coated gel silica, 28% carbon. (b) $Si_3N_4$ from coated gel silica, 36% carbon (c) $Si_3N_4$ from coated fumed silica, 28% carbon (d) $Si_3N_4$ from coated fumed silica, 36% carbon

however, disappeared when the initial carbon content was increased to 36 wt. % C. This whisker formation was either absent or insignificant in the $Si_3N_4$ powder produced from the gel silica. Also, the $Si_3N_4$ from gel silica is of slightly smaller particle size than that from fumed silica. All the produced $Si_3N_4$ powders shown in fig. 3 are in an as-produced condition, without any milling. The powder particles can also be seen to be equiaxed. The final powders from the coating process have a size range between 0.6 - 0.9µm, thereby significantly reducing the need to mill after synthesis.

(a) (b)

Fig 4. (a) $Si_3N_4$ seed powder, H C Stark, grade M11 (b) $Si_3N_4$ from mixed precursor with 36% carbon

Fig. 4 shows SEM micrographs of seed powder and powder from the mixed precursor with 36 wt. % C. The seed powder particles are irregular shaped with jagged edges and appear to have been significantly milled. Also, the size distribution is large for the seed particles while being narrow for the produced powders. The agglomeration in the powders produced by the coating process, from both gel and fumed silica, (Fig 3) appears to be loose agglomeration, unlike the strong agglomeration seen in the mixed powders (Fig 4).

## CONCLUSION

The carbon coating process is capable of producing high quality silicon nitride powders economically via carbothermal reduction. The coating step provides an even distribution and intimate contact between the solid reactants, eliminating the need for the long mixing times employed in the conventional process. It is capable of producing high yield and low oxygen content powders with relatively low carbon content in the initial precursor. The final powders have submicron particle size, with equiaxed particles and a narrow size range. Further experimental data and detailed explanations will be provided in an upcoming journal publication.

## ACKNOWLEDGMENT

This research was sponsored by the U.S. Department of Energy, Assistant Secretary for Energy Efficiency and Renewable Energy, Office of Industrial Technologies as part of the Advanced Industrial Materials Program under contract DE-AC05-84OR21400 with Lockheed Martin Energy Systems, Inc.

## REFERENCES

1. G. Schwier, G. Nietfeld and G. Franz, "Production and Characterization of Silicon Nitride Powders", *Matl. Sci. For.*, 47, 1-20 (1989)
2. S. C. Zhang and W. R. Cannon, "Preparation of Silicon Nitride from Silica", *J. Am. Cer. Soc.*, 67 (10), 691-695 (1984)
3. T. Licko, V. Figusch and J. Puchyova, "Synthesis of Silicon Nitride by Carbothermal Reduction and Nitriding of Silica: Control of Kinetics and Morphology", *J. Eur. Cer. Soc.*, 9, 219-230 (1992)
4. G. Glatzmaier and R. Koc, "Method for Silicon Carbide Production by Reacting Silica and Hydrocarbon gas", *U.S.Patent* # 5,324,494 (1994)
5. A. W. Weimer, G. A. Eisman, D. W. Susnitzky, D. R. Beaman and J. W. McCoy, "Mechanism and Kinetics of Carbothermal Nitridation Synthesis of α-Silicon Nitride", *J. Am. Cer. Soc.*, 80 (11), 2853-2863 (1997)
6. R. Koc and S. Kaza, "Synthesis of α-$Si_3N_4$ from Carbon Coated Silica by Carbothermal Reduction and Nitridation", To be published in *J. Eur. Cer. Soc.*

NEAR NET-SHAPED MAGNESIUM ALUMINATE SPINEL BY THE
OXIDATION OF SOLID MAGNESIUM-BEARING PRECURSORS

Pragati Kumar and Kenneth H. Sandhage
Department of Materials Science and Engineering
The Ohio State University, Columbus, OH 43210

ABSTRACT

The feasibility of fabricating near net-shaped spinel, $MgAl_2O_4$, by the oxidation of solid Mg-$Al_2O_3$-bearing precursors has been demonstrated. Dense disk-shaped and bar-shaped precursors were fabricated by the pressureless infiltration of molten Mg into porous (65-70% dense) $Al_2O_3$ preforms at 680-700°C. After solidification, some of the bar-shaped specimens were ground into rods of varied diameter. The shaped, solid Mg-$Al_2O_3$-bearing precursors were then oxidized in pure, flowing oxygen at 430°C/40 hours to 700°C/6 hours. The resulting mixtures of MgO and $Al_2O_3$ were converted into $MgAl_2O_4$ within 15 hours of annealing in oxygen at 1200°C. A subsequent 10 hour sintering treatment at 1700°C in flowing Ar yielded shaped spinel-bearing bodies with densities of 92.5% of theoretical. The dimensions of the final, sintered bodies were within 0.6% of those for the shaped Mg-$Al_2O_3$-bearing precursors. The phase content and microstructure at various stages of processing were evaluated with x-ray diffraction (XRD), scanning electron microscopy (SEM), and electron probe microanalysis (EPMA).

INTRODUCTION

Owing to its high melting point (2100°C), high hardness (16.1 GPa), resistance to chemical attack, and chemical compatibility with alumina, zirconia, and mullite, spinel ($MgAl_2O_4$) is an attractive matrix material for ceramic-matrix composites.[1,2,3] Monolithic, transparent spinel is also used in of optical applications, such as windows for pressure vessels and bulletproof vehicles.[2]

Spinel bodies are usually produced by ceramic powder-based processing. For example, spinel powder can be mixed with malleable organic material and then compacted and formed into a shaped green body (e.g., by injection molding, extrusion, etc.). After removal of the organic material by vaporization or pyrolysis at a modest temperature, the resulting porous spinel body is sintered at $\geq$ 1600°C to obtain a dense, shaped component.[4-6] Significant process optimization can be required during the organic burnout and sintering steps to avoid undesired defects within, and distortion of, the final ceramic body. Non-uniform or incomplete binder burnout can result in cracking or carbon contamination. Further, because the binder often occupies a significant volume fraction (20-40%) of the green body, the sintering shrinkage is often relatively large. As a result, dense ceramic parts produced by this conventional process will not retain the dimensions and, if densification is not uniform, the shape of the starting green body.

---

To the extent authorized under the laws of the United States of America, all copyright interests in this publication are the property of The American Ceramic Society. Any duplication, reproduction, or republication of this publication or any part thereof, without the express written consent of The American Ceramic Society or fee paid to the Copyright Clearance Center, is prohibited.

Recent work[7-12] has demonstrated that near net-shaped ceramic bodies can be fabricated by an exciting new process: the oxidation of solid, alkaline-earth (AE) metal-bearing precursors. AE elements (Mg, Ca, Sr, Ba) are ductile and low melting (650 to 840°C).[13] Hence, precursors containing AE metals can be formed into desired shapes by deformation processing or by melt infiltration into a porous ceramic preform. The molar volumes of AE metals tend to be larger than the molar volumes of the corresponding oxides (e.g., $V_m[Mg] > V_m[MgO]$). Such a volume reduction upon oxidation can be used to counter volume expansions associated with the oxidation of non-AE metals and/or with subsequent oxide-oxide reactions. Thus, by tailoring the precursor phase content, near net-shaped, all-ceramic bodies can be produced from AE-bearing precursors.

Consider the transformation of an equimolar $Mg-Al_2O_3$ precursor into $MgAl_2O_4$. The conversion of such a precursor can occur by: 1) the oxidation of Mg into MgO, and then 2) annealing of the resulting $MgO + Al_2O_3$ mixture to form $MgAl_2O_4$. The reduction in solid volume associated with magnesium oxidation can compensate for the increase in solid volume associated with the conversion of magnesia and alumina into spinel. As a result, the net volume change associated with spinel formation from an equimolar mixture of magnesium and alumina (reaction (1) below) is only 0.5%!

$$Mg(s) + Al_2O_3(s) \xrightarrow{O_2(g)} MgAl_2O_4(s) \quad\quad (1)$$

$$100 \cdot \Delta V/V_o = 100 \cdot \{V_m[MgAl_2O_4] - V_m[Mg] - V_m[Al_2O_3]\}/\{V_m[Mg] + V_m[Al_2O_3]\}$$
$$= 100 \cdot \{39.8 - 14.0 - 25.6\}/\{14.0 + 25.6\} = 0.5\%$$

($V_m[i]$ refers to the molar volume of species i, in $cm^3/mol$).

The theoretical dimensional change expected as a result of the transformation of such a $Mg-Al_2O_3$ precursor into spinel is only ~0.15% (assuming isotropic expansion, and equal pore fractions within the precursor and product bodies).

Shaped, dense $Mg-Al_2O_3$-bearing precursors to $MgAl_2O_4$ can be prepared by the infiltration of molten Mg into shaped, porous $Al_2O_3$ preforms.[8] Porous oxide preforms of desired shape can be fabricated by extrusion, pressing, slip casting, or other traditional powder processes. After infiltration and solidification of the magnesium, the solid precursor can be machined or ground into a more complicated shape. The ductile magnesium can endow the precursor with the green strength required for such machining; that is, the magnesium can serve the role of a binder. However, unlike fugitive organic binders, magnesium is retained in the precursor (as magnesia) upon oxidation. In other words, the magnesium binder is "burned in" upon oxidation. The amount of porosity in the oxidized precursor should, therefore, be significantly less than the porosity generated by organic binder burnout for a conventionally-processed body (i.e., for equivalent amounts of binder in both cases). Hence, less sintering can be required with the AE-metal-bearing precursor process, so that better control of the shape and size of the fired body can be achieved. A variety of monolithic ceramics and ceramic-bearing composite materials for electronic[14-17], magnetic,[18] structural[8,9,11,12] and biomedical[19] applications have been synthesized from solid, AE-metal-bearing precursors.

Other oxidation-based techniques developed for the syntheses of ceramics include the self-propagating high temperature synthesis (SHS)[20] and reaction-bonded aluminum oxide (RBAO)[21-24] methods. The SHS process involves the propagation of a reaction-induced thermal wave through a specimen. Although relatively little applied energy is required to fabricate ceramics with the SHS process, components produced by this method tend to be porous and do not tend to retain the dimensions of the precursor. In RBAO processing, an intimate mixture of $Al_2O_3$ and Al powder is formed into a shaped, porous green body. The volume expansion associated with the oxidation of aluminum is used to compensate for the initial porosity present in the green compact, so that relatively low net shrinkage is observed after oxidation and sintering. In the present method, however, the as-infiltrated Mg-Al$_2$O$_3$-bearing precursor is dense and more mechanically robust than a porous Al-Al$_2$O$_3$ precursor, so that stresses incurred upon machining/grinding are less likely to lead to macrocracking.

The objective of this study was to demonstrate the feasibility of fabricating dense and near net-shaped spinel bodies by the oxidation and subsequent annealing of melt-infiltration-derived, Mg-Al$_2$O$_3$-bearing precursors. The phases and microstructures produced upon infiltration, oxidation, and post-oxidation annealing were investigated.

EXPERIMENTAL PROCEDURE

(A) Precursor Preparation: A molten metal infiltration technique (Fig. 1) was used to prepare dense, shaped Mg-Al$_2$O$_3$ precursors. This technique consisted of the following steps:

1. *Al$_2$O$_3$ preform synthesis*: Porous Al$_2$O$_3$ preforms were fabricated by uniaxial pressing of Al$_2$O$_3$ powder. The Al$_2$O$_3$ powder, obtained from a commercial vendor (99.98% purity, Johnson Matthey, Inc., Ward Hill, MA), possessed a mean particle size of 7 microns, as determined with a particle size analyzer (Model SA-CP4, Shimadzu Corp., Kyoto, Japan). Uniaxial pressing was conducted in hardened-steel dies of circular (1 cm dia.) or rectangular (6 mm wide X 5 cm long) cross section using applied stresses of 800 or 100 MPa, respectively. The alumina disks and bars were then sintered in air at 1500°C for 5-12 hours so as to produce 65-70% dense preforms.

2. *Mg infiltration into Al$_2$O$_3$ preforms*: The porous Al$_2$O$_3$ preforms were infiltrated with molten Mg. The porous preform was placed between Mg disks (15 mm dia. X 3 mm thick, 99.95% purity, Johnson Matthey, Inc.) within a MgO crucible. An excess amount of Mg (molar Mg/Al$_2$O$_3$ ratio = 20-25) was used to allow for complete filling of open pores within the porous alumina preform. The crucible containing the Al$_2$O$_3$ preform and the Mg disks was then placed in a silica tube purged with flowing Ar. The melting and infiltration of Mg into the preform was conducted by thrusting the silica tube into a furnace preheated to 600°C. The furnace temperature was then ramped to 680 or 700°C at a rate of 6°C/min. The quartz tube was held at this temperature for $\leq$ 30 minutes, after which the power to the furnace was turned off, and the specimen was allowed to cool to room temperature. Most of the excess solid Mg surrounding the infiltrated disk was removed by cutting. The excess Mg close to the infiltrated disk surfaces was stripped from the disk by applying a modest shear stress. Light polishing with 1200 grit SiC-impregnated paper was then used to remove small, localized amounts of excess Mg adhering to the disk surface.

3. *Machining of Mg-Al$_2$O$_3$ bars:* Infiltrated precursor bars were machined into more complex shapes (see Fig. 9c) with the use of a lubricated, cubic BN grinding wheel (18 cm dia.) operating at a rotation rate of 3450 rpm.

(B) Heat Treatment: Isothermal oxidation of the Mg-Al$_2$O$_3$ precursor was conducted at 430 and 700°C in flowing oxygen for 40 and 6 hours, respectively. Post-oxidation annealing (for spinel formation) was conducted at 1200°C in flowing oxygen for 10-15 hours. Sintering was then conducted at 1700°C for 10 hours in a flowing Ar atmosphere. Fig. 2 shows the time-temperature profiles associated with various heat treatments.

(C) Characterization: X-ray diffraction (XRD), scanning electron microscopy (SEM), electron probe microanalysis (EPMA), and optical microscopy were used to characterize the specimens. XRD analyses were conducted with Cu-K$_\alpha$ radiation at room temperature on powder specimens produced by grinding with a stabilized-zirconia mortar and pestle. The ground powder was mixed with an x-ray transparent vacuum grease (Dow Corning, Inc., Midland, MI) and placed on a single crystal silicon substrate that had been ground so as to expose an irrational plane (i.e., so that silicon diffraction peaks would not be generated with the 2θ range examined). $Al_2O_3$ diffraction peaks were used as internal XRD standards. An external calibration method[25,26] was adopted for quantitative XRD analyses of the amounts of MgO, $Al_2O_3$, and $MgAl_2O_4$ within the samples. Specimens for optical and scanning electron microscopy were prepared by standard ceramographic techniques. The grain structure of sintered samples was revealed by thermal etching of polished surfaces for 4 hours at 1250°C in air. SEM and EDS analyses were conducted with a field emission gun microscope (Model XL-30, Philips Electronics N.V., Eindhoven, The Netherlands). The BSE and SE imaging were conducted on gold-coated specimens. EPMA/WDS analyses (Model SX-50, Cameca Instruments, Inc., Trumbull, CT) were conducted using a beam current of 20 nA and 10 kV with high-purity MgO and $Al_2O_3$ as standards.

Sample densities after various processing steps were determined using Archimede's method. The total porosity (open and closed) was evaluated with samples that were hermetically wrapped in a paraffin wax film of known density. Doubly-distilled water was used as the buoyant fluid. Archimede's measurements with and without the paraffin wax coating were used to evaluate the open porosity. Bulk density values were also obtained from dimensional and dry weight measurements. The flexural strengths of infiltrated Mg-$Al_2O_3$ precursor bars were evaluated with four-point bend tests conducted at room temperature as per ASTM standard C1161 (configuration A, inner span = 10 mm, outer span = 20 mm, cross-head speed = 0.2 mm/min.).[27]

RESULTS

The 65-70% dense $Al_2O_3$ preforms were infiltrated by molten Mg so as to obtain dense disk-shaped or bar-shaped Mg-$Al_2O_3$-bearing precursors. The closed porosity in the alumina preforms was only a few percent (e.g. 2.9% for the 70% dense preform). The molten Mg completely infiltrated the 1.5-1.8 mm thick disks of $Al_2O_3$ within 6 minutes at 680°C and the 5 mm thick bars within 30 minutes at 700°C under flowing Ar. No dimensional changes were detected after magnesium infiltration (Table I). The bulk densities of infiltrated precursors were measured to be 3.21 g/cm$^3$ which was equivalent to 95.8% of theoretical density (TD = 3.35 g/cm$^3$). The diffraction pattern obtained from an as-infiltrated precursor exhibited strong peaks for Mg and $Al_2O_3$, along with relatively weak peaks for MgO and $Mg_{17}Al_{12}$ (Fig. 3). The largest peak labeled Mg in Fig. 3 was shifted towards a higher 2θ value than expected for pure Mg. The secondary electron images of polished cross-sections of as-infiltrated samples (shown in Fig. 4) revealed interconnected networks of $Al_2O_3$ and Mg-rich phases. EPMA conducted on the metallic phase yielded a composition of 8.7 at% Al and 91.3 at% Mg. Four-point bend tests on bar-shaped infiltrated specimens yielded an average (8 samples) flexural strength of 88.2 MPa (72-107 MPa).

After oxidation for 40 hours at 430°C, the specimen weight increased by 6.9%. Further oxidation at 700°C for 6 hours resulted in an additional weight gain of 1.04% (i.e., a total weight gain of 7.94%). No additional weight change was detected with further heat treatment. A small amount (3.5 wt%) of MgO, in the form of loose powder, was observed near the specimens after the 700°C heat treatment. The microstructure of an as-oxidized sample (after 700°C) is shown in Fig. 5. The diffraction pattern obtained from the sample oxidized at 700°C (Fig. 6) revealed large diffraction peaks corresponding to MgO and $Al_2O_3$, along with smaller peaks for $MgAl_2O_4$. Diffraction peaks for Mg-Al solid solutions or Mg-bearing intermetallic compounds ($Al_3Mg_2$, $Al_{12}Mg_{17}$[13]) were not detected after the 700°C treatment.

The transformation to spinel was completed between 10 and 15 hours of annealing at 1200°C, as revealed by the absence of MgO diffraction peaks in Fig. 6c. Weak $Al_2O_3$ diffraction peaks

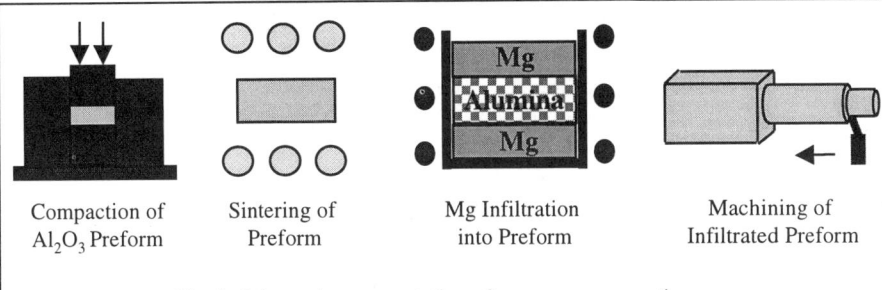

Fig. 1. Schematic representation of precursor preparation

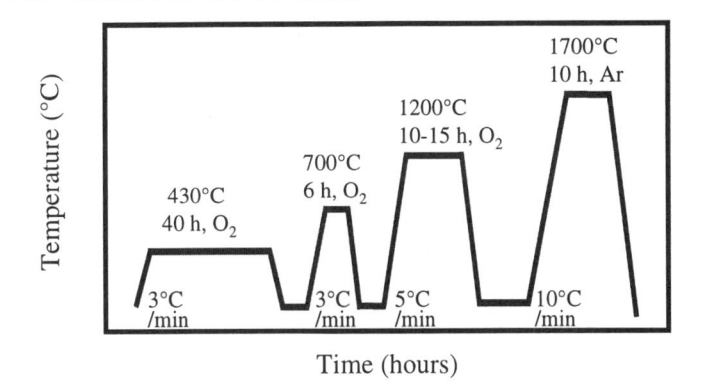

Fig. 2. Heat treatment schedule for Mg-Al$_2$O$_3$-bearing precursor

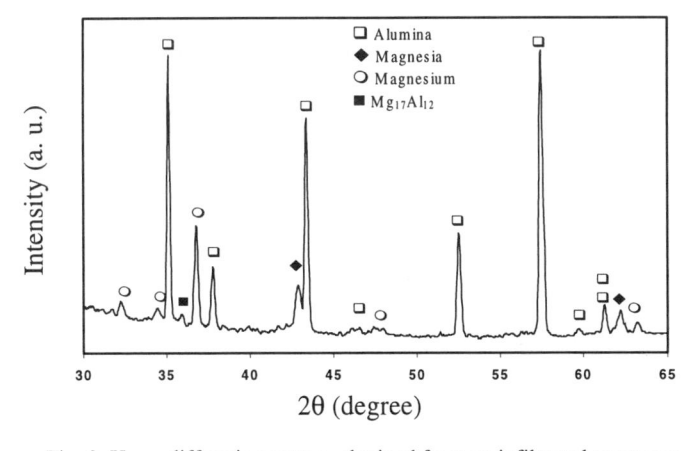

Fig. 3. X-ray diffraction pattern obtained from an infiltrated precursor

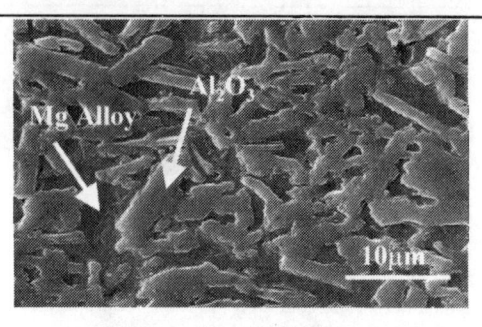

Fig. 4. SE image of a polished cross-section of an as-infiltrated precursor.

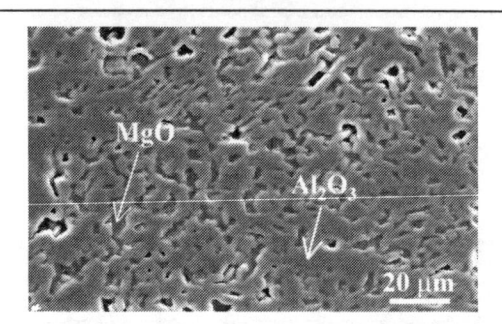

Fig. 5. SE image after oxidation at 700°C for 6 hours.

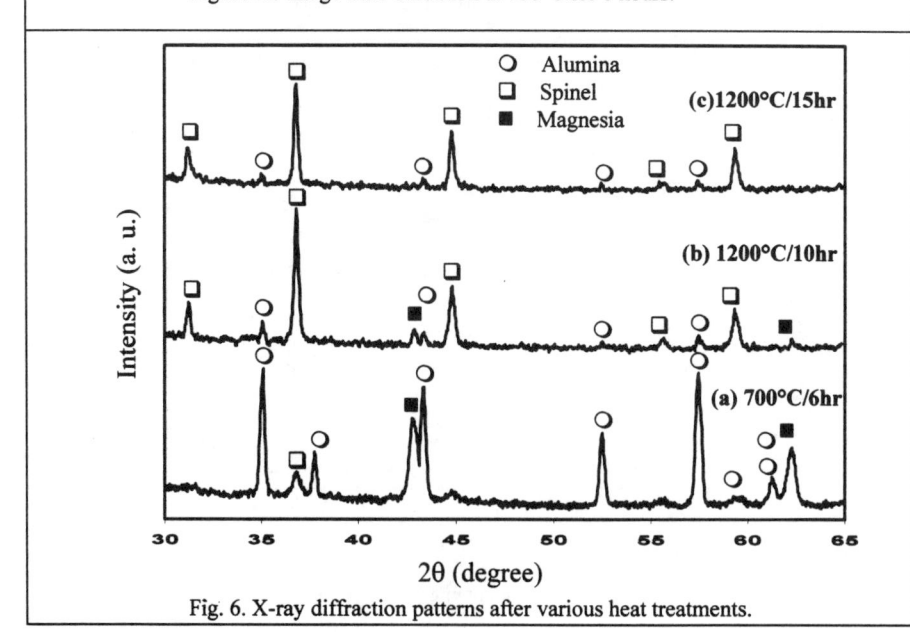

Fig. 6. X-ray diffraction patterns after various heat treatments.

were also observed in Fig. 6c, however. The microstructure of a sample annealed for 15 hours at 1200°C (Fig. 7) consisted of predominantly spinel with some isolated alumina particles and pores.

The 10 hour sintering treatment at 1700°C yielded samples with bulk densities of 3.33 g/cm$^3$, which corresponded to 92.5% of TD (3.60 g/cm$^3$). The amounts of the alumina and spinel phases in the sintered samples were 6.1 mol% (4.4 wt%) and 93.9 mol% (95.6 wt%), respectively (as determined by quantitative XRD analyses). Fig. 8 reveals the microstructure of the sintered spinel after thermal etching. Small, isolated Al$_2$O$_3$ particles were detected within a spinel matrix.

The disks and machined bars retained the shapes of the precursors after complete transformation (Fig. 9). Modest increases in dimensions were observed during transformation (Table I), with most of the increase occurring during spinel formation at 1200°C. After complete spinel formation at 1200°C and sintering at 1700°C, the dimensions of the disk and machined bar specimens returned to within 0.6% of the precursor values.

DISCUSSION

A. Precursor Preparation and Characterization

An equimolar mixture of Al$_2$O$_3$ and Mg corresponds to 64.6 vol% of Al$_2$O$_3$ and 35.4 vol% of Mg. Owing to such a relatively large ceramic content, deformation processing (extrusion, pressing, etc.) of an equimolar Mg-Al$_2$O$_3$ powder mixture was not thought to be a viable means of fabricating a high-density, shaped precursor body. However, because magnesium possesses a modest melting temperature (640°C[13]) and a high affinity for oxygen[28] (i.e., an affinity for wetting oxides), a melt-infiltration route (Fig. 1) was considered.

In order to produce an equimolar Mg-Al$_2$O$_3$ precursor by the infiltration of liquid Mg into a porous, rigid Al$_2$O$_3$ preform, the open porosity within the preform should be close to 35.4%. Lower or higher values of porosity should yield alumina-rich or magnesia-rich compositions, respectively, upon complete transformation. Since stoichiometric or alumina-rich compositions were preferred over magnesia-rich compositions (owing to the tendency of magnesia to undergo hydration and cracking, and to the thermal expansion mismatch between magnesia and spinel[29]), the alumina preforms of the present work were sintered to achieve open porosity values of 27-33 vol%. Such sintering also yielded rigid preforms. That is, sufficient necking occurred between Al$_2$O$_3$ particles that the capillary pressure generated upon Mg infiltration did not result in particle rearrangement and contraction (dimensional changes were not detected after infiltration).

The x-ray diffraction pattern obtained from an infiltrated precursor exhibited diffraction peaks corresponding to Mg, MgO, Al$_2$O$_3$, and Mg$_{17}$Al$_{12}$ phases (Fig. 3). The presence of MgO and Mg$_{17}$Al$_{12}$ peaks, and noticeable shifts in the positions of peaks labeled Mg in Fig. 3 (relative to peak positions expected for pure Mg), indicated that the following displacement reaction occurred between liquid Mg and solid Al$_2$O$_3$ during infiltration.

$$3\{Mg\} + Al_2O_3(s) \Rightarrow 3MgO(s) + 2\{Al\} \qquad (2)$$

where { } refers to a species in a liquid solution. The Al produced in this displacement reaction went into solution with liquid Mg at the infiltration temperature of 680-700°C. EPMA confirmed the presence of 8.7 at% Al in the metallic phase. As expected from the Mg-Al phase diagram,[13] the Mg-Al liquid solidified into two phases when cooled to room temperature: a Mg-Al solid solution and a small amount of Mg$_{17}$Al$_{12}$. The presence of a Mg-Al solid solution within infiltrated specimens was substantiated by the shifts observed for the Mg-like diffraction peaks towards higher 2θ values than expected for pure Mg (i.e., the lattice parameters of Mg-Al solid solutions are less than for pure Mg and decrease with increasing Al content[30]). The amount of Al in the Mg-

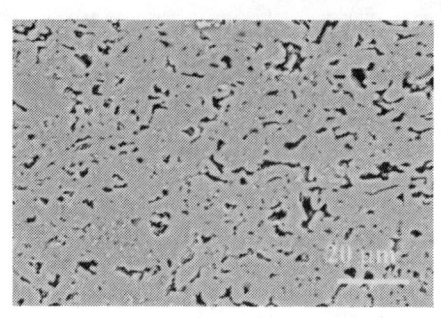

Fig. 7. BSE image of a polished cross-section of a sample annealed at 1200°C for 15 hours

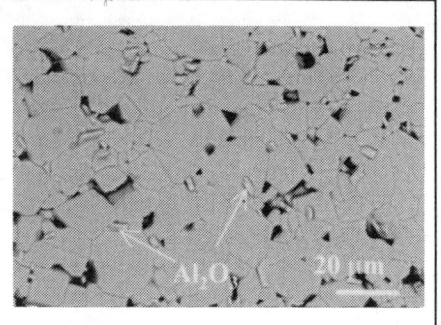

Fig. 8. BSE image of a spinel-rich sample sintered for 10 hours at 1700°C

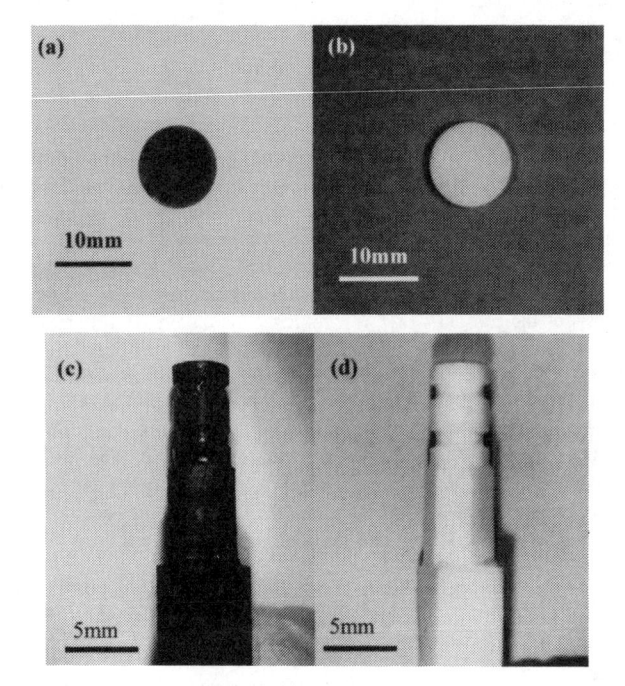

Fig. 9. Optical micrographs of (a) an as-infiltrated, disk-shaped sample, (b) same as in (a) - after sintering, (c) an infiltrated, machined preform and, (d) same as in (c) - after sintering.

Al solid solution was determined from the peak shift measurements to be 3.7 at%. Thus, the solidification of the Mg-Al melt can be expressed by the following reaction

$$10\{Mg_{0.913}Al_{0.087}\} => 0.0458Mg_{17}Al_{12} + 8.66<Mg_{0.963}Al_{0.037}> \quad (3)$$

where < > refers to a solid solution.

The ability to obtain complete pressureless infiltration of molten magnesium into the porous alumina preforms may have been enhanced by the observed displacement reaction (2). Indeed, other authors have shown that the pressureless penetration of molten aluminum into silica or silicate preforms was accompanied by a displacement reaction.[31,32] Prior published work with the PRIMEX process[33,34] has indicated that molten Al-Mg alloys can be pressureless infiltrated into porous $Al_2O_3$, if such infiltration is conducted in a $N_2$-bearing atmosphere. Successful infiltration was obtained in the present work with the use of a flowing Ar atmosphere.

Quantitative XRD analyses, coupled with mass and volume balance calculations, indicated that the precursors were comprised of 47.9 mol% $Al_2O_3$, 18.5 mol% MgO, 29.2 mol% Mg-Al solid solution and 4.4 mol% $Mg_{17}Al_{12}$. The theoretical density of this phase mixture is 3.35 g/cm³. Since the measured precursor density was 3.21 g/cm³, the infiltrated precursors possessed a porosity of 4.2%. Much of this porosity could be ascribed to the shrinkage of the molten Mg-Al metal during solidification. The shrinkage associated with the solidification of a Mg-8.7at% Al liquid at 680°C (i.e., reaction (3)) is 9.0%. Since this liquid comprised 27.3 vol% of the precursor, the expected porosity in the precursor due to solidification shrinkage was 2.5%. The remainder of the 4.2% porosity was due to the small amount of closed, non-infiltrated pores present in the $Al_2O_3$ preform.

Infiltrated, bar-shaped specimens possessed an average flexural strength of 88.2 MPa (72 to 107 MPa). Such a value of green strength is an order of magnitude higher than is typically achieved with conventionally pressed ceramic green bodies and is also higher than the green strengths reported for porous $Al-Al_2O_3$ precursors (20-50 MPa).[23,24] Such a high green strength value was a consequence of the continuous, percolative nature of the metallic Mg-Al alloy within, and the high density of, the infiltrated precursor (see Fig. 4).

B. Precursor Transformation

Owing to the occurrence of displacement reaction (2) during the infiltration of molten Mg into porous $Al_2O_3$, a metallic, Mg-rich Mg-Al liquid was produced at 680°C. The lowest solidus temperature for Mg-rich compositions in the Mg-Al system is 437°C.[13] Because the initial intent of this work was to oxidize a solid metal-bearing precursor, the initial oxidation heat treatment was conducted at 430°C. However, after 40 hours at this temperature, the weight gain expected for complete oxidation was not achieved. Hence, a subsequent treatment for 6 hours at 700°C was used to complete the oxidation of the magnesium and aluminum within the infiltrated specimens (as confirmed by weight gain measurements and XRD analyses). It is likely that, with further optimization of the oxidation cycle, complete oxidation could be achieved in less time.

A mixture of MgO and $Al_2O_3$ produced by the oxidation of a mixture of Mg-Al solid solution and $Mg_{17}Al_{12}$ should possess 16.2% less volume than the latter metallic mixture. However, very small positive dimensional changes were detected after the 700°C oxidation treatment (Table I). Hence, the oxidized specimens must have contained more porosity than the as-infiltrated precursors. Comparison of Figures 4 and 5 reveals that such an increase in internal porosity did indeed occur after the 700°C anneal. The pores generated upon oxidation were limited to dimensions less than or equal to those of the pores that had been present in the starting alumina precursor, however. Since little sintering of alumina occurred at 700°C, the alumina network in Fig. 5 looks similar to that in Fig. 4.

A small amount of spinel was detected after the 700°C anneal by XRD analyses. After further annealing for 10 hours at 1200°C, the solid-state reaction between magnesia and alumina yielded spinel as the major phase. After 15 hours at this temperature, spinel formation had been completed, as indicated by the absence of magnesia diffraction peaks (see Fig. 6c). A small amount of residual $Al_2O_3$ was still detected, however. Since some loose magnesia powder had separated from the specimens during the oxidation at 430-700°C, the presence of a small amount of residual alumina upon complete spinel formation was not surprising.

After the 15 hour anneal at 1200°C, a volume increase of 7.2% was detected (relative to the specimen volume after the 700°C treatment). This value was not far from the 7.9% increase in volume expected upon the formation of spinel from magnesia and alumina:

$$MgO(s) + Al_2O_3(s) => MgAl_2O_4(s) \qquad (4)$$

$$100 \cdot \Delta V/V_o = 100 \cdot \{V_m[MgAl_2O_4] - V_m[MgO] - V_m[Al_2O_3]\}/\{V_m[MgO] + V_m[Al_2O_3]\}$$
$$= 100 \cdot \{39.8 - 11.3 - 25.6\}/\{11.3 + 25.6\} = 7.9\%$$

Since a small amount of spinel was detected prior to the 1200°C anneal (see Fig. 6a), the slightly lower measured value of 7.2% was in reasonable agreement with the calculated volume change.

After sintering for 10 hours at 1700°C in flowing Ar, 92.5% dense specimens were produced. This sintering treatment resulted in a volume decrease of only 6.4% (relative to the specimen volume after the 1200°C treatment). A shaped green body comprised of spinel and organic binder would have exhibited a volume decrease of more than 4 times this value, upon sintering to the same density (i.e., for the case where the amount of organic binder equals the amount of Mg-Al alloy in the $Mg$-$Al_2O_3$-bearing precursors of the present work).

After the 1700°C anneal, the dimensions of the sintered disks and bars were within 0.6% of the precursor dimensions. As seen in the macroscopic optical images in Fig. 9, disk-shaped and machined bar specimens exhibited excellent shape/edge retention and were free of macrocracks after the 1700°C treatment. The before and after images in Fig. 9 confirm the near net-shape capability of the AE-metal-bearing precursor process.

Table I. Sample dimensions and cumulative variations (with respect to the porous alumina preform) after various processing steps.

| Processing Step | Diameter (d), mm | Thickness (t), mm | $\Delta d/d_o$ (%) | $\Delta t/t_o$ (%) | $\Delta V/V_o$ (%) |
|---|---|---|---|---|---|
| Preform | 9.81 | 1.61 | - | - | - |
| Infiltrated Preform | 9.81 | 1.61 | 0 | 0 | 0 |
| 700°C, 6 h | 9.83 | 1.62 | 0.20 | 0.62 | 1.03 |
| 1200°C, 15 h | 10.05 | 1.66 | 2.45 | 3.11 | 8.21 |
| 1700°C, 10 h | 9.86 | 1.62 | 0.51 | 0.62 | 1.65 |

CONCLUSIONS

(1) Molten magnesium was found to completely infiltrate 1.5-1.8 mm thick, porous alumina preforms within 6 minutes at 680°C (pressureless infiltration). 5 mm thick preforms were completely infiltrated within 30 minutes at 700°C.
(2) Bar-shaped $Mg$-$Al_2O_3$-bearing precursors possessed flexural strengths of 72-107 MPa (average = 88.2 MPa).
(3) Oxidation of the shaped $Mg$-$Al_2O_3$-bearing precursors was completed after heat treatment at 430°C for 40 hours and 700°C for 6 hours in pure, flowing oxygen.
(4) Spinel formation was completed within 15 hours at 1200°C in oxygen.

(5) Sintering in flowing argon for 10 hours at 1700°C yielded 92.5% dense, spinel-bearing bodies.
(6) Near net-shaped disks and machined bars were produced (i.e., the dimensions of the Mg-Al$_2$O$_3$-bearing precursor and final MgAl$_2$O$_4$-bearing bodies agreed to within 0.6%).

REFERENCES

[1] P. Hing, "Fabrication of Translucent Magnesium Aluminate Spinel and its Compatibility in Sodium Vapor," J. Mater. Sci., 11 1919-1926 (1976).
[2] Phase Diagrams in Advanced Ceramics, Ed. A. M. Alper, Academic Press, San Diego, Ca, pp. 23 (1995).
[3] Phase Diagrams for Ceramists, Ed. E. M. Levin, C. R. Robbins, H. F. McMurdie, Vol. 1, American Ceramic Society, Columbus, OH, Fig. 966, pg. 312 (1964).
[4] R. J. Bratton, " Sintering and Grain Growth Kinetics of MgAl$_2$O$_4$," J. Am. Ceram. Soc., 54 [3] 141-143 (1971).
[5] R. J. Bratton, " Translucent Sintered MgAl$_2$O$_4$," J. Am. Ceram. Soc., 57 [7] 283-286 (1974).
[6] J. T. Bailey and R. Russell, Jr., " Sintered Spinel Ceramics," Ceram. Bull., 47 [11] 1025-1029 (1968).
[7] K.H. Sandhage, " Electroceramics and Process for Making the Same," U.S. Patent No. 5,259,885, Nov. 9, 1993.
[8] K.H. Sandhage, "Processes for Fabricating Structural Ceramic Bodies and Structural Ceramic-Bearing Composite Bodies," U.S. Patent No. 5,318,725, June 7, 1994.
[9] S. M. Allameh, K. H. Sandhage, "The Oxidative Transformation of Solid, Barium-Metal-Bearing Precursors into Monolithic Celsian with a Retention of Shape, Dimensions, and Relative Density," J. Mater. Res., 13 [5] 1271-1285 (1998).
[10] H. J. Schmutzler, K. H. Sandhage, J. C. Nava, "The Fabrication of Dense, Shaped Barium Cerate by the Oxidation of Solid Metal-Bearing Precursors," J. Am. Ceram. Soc., 79 [6], 1575-1584 (1996)
[11] K. H. Sandhage, "The Fabrication of Alkaline-Earth-Bearing Ceramics by the Oxidation of Solid, Metal-Bearing Precursors," pp. 103-126 in Innovative Processing and Synthesis of Ceramics, Glasses, and Composites, Ceram. Trans., Vol. 85, Eds. N. P. Bansal, K. V. Logan, J. P. Singh, The American Ceramic Society, Westerville, OH, 1997
[12] K. H. Sandhage, S. M. Allameh, H. L. Fraser, "A Novel Solid Metal-Bearing Precursor (SMP) Route to Near Net-Shaped Alkaline-Earth Aluminosilicates," pp. 499-506 in Fourth Euro-Ceramics: Basic Science - Developments in Processing of Advanced Ceramics - I, Vol. 1, Ed. C. Galassi, Gruppo Editoriale Faenza Editrice, Faenza, Italy, 1995
[13] Moffatt's Handbook of Binary Phase Diagrams, Ed. J. H. Westbrook, Genium Publishing Corp., Schenectady, New York, 1992
[14] G.J. Yurek, J.B.Vander Sande, W.-X.Wang, and D.A.Rudman, "Direct Synthesis of Metal/Superconductor Oxide Composite by Oxidation of a Metallic Precursor," J. Electrochem. Soc., 134[10] 2635-36 (1987).
[15] K. Matsuzaki, A.Inone, and T.Masumoto, "High-T$_c$ Superconductor Produced by Oxidation of Melt Spun Ag-Ho-Ba-Cu Alloy Ribbon," Jpn. J. Appl. Phys., 27[2] L195-L198 (1988).
[16] K.H. Sandhage, W.Carter, L.Masur, C.Joshi, H.Hsu, and G.J.Yurek, "Synthesis of Ba-Pb-Bi-O/Ag Superconducting Composite by the Oxidation of a Ba-Pb-Bi-Ag precursor," Physica C, 177, 95-100 (1991).

[17] K.H. Sandhage, " The Preparation of Superconducting $YBa_2Cu_3O_{7-y}$/Ag Microlaminates by an Oscillating Oxidation Scheme," J. Electrochem. Soc., 139[6] 1661-1671 (1992).

[18] G. A. Ward, K. H. Sandhage, "Synthesis of Barium Hexaferrite by the Oxidation of a Metallic Ba-Fe Precursor," J. Am. Ceram. Soc., 80 [6] 1508-16 (1997)

[19] E. Saw, K. H. Sandhage, P. K. Gallagher, A. S. Litsky, "Synthesis of Hydroxyapatite by the Oxidation of Solid, Metal-Bearing Precursors," pg. 328 in Trans. Fifth World Biomaterials Congress, Univ. Toronto Press, Toronto, Canada, 1996

[20] Z. A. Munir, " Synthesis Of High Temperature Materials By Self-Propagating Combustion Methods," Ceramic Bulletin, V 67 [2], 342-49, 1988.

[21] N. Claussen, T Le, S. Wu, "Low-Shrinkage Reaction-Bonded Alumina," J. Eur. Ceram. Soc., 5, 29-35 (1989).

[22] S. Wu and N. Claussen, "Fabrication and Properties of Low-Shrinkage Reaction-Bonded Mullite," J. Am. Ceram. Soc., 74, 2460-2463 (1991).

[23] N. Claussen, N. A. Travitzky, S. Wu, "Tailoring of Reaction-Bonded $Al_2O_3$ (RBAO) Ceramics," Ceram. Eng. Sci. Proc., 11, 806-20 (1990).

[24] S. Wu, D. Holz, N. Claussen, "Mechanisms and Kinetics of Reaction-Bonded Aluminum Oxide Ceramics," J. Am. Ceram. Soc., 76 [4] 970-80 (1993).

[25] F. H. Chung, "Quantitative Interpretations of X-Ray Diffraction Patterns of Mixtures. I. Matrix Flushing Method for Quantitative Multicomponent Analysis," J. Appl. Crystallog., 7, 519 (1974)

[26] F. H. Chung, "Quantitative Interpretations of X-Ray Diffraction Patterns of Mixtures. III. Simultaneous Determination of a Set of Reference Intensities," J. Appl. Crystallog., 8, 17 (1975)

[27] Standard Test Method for Flexural Strength of Advanced Ceramics at Ambient Temperature, 1998 Annual book of ASTM standards, Vol. 15.01, pp 304-309, ASTM, West Conshohocken, PA.

[28] L. B. Pankratz, J. M. Stuve, N. A. Gokcen, Thermodynamic Data for Mineral Technology, Bulletin 677, U.S. Bureau of Mines, 1984

[29] Y. S. Touloukian, R. K. Kirby, R. E. Taylor, T. Y. R. Lee, Thermophysical Properties of Matter: Thermal Expansion (Nonmetallic Solids), (Plenum Press, New York, NY, 1975), Vol. 13, pp. 176, 288, 479

[30] J. L. Murray, "The Magnesium-Aluminum Phase Diagram," Bull. Alloy Phase Diagrams, 3 [1] 60-74 (1982).

[31] M. C. Breslin, J. Ringnalda, L. Xu, M. Fuller, J. Seeger, G. S. Daehn, T. Otani, and H. L. Fraser, "Processing, Microstructure, and Properties of Co-Continuous Alumina-Aluminum Composites," Mater. Sci. Eng. A, 195 [1-2] 113-119 (1995).

[32] R. E. Loehman, K. Ewsuk, A. P. Tomsia, " Synthesis of Alumina-Al composites by Reactive Metal Penetration," J. Am. Ceram. Soc., 79 [1] 27-32 (1996).

[33] M. K. Aghajanian, M. A. Rocazella, J. T. Burke, S. D. Keck,"The Fabrication of Metal Matrix Composites by a Pressureless Infiltration Technique," J. Mater. Sci., 26 447-454 (1991).

[34] M. K. Aghajanian, A. S. Nagelberg, Method of Forming a Metal Matrix Composite by a Spontaneous Infitration Technique, U. S. Patent Number 5,456,306, October 10, 1995.

# THE DISPLACIVE COMPENSATION OF POROSITY (DCP) METHOD FOR FABRICATING DENSE OXIDE/METAL COMPOSITES AT MODEST TEMPERATURES WITH SMALL DIMENSIONAL CHANGES

Kirk A. Rogers, Pragati Kumar, Ramazan Citak, and Ken H. Sandhage,
Department of Materials Science and Engineering,
The Ohio State University, Columbus, OH 43210

## ABSTRACT

Dense oxide/metal composites have been produced by the displacive compensation of porosity (DCP) method. Mg-bearing liquids were infiltrated (pressureless) at 900-1000°C into porous oxide-bearing ($Al_2O_3$ or $NiAl_2O_4$) preforms. Upon infiltration, the Mg in the liquid underwent a displacement reaction with the oxide in the preform. The resulting oxide product (MgO or $MgAl_2O_4$) possessed a larger volume than the starting oxide. For the case of $NiAl_2O_4$-bearing preforms, the displaced Ni reacted with Al in the liquid and solid Fe in the preform to produce a solid Fe-Ni-Al alloy (i.e., reaction-induced solidification). The increased solid volume (from new oxide or new oxide and metal) generated by these reactions resulted in pore filling, so that dense bodies were produced with minimal dimensional changes (i.e., densification without sintering). Co-continuous $MgAl_2O_4$/Fe-Ni-Al-bearing composites possessed ave. room-temperature flexural strength and fracture toughness values of 398 MPa and 13.2 MPa·m$^{1/2}$, respectively.

## INTRODUCTION

In the decade following the second world war, a significant amount of research was conducted throughout the world to develop high-temperature metal/oxide composites to facilitate the development of jet engines with elevated temperature capability and, therefore, higher efficiency [1-9]. Among the composite systems examined were $Cr/Al_2O_3$ and $Cr-Mo/Al_2O_3$ [1-4,7,9]. Cr and Cr-based alloys were considered to be attractive reinforcements owing to high melting temperatures, good resistance to oxidation, and the chemical compatibility of $Cr_2O_3$ (the oxide formed on Cr or Cr-based alloys) with $Al_2O_3$ [1-4,7,9]. Such composites were fabricated by conventional powder processing routes; that is, powder mixtures were compacted and formed into shaped green bodies that were then densified by high-temperature, pressureless sintering or by hot pressing [1-4,9]. The shrinkage accompanying the densification of such composites resulted in the loss of the dimensions of the starting green bodies [2,3]. If the sintering shrinkage was non-uniform, then the shape of the green body would also be lost. Nonetheless, co-continuous $Cr/Al_2O_3$ and Cr-$Mo/Al_2O_3$ composites produced in this manner possessed fracture strengths ranging from about 350 to 550 MPa (i.e., for composites comprised of about 40 to 94 vol% $Al_2O_3$) [1-3,9].

Over the past decade, considerable research effort has continued to be expended on the fabrication and characterization of oxide/metal composites [10-38]. As during the 1950s, much of the recent work has been focused on alumina-bearing composites [12-38]. Although powder-based processing similar to that reported in the 1950s continues to be used to fabricate alumina/metal composites [12-16], a variety of reaction-based processing routes have also been examined over the past decade. The types of reactions involved in such processes have included: gas-solid reduction reactions, solid-solid displacement reactions, solid-liquid displacement reactions ($C^4$, RMP, 3A), and liquid-gas oxidation reactions (e.g., DIMOX[TM]) [12,13,17-38]. Unlike the gas-solid reduction or solid-solid displacement reaction methods, the $C^4$, RMP, 3A, and DIMOX[TM] approaches have yielded near net-shaped alumina/metal composites [25-38]. However, the published work on the $C^4$, RMP, and DIMOX[TM] methods has not been aimed at the direct fabrication of alumina-bearing composites reinforced with high-melting metallic phases (i.e., $Al_2O_3/Al$ composites were synthesized) [25-32,36-38]. The 3A process, on the other hand, has been used to directly synthesize alumina-bearing composites with high-melting aluminide reinforcements [33-35].

In this paper, a novel displacement-reaction-based process has been used to fabricate dense, shaped oxide/metal composites at modest temperatures (900-1000°C). This process, known as the DCP (Displacive Compensation of Porosity) method [39], involves: 1) the infiltration of a liquid metal into a shaped, porous oxide preform, and 2) a displacement reaction between the liquid metal and the solid oxide that generates *a larger volume of solid oxide than is consumed* (i.e., unlike the displacement reactions in the C[4], RMP, and 3A processes). The increase in solid volume associated with the displacement reaction allows the open pores present in the original oxide preform to become filled, so that a dense oxide/metal composite is produced (i.e., without the need for a subsequent high-temperature sintering step). Unlike the C[4], RMP, and 3A processes, the volume fraction of ceramic present in the final composite is larger than the ceramic volume fraction of the starting, porous preform. Hence, composites with relatively high ceramic contents can be fabricated by the DCP method. In addition, if the metallic liquid and the ceramic preform are properly alloyed, then the displaced metallic element can react with the remaining components of the liquid to yield an all-solid body at the reaction temperature [39]. That is, composites reinforced with high-melting metal alloys can be produced by reaction-induced solidification during the DCP transformation. With proper tailoring of the liquid and preform compositions and the preform density, composites with a wide range of oxide and metal contents can be fabricated [39].

The purpose of this paper is to demonstrate the feasibility of fabricating 2 types of oxide/metal composites by the DCP method: lightweight MgO/Mg-Al composites with a high oxide content, and higher-melting, co-continuous $MgAl_2O_4$/Fe-Ni-Al-bearing composites.

THE DISPLACIVE COMPENSATION OF POROSITY (DCP) PROCESS

The DCP processing route consists of the following steps [39]:

1) Preparation of a shaped, porous, rigid preform containing an oxide (reactant) phase
2) Infiltration of the porous preform with a metallic liquid
3) Reaction of the metallic liquid with an oxide in the preform so as to yield an oxide product that possesses a larger volume than the oxide reactant

In the DCP process, displacement reactions are chosen that lead to an increase in solid oxide volume. For example, consider the following reactions:

$$3\{Mg\} + Al_2O_3(s) => 3MgO(s) + 2\{Al\} \qquad (1)$$

$$3\{Mg\} + 4Al_2O_3(s) => 3MgAl_2O_4(s) + 2\{Al\} \qquad (2)$$

where { } refers to a liquid solution. Both of these reactions are strongly favored from a thermodynamic standpoint at 900-1000°C (e.g., $\Delta G°_{rxn(1)}(1000°C)$ = -112.7 kJ/mole and $\Delta G°_{rxn(2)}(1000°C)$ = -192.3 kJ/mole [40]). Both reactions also lead to increases in oxide volume. The volume of 3 moles of MgO is 31.9% larger than the volume of 1 mole of $Al_2O_3$, whereas the volume of 3 moles of $MgAl_2O_4$ is 16.6% larger than the volume of 4 moles of $Al_2O_3$ (the molar volumes of MgO, $Al_2O_3$, and $MgAl_2O_4$ are 11.25, 25.58, and 39.76 cm$^3$/mole, respectively [41]). Since such displacement reactions can occur at all available liquid/solid interfaces, and since the liquid occupies the prior pore volume of the alumina preforms, these volume increases will lead to a filling of the prior pore spaces. That is, a reaction-induced densification will occur. If the reaction-induced increase in solid volume is similar to the pore volume in the starting preform, then a low porosity body with a high ceramic content can be produced with minimal dimensional changes.

By alloying the metallic liquid and the oxide preform, displacement reactions can be tailored to yield composites with multicomponent metallic and ceramic phases. Consider, for example, the following reactions:

$$4/3\{Mg_{3/4}Al_{1/4}\} + 8Fe + NiAl_2O_4 \Rightarrow MgAl_2O_4 + 8[Fe_1Ni_{1/8}Al_{1/24}] \quad (3)$$

$$16/3\{Mg_{3/4}Al_{1/4}\} + 8Fe + NiAl_2O_4 \Rightarrow 4MgO + 8[Fe_1Ni_{1/8}Al_{5/12}] \quad (4)$$

where [ ] refers to a solid solution. For these reactions, the porous preform would consist of a solid mixture of 59 vol% Fe and 41 vol% $NiAl_2O_4$, whereas the liquid would consist of a solution of 75 at% Mg and 25 at% Al (the liquidus temperature of this Mg-Al composition is 490°C [42]). The Fe-Ni-Al alloys produced by reactions (3) and (4) are considerably higher melting (i.e., solidus temperatures of ≈1410°C and ≈1380°C, respectively [43]) than the Mg-Al liquids produced by reactions (1) and (2). Such Fe-Ni-Al alloys are produced by the reaction of the reduced element(s) (Ni or Ni and Al) from $NiAl_2O_4$ with the solid Fe in the preform and the remaining Al in the melt. Since the combined volumes of the oxide (1 mole of $MgAl_2O_4$ or 4 moles of MgO) and Fe-Ni-Al products in reactions (3) and (4) are larger than the combined volumes of 8 moles of Fe and 1 mole of $NiAl_2O_4$, the pores in Fe-$NiAl_2O_4$ preforms can be filled at the reaction temperature by the increase in volume associated with the formation of these reaction products. The difference between reactions (3) and (4) is the extent of reduction of the nickel aluminate; that is, only the nickel oxide in nickel aluminate is reduced in reaction (3), whereas both nickel oxide and alumina are reduced in reaction (4). By tailoring the ratio of Mg-Al liquid to $NiAl_2O_4$ (i.e., by controlling the pore volume in the preform), the displacement reaction can result in the selective or complete reduction of $NiAl_2O_4$. The composites produced by reactions (3) and (4) should consist of 41 and 44 vol% oxide, respectively [44]. In general, composites with a wide range of oxide and metal contents can be fabricated by varying the fractions of metal and porosity in the preform, as well as the compositions of the liquid and solid phases (e.g., if the molar Fe:$NiAl_2O_4$ ratio within the preform were 4:1, then reactions of the type (3) and (4) would yield composites with 59 and 61 vol% oxide, respectively).

EXPERIMENTAL PROCEDURES

The DCP process was used to convert: 1) porous $Al_2O_3$ preforms into dense MgO/Mg-Al-bearing composites, and 2) porous Fe + $NiAl_2O_4$ preforms into dense $MgAl_2O_4$/Fe-Ni-Al-bearing composites.

Porous Preform Fabrication: Bar-shaped (5.0 cm X 7.0 mm X 6.0 mm), porous $Al_2O_3$ preforms were prepared by uniaxial pressing of $Al_2O_3$ powder (99.98% purity, 7 μm ave. size, Johnson Matthey, Inc., Ward Hill, MA) followed by partial sintering at 1500°C for 4-8 h in air to achieve relative densities in the range of 55-71%. Bar-shaped, preforms of Fe and $NiAl_2O_4$ were also prepared by uniaxial pressing. The $NiAl_2O_4$ in such preforms was produced by milling an equimolar mixture of Ni (99.99% purity, ≤ 3.0 μm size, Johnson Matthey, Inc.) and $Al_2O_3$ powder (same as above) for 1 h in a high-energy vibratory ball mill (Model 8000 mixer/mill, SPEX Industries, Edison, NJ) and then annealing the mixture for 10 h at 1500°C in air. The $NiAl_2O_4$ powder was then mixed with Fe powder (99% purity, 10 μm ave. size, Aldrich Chemical, Milwaukee, WI) in a 1:8 molar ratio. The $NiAl_2O_4$/Fe mixtures were milled for 15 min, pressed, and then annealed for 10 h at 1325°C in a flowing argon atmosphere to yield relative densities of 84-90.

DCP Processing of MgO/Mg-Al-bearing Composites: A given porous $Al_2O_3$ preform was placed between semi-circular disks of Mg (1.5 cm dia. X 5 mm thick, 99.95% purity, Johnson Matthey, Inc.) within a 1020 steel tube, which was then welded shut in air. The molar Mg:$Al_2O_3$ ratio inside the tube was 20-25:1. The sealed steel tube was then annealed in a flowing argon atmosphere at 1000°C for 10-15 h. After cooling to room temperature, the tube was cut open and the transformed sample was removed.

DCP Processing of MgAl$_2$O$_4$/Fe-Ni-Al-bearing Composites: A given porous NiAl$_2$O$_4$/Fe preform was placed under a similarly-shaped bar comprised of 75 at% Mg + 25 at% Al. The latter bar was produced by uniaxial pressing of a mixture of Mg flakes (99.99% purity, ≤ 1 cm length, Johnson Matthey, Inc.) and Al powder (99% purity, 20 μm ave. size, Johnson Matthey, Inc.). The assembly was placed in a Ni crucible and then annealed for 0.5 h at 700°C in a flowing N$_2$ atmosphere. Under these conditions, the Mg-Al bar melted and infiltrated into the porous NiAl$_2$O$_4$/Fe preform. The infiltrated bar was placed within a 1020 steel tube, that was then welded shut in air. The sealed tube was annealed for 10 h at 900°C within a flowing Ar atmosphere. After cooling, the transformed sample was removed from the tube.

Composite Characterization: Microstructural characterization was conducted on polished composite cross-sections with the use of a field emission gun SEM (Model XL-30, Philips Electronics N. V., Eindhoven, The Netherlands) equipped with a Si/Li detector (Edax International, Mahwah, NJ). Archimede's density measurements were conducted with distilled water as the buoyant fluid. Four-point bend tests were conducted at room temperature as per ASTM standard C1161 (configuration A, inner span = 10 mm, outer span = 20 mm, cross head speed = 0.2 mm/min) to obtain fracture strength values. Prior to testing, the composite bars were ground to dimensions of 1.5 mm X 2.0 mm X 25 mm and polished to 1 micron diamond finish. Toughness measurements were conducted using the ASTM standard PS-070-97. In this case, polished bars with dimensions of 30 mm X 4 mm X 3 mm were indented with a standard Knoop microhardness indentor using a load of 15 Kg and a specimen tilt angle of ≈1 degree. The indented zone and the associated region of high residual stress were then removed by dry polishing with SiC-impregnated (600 grit) paper. Four point bending with the ASTM C1161 standard configuration was then used to fracture the specimens. Indentation toughness measurements were also conducted on specimens polished to a 1 μm finish [45]. The hydration resistance of each type of composite specimen was evaluated according to ASTM standard C492.

RESULTS AND DISCUSSION

Microstructural Characterization of MgO/Mg-Al-bearing Composites: Fracture and polished cross-sections of a MgO/Mg-Al specimen produced by the DCP transformation of a 55% dense $Al_2O_3$ preform are shown in Figs. 1a and b, respectively. After transformation, the composite specimens were quite dense ($\geq 98\%$ of TD) and consisted primarily of MgO with a lesser amount of fine, well-dispersed Mg-Al alloy. Image analyses revealed that the specimen of Fig. 1b consisted of 74 vol% MgO and 26 vol% Mg-Al alloy. Composites containing even higher fractions of MgO were obtained upon conversion of $Al_2O_3$ preforms with less than 45% porosity. For example, a 71% dense $Al_2O_3$ preform was converted into a composite containing 88 vol% MgO. The composite bodies retained the shapes and dimensions (to within 1-1.7%) of the starting porous preforms. The small dimensional changes, high density values, and high ceramic contents achieved upon conversion indicated that the pores in the starting preforms were effectively filled by the MgO product phase. These results demonstrate the feasibility of using the DCP method to fabricate dense, lightweight, oxide-rich (71-88% oxide) bodies with small dimensional changes at a modest temperature (1000°C). In order to determine whether co-continuous composites containing higher-melting metals could be fabricated, the melt and the preform were both alloyed with other components, as discussed below.

Microstructural Characterization of $MgAl_2O_4$/Fe-Ni-Al-bearing Composites: Back-scattered electron images of polished cross-sections of a composite specimen are shown in Figs. 1c and 1d. These images were obtained from a specimen produced by the DCP transformation of a preform with a molar Fe:$NiAl_2O_4$ ratio of 8:1. The transformed specimen consisted of a well-dispersed mixture of metal (bright) and oxide (grey) phases. An interesting, duplex microstructure was observed for the metallic phase; that is, the metal was present as a continuous matrix phase and as discrete particles (a few microns in size) encased within the oxide-bearing regions. Image analyses indicated that this composite specimen consisted of 42±2 vol% oxide and 58±2 vol% metal. EDS analyses indicated that the oxide phase was enriched in magnesium, whereas magnesium could not be detected in the metallic phase. As expected from the strong thermodynamic driving forces for reactions (3) and (4) [40], magnesium was completely oxidized by the displacement reaction within this

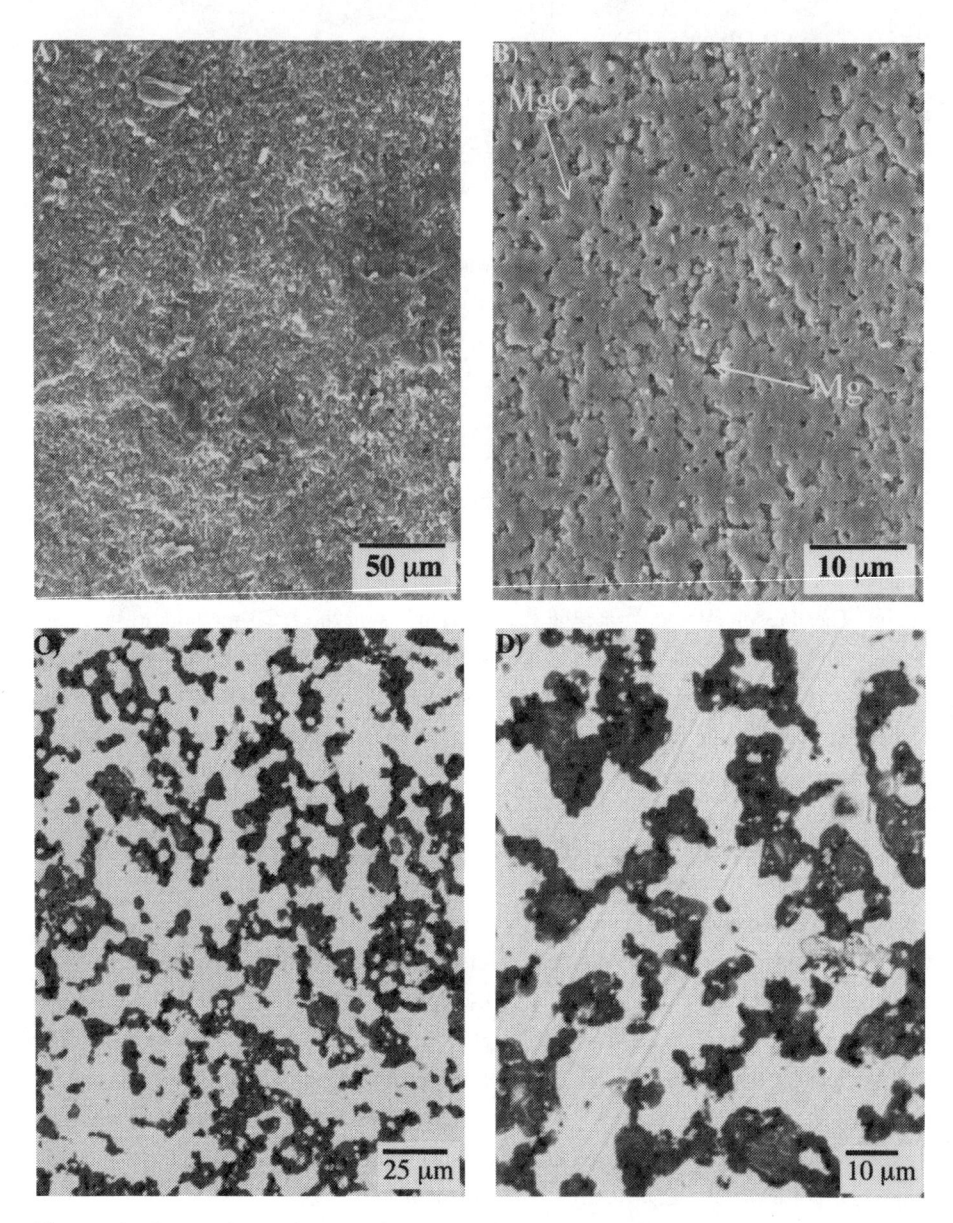

Figure 1. Secondary electron images of DCP-derived composite specimens. A) Fracture Section of a MgO/Mg-Al composite; B) Polished and etched section of the composite in A); C),D) MgAl$_2$O$_4$/Fe-Ni-Al-bearing composite

specimen. Some, but not all, of the MgO-bearing grains were also enriched in aluminum, which indicated that a mixture of $MgAl_2O_4$ and MgO had been produced. Hence, some of the alumina in the nickel aluminate must have been reduced by the magnesium in the liquid. Weight gain measurements after infiltration of the Mg-Al liquid into the porous preform at 700°C (prior to the DCP reaction at 900°C) indicated that the molar $Mg:NiAl_2O_4$ ratio of the infiltrated preform was ≈1.3:1; that is, more Mg than required to complete reaction (3), but less than that required to complete reaction (4), had infiltrated into the specimen. EDS analyses indicated that the metallic phase consisted of an Fe-Ni-Al alloy with an average composition of 78.3±6.6at% Fe, 7.6±3.8 at% Ni, 14.1±2.9 at% Al. The composition and amount of the Fe-Ni-Al alloy in this specimen were between the values expected for reactions (3) and (4), which again was consistent with the reduction of some of the alumina in the starting $NiAl_2O_4$. Very small dimensional changes were observed upon conversion of $NiAl_2O_4$/Fe preforms into $MgAl_2O_4$/Fe-Ni-Al-bearing composites. For example, the length of a particular preform bar in the partially-sintered, as-infiltrated, and completely-transformed states remained at 2.72±0.02 cm.

<u>Strength, Toughness, and Hydration Resistance</u>: The $MgAl_2O_4$/Fe-Ni-Al-bearing composites exhibited average room-temperature flexural strength and fracture toughness values of 398 MPa and 13.2 MPa·m$^{1/2}$, respectively. Lower average values of flexural strength and toughness, 255 MPa and 3.8 MPa·m$^{1/2}$, were obtained from the MgO/Mg-Al-bearing composites. The latter modest values were not surprising, given the relatively high amount of oxide in the MgO/Mg-Al-bearing composites. The weight changes detected for $MgAl_2O_4$/Fe-Ni-Al-bearing and MgO/Mg-Al-bearing specimens after exposure to water vapor at 71°C for 24 h (as per ASTM standard C492) were <0.1%. Such excellent hydration resistance, particularly for the MgO/ Mg-Al specimens, was consistent with the high densities of these composites.

CONCLUSIONS

The feasibility of fabricating dense MgO/Mg-Al-bearing and $MgAl_2O_4$/Fe-Ni-Al-bearing composites at ≤ 1000°C by the DCP method has been demonstrated. For both types of composites, a Mg-bearing liquid was infiltrated (pressureless) into porous, oxide-bearing preforms. The magnesium then underwent a volume-increasing displacement reaction with the solid phase(s) in the preform. Lightweight

MgO/Mg-Al-bearing composites were produced by the reaction of Mg(l) with $Al_2O_3$ preforms. The MgO produced by this reaction filled the pores within the preforms and yielded dense composites with relatively high oxide contents (up to 88 vol% oxide). Higher-melting, co-continuous $MgAl_2O_4$/Fe-Ni-Al-bearing composites (42 vol% oxide, 58 vol% metal) were synthesized by the reaction of an Mg-Al(l) with $NiAl_2O_4$ and Fe. In this case, the Ni liberated by the displacement reaction reacted with Al and Fe to produce a Fe-Ni-Al alloy ($T_m \approx 1400°C$). The co-continuous $MgAl_2O_4$/Fe-Ni-Al-bearing composites exhibited average flexural strength and fracture toughness values of 398 MPa and 13.2 MPa·m$^{1/2}$, respectively. Average strength and toughness values of 255 MPa and 3.8 MPa·m$^{1/2}$, respectively, were obtained for the high-oxide-content MgO/Mg-Al composites. Both types of composites exhibited excellent hydration resistance. These DCP-derived composites retained the shapes and dimensions (to within a few percent) of the starting porous preforms.

REFERENCES

1. R. Blackburn, T. S. Shevlin, "Fundamental Study and Equipment for Sintering and Testing of Cermet Bodies: V, Fabrication, Testing and Properties of 30 Chromium-70 Alumina Cermets," *J. Am. Ceram. Soc.*, **34** [11] 527-31 (1951)
2. T. S. Shevlin, "Fundamental Study and Equipment for Sintering and Testing of Cermet Bodies: VI, Fabrication, Testing and Properties of 72 Chromium-28 Alumina Cermets," *J. Am. Ceram. Soc.*, **37** [3] 140-45 (1954)
3. T. S. Shevlin, C. A. Hauck, "Fundamental Study and Equipment for Sintering and Testing of Cermet Bodies: VII, Fabrication, Testing and Properties of 34 $Al_2O_3$-66 Cr-Mo Cermets," *J. Am. Ceram. Soc.*, **38** [12] 450-454 (1955)
4. L. J. Cronin, "Refractory Cermets," *Ceram. Bull.*, **30** [7] 234-38 (1951)
5. J. H. Westbrook, "Metal-Ceramic Composites" *Ceram Bull.*, **31** [6] 205-08 (1952)
6. J. Graham, J. H. Weymouth, L. S. Williams, "Fundamental and Technological Aspects of Infiltrated Oxide Cermets," *J. Aust. Inst. Metall.*, **8** [3] 280-96 (1963)
7. A. E. S. White, F. K. Earp, T. H. Blakeley, J. Walker, "Metal-Ceramic Bodies," *J. Inst. Metals*, **22**, 880-83 (1955)
8. E. Glenny, T.A. Taylor, "The High-Temperature Properties of Ceramics and Cermets," *Powder. Metall.*, **1-2**, 189-226 (1958)
9. J. B. Huffadine, L. Longland, N. C. Moore, "The Fabrication and Properties of Chromium-Alumina and Molybdenum-Chromium-Alumina Cermets," *Powder. Metall.*, **1-2**, 235-52 (1958)
10. L. S. Sigl, P. A. Mataga, B. J. Dalgleish, R. M. McMeeking, A. G. Evans, "On the Toughness of Brittle Materials Reinforced With a Ductile Phase." *Acta Metall.*, **36** [4] 945-953 (1988)
11. A. G. Evans, D. B. Marshall, "The Mechanical Behavior of Ceramic Matrix Composites." *Acta Metall.*, **37** [10] 2567-83 (1989)

12. W. H. Tuan, R. J. Brook, "The Toughening on Alumina with Nickel Inclusions," *J. Euro. Ceram. Soc.*, **6**, 31-7 (1990)
13. W. H. Tuan, R. J. Brook, "Processing of Alumina/Nickel Composites," *J. Euro. Ceram. Soc.*, **10**, 95-100 (1992)
14. W. B. Chou, W. H. Tuan, S. T. Chang, "Preparation of NiAl Toughened $Al_2O_3$ by Vacuum Hot Pressing," *Trans. Brit. Ceram. Soc.*, **95** [2] 71-74 (1996)
15. K. B. Alexander, H. T. Lin, H. H. Schneibel, P. F. Becher, "Fabrication and Properties of Alumina Matrix Composites Containing Nickel Aluminide Reinforcements," pp. 877-882 in *Processing and Fabrication of Advanced Materials III*, Ed. V. A. Ravi, T. S. Srivatsan, J. J. Moore, TMS, 1994
16. X. Sun, J. A. Yeomans, "Microstructure and Fracture Toughness of Ni Particle Toughened Alumina Matrix Composites," *J. Mater. Sci.*, **31**, 875-880 (1996)
17. C. A. Handwerker, T. J. Froecke, J. S. Wallace, U. R. Kattner, R. D. Jiggets, "Formation of Alumina Chromia-Chromium Composites by a Partial Reduction Reaction," *Mater. Sci. Eng.*, **A195**, 89-100 (1995)
18. R. Subramanian, E. Ustündag, S. L Sass, R. Dieckmann. "In-Situ Formation of Metal-Ceramic Microstructure by Partial Reduction Reactions," *Solid State Ionics*, **75**, 241-255 (1995)
19. E. Ustündag, P. Ret, R. Subramanian, R. Dieckmann, S. L. Sass. "In Situ Metal-Ceramic Microstructures by Partial Reduction Reactions in the Ni-Al-O System and the Role of $ZrO_2$," *Mater. Sci. Eng.*, **A195**, 39-50 (1995)
20. E. Ustundag, R. Subramanian, R. Dieckmann, S. L. Sass, "*In Situ* Formation of Metal-Ceramic Microstructures in the Ni-Al-O System by Partial Reduction Reactions," *Acta Metall. Mater.*, **43** [1] 383-389 (1995)
21. C. H. Henager, Jr., J. L. Brimhall, "Solid State Displacement Reaction Synthesis of Interpenetrating-Phase Ni-Al/$Al_2O_3$ Composites," *Scripta Met. Mater.*, **29**, 1597-1602 (1993)
22. S. A. Jones, J. M. Burlitch, "In Situ Formation of Composites of Alumina with Nickel and with Nickel Aluminide," *Mater. Lett.*, **19**, 233-235 (1994)
23. S. A. Jones, J. M. Burlitch, E. Ustundag, J. Yoo, A. T. Zehnder, "Nickel-Alumina Composites: *In Situ* Synthesis by a Displacement Reaction and Mechanical Properties," pp. 53-58 in *Mater. Res. Soc. Symp. Proc.*, Vol. 365, Materials Research Society, Warrendale, PA, 1995
24. D. W. Song, R. Subramanian, R. Dieckmann, "Displacement Reactions in the Ni-Al-O System Resulting in Periodic Layer Structures," pp. 59-64 in *Mater. Res. Soc. Symp. Proc.*, Vol. 365, Materials Research Society, Warrendale, PA, 1995
25. M. C. Breslin, "Process for Preparing Ceramic-Metal Composite Bodies," *U. S. Patent No. 5,214,011,* May 25, 1993
26. G. S. Daehn, B. Starck, L. Xu, K. F. ElFishawy, J. Ringnalda, H. L. Fraser, "Elastic and Plastic Behaviour of Co-Continuous Alumina/Aluminum Composites," *Acta Mater.*, **44**, 249-261 (1996)
27. M. C. Breslin, J. Ringnalda, L. Xu, M. Fuller, J. Seeger, G. S. Daehn, T. Otani, H. L. Fraser, "Processing, Microstructure, and Properties of Co-Continuous Alumina-Aluminum Composites," *Mater. Sci. Eng.*, **A195**, 113-119 (1995)
28. M. C. Breslin, J. Ringnalda, J. Seeger, A. L. Marasco, G. S. Daehn, H. L. Fraser, "Alumina/Aluminum Co-Continuous Ceramic Composite ($C^4$) Materials Produced by Solid/Liquid Displacement Reactions: Processing Kinetics & Microstructures," *Ceram. Eng. Sci. Proc.*, **15** [4] 104-112 (1994)

29. W. Liu, U. Koster, "Microstructures and Properties of Interpenetrating Alumina/Aluminum Composites Made by Reaction of $SiO_2$ Glass Preforms with Molten Aluminium, *Mater. Sci. Eng.*, **A210**, 1-7 (1996)
30. W. G. Fahrenholtz, K. G. Ewsuk, D. T. Ellerby, R. E. Loehman, "Near-Net-Shape Processing of Metal-Ceramic Composites by Reactive Metal Penetration," *J. Am. Ceram. Soc.*, **79** [9] 2497-99 (1996)
31. R. E. Loehman, K. G. Ewsuk, A. P. Tomsia, "Synthesis of $Al_2O_3$-Al Composites by Reactive Metal Penetration," *J. Am. Ceram. Soc.*, **79** [1] 27-32 (1996)
32. K. G. Ewsuk, S. J. Glass, R. E. Loehman, A. P. Tomsia, W. G. Fahrenholtz, "Microstructure and Properties of $Al_2O_3$-Al(Si) and $Al_2O_3$-Al(Si)-Si Composites Formed by In Situ Reaction of Al with Aluminosilicate Ceramics," *Metall. Mater. Trans.*, **27A** [8] 2122-2129 (1996)
33. D. E. Garcia, S. Schicker, J. Bruhn, R. Janssen, N. Claussen, "Synthesis of Novel Niobium Aluminide-Based Composites," *J. Am. Ceram. Soc.*, **80** [9] 2248-52 (1997)
34. N. Claussen, D. E. Garcia, R. Janssen, "Reaction Sintering of Alumina-Aluminide Alloys (3A)," *J. Mater. Res.*, **11** [11] 2884-2888 (1996)
35. S. Schicker, D. E. Garcia, J. Bruhn, R. Janssen, N. Claussen, "Reaction Processing of $Al_2O_3$ Composites Containing Iron and Iron Aluminides," *J. Am. Ceram. Soc.*, **80** [9] 2294-2300 (1997)
36. M. S. Newkirk, A. W. Urquhart, H. R. Zwicker, E. Breval, "Formation of Lanxide™ Ceramic Composite Materials," *J. Mater. Res.*, **1** [1] 81-89 (1986)
37. M. S. Newkirk, H. D. Lesher, D. R. White, C. R. Kennedy, A. W. Urquhart, T. D. Claar, "Preparation of Lanxide™ Ceramic Matrix Composites: Matrix Formation by the Directed Oxidation of Molten Metals," *Ceram. Eng. Sci. Proc.*, **8** [7-8] 879-885 (1987)
38. A. W. Urquhart, "Novel Reinforced Ceramics and Metals: A Review of Lanxide's Composite Technologies," *Mater. Sci. Eng.*, **A144**, 75-82 (1991)
39. P. Kumar, K. H. Sandhage, "Method for Fabricating Shaped Monolithic Ceramics and Ceramic Composites Through the Displacive Compensation of Porosity and Ceramics and Composites Made Thereby," *U.S. Patent Application No. 60/083,534,* 1998.
40. I. Barin, *Thermochemical Data of Pure Substances* (VCH Verlagsgesellschaft, Weinheim, Germany, 1989), pp. 43, 48, 148, 868, 869, 1067, 1069, 1427
41. *JCPDS Card File* Numbers 6-696 for Fe, 45-946 for MgO, 43-1484 for $Al_2O_3$, 21-1152 for $MgAl_2O_4$, and 10-339 for $NiAl_2O_4$.
42. J. L. Murray, "The Al-Mg (Aluminum-Magnesium) System," *Bull. Alloy Phase Diag.*, **3** [1] 60-74 (1982)
43. *Handbook of Ternary Alloy Phase Diagrams*, Vol. 3, Ed. P. Villars, A. Prince, H. Okamoto, ASM International, Materials Park, OH, 1995
44. W. B. Pearson, *A Handbook of Lattice Spacings and Structures of Metals and Alloys,* Pergamon Press, New York, NY, 1958, pp. 347-351
45. B. R. Lawn, A. G. Evans and D. B. Marshall, "Elastic/Plastic Indentation Damage in Ceramics: The Median/Radial Crack System," *J. Am. Ceram. Soc.*, **63** [9-10] 574-81 (1980)

INNOVATIVE PROCESSING FOR NON-OXIDES BY ADDING SMALL AMOUNTS OF ELEMENTS

Katsutoshi Komeya and Takeshi Meguro
Department of Materials Chemistry,
Yokohama National University
Hodogayaku, Yokohama 240, Japan

ABSTRACT

In the practical use of ceramics, the increase of reliability and the cost reduction are mostly required. In particular, the cost problem is the critical item to be solved for the expansion of applications in non-oxide ceramics. To overcome of this problem, the discovering of the new process which can be conducted at lower temperatures is strongly required. It is well known that small amounts of elements strongly affect reaction rates and microstructures. Some of them have been very useful for the creation of novel synthesis and processing of ceramics, especially silicon nitride and aluminum nitride. Based on this concept we have been studying the development of both nitrides. In this paper, we introduce the background and some examples from our recent work, and discuss the effect of small amounts of elements on the reaction behavior and microstructure control.

INTRODUCTION

More than 40 years has passed since non-oxide ceramics were introduced as new high-temperature materials in the latter half of the 1950s. During that period, several materials including aluminum nitride with high thermal conductivity and silicon nitride with high strength materials were developed and have been put to practical use as new ceramic components. Collins [1] reported in 1955 that silicon nitride as structural ceramics can be obtained through reaction sintering, and Deeley et al. [2] reported in 1961 that high-density sintered bodies can be obtained through hot pressing of $Si_3N_4$-MgO. Thereafter, high density and high strength were achieved simultaneously using rare earth oxides as sintering aids, as discovered by the authors [3][4] and others [5]. Materials development was then further advanced through the development of the gas-pressure sintering process [6]. A glow plug was subsequently put into practical use in 1981 and a turbocharger in 1985. Although silicon nitride was developed through these historical processes, there are still many problems left unsolved. Among them, the problem of insufficient reliability in terms of strength and toughness is considered to be the largest one. CMC, MMC, FGM, nano-composite, synergy ceramics and etc. have thus been developed to replace monolithic materials, and research and development of these materials is now in progress.

Aluminum nitride has also been developed along a historical flow similar to the above. After much effort, suitable sintering aids for the densification, pure fine grain powders and processes for high thermal conductivity were developed [7][8][9]. The first practical use was as a heat dissipation substrate of the thyristor for electric trains, thereafter high thermal conductive substrates and packages for IC/ISI have been developed in these years. Furthermore, the use of aluminum nitride powder as a filler for high thermal conductive resins is being examined. More recently, aluminum nitride is also being developed as a structural part for semiconductor process, utilizing its excellent resistance to halogen and plasma.

However, a point that must be remembered regarding the creation of a new material series is that those materials manufactured through a high-grade or complex process are generally more expensive. Today, it is believed that, judging from the fact that silicon nitride and aluminum nitride ceramics have excellent properties, the cause of the slow growth of non-oxide ceramics lies in their poor cost competitiveness. Although composite materials such as CMC are important as future structural materials, the authors consider that it is necessary to resolve both problems of reliability and cost in monolithic silicon nitride and aluminum nitride ceramics in near future. As a basic approach for that purpose, this research focuses on utilizing small amounts of elements as an effective means for promoting reactions and for microstructure control instead of treating them as harmful contaminants.

CONCEPT

Non-oxide ceramics such as silicon nitride and aluminum nitride exhibit high covalent bonding. Due to this basic property, these materials are considered to have poor sinterability and poor reactivity with other substances. However, as a small amount of oxygen is contained in these non-oxide substances, some nitride/oxide systems change into oxide/oxide systems and reciprocal reactions will be expedited easily. The phenomenon of densification of substances containing AlN-$Y_2O_3$ promoted by oxygen contaminants in aluminum nitride is a typical example of such reactions. However, contamination with small amounts of components may also lead to property deterioration of the sintered body. Also, in the sintering aluminum nitride with rare earth oxides and with alkali earth compounds, which were invented by the authors [7][8], densification may be difficult to obtain unless there are oxygen contaminants on the surface and inside the aluminum nitride powder. Further, the grain morphology in the sintered specimens with these sintering aides changes as they contain contaminants. The commercialized silicon nitride ceramics have been also developed by utilizing these phenomena. In other words, the progress and development of silicon nitride and aluminum nitride achieved so far are nothing more than the effective utilization of their respective small amounts of components. It is considered that the control of the microstructure through the promotion of reactions using small amounts of components in this way will not only improve the reliability of monolithic ceramics, which have been developed conventionally, but will also enable providing those industrial materials with wide applications. We have conducted our research for the past 10 years based on the above concept. Examples of researches focused on the utilization of small amounts of components are listed in Table 1 together with typical previous reports. Some results of the researches performed by the authors and others are briefly described in the following section.

Table 1 Small amounts of elements for synthesis and processing of non-oxides.

| Materials | Contents | Small amount of elements | Effect |
|---|---|---|---|
| $Si_3N_4$ | Nitridation of Si | Fe | Reaction promotion |
| | Post reaction sintering | Fe | Reaction promotion |
| | $SiO_2$ reduction-nitridtion | $Si_3N_4$, Fe | Reaction promotion, size & shape control |
| | Pressureless sintering | MgO, $Y_2O_3$, $Yb_2O_3$, $Y_2O_3$-$Al_2O_3$ | Densification, high strength |
| | $Si_3N_4$-$Y_2O_3$-$Al_2O_3$-$TiO_2$ | $TiO_2$ | Densification, wear resist. |
| | $Si_3N_4$-$Y_2O_3$-$Al_2O_3$-$SiO_2$ | $SiO_2$ | Oxidation resist. |
| | $Si_3N_4$-$Y_2O_3$-$Al_2O_3$-Seed($Si_3N_4$) | Seed($Si_3N_4$) | Grain structure control, toughness |
| | $Si_3N_4$-$Y_2O_3$-AlN-$HfO_2$ | $HfO_2$ | Grain boundary Crystallization |
| | $Si_3N_4$-$Y_2O_3$-AlN-$HfO_2$ | Cl, F | High temperature strength degradation |
| AlN | Nitridation of AlN alloys | Li, Ca | Reaction promotion |
| | $Al_2O_3$ reduction nitridation | Alkali earth & rare earth comp. | Reaction promotion, size & shape control |
| | Pressureless sintering | Alkali earth comp. Rare earth comp. | Densification, high thermal conductivity |
| | AlN-$Y_2O_3$ | O | Densification |
| | Low temperature sintering | CaO-$Y_2O_3$, YF | Densification at low temperatures |
| SiC | Pressureless sintering | B, $Al_2O_3$, $Y_2O_3$-$Al_2O_3$ | Densification |
| Composites | Oxidation of Alloys | Alloying elements | New fabrication process |

## EFFECT OF SMALL AMOUNTS OF ELEMENTS ON SYNTHESIS AND PROCESSING OF ALUMINUM NITRIDE

Effect of Various Additives on Synthesis of Aluminum Nitride by Carbothermal Reduction-Nitridation of Alumina

The effect of additives on the synthesis of aluminum nitride by the carbothermal reduction-nitridation of alumina was investigated, and consequently confirmed that alkali earth compounds such as calcium fluoride and rare earth compounds such as yttria promoted the reaction rate (Fig.1) [10], and controlled, the size and shape of the products.

Upon further study, it was recognized that the nitridation rate was dependent on the synthesis method and the particle size of raw alumina powder. Full nitridation was achieved after 20 to 30min in the specimens with calcium fluoride with 3wt% fired at 1450°C, whereas in the specimen without additives 120min was required for 100% nitridation under the same conditions. Equiaxial aluminum nitride particle size and shape were observed in the products. It is especially meaningful that, even though various particle sizes of alumina were used, aluminum nitride particles synthesized under the same condition indicated almost the same particle size with uniaxial shape and narrow particle size distribution [11]. This phenomenon suggests that a factor determining the morphology of resulting aluminum nitride is in the process of nitridation of the intermediate compounds produced by the reaction of alumina with calcium fluoride. Since the products were composed of aluminum nitride and calcium fluoride, they could be easily densified without additional sintering aids. In other alkali earth compounds and rare earth compounds addition, similar phenomena were observed.

Direct Nitridation of Al-Li and Al-Ca Alloys

The reaction of pure aluminum powder compact with nitrogen takes place only on the surface of aluminum particles. Recently, Lanxide Co., Ltd. discovered new synthesis processes of ceramic composites through the chemical reaction of aluminum-alloys, which has been known as the DIMOX process [12]. Then Scholz et al. [13] reported synthesis of aluminum nitride by the nitridation of the Al-Li and Al-Si-Mg as a new approach. We also studied these works, and consequently the aluminum-alloys such as Al-Li and Al-Ca completely converted to aluminum nitride in very short time after being fired at 1000°C in nitrogen flow (Fig.2) [14]. This nitridation promotion was considered to be based on the oxygen concentration in the reactant gas. It was confirmed by further experiments that the nitridation ratio exceeded 90% in nitrogen containing

Fig. 1 Nitridation rate in the carbothermal reduction-nitridation of alumina powder with various additives: fired at 1450°C in $N_2$.

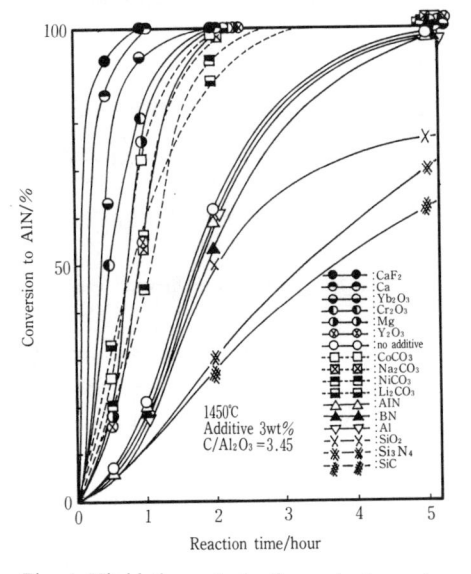

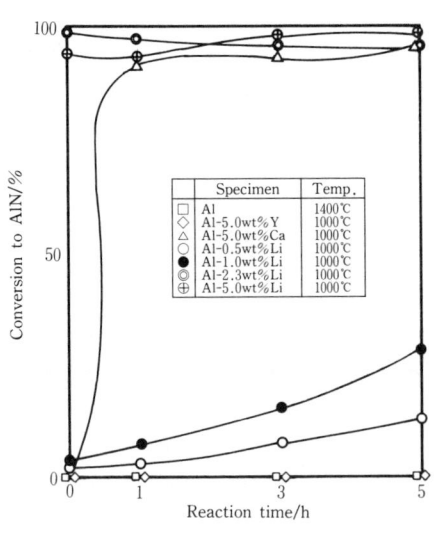

Fig. 2 Nitridation rate of various aluminum alloys fired at 1000°C and 1400°C in $N_2$.

oxygen of 1150 and 250ppm. The reason of such nitridation phenomenon in the aluminum-alloys is considered to be that lithium near the surface of the alloy is changed to $Li_2O$ and/or $LiAlO_2$ by the reaction with oxygen at around 1000°C. As a result, lithium near the surface decreases, and then lithium inside the alloy moves toward the surface under the influence of the lithium concentration gradient and then further the reaction occurs.

# EFFECT OF SMALL AMOUNTS OF ELEMENTS ON SYNTHESIS AND PROCESSING OF SILICON NITRIDE

Synthesis of Silicon Nitride Powder by Carbothermal Reduction-Nitridation of Silica with Silicon Nitride Seed

The effects of addition of silicon nitride seeds on the particle size and α content in the resulting silicon nitride powder were investigated. The addition of small amounts of seeds caused a steep increase in the α content of the resulting silicon nitride powder regardless of the crystalline phases of the silicon nitride seeds (Fig.3) [15]. As the amounts of seeds increased, the powder began to reflect the original crystal form. The use of the seed prepared through a pulverizing process yielded very fine silicon nitride particles of 0.5 to 0.6μm, which were equal to the mean diameter of the seed powder (Fig.4) [15]. It is inferred that this was due to the existence of a large numbers of very fine particles in the seed powder. The seed by silicon-imide decomposition method produced relatively large particles because of the lack of very fine particles. The relation between resulting particle size (D) and the number of seeds could be described by the equation $D = k (1/N)^{1/3}$, when N is the number of seeds and k is a constant.

Hafnia Addition on Sintering Behavior of the System $Si_3N_4$-$Y_2O_3$

The high temperature strength of silicon nitride ceramics depends on the characteristics of grain boundary phases containing additives. As an approach, the grain boundary phase crystallization methods were developed to improve high temperature strength. One method is the heat-treatment of the sintered bodies at around 1400°C. The other method is the

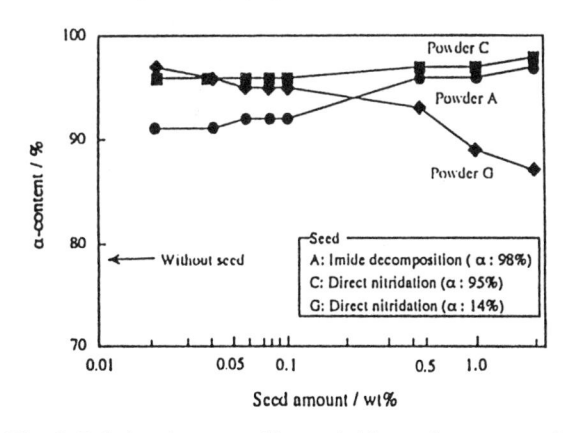

Fig. 3 Relation between silicon nitride seed amount and α content in the silicon nitride powder produced by carbothermal reduction-nitridation of silica: fired at 1450°C for 2 h in $N_2$.

search for the adaptable compositions by which the grain boundary crystallization proceeds after the densification by liquid phase sintering.

The hafnia addition to the $Si_3N_4$-$Y_2O_3$-AlN was investigated as a latter method [16]. From the results as shown in Fig.5, it was confirmed that hafnia promoted the densification in the initial stage, in which the densification behaviors differed below and above 1850°C. Below 1850°C, the maximum densities appeared in the hafnia addition amount of 1 to 1.5wt%, and in the region of amount of greater than 1.5wt% the densities decreased with increasing hafnia content. The densities of the specimens fired at 1850 and 1900°C increased with increasing hafnia amount added, and they shifted on the same level or an increment somewhat over 1.5wt% of hafnia. This phenomenon might be based on the phase relation of the system $HfO_2$-$Al_2O_3$-$SiO_2$. The XRD profiles of the

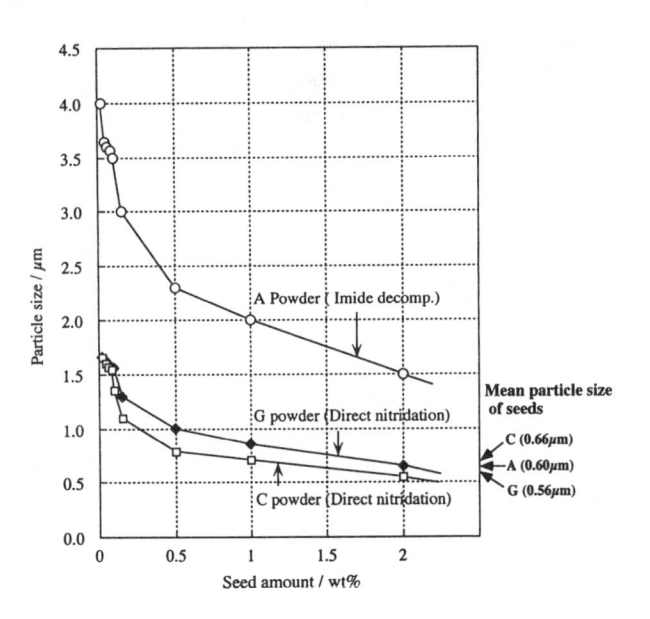

Fig. 4 Particle size of the products for silicon nitride seed amount produced by carbothermal reduction–nitridation of silica with silicon nitride seed: fired at 1450°C for 2 h in $N_2$.

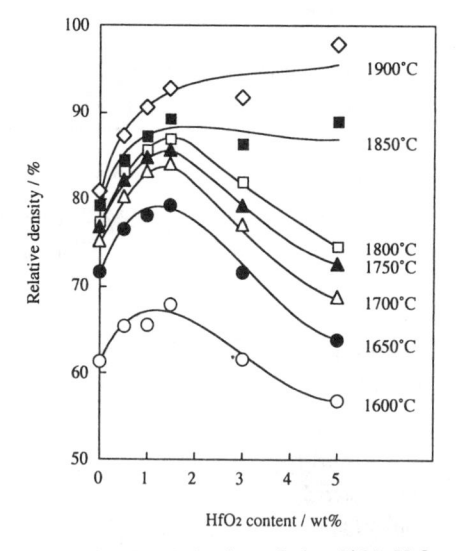

Fig. 5 Densification behavior of the $Si_3N_4$-$Y_2O_3$ system for hafnia addition amount as a parameter of firing temperature.

specimens without hafnia fired at 1900°C showed no existence of crystalline phase except $Si_3N_4$, indicating that the grain boundary was in a glass phase. However, in the specimens with hafnia some crystalline phases such as $Y_2O_3 \cdot 2HfO_2$ and $5Y_2O_3 \cdot Al_2O_3 \cdot Si_3N_4$ were observed, which shows the crystallization of the grain boundary phase occurred due to the addition of hafnia. The crystallization promotion was estimated as follows: hafnia promoted densification by the formation of liquid phases in the initial stage. Through the densification promotion, Al, O and Y dissolved in the silicon nitride grains to produce $\alpha$ and/or $\beta$ sialons, and hafnia was precipitated as isolated grains from glass phases, which was confirmed by TEM analysis. It is considered that the crystallization of the remained compositions existing in the grain boundary phase took place in the final stage.

Effect of Iron on the Post-Reaction Sintering of Silicon Nitride

Reaction sintering of silicon and $Si-Y_2O_3-Al_2O_3$ in the $N_2-5\%H_2$ gas flow was promoted by the small amount of Fe addition, of about 0.1wt%. Alpha content increased with the increase in the amount of iron added in both specimens, which might be based on the higher partial pressure of silicon monoxide during firing because of rapid reduction rate of nitridation. This fact means that the densification accompanying the formation of elongated grain structure will be achieved at high firing temperature.

CONCLUSIONS

Although silicon nitride and aluminum nitride ceramics attract the highest attention among non-oxides, they still have problems of reliability and cost. For reliability, various examinations have been performed using a new process for hyper organized microstructure control, and good results are readily obtained. However, if expensive materials and/or complex processes are used, it will generally lead to higher cost, so further breakthroughs will be required to implement new technology. This means that there is no better way other than manufacturing high-grade materials with controlled microstructures using a simple process. We consider that the utilization of small amounts of components, as reported in this paper, can be an attractive approach considering the fact that it promotes the rate of reactions and enables microstructure control as required. Some of the examples of the processes shown here are expected to contribute to the cost reduction since they are based on the processes that have been implemented conventionally. There are many examples of success where these ideas are used unconsciously. However, we consider that there is too strong a trend for new materials and new technologies to be sought with existing problems left unsolved. We will endeavor to develop and improve the targeted materials as well as to discover new processes by performing research based on this concept positively and systematically in the future.

# REFERENCES

[1] J. F. Collins and R. W. Gerby, *J. of Meter.*, 7, 612(1955).

[2] G. G. Deeley, J. J. M. Herbert and N. C. Moore, *Powder Metallurgica*, 8, 145 (1961).

[3] A. Tsuge, O. Kudo and K. Komeya, *Journal of the American Ceramic Society*, 57 [6] 269-70 (1974).

[4] A. Tsuge, K. Nishida and M. Komatsu, *Journal of the American Ceramic Society*, 58 [7-8] 323-26 (1975).

[5] G. E. Gagga, *Journal of American Ceramic Society*, 56, 662(1973).

[6] M. Mitomo, *Journal of Materials Science*, 11, 1103(1976).

[7] K. Komeya and H. Inoue, *Trans. and J. of the Brit. Ceramic Soc.*, 70 [3] 107-14 (1971).

[8] K. Komeya, H. Inoue and A. Tsuge, *Yogyo-kyokai-shi*, 89(6), 330-36 (1981).

[9] K. Shinozaki et al., *Proc. of the Annual Meeting of the Ceram. Soc. of Japan*, 3A60 (1985).

[10] K. Komeya, E. Mitsuhashi and T. Meguro, *Journal of the Ceramic Society of Japan*, 101 [4] 377-82 (1993).

[11] K. Komeya, T. Ide and T. Meguro, Proceedings of International Symposium on Aluminum Nitride Ceramics, Tokyo, March 8-11, 1998.

[12] M. S. Newkirk, A. W. Urquhart, H. R. Zwicker and E. Breval, *J. of Mater. Res.*, 1 [1] 81 (1986) and M. S. Newkirk, H. D. Lesher, D. R. White, C. R. Kennedy, A. W. Urquhart and T. D. Claar, *Ceram. Eng. Sci. Proc.*, 8 [7-8] 879 (1987).

[13] H. Scholz and P. Greil, *Journal of Materials Science.*, 26, 669-77 (1991).

[14] K. Komeya, N. Matsukaze and T. Meguro, *Journal of the Ceramic Society of Japan*, 101 [12] 1319-23 (1993).

[15] I. H. Kang, K. Komeya, T. Meguro, M. Naito and O. Hayakawa, *Journal of the Ceramic Society of Japan*, 104 [12] 471-475 (1996).

[16] K. Komeya, M. Komatsu, T. Kameda, Y. Goto, and A. Tsuge, *Journal of Materials Science*, 26, 5513-16 (1991).

# Sol-Gel Synthesis

# APPLICATION OF SOL-GEL CONCEPTS TO SYNTHESIS OF NON-OXIDE CERAMICS

Prashant N. Kumta, Jin Yong Kim and Mandyam A. Sriram
Carnegie Mellon University, Pittsburgh, PA 15213

## ABSTRACT

Metal alkoxides have been used extensively for synthesizing oxide ceramics and glasses. Very little is known about their use for synthesizing non-oxide sulfide and nitride ceramics. In this paper the applicability of metal alkoxides for synthesizing non-oxide ceramics has been demonstrated. Two reactive metal systems, Al and Ti have been selected to demonstrate this applicability. Powders in the Al-O-N system were synthesized by reacting aluminum tri-*sec*-butoxide with anhydrous hydrazine in acetonitrile at 80°C. The resultant precipitate was dried and heat-treated in argon, nitrogen and ammonia atmospheres at various temperatures. The phases obtained from the different heat-treatments were examined using X-ray diffraction, while chemical analysis was conducted to analyze the chemical composition of the precursors. Fourier transform infrared (FTIR) was also used to investigate the structural changes occurring in the precursors during heat-treatment. At the same time, scanning electron microscopy (SEM) and transmission electron microscopy (TEM) were used to analyze the morphology and particle sizes of the fine powders. Results of these studies demonstrate the potential of the new sol-gel based chemical approach for synthesizing various powders in the Al-O-N system. Similarly, the potential of metal alkoxides for synthesizing transition metal sulfides has been demonstrated by reacting titanium alkoxide with hydrogen sulfide at room temperature. Replacement of the alkoxy groups with hydrosulfide species result in an alkoxy-sulfide precursor that transforms to yield $TiS_2$. Results of these studies are discussed in the context of the oxide sol-gel process so well known for the synthesis of oxide ceramics.

## INTRODUCTION

Traditional solution sol-gel approaches have been exploited very well for synthesizing a myriad of oxide glasses, glass-ceramic, refractory oxide fibers, powders, and thin films.[1] However, the applicability of similar chemical routes particularly using metal alkoxides for synthesizing non-oxides such as sulfides and nitrides remains somewhat unexplored mainly due to the extreme problems related to the chemistry of the precursors. A major problem is the control of the molecular structure and the inherent sensitivity of the precursors towards attack by OH moieties. Thus, no significant efforts have been made at directly replacing any of the alkoxy groups in the alkoxide by non-oxygen related species to form precursors for synthesizing non-oxide glasses, glass-ceramics and composites. In order to circumvent the difficulties associated with chemical stability and reactivity, much of the work has therefore focused on generating sol-gel based oxide precursors which have then been reacted in suitable atmospheres to form the corresponding non-oxide materials.[2,3] This approach although successful in yielding the desired non-oxide, unfortunately is not very conducive for controlling the morphology and microstructure. Moreover, control of composition and the related defect chemistry becomes almost a distant reality due to the high temperatures needed and the accompanying grain growth, agglomeration and temperature related instabilities leading to decomposition. The incorporation of the non-oxygen anion and the consequent formation of the metal-non-oxygen anion bond at the precursor stage itself would prove to be of significant benefit particularly in terms of not only controlling the composition but also the defect chemistry and microstructure of the non-oxide ceramic.

Metal alkoxides in a sense, would be good candidate starting source materials for initiating any of the reactions leading to the formation of metal-non-oxygen bonds in the precursor. This is because of the susceptibility of the metal centers in the alkoxides towards nucleophilic attack (by $H_2O$, $H_2S$, RSH and $R_2NH$) making them favorable for synthesizing non-oxide ceramics which otherwise prove to be quite challenging. We have demonstrated this unique feature of alkoxides for synthesizing $La_2S_3$.[4,5] In the present paper, we extend the applicability of metal alkoxides by demonstrating its potential for synthesizing non-oxide nitrides and sulfides of reactive metals such as Al and Ti respectively. Accordingly, we have shown that the alkoxide can be reacted in solution with molecular nitriding agents such as hydrazine ($N_2H_4$) to facilitate the formation of condensed species containing Al-N linkages in the solid precursor leading to the formation of oxynitride (Al-O-N) glasses containing AlN ceramic particles. Hydrazine is a strong nucleophile and has a tendency to form adducts with several metalorganic compounds including Al, Ti and Zr alkoxides.[6] However, these reactions have not been investigated further with regards to the formation of nitride and oxynitride glasses and ceramics. Similarly, we have shown that the alkoxide can be reacted with sulfidizing agents such as $H_2S$ to form alkoxy sulfide precursors that transform to the crystalline sulfide when heat treated in a sulfidizing atmosphere.

In the present study, we have investigated the reaction of aluminum tri-*sec*-butoxide [$Al(OC_4H_9)_3{}^s$] with $N_2H_4$ in a pure acetonitrile solution and have studied the pyrolysis of the precursors formed, in argon, nitrogen and ammonia atmospheres. X-ray diffraction, Fourier transform infrared (FTIR) spectroscopy have been used to both, identify the structure and various molecular linkages in the evolved phases, and understand the molecular changes occurring in these precursors during nitridation. The details of this reaction and the subsequent processes leading to the formation of the nitrides have been described and discussed. In order to confirm the nitriding ability of hydrazine, water was also introduced into the reaction along with hydrazine. The influence of water on the nitriding ability of water has been studied and reported in the present work. Similarly, an extension of the process for synthesizing transition metal sulfides such as $TiS_2$ has been outlined. Specifically, the reaction of Ti-isopropoxide [$Ti(OC_3H_7)_4$] with $H_2S$ has been explored and its transformation to crystalline $TiS_2$ has been reported. The mechanisms involved in the formation of the precursor and its structural details are discussed in relation to the well known oxide sol-gel process.

EXPERIMENTAL PROCEDURE

The experimental procedures for the two systems are described in two sections.

A. Synthesis of Aluminum Nitride

*1. Powder Synthesis and Heat-Treatment:*

Two different processes were used to synthesize the solid precursors. The first process called "Process-I" consisted of reacting aluminum tri-*sec*-butoxide with anhydrous hydrazine in an anhydrous acetonitrile solution in the absence of air and moisture. The second process, called

"Process-II" involves the execution of the same reactions with the addition of controlled amounts of water once again in the absence of air to analyze the role of hydrazine as a nitriding agent. Fig. 1 shows the procedures employed for synthesizing the precursors using both process-I and process-II.

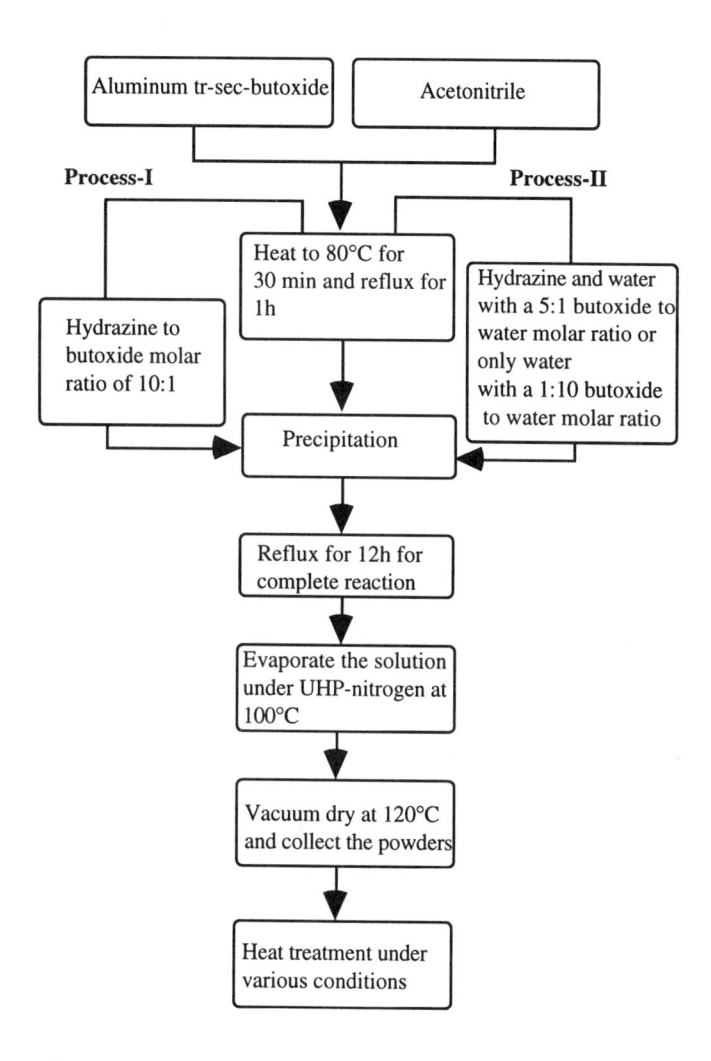

Fig. 1 Flow sheet showing the procedure for Process-I and Process-II.

Process-I

All the reactions in this process were conducted in the absence of air and moisture. Aluminum tri-sec-butoxide, (Al(OBu$^s$)$_3$, [Al(OC$_4$H$_9$)$_3$$^s$]: Aldrich, 97%) was dissolved in acetonitrile (MeCN, [CH$_3$CN]: HPLC grade) and heated to 80°C for 30 min. Anhydrous hydrazine (N$_2$H$_4$: Aldrich, 98%) was then carefully added to the solution to achieve a hydrazine to butoxide molar ratio of 10:1. After few minutes, a turbid solution resulted due to the precipitation of a solid. The reaction was allowed to continue at the same temperature for 12 h and then the liquid was distilled under an environment of flowing ultra-high-purity nitrogen (UHP-N$_2$). The precipitate was dried under vacuum at 120°C for 12 h, while taking proper precaution against exposure to air and moisture. The dried precursor was then heat-treated at 800°C, 1000°C, 1200°C and 1300°C under either flowing UHP-Ar or UHP-N$_2$ or NH$_3$. It should be mentioned that hydrazine is a toxic chemical and its vapors could be explosive of brought in contact with air and hence extreme caution must be exercised while handling it.

Process-II

In this second approach, water was added either with or without hydrazine to induce precipitation. Thus two solid precursors were generated, one using pure excess deionized (DI) water and the other utilizing a mixture of water and anhydrous hydrazine. These precursors are referred to as 'WT' and 'HZ+WT' in all the later discussions to follow. Water as mentioned earlier was selected in this study to initiate the hydrolysis reaction and thereby observe its effect on the nitridation reaction of hydrazine. In addition, the competing hydrolysis reaction of water also helps to demonstrate the flexibility of this process to synthesize oxide, oxynitride and nitride powders. A series of these reactions were essentially conducted to serve as control experiments to test the role of hydrazine as a nitriding agent in its reaction with the alkoxide. Thus similar to process-I, the alkoxide was first dissolved in anhydrous acetonitrile and heated to 80°C for 30 min. Hydrazine was then added, maintaining the same molar ratio of hydrazine to butoxide (10:1) that was used in Process-I. The 'HZ+WT' precursor was then generated by introducing a small amount of DI water enough to partially hydrolyze the alkoxide. The molar ratio of water to butoxide that was used in this process was 1:5. The third type of precursor (WT) was synthesized by adding excess water alone to the solution of the pure alkoxide in acetonitrile to initiate the well known hydrolysis and condensation reactions similar to the solution sol-gel process. A solid alumina gel was therefore obtained by the addition of water[7]. A water to butoxide ratio of 10:1 was used to generate this precursor. In all the processes involving water additions, the reaction was allowed to continue at the same temperature of 80°C for 12h after the addition of water, followed by distillation of the liquid under flowing UHP-N$_2$. The resulting precipitates were then dried under vacuum at 120°C.

The dried precursors were subjected to various heat-treatment conditions in different atmospheres. The complete heat-treatment profile used to generate the desired ceramic powders are shown in Table 1. The table also shows the nomenclature followed for naming each heat-treated sample. The as-prepared precursors derived from process-I heat-treated in ultrahigh-purity argon, nitrogen and anhydrous ammonia are referred to as 'Ar', 'N$_2$', and 'NH$_3$' respectively. The heat-treatments for each atmosphere were conducted at 800°C, 1000°C, 1200°C and 1300°C (the numbers to right in the sample nomenclature represent the temperatures). The temperature of 1300°C was selected to ensure complete crystallization and transformation of the resultant ceramic non-oxide or oxide phases.

On the other hand, the two types of precursors generated using process-II namely, 'HZ+WT' and 'WT', were subjected to heat treatments only in flowing anhydrous ammonia at various temperatures. These samples are therefore referred to as 'HZ+WT-NH$_3$' and 'WT-NH$_3$' respectively. Similar to the samples derived from process-I, these precursors were also heat-treated in ammonia at final temperatures of 800°C, 1000°C and 1300°C, respectively.

Table 1 Heat-treatment schedule

| Sample | Atmosphere | Temperature(°C) | Duration (h) | Heating rate (°C/min) |
|---|---|---|---|---|
| Ar-800 | Ar | | | |
| N$_2$-800 | N$_2$ | 800 | 10 | 2 |
| NH$_3$-800 | NH$_3$ | | | |
| Ar-1000 | Ar | | | |
| N$_2$-1000 | N$_2$ | 1000 | 10 | 2 |
| NH$_3$-1000 | NH$_3$ | | | |
| Ar-1200 | Ar | | | 5 up to 800°C |
| N$_2$-1200 | N$_2$ | 1200 | 10 | 1 up to 1200°C |
| NH$_3$-1200 | NH$_3$ | | | |
| Ar-1300 | Ar | | | 5 up to 800°C |
| N$_2$-1300 | N$_2$ | 1300 | 15 | 1 up to 1300°C |
| NH$_3$-1300 | NH$_3$ | | | |
| HZ+WT-NH$_3$-800 | | 800 | 10 | 2 |
| HZ+WT-NH3-1000 | NH$_3$ | 1000 | 10 | 2 |
| HZ+WT-NH3-1300 | | 1300 | 15 | 5 up to 800°C |
| | | | | 1 up to 1300°C |
| WT-NH3-800 | | 800 | 10 | 2 |
| WT-NH3-1000 | NH$_3$ | 1000 | 10 | 2 |
| WT-NH$_3$-1300 | | 1300 | 15 | 5 up to 800°C |
| | | | | 1 up to 1300°C |

* All samples were furnace-cooled.

## 2. Characterization

All the heat-treated powders were studied for their phase evolution and crystallization characteristics using X-ray diffraction, employing a Rigaku θ/θ diffractometer. X-ray analysis was not conducted on the as-prepared precursors owing to its reactivity and potential hazards on exposure to moisture. Infrared (IR) spectroscopic analyses were conducted on all the heat treated powders generated using the above process. Absorption infrared spectra (using a Mattson Galaxy Series 5000 FTIR spectrometer) were collected on the precursors (Ar-800, N$_2$-800, NH$_3$-800, Ar-1000, N$_2$-1000, and NH$_3$-1000) in the spectral range of 400 ~ 4000 cm$^{-1}$, using the KBr-pellet technique. Identical amounts of powders were used in all the KBr pellets in order to make comparisons of the different spectra. These six heat-treated powders were also chemically analyzed by Galbraith Laboratories (Knoxville, TN) for Al, N, C and H contents. Scanning electron microscopy (SEM) was used to observe the morphology of all the as-prepared and heat-treated powders employing a CamScan scanning electron microscope. The powders obtained after heat-treatment in NH$_3$ were also observed under a transmission electron microscope (TEM, JEOL 1200 CX) in order to assess both, the particle size and the crystalline structure.

## B. Synthesis of Titanium Sulfide

### 1. Powder Synthesis and Heat-Treatment

Several different reactions of Ti-alkoxide with inorganic and organic sulfidizing agents were explored for synthesizing $TiS_2$ and the details regarding these processes have been reported elsewhere[8]. However, in this paper only the reaction of Ti-isopropoxide with $H_2S$ is described and discussed. In all these experiments, titanium-isopropoxide was used as received from Johnson Matthey, technical grade $H_2S$ was used from Matheson gases and benzene was used after distilling over sodium chips and molecular sieves. All manipulations of the reagents during the reaction as well as handling of the precursors were conducted in the absence of air. The precursors were therefore synthesized by reacting $Ti(OPr^i)_4$ with $H_2S$. This was done by dissolving the alkoxide in anhydrous benzene, and bubbling $H_2S$ through it at room temperature. A black precipitate was observed within a minute, but the bubbling was continued for ten minutes in order to ensure complete reaction. After completion of the reaction, the reaction vessel was sealed and kept isolated for twelve hours before collecting the precipitate. The precipitates were collected in a soxhlet extractor and then washed thoroughly with benzene. The filtrate from the reaction was dark brown in color. The collected powders were dried for three hours in a vacuum oven at 40°C after which they were perceived to be extremely air-sensitive.

### 2. Characterization

Elemental chemical analyses were conducted on the dried powders by Galbraith Laboratories Inc. (Knoxville, TN). The precursor powders were also heat treated in flowing $H_2S$ at 600, 700 and 800°C for a period of six hours. X-ray diffractograms (θ/θ diffractometer, Rigaku, Tokyo, Japan) were collected on the as prepared precipitates and the powders obtained after heat treatment at each of the temperatures. In order to clearly identify the molecular reactions occurring in solution, it was decided to characterize the solid and liquid precursors. An infrared (IR) spectrum was obtained from the powder in a KBr pellet using a Fourier transform infrared (FTIR) spectrometer (113V, Brucker Instruments, Billerica MA, equipped with a mercury cadmium telluride detector), in the 4000 to 600 $cm^{-1}$ wave number range. The transmission spectrum in the spectral range of 500 to 200 $cm^{-1}$ was collected using a CsI pellet in a Biorad FTIR spectrometer equipped with a DTGS detector and CsI optics. The dark brown filtrate was distilled and gas chromatography (GC) (5830A Hewlett Packard, Avondale PA) was performed on the distillate to identify the products of the reaction. After ensuring that all the excess benzene was distilled off, chemical analysis was also performed on the dark brown liquid. In addition, the liquid was analyzed employing electron impact mass spectroscopy (Model 7070, VG Analytical, Manchester, England).

## RESULTS AND DISCUSSION

The results of the study will be described and discussed in two sections. Section A will discuss the results of the work conducted on the synthesis of aluminum nitride while section B will describe the results of the study on the synthesis of titanium sulfide.

## A. Synthesis of Aluminum Nitride

### 1. Phase Evolution of the Heat-Treated Precursors

Fig. 2 shows the XRD traces collected on the heat-treated powders in Ar. The precursor heat-treated in Ar at 800°C for 10 h remains amorphous (see Fig. 2(a)), while exhibiting the onset of crystallization after heat-treatment in Ar at 1000°C for 10 h (Fig. 2(b)). Heat-treatment of the precursor in Ar at 1200°C for 10 h however yields crystalline phases corresponding to AlN, α-$Al_2O_3$ and γ-$Al_2O_3$ or γ-AlON (see Fig. 2(c)). The apparent structural similarity between γ-$Al_2O_3$ and γ-AlON makes it difficult to clearly distinguish the phase at 1200°C. Powders obtained after heat-treating the as-prepared precursors in Ar at 1300°C for 15 h show the appearance of α-$Al_2O_3$ along with increase in the intensity of peaks corresponding to AlN in comparison to the phase denoted as 'γ' (see Fig. 2(d)). This result clearly suggests that the phase labeled as 'γ' probably undergoes decomposition to AlN and α-$Al_2O_3$ at 1300°C. This observation combined with the fact that γ-AlON is unstable beyond 1000°C, and is known to decompose to α-$Al_2O_3$ and AlN, lends credence to our assumption that the 'γ' phase is indeed γ-AlON. The formation of AlN after heat-treatment in Ar is a reflection of the possible replacement of some of the alkoxy groups with hydrazine to generate alkoxy-hydrazide species. Preliminary analysis of the liquid solution obtained after reaction of the alkoxide with hydrazine using gas chromatography shows the presence of butanol suggesting the formation of these species. These results are still in progress and will be reported elsewhere[9]. However, Bains et al.[6] have already reported on the generation of adducts between Al-butoxide and $N_2H_4$. Thus it is possible that these adducts form in addition to replacement of some alkoxy groups. The presence of these hydrazide species attached directly to Al as well as the formation of adducts then lead to the formation of direct Al-N linkages which convert to AlN during heat-treatment. Nevertheless, the presence of α-$Al_2O_3$ and γ-AlON along with AlN after heat-treatment at 1300°C in Ar suggests that the unreacted butoxy groups in turn convert to α-$Al_2O_3$, while reacting with nitrogen in presence of carbon to form γ-AlON during heat-treatment in Ar.

The XRD traces collected on the $N_2$-treated powders are shown in Fig. 3. Similar to the Ar-treated samples, the $N_2$-treated samples appear to be amorphous at 800°C. However, at 1000°C, the powder exhibits different features compared to the sample heat-treated in Ar at the same temperature (compare Fig. 3(b) with Fig. 2(b)). The $N_2$-treated sample clearly shows the presence of single phase of γ-AlON, while the Ar-treated powder shows broad peaks implying the initiation of a crystalline phase that is at present unidentifiable although it could be related to the nucleation and growth of AlN. This observation appears to be related to the reaction of the as-prepared precursor with $N_2$ which is extremely unlikely during Ar-treatments. The as-prepared precursor when heat-treated in $N_2$ at 1200°C for 10 h, results in the formation of AlN and α-$Al_2O_3$. In other words, the AlON phase which forms in $N_2$ at 1000°C decomposes into AlN and α-$Al_2O_3$ at 1200°C. At 1300°C, the intensities of peaks corresponding to AlN and α-$Al_2O_3$ increase due to further decomposition of the AlON phase which is known to be unstable beyond 1000°C.

Heat-treatment of the precursors in $NH_3$ however, as expected reveal an entirely different phase evolution behavior in comparison to the Ar or $N_2$-treated samples. Fig. 4 shows the XRD traces collected on the precursors heat-treated in $NH_3$. The powder heat-treated in $NH_3$ at 800°C for 10 h exhibits broad peaks corresponding to AlN (see Fig. 4(a)), while the powders remained largely amorphous after heat-treatment in Ar or $N_2$ (Figs. 2(a) and 3(a)). The powder continues to show peaks characteristic of single phase of crystalline AlN after heat-treatment in $NH_3$ at 1000°C for 10 h (see Fig. 4(b)). Continued heat-treatment in $NH_3$ at 1200°C for 10 h shows the growth in the intensities of AlN peaks as expected, while a small peak, characteristic of crystalline 'γ' phase (which is probably γ-AlON), is also seen to appear. This suggests that, although the precursors

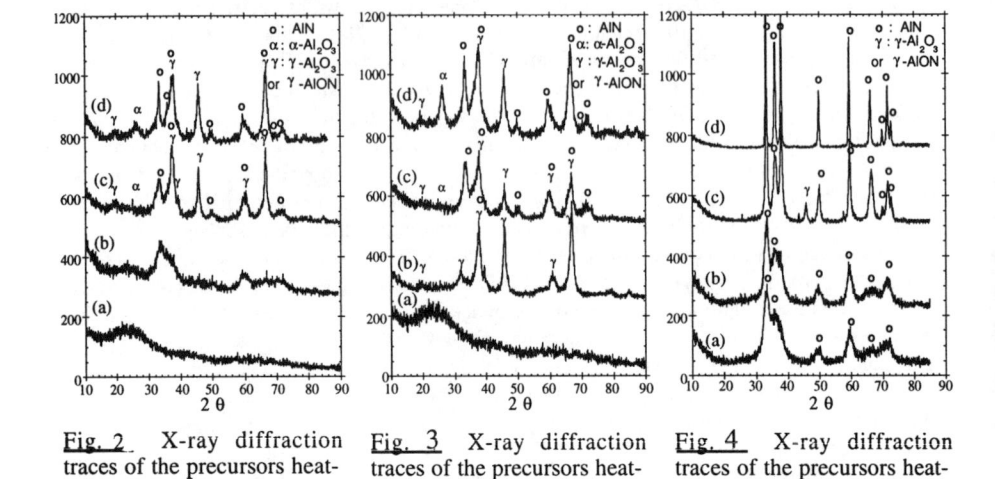

Fig. 2 X-ray diffraction traces of the precursors heat-treated in UHP-Ar: (a) Ar-800, (b) Ar-1000, (c) Ar-1200, and (d) Ar-1300.

Fig. 3 X-ray diffraction traces of the precursors heat-treated in UHP-N$_2$: (a) N$_2$-800, (b) N$_2$-1000, (c) N$_2$-1200, and (d) N$_2$-1300.

Fig. 4 X-ray diffraction traces of the precursors heat-treated in NH$_3$: (a) NH$_3$-800, (b) NH$_3$-1000, (c) NH$_3$-1200, and (d) NH$_3$-1300.

show to a large extent the formation of single phase of AlN after heat-treatment in NH$_3$ at 800°C and 1000°C, there is still some amorphous oxide phase formed due to the residual unreacted butoxy groups. These groups retained in the as-prepared precursor react with NH$_3$ at 1200°C to form the 'γ' phase. This crystalline phase labeled as 'γ' however disappears after heat-treatment in NH$_3$ at 1300°C for 15 h and single phase of AlN is obtained (refer to Fig. 4(d)). Similar heat-treatments of the precursor at 1300°C for 15 h in Ar and N$_2$ showed a mixture of AlN, α-Al$_2$O$_3$ and γ-AlON. Therefore, it can be concluded that NH$_3$-treatment of the precursor is beneficial for the formation of crystalline AlN while also facilitating the removal of residual butoxy groups from the as-prepared precursor. The formation of adducts and alkoxy-hydrazide species in the precursor stage itself by the reaction of the alkoxide with N$_2$H$_4$ therefore helps to enhance the kinetics of the subsequent reactions during heat treatment leading to formation of crystalline AlN. This is because of the presence of Al-N type linkages as well as the relative ease with which rearrangements of the molecules occur thereby facilitating the nucleation and growth of the AlN phase.

## 2. Structural Characterization Using FTIR Spectroscopy Techniques

Infrared absorption spectra obtained on the argon and nitrogen treated 'Ar-800' and 'N$_2$-800' samples are shown in Fig. 5. Overall, the IR spectra collected on both the 'Ar-800' and 'N$_2$-800' samples are similar. In both cases, an intense vibration at 2160 cm$^{-1}$ is observed. This absorption is due to the stretching vibration of the C≡N triple bond ($v_{C≡N}$)[10,11], suggesting the presence of terminal nitrogen bonds on the carbon residue within the structure obtained after pyrolysis. There are also peaks at 1530 cm$^{-1}$ and 1330 cm$^{-1}$ in both samples. The absorption at 1330 cm$^{-1}$

corresponds to carbon vibrations in disordered graphite (indicated as 'D' in Fig. 5)[12,13] and the one at 1530 cm$^{-1}$ corresponds to the carbon vibrations in a graphitic carbon network modified by nitrogen atoms (indicated as 'G' in Fig. 5).[13] The spectra in the range between 750 cm$^{-1}$ and 500 cm$^{-1}$ correspond to Al-related linkages. Both the 'Ar-800' and 'N$_2$-800' powders reveal broad peaks at 750 cm$^{-1}$ and 540 cm$^{-1}$, a small peak at 700 cm$^{-1}$, and a shoulder at 600 cm$^{-1}$. The broad peak centered at 750 cm$^{-1}$ corresponds to both Al-N and Al-O vibrations (Al-O vibration is observed at ≈750 cm$^{-1}$ for γ-Al$_2$O$_3$ and γ-AlON)[14,15], while the low intensity peak centered at 700 cm$^{-1}$ corresponds to the Al-N linkage.[11] The shoulder shown at 600 cm$^{-1}$ and the broad peak centered at 540 cm$^{-1}$ correspond to Al-O and pseudo-γ-Al$_2$O$_3$, respectively.[14] The presence of Al-N linkage in the powder obtained after heat-treating the precursor in Ar is an indication of the reaction of the alkoxide with N$_2$H$_4$. The fact that the powder heat-treated in N$_2$ at 800°C (N$_2$-800) exhibits a spectra almost identical to the Ar-treated sample (Ar-800) suggests that there is no significant reaction between the as-prepared precursor and N$_2$. Analysis of the IR finger prints from both these spectra however suggest the influence of N$_2$H$_4$ and its possible reaction with the alkoxide in solution. In other words, the adducts and the alkoxy-hydrazide species formed by the reaction of alkoxide and hydrazine in solution undergo reconstruction during subsequent heat-treatment resulting in the formation of Al-N type molecular linkages.

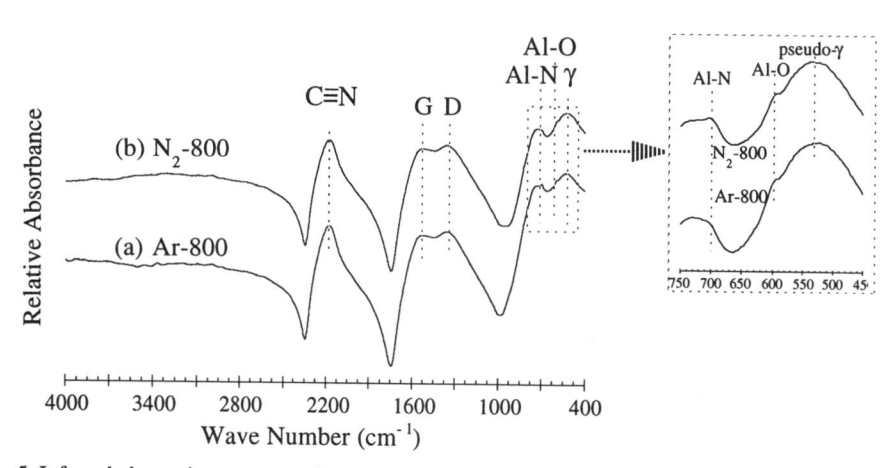

Fig. 5 Infrared absorption spectra collected on the precursors heat-treated at 800°C in Ar and N$_2$: (a) Ar-800 and (b) N$_2$-800.

Fig. 6 shows the IR absorption spectra obtained on the 'Ar-1000' and 'N$_2$-1000' samples. Similar to the samples heat-treated at 800°C, both the 'Ar-1000' and 'N$_2$-1000' samples reveal an intense vibration at 2160 cm$^{-1}$ corresponding to the stretching vibration of the C≡N triple bond (νC≡N) and peaks at 1530 cm$^{-1}$ and 1330 cm$^{-1}$ which correspond to carbon vibrations in graphitic carbon and the carbon vibrations in disordered graphite network, respectively. This result clearly indicates that it is difficult to remove carbon in Ar and N$_2$ due to the inert nature of these gases. Some differences however occur in the range of 750 ~ 500 cm$^{-1}$, related to Al-N and Al-O linkages. Compared to the samples heat-treated at 800°C, the powders heat-treated at 1000°C in either Ar or N$_2$ exhibit a collapse of the peak centered at 540 cm$^{-1}$ corresponding to pseudo-γ-Al$_2$O$_3$. This result indicates that the unstable linkages related to γ-Al$_2$O$_3$ tend to disappear during heat-treatment at the higher temperature of 1000°C. Another distinct feature exhibited by the samples heat-treated at 1000°C is the presence of a shoulder at 600 cm$^{-1}$ corresponding to Al-O linkages. The intensity of this peak observed in the N$_2$-treated sample (N$_2$-1000) is relatively smaller compared to the precursor heat-treated in Ar at 1000°C (Ar-1000: refer to Figs. 6(a) and 6(b)). This suggests a possible nitridation reaction between the precursor and N$_2$ when heat-treated at 1000°C, while this reaction was not observed in the precursor heat-treated at 800°C.

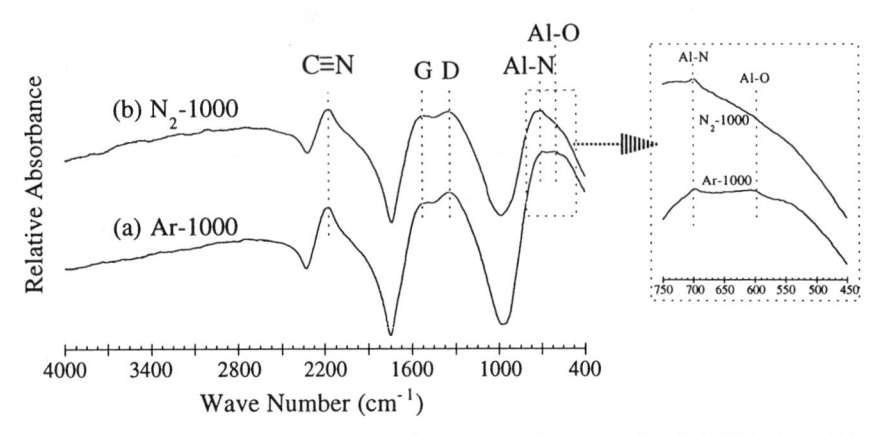

Fig. 6 Infrared absorption spectra collected on the precursors heat-treated at 1000°C in Ar and N$_2$: (a) Ar-1000 and (b) N$_2$-1000.

The IR absorption spectra collected on the precursors heat-treated in NH$_3$ at 800°C and 1000°C are shown in Fig. 7. The precursors heat-treated in NH$_3$ at 800°C and 1000°C reveal several characteristic vibrations. Thus, N-H stretching vibration (3440 cm$^{-1}$)[16-18], stretching vibration of the C≡N bond ($v_{C≡N}$: 2160 cm$^{-1}$)[10,11], C=C stretching vibration ($v_{C=C}$: 1625 cm$^{-1}$)[19], carbon vibrations representative of graphitic carbon network (G: 1530 cm$^{-1}$) and disordered graphite (D: 1330 cm$^{-1}$)[12,13], and Al-N stretching vibration (700 cm$^{-1}$)[14] are observed. In comparison to the Ar and N$_2$-treated samples at both 800°C and 1000°C (Ar-800, N$_2$-800, Ar-1000, and N$_2$-1000), the precursors heat-treated in NH$_3$ (NH$_3$-800 and NH$_3$-1000) show very different features. First, the stretching vibration mode of N-H bonds is observed in the NH$_3$-treated sample due to the reaction of the precursor with NH$_3$. Second, the peaks related to carbon such as '$v_{C≡N}$' (2160 cm$^{-1}$), 'G' (1530 cm$^{-1}$) and 'D' (1330 cm$^{-1}$) show very low intensities owing to the well known ability of NH$_3$ to remove carbon.[20] A third distinct feature is the strong intensity of the peak centered at 700 cm$^{-1}$ corresponding to the Al-N stretching vibration. This result can be expected since X-ray diffraction analysis shows the formation of crystalline AlN after heat-treatment in NH$_3$ at 800°C, while samples heat-treated in Ar and N$_2$ under identical conditions reveal amorphous patterns. The peaks related to Al-O linkages [Al-O (600 cm$^{-1}$) and pseudo-γ-Al$_2$O$_3$ (540 cm$^{-1}$) vibrations] have almost disappeared and only remain as shoulders of very low intensity. These results indicate that NH$_3$ reacts with carbon in the as-prepared precursors and facilitates the formation of Al-N linkages which crystallize into AlN. The intensity of the peak corresponding to Al-N vibrations increases and correspondingly, intensities of the peak related to carbon ['$\nu_{C\equiv N}$' (2160 cm$^{-1}$), 'G' (1530 cm$^{-1}$) and 'D' (1330 cm$^{-1}$)] decrease with increase in the temperature of the NH$_3$-treatment.

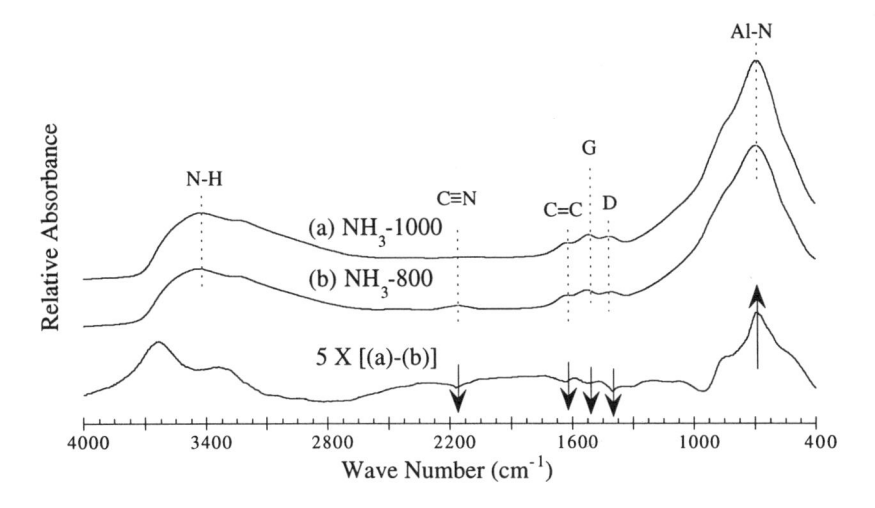

Fig. 7 Infrared absorption spectra collected on the precursors heat-treated in NH$_3$: (a) NH$_3$-1000, (b) NH$_3$-800, and (c) the difference between these two spectra (Scale of the difference spectrum has been expanded by a factor of 5).

Thus based on the infrared studies conducted on the precursors heat treated in argon, nitrogen and ammonia, it can be concluded that there is clearly evidence suggesting the presence of Al-N type linkages after heat treatment at 800°C. This evidence is also strikingly clear in the precursor heat treated in an inert argon atmosphere. The presence of Al-N linkages in a sample that is X-ray amorphous at a temperature of 800°C shows the influence of the nitriding role of hydrazine. The incorporation of nitrogen via both replacement of alkoxy groups and the formation of adducts certainly helps to facilitate the nucleation and growth of AlN. Detailed gas chromatography has been conducted on the liquid product to confirm the replacement of alkoxy groups via hydrazide species in solution due to the reaction of the alkoxide with hydrazine. Similarly magic angle sample spinning nuclear magnetic resonance spectroscopy has been conducted on the solid precursors to analyze the changes in the coordination environment around Al during heat treatment. The results of these studies are currently in progress and will be published elsewhere[9].

Results of the elemental chemical analysis conducted on the heat treated precursors are shown in Table 2. These results also confirm the trend revealed by the FTIR analysis. Compared to the precursors heat-treated in Ar and N$_2$, NH$_3$-treated samples show much lower C contents due to the reducing nature of NH$_3$ and its known ability to remove carbon.[20] The C content decreases with increase in the temperature of NH$_3$-treatment as expected, and also confirmed by the results of the IR analysis. In the case of the powders heat-treated in N$_2$ (N$_2$-800), N/C and N/Al ratios are slightly higher in comparison with the Ar-treated samples (Ar-800) and this tendency increases with increase in the temperature (compare N$_2$-1000 with Ar-1000 in Table 2). This result is also in agreement with the results of the IR analysis. The IR analysis results discussed above clearly shows the reaction of the precursor with N$_2$ at the elevated temperature of 1000°C. Another observation that can be made from the results of the chemical analysis is that the heat-treatment in NH$_3$ also yields a product with a significantly lower N content. This reduction (along with the nature of bonding of C, H and N in the powders) can be explained on the basis of the IR analysis of these powders. In other words, the precursors heat-treated in Ar and N$_2$ contain a larger amount of nitrogen which is mainly bonded to carbon in the form of C≡N bonds as shown in Figs. 5 and 6. These nitrile bonds most probably originate from acetonitrile which is used as a solvent in these reactions. However, although the nitrile (C≡N) species are present in the precursors, they are easily removed along with carbon during NH$_3$-treatment. Since the NH$_3$-treated samples show a much stronger peak in the IR spectra (Fig. 7) corresponding to the Al-N linkage, it can be concluded that a majority of nitrogen in the powders heat-treated in Ar and N$_2$ remains mainly bonded with carbon as C≡N bonds, while only a small fraction of nitrogen in the precursors heat-treated in NH$_3$ is bonded to carbon (see Fig. 7 which shows the small intensity of C≡N vibration mode). Thus, NH$_3$ reacts with carbon in the as-prepared precursors and helps in the formation of Al-N linkages.

Table 2 Chemical analysis of the heat-treated powders

| Sample | Carbon* | Hydrogen* | Nitrogen* | Aluminum* | N/Al** | N/C** |
|---|---|---|---|---|---|---|
| Ar-800 | 28.21 | <0.5 | 31.23 | 26.87 | 2.24 | 0.95 |
| N$_2$-800 | 28.23 | <0.5 | 31.36 | 26.37 | 2.29 | 0.96 |
| NH$_3$-800 | 0.92 | 1.89 | 12.10 | 41.78 | 0.56 | - |
| Ar-1000 | 27.62 | <0.5 | 30.54 | 25.37 | 2.23 | 0.95 |
| N$_2$-1000 | 27.17 | <0.5 | 31.61 | 25.19 | 2.42 | 0.99 |
| NH$_3$-1000 | 0.84 | 1.68 | 13.68 | 42.62 | 0.61 | - |

\* All the analyses are in wt%.
\*\* N/Al and N/C are listed in molar ratios.

The formation of Al-N linkages in the precursor helps in the nucleation and crystallization of AlN during NH$_3$-treatments. On the other hand, Ar and N$_2$ do not remove carbon significantly due to the inert nature of these gases even though there is a slight reaction of the precursor with N$_2$. However, as is clear from the X-ray data in Figs. 2 and 3, although there is almost no removal of carbon, the adduct formation of Al-alkoxide with N$_2$H$_4$ as well as the replacement of some of the alkoxy groups with hydrazide species helps in the incorporation of nitrogen into the amorphous precursor. The presence of carbon inhibits the structural rearrangement until 1200°C where formation of AlN can be seen. Observation of AlN in the XRD data at 1200°C, therefore, confirms that the weak absorption peak centered at 700 cm$^{-1}$ seen in both the Ar and N$_2$-treated samples (Figs. 5 and 6) at 800°C and 1000°C is indeed due to Al-N bonds. Thus, it is possible that the AlN phase has already nucleated at 800°C, but the presence of carbon prevents its growth until a temperature of 1200°C. These observations are largely speculative at this point, although the formation of Al-N linkages and the growth of AlN is clear.

## 3. Microstructure of the As-Prepared and Heat-Treated Powders

The microstructure of the as-prepared and heat treated powders was observed using a scanning electron and transmission electron microscope. This was done to acquire some information and obtain an insight into the size and distribution of the amorphous particles as well as the crystalline particles embedded in the amorphous matrix. Similarly, by conducting microscopy on the heat treated powders, it would be possible to observe the progressive growth in the AlN crystallites with heat treatment in ammonia from 800°C to 1200°C.

Fig. 8 SEM micrographs showing the morphology of the as-prepared and heat-treated powders: (a) As-prepared precursor (b) Ar-800, (c) $N_2$-800, and (d) $NH_3$-800.

Innovative Processing/Synthesis: Ceramics, Glasses, Composites II

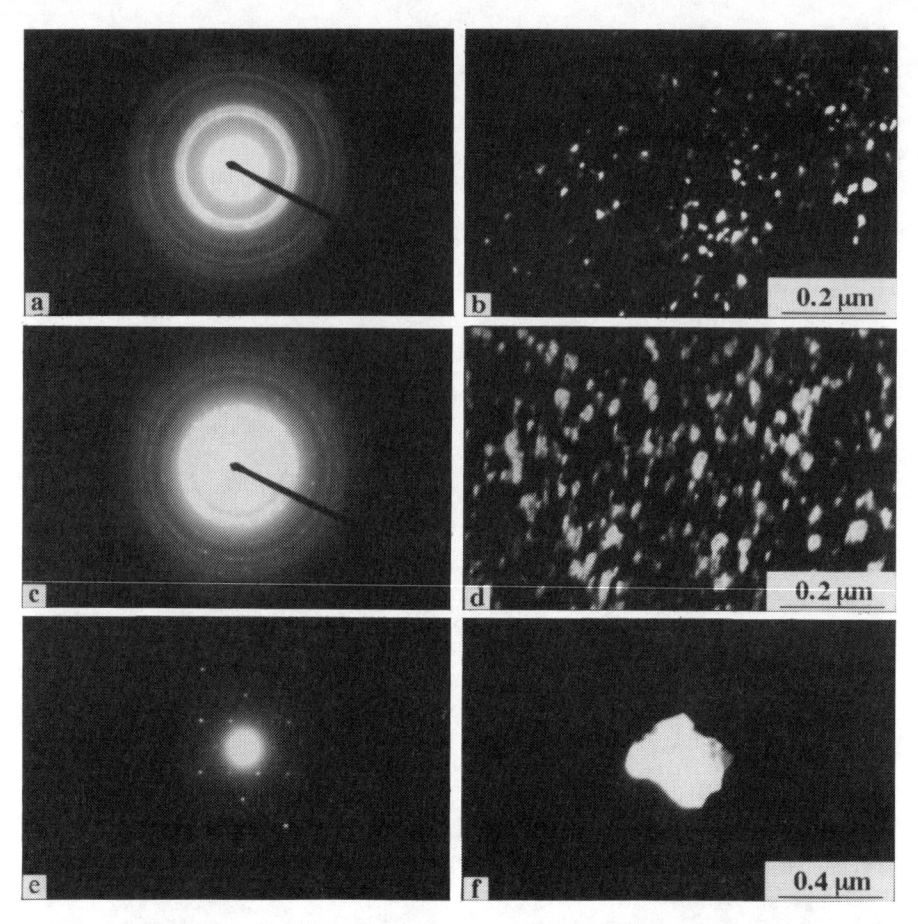

Fig. 9 TEM micrographs of powders heat-treated in $NH_3$ showing the diffraction pattern (DP) and the morphology of the AlN crystallites: (a) DP collected on $NH_3$-800, (b) dark field (DF) image obtained from the brightest ring in (a), (c) DP collected on $NH_3$-1000, (d) DF image obtained from the brightest ring in (c), (e) DP collected on $NH_3$-1300, and (f) DF image obtained from the spot pattern in (e). All the ring and spot patterns can be indexed to single phase of AlN.

Fig. 8 shows the SEM micrographs of the as-prepared precursor and heat-treated powders. The as-prepared precursor powder consists of particles less than 0.5 µm in size (see Fig. 8(a)). Figs. 8(b) and 8(c) show the SEM micrographs which represent the fracture surfaces of the partially fused clumps of amorphous powders obtained after heat-treatment at 800°C in Ar and $N_2$, respectively. The powders heat-treated in Ar and $N_2$ show the conchoidal fracture surfaces characteristic of the 'glassy' phases present in both cases. The morphology of the powders obtained after heat-treatment of the precursors in $NH_3$ at 800°C is shown in Fig. 8(d). After heat-treatment in $NH_3$ at 800°C ($NH_3$-800), the morphology of the powders remains essentially unchanged and is similar to the morphology of the as-prepared precursors shown in Fig. 8(a), although there is growth of the AlN phase as shown by the XRD results (Fig. 4). On the other hand, the powders heat-treated in Ar and $N_2$ at 800°C appear to have undergone fusion due to the possible viscous sintering of the precursor to some extent. These amorphous particles therefore reveal the conchoidal fracture surface characteristic of the brittle 'glassy' phase.

The SEM micrographs clearly show the differences in the particle size and morphology of the precursor when heat treated in the inert atmospheres of argon and nitrogen, and reactive atmosphere of ammonia. An interesting observation is the retention of the morphological features of the precursor seen in the as-prepared stage even after heat treatment in ammonia at 800°C and growth of the AlN phase. This aspect could be extremely useful in the design of materials for catalytic applications wherein morphology and particle size are important factors affecting the surface area and catalytic activity. In order to observe the morphology and size of the crystalline AlN particles present in the $NH_3$-treated powders as identified by the XRD traces (refer to Fig. 4), transmission electron microscopy (TEM) was conducted on these powders. Fig. 9 shows the diffraction patterns and the corresponding dark field images obtained on the powders at each stage of the heat treatment. Thus both the diffraction patterns and the dark field images were obtained from the precursors heat-treated in $NH_3$ at 800°C ($NH_3$-800), 1000°C ($NH_3$-1000), and 1300°C ($NH_3$-1300). Although these three samples show the presence of single phase of AlN in the X-ray diffraction (see Fig. 4(a), 4(b) and 4(d)), the morphology and more importantly, the size of the AlN particles is quite different as revealed by the TEM micrographs. The powder heat-treated in $NH_3$ at 800°C shows the diffraction pattern consisting of rings corresponding to polycrystalline AlN (see Fig. 9(a)). Dark field image collected on the brightest ring of this diffraction pattern exhibits a homogeneous distribution of nanocrystalline AlN particles less than 10 nm in size (Fig. 9(b)). The precursor heat-treated in $NH_3$ at 1000°C also reveal the ring diffraction pattern characteristic of polycrystalline AlN phase (see Fig. 9(c)). Nanocrystalline particles of AlN are also observed in this precursor (see Fig. 9(d)) at this stage, although there is an increase in both the particle size and volume fraction of crystalline particles in comparison to the precursor heat-treated at 800°C. This is expected due to the removal of carbon promoting subsequent formation and growth of AlN.

The presence of the nanocrystalline AlN phase in the glassy matrix is in agreement with the XRD results of the powders heat treated at 800°C and 1000°C which show the broad peaks corresponding to a single phase of AlN. The diffraction pattern obtained on the precursor heat-treated at 1300°C for 15 h in $NH_3$ reveals spots corresponding to the crystalline AlN phase and the dark field image shows that the AlN particle is about 0.5 µm in size (see Figs. 9(e) and 9(f)). The homogeneous distribution of AlN particles observed in the samples labeled as '$NH_3$-800' and '$NH_3$-1000' indicates that a nanocrystalline AlN phase has been internally nucleated due to the presence of Al-N bonds in the as-prepared precursor created by the reaction of Al-butoxide with $N_2H_4$ resulting in the formation of adducts and solid precursors containing alkoxy-hydrazide species. If crystallization resulted primarily due to the reaction between the butoxide and $NH_3$, the AlN phase should have nucleated from the surface of the particles. This result therefore implies that $N_2H_4$

does help in the formation of Al-N bonds in the as-prepared precursor due to its reaction with the butoxide to form both adducts and partially replaced butoxide to yield butoxy-hydrazide species.

The alkoxy-hydrazide groups and adducts subsequently undergo dissociation and reconstruction thereby promoting further replacement of butoxy groups with Al-N type linkages during subsequent post heat-treatments. As shown in the XRD traces, the precursor heat-treated in $NH_3$ at 1300°C reveals the formation of single phase of crystalline AlN (see Fig. 4(d)) while the powders heat-treated in Ar and $N_2$ at the same temperature show the presence of oxide, oxynitride and nitride (see Figs. 2(d) and 3(d)). This result suggests that $NH_3$ helps to remove carbon thereby accelerating the kinetics of replacement of the unreacted butoxy groups with bonded nitrogen which otherwise would form oxide phases during heat-treatment in inert gases such as Ar and $N_2$. Therefore, it can be concluded that the formation of single phase of AlN is due to a combination of generation of adducts $[Al(OC_4H_9)_3{}^s{\cdot}xN_2H_4]$ and the formation of partially replaced butoxy-hydrazide species $[Al(OC_4H_9)_{3-x}(NHNH_2)_x]$ in the as-prepared precursor via reaction of the alkoxide with hydrazine. Reconstructive association of nitrogen together with the beneficial role played by $NH_3$ in the removal of the residual carbon, facilitates the replacement of the residual butoxy groups with Al-N linkages. Thus, the present sol-gel based approach demonstrates an excellent application of metal-alkoxides to generate non-oxide nitrides of reactive elements such as Al in addition to sulfides. Moreover, as shown by the TEM dark field image in Fig. 9(b), the present approach allows the unique possibility of synthesizing nanocrystalline AlN trapped in an oxynitride glass matrix. The generation of such nanocrystalline non-oxide glass-ceramic composites could have some unique electronic, catalytic, and electrochemical applications

B. Synthesis of Titanium Sulfide

*1. Synthesis of Precursors*

The reaction of Ti-alkoxide with hydrogen sulfide generated both solid and liquid products. Since the sulfidization reaction in solution has hardly been studied, an attempt was made to investigate the reaction of the alkoxide with hydrogen sulfide in greater detail. Accordingly, both the liquid and solid products of the reaction were analyzed using various analytical techniques. Chemical analysis of the as-precipitated powder from this reaction (dried at 40°C) showed a S/Ti molar ratio of about 1.4 and O/Ti molar ratio of 1.2, see Table 3. The IR spectrum collected on the precipitate from 2000 to 900 $cm^{-1}$, shown in Figure 10, and 500 to 200 $cm^{-1}$ (see the inset in Figure 10) indicate the existence of isopropoxy groups. The doublet seen at 1377 $cm^{-1}$ and 1360 $cm^{-1}$ is representative of the characteristic *gem*-dimethyl structure of the isopropoxy group[21,22]. In addition, absorptions characteristic of isopropoxy groups bonded to Ti are observed at 1160, 1127 and 1013 $cm^{-1}$ [22], while the Ti-S vibrations can be seen centered around 300 $cm^{-1}$ [23]. The IR and chemical analyses results indicate that the reaction in solution leads to the incorporation of sulfur by a possible replacement of the alkoxy groups attached to the titanium center similar to the reaction of Al-butoxide with hydrazine described above.

In order to understand the mechanism of the replacement reaction responsible for the incorporation of sulfur, the liquid was distilled and analyzed using GC, which showed that isopropanol was liberated during the reaction. This, in combination with the IR results therefore suggests a replacement of the isopropoxy (-$OC_3H_7$) groups by thiol (-SH) groups from $H_2S$, resulting in the liberation of isopropanol from the alkoxide. However, the replacement is not complete as indicated by the presence of isopropoxy groups in the solid formed due to the reaction. The attack of the alkoxy groups by the thiol species would form the basis of a thiolysis reaction very similar to the hydrolysis reactions seen in the sol-gel process.

Table 3    Chemical Analysis of Precursors obtained from the thio-sol-gel reaction

| Material | Ti | S | C | H | O* | S/Ti† |
|---|---|---|---|---|---|---|
| Precursor | 33.6 | 32.3 | 17.3 | 3.5 | 13.3 | 1.4 |

* oxygen obtained from balance
† molar ratio of sulfur to titanium in the precursor.

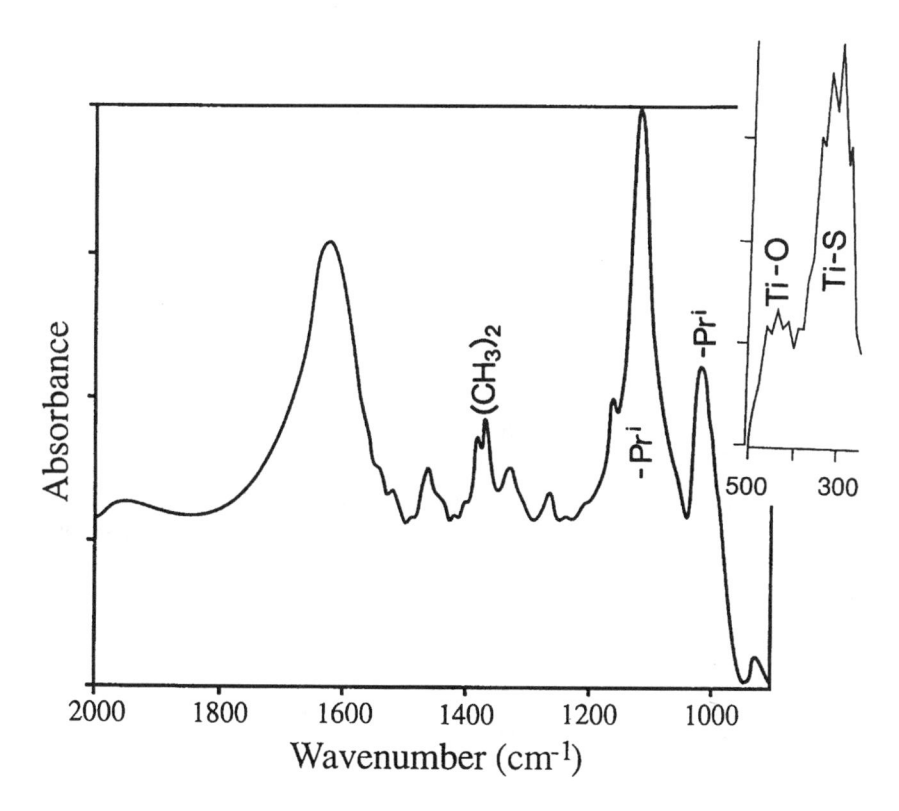

Fig. 10 Infrared spectrum collected on the precipitate obtained from the reaction of titanium isopropoxide with $H_2S$ showing the characteristic absorptions of Ti-OPr$^i$ and Ti-S bonds (inset).

Innovative Processing/Synthesis: Ceramics, Glasses, Composites II

$$H_2S + Ti(OR)_4 \longrightarrow H_2S \rightarrow Ti(OR)_4 \longrightarrow (HS)(OR)_3Ti \leftarrow OH \longrightarrow Ti(OR)_3SH + ROH$$
$$\phantom{XXXXXXXXXXXXXXXXXXXXXXXXXXXXXXXXX}|$$
$$\phantom{XXXXXXXXXXXXXXXXXXXXXXXXXXXXXXXXX}R$$

The overall sulfidization reaction can therefore be written as:

$$Ti(OR)_4 + nH_2S \longrightarrow Ti(OR)_{4-n}(SH)_n + nROH \qquad (1)$$
$$n \leq 4$$

From the above, it can be seen that the thiolysis reaction of titanium isopropoxide leads to the formation of alkoxy-thiol species. The formation of the black precipitate is indicative of further condensation-polymerization of the alkoxy-thiol species by the liberation of $H_2S$ represented below:

$$2p\{Ti(OR)_{4-n}(SH)_n\} \longrightarrow \{(RO)_{4-n}(SH)_{n-1}Ti\text{-}S\text{-}Ti(SH)_{n-1}(OR)_{4-n}\}_p + p\,H_2S \qquad (2)$$
$$n \leq 4, p \text{ is variable and } R = C_3H_7.$$

The condensation reaction (2) has been verified by estimating the isopropanol liberated using quantitative gas chromatography and correlating it with calculated amounts derived from the chemical analyses of the solid precipitate and the liquid product[24,25]. The formation of an azeotrope in the benzene-isopropanol system greatly simplified the quantitative analysis of the liquid products of the reaction. The results indicated the extent to which the thiolysis reaction (1) had occurred, and also indicated that all the thiol groups condensed to form Ti-S-Ti linkages in the solid according to (2) (i.e. partial thiolysis and complete condensation of the thiol groups).

Chemical analysis conducted on the liquid product showed a S/Ti molar ratio of about 0.07. In addition, the mass fragmentation pattern of the liquid was identical to that of titanium isopropoxide, implying that the dark liquid was in fact the alkoxide, which had not undergone significant reaction with $H_2S$. Some of the alkoxide molecules remained unreacted, possibly due to (a) association in inert solvents like benzene (titanium isopropoxide is known to have an average molecular complexity of 1.4 in benzene[26]), or (b) due to the formation of partially reacted soluble oligomers which do not undergo any further reaction in benzene because of steric hindrance to nucleophilic attack by $H_2S$. Experiments conducted in acetonitrile, (a coordinating solvent which is known offer a better medium for dissociation of $H_2S$), have shown more than five-fold increment in product yield, providing support to the hypothesis.

The above results suggest that competition between thiolysis and condensation reactions plays an important role in controlling the amount of sulfur incorporated into the precursor. Accordingly, significant variation in the structure of the precursors could be envisaged by affecting changes in the thiolysis and condensation reactions. This could be achieved in a number of ways. For example, by changing or modifying the reactants, using different solvent systems, etc. Modification of the alkoxide by using organic acids containing sulfur such as sulfonic acids have been studied. The results of such a modification reaction are very similar to the modification of the alkoxides with organic carboxylic acids such as acetic acid in the sol-gel studies to form oxides. A slower kinetics of the ensuing sulfidization reaction is seen. These results have been published elsewhere[24] . Similarly, the use of more polar aprotic solvents such as acetonitrile have been utilized as the medium for solubilizing the alkoxides and reacting with hydrogen sulfide. Acetonitrile provides a more facile medium for dissociation of hydrogen sulfide since the negatively charged nitrogen bearing nitrile groups shield the $H^+$ ions thereby allowing $SH^-$ species to attack the alkoxy groups. The results of these studies are also published elsewhere and will not be mentioned in this short review.[25] The use of such modifying agents and changes in the solvent have a significant impact on the molecular structure of the sulfidized precursor. These changes could have an impact on the temperature of formation of $TiS_2$, the crystallite size and the evolved morphology of the sulfide.

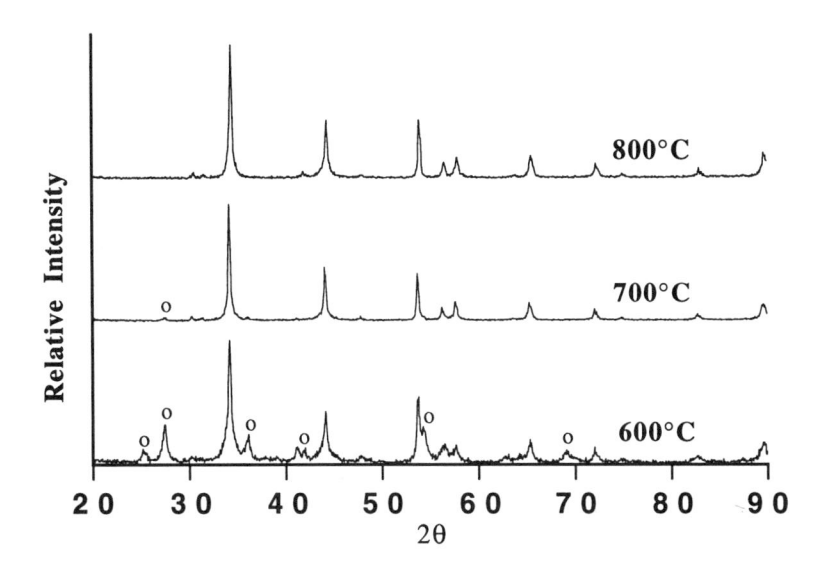

**Fig. 11** X-ray diffraction patterns (using Cu-K$_\alpha$ radiation) collected on samples obtained from the reaction of Ti-isopropoxide and H$_2$S, heat treated for 6 hours in flowing H$_2$S at 600, 700 and 800°C. The "O" marks indicate the position of oxygen peaks. All other peaks are due to hexagonal TiS$_2$.

2. *Conversion of the Precursor to Crystalline Sulfide*

The precipitated precursors obtained from the reaction between the isopropoxide and hydrogen sulfide are amorphous and the x-ray diffraction patterns collected on the heat treated powders obtained from the reaction is shown in Figures 11. The figure clearly shows that crystalline TiS$_2$ (hexagonal phase) is seen to form after heat treatment at 600°C for six hours in an H$_2$S environment. However, crystalline TiO$_2$ phases (both rutile and anatase ) are also formed at this temperature, an evidence of the presence of unreplaced alkoxy groups that condense to form Ti-O-Ti bonds:

$$\text{Ti-OR} + \text{RO-Ti} \longrightarrow \text{Ti-O-Ti} + \text{hydrocarbons} \qquad (3)$$

When heat treated at 700°C, the XRD pattern obtained shows an oxide peak at a relatively low intensity. However, we have observed that single phase TiS$_2$ could be synthesized when the precursor is subjected to a ten hour heat treatment at 700°C. Upon heat treatment at 800°C in flowing H$_2$S, the precursor clearly transform to TiS$_2$ in 6 h. In addition, control experiments conducted for comparison, by heat treating sol-gel derived TiO$_2$ in flowing H$_2$S at 800°C under identical conditions showed TiO$_2$ as the major phase in the XRD pattern. This implies the enhanced kinetics of the high temperature reaction due to incorporation of S at the precursor stage. The kinetics of the ensuing sulfidization reaction during the subsequent gas phase thiolysis reaction has a strong influence on the defect chemistry, composition and morphology of titanium sulfide. This control of the composition and morphology has a profound implication on the electrochemical properties of the sulfide. These results have already been published elsewhere and will not be discussed here since the objective of this review is to demonstrate the applicability of metal alkoxides for synthesizing non-oxide sulfide and nitrides[8, 24, 25].

This study clearly demonstrates the flexibility of alkoxides to synthesize $TiS_2$. This is possible not only by direct reaction of the alkoxide with hydrogen sulfide but also by modifying the alkoxide as explained above, and also by using various sulfidizing agents to generate the intermediate precursors[25]. All these different reactions result in unique morphologies and microstructures of the sulfide which could exhibit very different surface characteristics affecting the properties of the material, for example the kinetics of the intercalation reaction in the case of $TiS_2$ and consequently the electrochemical performance of $TiS_2$ as a cathode material. Electrochemical studies of thio-sol-gel synthesized $TiS_2$ have been conducted, the results of which have been published elsewhere[27]. The results do indicate a significant influence of the morphology and microstructure on the electrochemical utilization of $TiS_2$ cathodes.

CONCLUSIONS

The application of metal alkoxides for synthesizing non-oxide nitrides and sulfides of reactive metals such as the early transition metals and group III elements respectively has been shown. It has been demonstrated very clearly in the two cases that the reactions proceed along the lines similar to the solution sol-gel process well known for synthesizing oxide glasses and ceramics. The reaction of hydrazine with Al tri-*sec*–butoxide in acetonitrile causes the precipitation of a solid precursor. Pyrolysis of this precursor to 800°C in UHP-Ar and UHP-$N_2$ resulted in amorphous powders, while the $NH_3$-treated powders showed the formation of nanocrystalline ($\approx$ 10 nm) AlN at the same temperature. Chemical analyses revealed that the amorphous powder obtained after pyrolysis in $NH_3$ had lower C and N contents in comparison with that pyrolyzed in Ar and $N_2$. At the same time, IR spectroscopy indicated that some of the nitrogen was bonded to carbon in the residue obtained after pyrolysis in Ar and $N_2$. The presence of Al-N bonds was detected in the powders pyrolyzed in the three atmospheres using IR spectroscopy. Heat-treatment of this precursor at 1300°C in $NH_3$ for 15 h resulted in single phase of hexagonal AlN, while the precursors heat-treated in Ar and $N_2$ exhibited a mixture of AlN, $\alpha$-$Al_2O_3$ and $\gamma$-AlON. The formation of AlN in inert atmospheres is due to the formation of Al-N linkages during heat-treatment, which originate from the adducts and partially nitrided butoxy-hydrazide species formed by virtue of the reaction between Al-butoxide and hydrazine in solution.

In the case of sulfides, the reaction of Ti-isopropoxide with hydrogen sulfide results in the formation of an ispropoxy-sulfide precursor via a thiolysis and condensation analogous to the hydrolysis and condensation reaction seen in the oxide sol-gel process. The partially sulfided alkoxy-sulfide upon subsequent heat treatment in hydrogen sulfide leads to the formation of the crystalline $TiS_2$. Thus the mechanisms are identical to that of the oxide solution sol-gel process wherein the gel generated by the hydrolysis and condensation reaction is transformed to the oxide ceramic via heat treatment in air. The sol-gel based approach described herein, therefore, offers excellent potential for the synthesis of nanocrystalline nitrides and oxynitrides of oxophylic metals such as Al. At the same time, the approach also demonstrates the unique ability to synthesize glass-ceramics in the Al-O-N system containing nanocrystalline nitrides. Similarly, the approach offers a unique and novel methodology to form sulfides of reactive transition metals. The technique is currently being explored for synthesizing other potentially useful nitrides and sulfides.

ACKNOWLEDGMENTS

PNK, JYK and MAS would like to acknowledge the donors of the Petroleum Research Fund, administered by the American Chemical Society for support of this research (Grant # PRF 25507-G3), the National Science Foundation (Grant DMR 9301014, Ceramics Program), a Research Initiation Award (RIA) from the National Science Foundation (Grants CTS–9309073 and CTS-9700343).

REFERENCES

1. C.J. Brinker and G.W. Scherer, *Sol-Gel Science* (Academic Press, Boston, MA) 1990.
2. T.H. Elmer and M.E. Nordberg, *J. Am. Ceram. Soc.* 50 (1967) 275.
3. C.J. Brinker and D.M. Haaland, *J. Am. Ceram. Soc.* 66 (1983) 758.
4. P.N. Kumta and S.H. Risbud, *J. Mater. Res.* 8 (1993) 1394.
5. P.N. Kumta and S.H. Risbud, *J. Mat. Sc,* 29 (1994) 1135.
6. M.S. Bains, D.C. Bradley, *Canadian J. Chem.* 40 (1962) 1350.
7. B.E. Yoldas, *J. Mater. Sci.* 10 (1975) 1856.
8. M.A. Sriram, 'Thio-Sol-Gel Synthesis, Structural and Electrochemical Characterization of TiS$_2$', Ph.D. Thesis (1996).
9. J.Y. Kim, P.N. Kumta, B.L. Phillips, and S.H. Risbud, submitted to *J. Phys. Chem.* (1998).
10. H. Endress, in *Comprehensive Coordination Chemistry - Vol.2 - Ligands*, eds. G. Wilkinson, R.D. Gillard and J.A. McCleverty (Pergamon Press, Oxford,UK 1987) p. 261
11. .X.-H. Han and B.J. Feldman, *Solid State Commun.*, 65 (1988) 921.
12. H.-C. Tsai and D.B. Bogy, *J. Vac. Sci. Technol.* A5 (1987) 3287.
13. L. Maya, R.C. Cole and E.W. Hagarman, *J. Am. Ceram. Soc.*, 74 (1991) 1686.
14. M. Hoch, T. Vernardakis and K.M. Nair, *Sci. Ceram.*, 10 (1980) 227.
15. J. Sappei, D. Goeuriot, F. Thevenot, Y. Laurent, J. Guyader and P. L'Haridon, *Euro. Ceram. Soc.*, 8 (1991) 257.
16. L. Sacconi and A. Sabatini, *J. Inorg. Nucl. Chem.* 25 (1963) 1389.
17. A. Earnshaw, L.F. Larkworthy, and K.S. Patel, *Z. Anorg. Allg. Chemie.* 334 (1964) 163.
18. W.G. Paterson and M. Onyszchuk, *Canad. J. Chem.* 41 (1963) 1872.
19. D.R. McKenzie, R.C. McPhedran, and N. Savvides, *Thin Solid Films* 108 (1983) 247.
20. S. Colque and P. Grange, *J. Mater. Sci. Lett.* 13 (1994) 621.
21. J.V. Bell, J. Heisler, H. Tannenbaum and J. Goldenson *Anal. Chem.* 25[11], 1720 (1953).
22. C.T. Lynch, K.S. Mazdiyasni, J.S. Smith and W.J. Crawford *Anal. Chem.* 36[12], 2332 (1964).
23. R. R. Chianelli and M. B. Dines *Inorg. Chem.* 17[10], 2758 (1978).
24. M.A. Sriram and P.N. Kumta *J.Am. Ceram. Soc.* 77[5], 1381 (1994).
25. M.A. Sriram and P.N. Kumta *Mat. Sci. & Eng.* B33, 140 (1995).
26. D.C. Bradley, R.C. Mehrotra and W. Wardlaw *J. Chem. Soc.* 5020 (1952).
27. M.A. Sriram and P.N. Kumta, 'Electrochemical Characterization of Thio-Sol-Gel Derived Titanium Sulfide Powders with Unique Morphologies', in *Role of Ceramics in Advanced Electrochemical Systems*, Eds. P.N. Kumta, G.S. Rohrer and U. Balachandran, Ceramic Transaction, 65, Proceedings of the 97th annual meeting of the American Ceramic Society, Cincinnati, OH, Apr.30 to May 3, 1995, 163 (1996).

# COMPOSITE ALUMINA SOL-GEL CERAMICS

Tom Troczynski and Quanzu Yang
*Department of Metals and Materials Engineering*
*University of British Columbia*
*Vancouver, BC, Canada*

## ABSTRACT

A sol-gel based processing route has been developed to fabricate alumina Composite Sol-Gel (CSG) ceramic by dispersing calcined alumina powder in alumina sol. The research results indicate that the alumina sol acts as a dispersant and sintering accelerator for the calcined alumina. At the sintering stage, the sol-gel derived phase bonds the alumina particles and determines the microstructure and properties of CSG. The sintering kinetics and final microstructure of CSG is strongly affected by MgO content in the sol-gel derived matrix phase.

## INTRODUCTION

Sol-gel matrix composites reinforced with ceramic particles or whiskers receive much attention as high performance materials, for a wide range of engineering applications [1,2,3]. The potential advantages of sol-gel processing for ceramic composites are fine scale mixing and low densification temperature, leading ultimately to improved properties. For example, composites of $\alpha$-$Al_2O_3$ seeded sol-gel-derived alumina-zirconia were fabricated with submicrometer alumina grains and small intergranular zirconia particles of average grain size of 0.4 μm [4]. The sol-gel composite processing typically involves organometallic compounds as an oxide precursor, such as tetraethylorthosilicate [6], aluminum isopropoxide [7], or zirconium isopropoxide [8]. The different components of sol may be tailored so they do not react with each other to form new components. Variety of solid phases, such as fine powders or fibers, can be dispersed into a sol before gelation, leading to a composite with good homogeneity and intimate contact between the components. The composite slurry is typically dried at 25-250°C, and sintered at temperatures several hundred °C lower than the counterpart calcined ceramics. The main objective of the present research was to investigate the microstructure of ceramic-composites, wherein both sol-gel matrix and the dispersed phase is alumina. A minor amount of sintering additive (MgO) was used to effect densification kinetics of the composite. The goal is to study the effects of the process variables on porosity, microhardness and microstructure of the resulting alumina ceramics.

## EXPERIMENTAL

Calcined alumina A-16 (Alcoa Industrial Chemicals, Pittsburgh, PA, USA) was dispersed in alumina sol. A water-soluble precursor of magnesia was added to the alumina sol to control sintering of the composite sol-gel ceramics. The concentration of magnesia in the alumina sol was varied from 0 to 2 mole%. The particle size distributions in the sol was monitored (Horiba CAPA 700, Irvine, CA, USA ). The microstructures of the composite sol-gel ceramics were examined using SEM, and the microhardness using 5 and 10 N loads applied for 10 seconds. The open porosity of CSG was measured by impregnating the samples in deionized water at 25°C and $10^{-3}$ Torr vacuum for 6 hours.

## RESULTS AND DISCUSSION

### Alumina Sol-Gel Matrix Phase

For the hydrolysis mole ratio $[OH^-]/[Al^{3+}]$ between 0.5 and 2.5, alumina sol particles (clusters) have approximate composition $AlO_4Al_{12}(OH)_{24}(H_2O)_{12}^{+7}$ [11]. The clusters have central tetrahedrally coordinated aluminum $[AlO_4]$, surrounded by twelve edge-linked octahedrally coordinated aluminum $[AlO_6]$ [11]. The sets of three bonded octahedra form $Al_3O_{13}$ units that are interlined by double OH bridges [12]. The average size of such sol clusters is approximately 1-2 nm. The cluster structure and interactions are stable for pH <5.5, demonstrated through constant viscosity of the sol in this range, Figure 1.

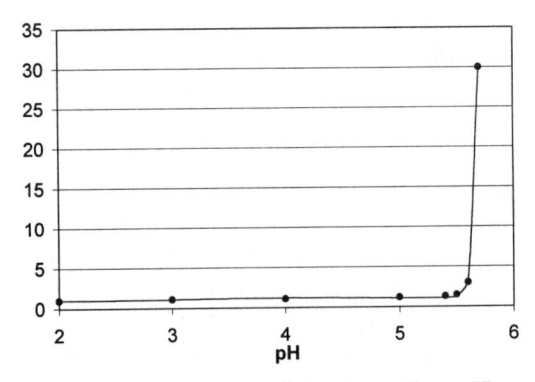

**Figure 1.** Viscosity of the alumina sol vs pH

When the pH exceeds 5.6, the viscosity rapidly increases from 1.2 cps to more than 30 cps, Figure 1. This is because a major rearrangement of the sol cluster structure takes place for the hydrolysis mole ratio at 2.5<[OH$^-$]/[Al$^{3+}$]<3.0, i.e. when gelation begins. It is believed that AlO$_4$Al$_{12}$(OH)$_{24}$(H$_2$O)$_{12}^{+7}$ is transformed during gelation into pseudobehmite (AlOOH) because of the loss of the tetrahedrally coodinated aluminum [13]. The formation of gels is by the stepwise linkage of AlOOH dimers into double chains to form 3D-network [11]. This causes an increase in the polycondensation reaction rate, and gelation time becomes very short (1-5min). The following phase transformations take place upon subsequent heat treatment of the resulting hydroxides [1]:

$$\gamma\text{AlOOH} \xrightarrow{450°C-550°C} \gamma\text{Al}_2\text{O}_3 \xrightarrow{850°C}$$

$$\delta\text{Al}_2\text{O}_3 \xrightarrow{1050°C} \theta\text{Al}_2\text{O}_3 \xrightarrow{1200°C} \alpha\text{Al}_2\text{O}_3$$

**Sol-Gel Composite**

The particle size distribution of the calcined alumina, dispersed in alumina sol, is shown in Figure 2. The powder has a median particle diameter of 0.31 μm and a specific surface area of 5.478 m$^2$/g.

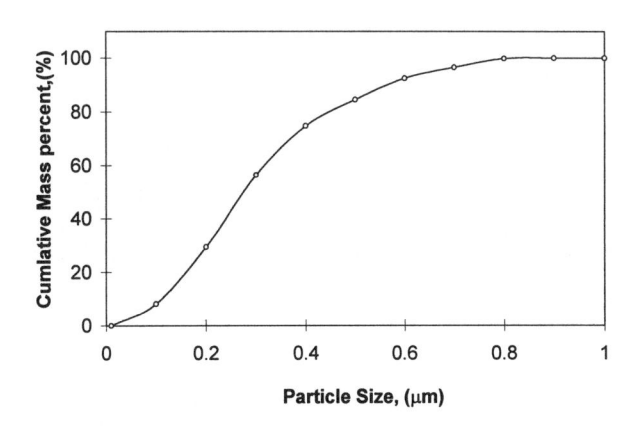

**Figure 2.** Particle size distribution of the calcined alumina dispersed in alumina sol

The calcined alumina / alumina sol slurry is stabilized because the positively charged (+7) small size (~1 nm) colloidal clusters interact with the large Al$_2$O$_3$ particles (~300 nm) and increase the electrostatic repulsion force. During the gelation process, the hydrolyzed surface of alumina particles (=AlOH) enters a polycondensation reaction with the alumina sol clusters:

$$=AlOH + AlOOH \rightarrow =Al-O-Al= + H_2O$$

The anticipated structure resulting from this reaction is illustrated in Fig. 3. It is expected that both hydrogen bonds and ionic/covalent bonds are formed at the interface of alumina particles/sol-gel matrix.

During the drying stage, removal of the solvent condenses alumina gel towards a continuous matrix providing mechanical strength to a presintered body. When the composite green bodies were dried at 300 °C, almost all excess solvents escaped through interconnected porosity, accompanied by approximately 60% weight loss and 25% volume reduction. Figure 4 illustrates the typical weight loss of the composite sol-gel with 86.2 vol% calcined alumina and of pure sol-gel phase as a function of temperature.

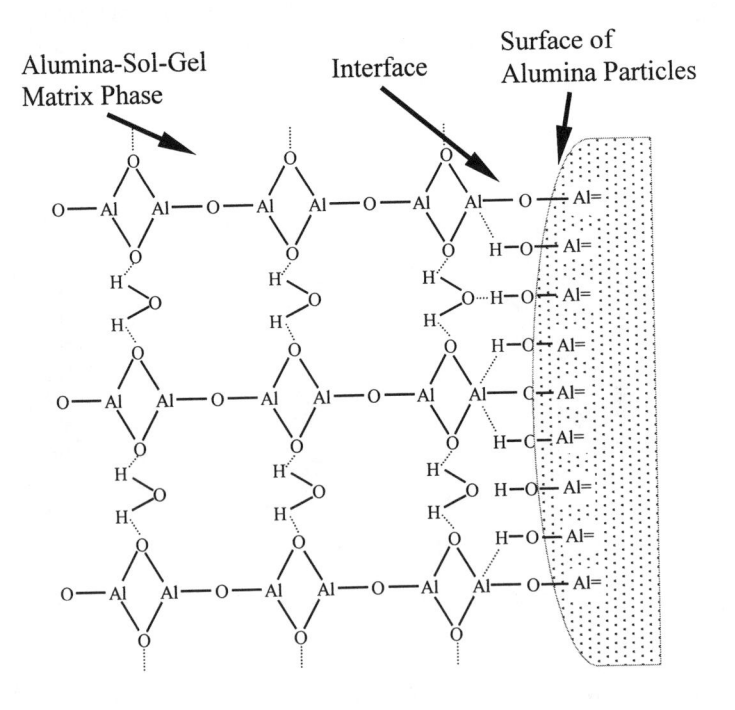

**Figure 3.** Schematic of the anticipated interaction of the alumina sol with hydrolyzed alumina particles

Heat treatment between 500 and 550°C causes rapid de-hydration of gel as AlOOH transforms to γ-Al$_2$O$_3$. This results in a 15.3 wt% loss for pure alumina sol-gel phase and 2.2 wt% loss for the composite sol-gel with 86.2 vol% calcined alumina, region II in Figure 4.

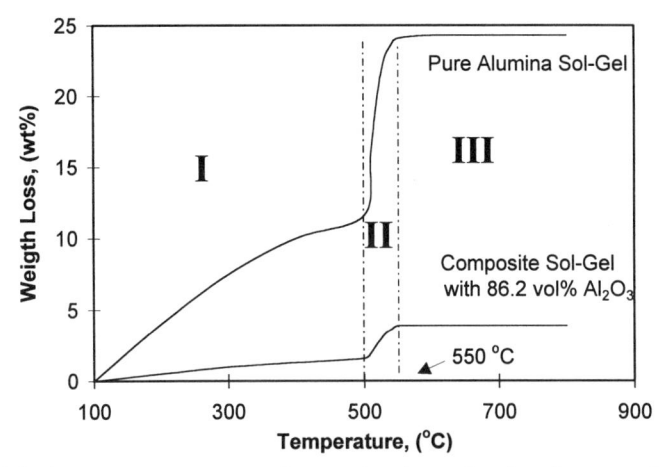

Figure 4. Weight loss vs temperature of pure alumina sol-gel and composite sol-gel with 86.2 vol% calcined alumina after drying at 100 °C for 20 hours

**Properties of CSG**

Figure 5 shows the microhardness of the alumina-alumina CSG without additives, as a function of the sintering temperature. The microhardness of pure sol-gel derived alumina (0 vol% calcined phase), although relatively higher at low sintering temperatures, is eventually lower than that of the composite sol-gel after sintering at 1400 °C for 3 hours. This is because the pure sol-gel derived alumina microcracks during sintering due to large densification strain. The final microhardness increases with adding calcined alumina as secondary phase, to reach a maximum of 17.5 GPa at about 86 vol%, and decreases thereafter. The microhardness of the control sample of pure calcined alumina (100 vol% calcined alumina line in Fig. 5) sintered in parallel with CSG, reached 12.4 GPa at 1400 °C. This suggests a strong effect of the sol-gel derived phase on the sinterability of the composite. CSG hardness is inversely proportional to porosity at different sintering temperatures, as illustrated in Figure 6.

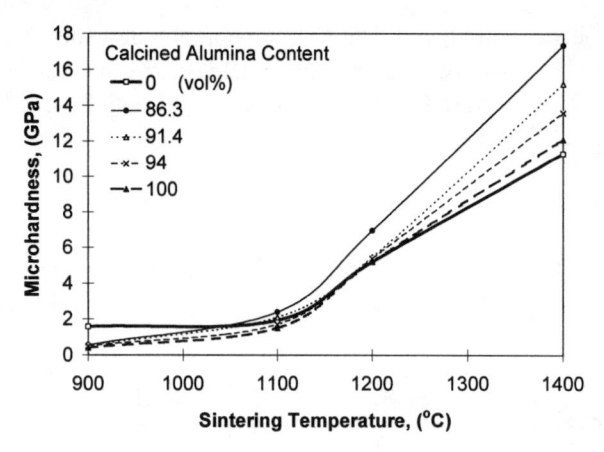

**Figure 5.** Microhardness (HV$_{1Kg}$) vs sintering temperature for alumina-alumina CSG

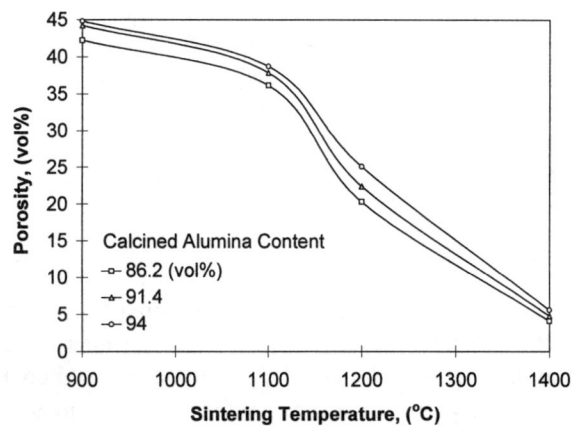

**Figure 6.** Porosity vs sintering temperature for alumina-alumina CSG

**Effect of MgO on Sinterability of Alumina-Alumina CSG**

Figures 7 and 8 show variation of microhardness and porosity of CSG as a function of MgO content in alumina sol-gel derived phase. The microhardness of CSG increases and the porosity of GSG decreases with increasing MgO content in alumina sol-gel phase, as anticipated [14]. Surprisingly however, the microhardness of CSG with 2 mol% MgO in the sol-gel matrix phase appears to be in excess of 20 GPa, the value approaching that of a single crystal alumina. Porosity of these CSG is less than 1 vol%, Fig. 8.

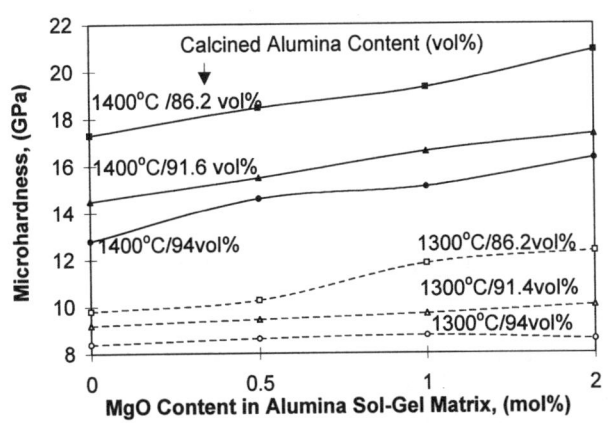

**Figure 7.** Microhardness ($HV_{1Kg}$) vs MgO content in alumina sol-gel matrix phase, for CSG sintered at 1300 °C and 1400°C

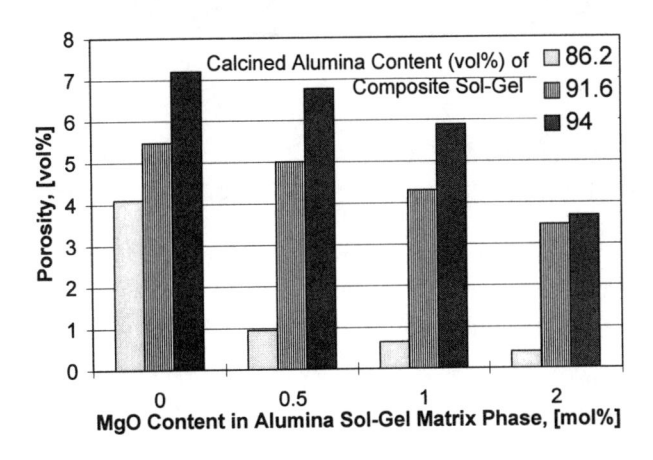

**Figure 8.** Porosity vs MgO content in alumina sol-gel matrix phase, for CSG sintered at 1400°C

## SEM Micrographs of CSG

Figure 9 illustrates microstructure of 86.2 vol% calcined alumina CSG sintered at 1300 °C and 1400°C for 3 hours. Two kinds of pores can be identified in these CSG. The "small" pores (< 0 μm), at the triple junctions of the grains, shrink rapidly at increased sintering temperature and with increased content of the sol-gel derived matrix phase. The "large" pores (~1 μm) are produced through imperfect consolidation of the CSG slurry. Most of the "large" pores are open and remain interconnected until CSG is sintered at 1400 °C. The average grain size at 1300 °C is about 0.6 μm, twice the starting calcined alumina particle size. The average grain size at 1400 °C is about 1.2 μm.

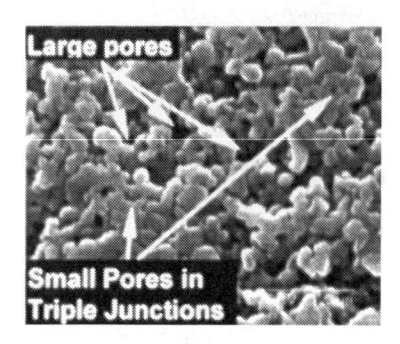

a        b

**Figure 9.** SEM micrograph of CSG with 86.2 vol% calcined alumina sintered at (a) 1300°C and (b) 1400°C for 3 hours. Bar=2 μm

Figure 10 shows the microstructures of of 86.2 vol% calcined alumina CSG with different MgO contents in sol-gel matrix phase, fired at 1400 °C for 3 hours. The average pore size and content decrease with increasing MgO in the sol-gel matrix phase.

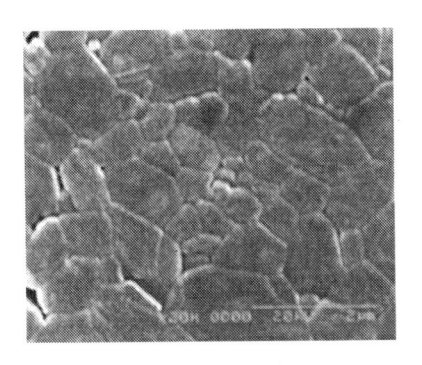

a

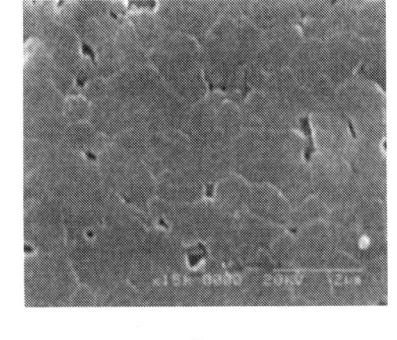

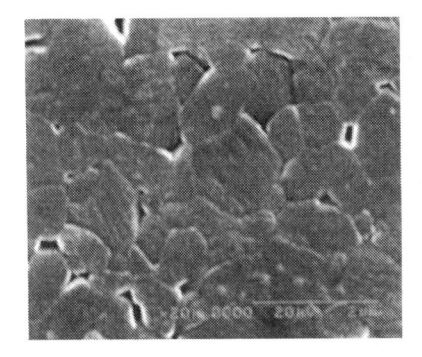

b

c

**Figure 10.** SEM micrograph of CSG with 86.2vol% calcined alumina sintered at 1400°C for 3 hours, with different MgO contents in the sol-gel matrix: (a) 0.5 mol% MgO, (b) 1.0 mol% MgO, and (c) 2.0 mol% MgO. . Bar=2 μm

## CONCLUSIONS

Alumina composite sol-gel (CSG) ceramics has been produced by dispersing calcined alumina powder in alumina sol. It has been found that the alumina sol acts as a dispersant and sintering accelerator for the calcined alumina. The sol-gel matrix phase of CSG bonded to calcined alumina forms a strong 3-D network. The microhardness and porosity of CSG depends on the sintering temperature and sol-gel phase content. When sol-gel matrix phase is 13.8 vol%, the microhardness reaches the maximum value (17.5 GPa) for pure alumina-alumina CSG. The porosity decreases with increasing sol-gel matrix phase. The microhardness of composite sol-gel with MgO increases to more than 20 GPa and porosity decreases to less than 1vol% at a sintering temperature of 1400 °C. The sintering kinetics and final microstructure of CSG is strongly affected by MgO content in the sol-gel derived matrix phase.

# REFERENCES

1. Borrow, T. Edward, M.S. US Patent (5,585,136),1996
2. Martin Sternitzke, J. of the European Ceramic Society 17 (1997) 1061-1082
3. Y. Xu, A. Nakahira, and K. Niihara, J. of the Ceramic Society of Japan . Int. Edition, Vol.102 No. 3 pp312-315, 1994.
4. D.C. Bradley, R.C. Metrotra, and D.P. Gaur, "MetalAlkoxides" Academic Press, London, 1978.
5. B.E. Yoldas, Ceramic Bulletin, Vol 54, No.3 pp289-299 (1975).
6. K. Kamiya, S.Sakka, and Y. Tatemichi, J. Mat. Sci. 15, 1765(1980).
7. P.F. Becher, J. Am. Ceram. Soc., Vol 72 pp255-269(1991).
8. M. Harmer, H.M. Chen, and G.A. Miller, J. Am. Ceram. Soc., Vol.75 pp1715-1728 (1992)
9. V.V. Srdic and L. Radonjic, J. Am. Ceram. Soc. 80[8] 2056-60(1997).
10. S. Jiao, M.L. Jenkins and R.W. Davidge, Acta Mater., Vol.45, No. 1 pp149-156, 1997
11. Susan M. Bradley, Ronald A. Kydd, and Russell F. Howe, Journal of Colloid and Interface Science, 159, 405-412(1993)
12. J.W. Akitt and J.M. Elders, J. Chem. Soc. Faraday Trans. 1, 1987,83,1725-1730
13. G. Fu and L.F. Nazar, Chem. Mater. 1991, 3, 602-610.
14. W.E. Lee and W.M. Rainforth, Ceramic Microstructures, pp266-289, 1994, Chapman&Hall, London

# AQUEOUS SOL-GEL METHOD FOR THE SYNTHESIS OF NANO-SIZED CERAMIC POWDERS

Panchanan Pramanik and Narendra Nath Ghosh
Department Of Chemistry
Indian Institute Of Technology
Kharagpur, 721 302, India

## ABSTRACT
A series of nanosized multicomponent ceramic powders has been synthesized using aqueous sol-gel method. In this method metal formates and precipitated silica were used as precursor compounds instead of metal alkoxides and water was used as reaction medium instead of the commonly used solvent alcohol. The gels were prepared by mixing precipitated silica and aqueous solutions of metal formates. The gels prepared by this method were calcined at different temperatures and characterized by X-ray powder diffraction, Infrared spectroscopy, differential thermal analysis, thermogravimatric analysis and transmission electron microscopy. This aqueous sol-gel method is shown to provide homogeneous and nono-sized ceramic powders at effectively low temperatures and offers the potential of technically simpler and cost effective route than other reported sol-gel methods.

## INTRODUCTION
The design of advanced ceramics depends on the availability of powders with outstanding properties in terms of composition, purity and size distribution. Sol-gel processes allow the synthesis of powders to have a more elaborate structure. Sol-gel refers to a processing in a liquid medium to obtain a solid matter which does not settle under gravity - that is to say which does not precipitate. This solid matter can be composed of single solid network spreading throughout the liquid matrix, which is the definition of gel. The application of sol-gel technology to the production of ceramics and glasses is an area of intense interest[1-6]. The sol-gel

technique has been used by Roy and co-workers for the preparation of highly homogeneous glasses and ceramics using different metal-alkoxides[7,8]. As this procedure allows the mixing of precursors at molecular level, there is a better control over the whole process, facilitating synthesis of "tailor-made" materials. Based on the knowledge of sol-gel conversions, it is possible to prepare fibers, films and composites[9-12]. Multicomponent alkoxides have been used to prepare a wide variety of ultrafine and high-purity powders which are difficult to prepare by conventional ceramic processing. However, the sol-gel transformations involving a mixture of single metal alkoxides lead to microscopic inhomogeneities in the different resultant gel and oxides due to different rate of hydrolysis of various alkoxides. To overcome this limitation, several approaches have been attempted including matching of hydrolysis rates by chemical modifications with chelating ligands, or synthesis of multication alkoxides or partial prehydrolysis of an alkoxide[13-16].

As all the above approaches are complex in nature, the authors have initiated to simplify the all-alkoxide sol-gel method[17-23]. The objective of the present work is the development of an efficient and cost-effective sol-gel route for the preparation of nano-sized multicomponent ceramic powders. The sol-gel method has been modified by using metal formates as precursor compounds instead of metal alkoxides, and water as reaction medium instead of alcohol. In this paper synthesis of a series of multicomponent ceramic powders (i) $3Al_2O_3$-$2SiO_2$ (ii) $3Al_2O_3$-$2SiO_2$.$ZrO_2$ (iii) $Li_2O.Al_2O_3.2SiO_2$ (iv) $Li_2O.Al_2O_3.2SiO_2.ZrO_2$ (v) $Li_2O.Al_2O_3.4SiO_2$. (vi) $Li_2O.Al_2O_3.4SiO_2.ZrO_2$ (vii) $2MgO.2Al_2O_3.5SiO_2$ (viii) $2MgO.2Al_2O_3.5SiO_2.ZrO_2$ (ix) $ZrO_2.SiO_2$ using this aqueous sol-gel method has been described.

EXPERIMENTAL PROCEDURES

The starting materials were aluminium nonahydrate (>98.5%) (SD Fine Chemicals Ltd., India), lithium carbonate (>99%) (BDH Chemicals Ltd., India) zirconium oxychloride octahydrate (>99% Aldrich Chemicals), magnesium nitrate (>99%) (BDH Chemicals Ltd., India), and sodium silicate (>99%) (BDH Chemicals Ltd., India). Freshly precipitated aluminium hydroxide, magnesium hydroxide and zirconium hydroxide were prepared by adding ammonium hydroxide to the aqueous solutions of aluminium nitrate, magnesium nitrate and zirconium oxychloride respectively. The solutions were filtered and the precipitates were washed with distilled water several times. These hydroxides were then reacted with aqueous formic acid solution (50%) to give the corresponding metal formate solutions. Precipitated silica was prepared by adding aqueous solution of ammonium nitrate to the aqueous solution of sodium silicate. It was dried over a water bath for 5h. The precipitate was washed with dilute solution of nitric acid followed by distilled water for several times. The concentration of alkali metal ions in this precipitated silica was dropped to <100 ppm. The precipitated silica was mixed with aqueous solutions of metal formates in the requisite molar ratio. The mixture was stirred using a magnetic stirrer until the formation of gels. The specific conditions and experimental details are listed in Table I.

Table I. Experimental details of the preparation of gels.

| Target composition | Sample | SiO$_2$ | Zrf | Alf | Lif | Mgf | t$_{gel}$ |
|---|---|---|---|---|---|---|---|
| 3Al$_2$O$_3$-2SiO$_2$ | MZ0 | 1 | - | 3 | - | - | 10 |
| 3Al$_2$O$_3$-2SiO$_2$ with 25 mol% ZrO$_2$ | MZ25 | 1 | 0.25 | 3 | - | - | 9 |
| 3Al$_2$O$_3$-2SiO$_2$ with 35 mol% ZrO$_2$ | MZ35 | 1 | 0.35 | 3 | - | - | 9 |
| 3Al$_2$O$_3$-2SiO$_2$ with 50 mol% ZrO$_2$ | MZ50 | 1 | 0.50 | 3 | - | - | 8 |
| Li$_2$O-Al$_2$O$_3$-2SiO$_2$ | EZ0 | 1 | - | 1 | 1 | - | 4.5 |
| Li$_2$O-Al$_2$O$_3$-2SiO$_2$ with 5 mol% ZrO$_2$ | EZ5 | 1 | 0.05 | 1 | 1 | - | 6 |
| Li$_2$O-Al$_2$O$_3$-2SiO$_2$ with 10 mol% ZrO$_2$ | EZ10 | 1 | 0.10 | 1 | 1 | - | 6 |
| Li$_2$O-Al$_2$O$_3$-2SiO$_2$ with 15 mol% ZrO$_2$ | EZ15 | 1 | 0.15 | 1 | 1 | - | 6 |
| Li$_2$O-Al$_2$O$_3$-4SiO$_2$ | SZ0 | 2 | - | 1 | 1 | - | 4 |
| Li$_2$O-Al$_2$O$_3$-4SiO$_2$ with 5 mol% ZrO$_2$ | SZ5 | 2 | 0.05 | 1 | 1 | - | 3 |
| Li$_2$O-Al$_2$O$_3$-4SiO$_2$ with 10 mol% ZrO$_2$ | SZ10 | 2 | 0.10 | 1 | 1 | - | 4 |
| Li$_2$O-Al$_2$O$_3$-4SiO$_2$ with 15 mol% ZrO$_2$ | SZ15 | 2 | 0.15 | 1 | 1 | - | 4 |
| 3MgO-2Al$_2$O$_3$-5SiO$_2$ | CZ0 | 5 | - | 4 | - | 3 | 2 |
| 3MgO-2Al$_2$O$_3$-5SiO$_2$ with 5 mol% ZrO$_2$ | CZ5 | 5 | 0.05 | 4 | - | 3 | 3 |
| 3MgO-2Al$_2$O$_3$-5SiO$_2$ with 15 mol% ZrO$_2$ | CZ15 | 5 | 0.15 | 4 | - | 3 | 5 |
| 10ZrO$_2$-90SiO$_2$ | 10ZS | 9 | 1 | - | - | - | 7 |
| 30ZrO$_2$-70SiO$_2$ | 30ZS | 3 | 7 | - | - | - | 8 |
| 50ZrO$_2$-50SiO$_2$ | 50ZS | 1 | 1 | - | - | - | 8 |

Zrf: Zirconium formate; Alf: Aluminium formate; Lif: Lithium formate; Mgf: Magnesium fomate.
t$_{gel}$: Time of formation of gels in hours.

The gels were dried over a water bath and ground to powders. These powders were then calcined at different temperatures.

The crystalline phases of the calcined powders were identified by X-ray diffraction (XRD) using a Phillips X-ray diffractometer PW 1840 and using CuK$_\alpha$ radiation. Infrared spectra were recorded using a Perkin Elmer 883 spectrophotometer. The IR samples were prepared using the KBr pellet method. Thermogravimetric analysis (TGA) and differential thermal analysis (DTA) were carried out at heating rate of 10K min$^{-1}$ in air using a Shimadzu thermal analyzer DT-40. Electron microscopic examination of powders were carried out by transmission electron microscope (TEM) using a Phillips CM-12.

RESULTS AND DISCUSSIONS

X-ray powder diffraction patterns were taken for all the samples calcined at different temperatures and an example shown is shown in Fig. 1.

From the XRD patterns it was observed that:
(i) All the dried gel powders were amorphous in nature.
(ii) XRD peaks corresponding to mullite were observed when MZO was calcined for 1h at 1200°C. In MZ25, MZ35, and MZ50 tetragonal zirconia formed along with mullite on calcination at 1200°C or higher temperatures.
(iii) For sample EZ0, $\alpha$ and $\beta$-eucryptite phases were formed on calcination for 2h at 600°C. Tetragonal zirconia appeared along with $\alpha$ and $\beta$-eucryptite when EZ5, EZ10, and EZ15 were calcined for 2h at 700 and 800°C.
(iv) XRD peaks corresponding to $\beta$-spodumene phase were observed when sample SZ0 was calcined for 2h at 700°C. When the calcination temperatures were 800 and 900°C SZ5, SZ10 and SZ15 showed XRD peaks corresponding to $\beta$-spodumene and tetragonal zirconia phase.
(v) The sample CZ0 produced $\alpha$-cordierite phase when calcined at 1200°C. Samples CZ5 and CZ15 showed the characteristic XRD peaks of zircon and tetragonal zirconia together with $\alpha$-cordierite phase on calcination at 1300°C.
(vi) In the samples 30ZS and 50ZS the tetragonal zirconia phase was present after calcination at 700 and 1000°C for 6h. When the calcination temperature was 1200°C, only tetragonal zirconia was formed with 10ZS but with 30ZS and 50ZS monoclinic $ZrO_2$ was formed along with tetragonal zirconia phase.

The observed IR frequencies of all the materials are in good agreement with the values reported in literature. The most striking features of the IR spectra on increase of the calcination temperatures were as follows:
(i) The principal absorption band of the formate group at 1380 cm$^{-1}$ was observed in the IR spectra of all gel powders dried at 100°C. The intensity of this band diminished to zero when the gel powders were calcined at 500°C or higher temperatures.

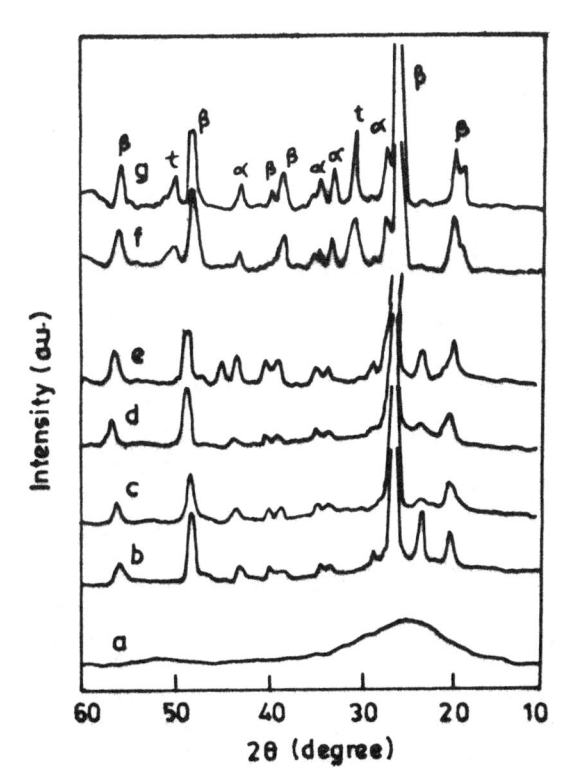

**Fig. 1.** XRD patterns of the samples after calcination (a) EZ15 at 500°C, (b) EZ5 at 600°C, (c) EZ15 at 600°C, (d) EZ10 at 600°C, (e) EZ5 at 700°C, (f) EZ10 at 800°C and (g) EZ15 at 800°C.
α: α-eucryptite; β: β- eucryptite; t: tetragonal zirconia.

(ii) The IR spectra of MZ0 calcined at 1200°C showed the characteristic bands of mullite[24] at ~1075, 1125, 814, 750, 560 and 450 cm$^{-1}$. The samples MZ25, MZ35 and MZ50 showed the characteristic IR bands of mullite along with tetragonal zirconia when the calcination temperature was 1300°C.

(iii) alcination at 600°C or higher temperatures of EZ0 showed the IR bands at 1015, 755, 720 and 705 cm$^{-1}$ which are characteristic bands of eucryptite[24]. The characteristic bands of tetragonal zirconia along with the bands of eucryptite were observed when EZ5, EZ10 and EZ15 were calcined at 700°C.

(iv) The characteristic IR bands of β-spodumene[24] at 1017, 765 and 560 cm$^{-1}$ were observed when the samples SZ0, SZ5, SZ10 and SZ15 were calcined for 2h at

800°C. The composition SZ5, SZ10 and SZ15 also showed the IR bands corresponding to the tetragonal zirconia.

(v) The spectra of CZ0 showed the characteristic bands of cordierite[24] on calcination at 1300°C. The characteristic bands of cordierite, zircon and tetragonal zirconia were observed when CZ5 and CZ15 were calcined at 1300°C.

(vi) The compositions 10ZS, 30ZS and 50ZS showed the characteristic IR bands of tetragonal zirconia[24] at 600cm$^{-1}$ when the calcination temperature was 1200°C and 50ZS also showed the characteristic bands of monoclinic $ZrO_2$ at 732 and 420 cm$^{-1}$.

Table II. Average particle size (± 10nm) of the powders calcined at different temperatures measured by TEM.

| Sample | \multicolumn{6}{c}{Average particle size (nm)} |
|---|---|---|---|---|---|---|
| | 800°C | 900°C | 1000°C | 1100°C | 1200°C | 1300°C |
| MZ0 | - | - | 90 | 110 | 125 | 135 |
| MZ25 | - | - | 100 | 120 | 130 | 145 |
| MZ35 | - | - | 110 | 125 | 135 | 150 |
| MZ50 | - | - | 115 | 125 | 135 | 160 |
| EZ0 | 165 | 205 | - | - | - | - |
| EZ5 | 185 | 225 | - | - | - | - |
| EZ10 | 175 | 210 | - | - | - | - |
| EZ15 | 200 | 245 | - | - | - | - |
| SZ0 | 170 | 200 | - | - | - | - |
| SZ5 | 200 | 310 | - | - | - | - |
| SZ10 | 210 | 265 | - | - | - | - |
| SZ15 | 245 | 285 | - | - | - | - |
| CZ0 | - | - | 105 | 115 | 120 | 125 |
| CZ5 | - | - | 110 | 115 | 125 | 130 |
| CZ15 | - | - | 115 | 120 | 130 | 140 |
| 10ZS | - | - | 10 | 20 | 30 | |
| 30ZS | - | - | 15 | 25 | 35 | |
| 50ZS | - | - | 25 | 40 | 45 | |

The average particle sizes of all the materials calcined at different temperatures as measured by TEM are listed in Table II. Some examples are shown in Fig. 2. It was observed the particles were nanosized with a narrow size distribution. The average particle size increased with increase in calcination temperatures.

CONCLUSION

A series of multicomponent ceramic powders were prepared by using the aqueous sol-gel method. Here metal formates and precipitated silica were used as precursor compounds instead of metal alkoxides and water was used as the reaction medium

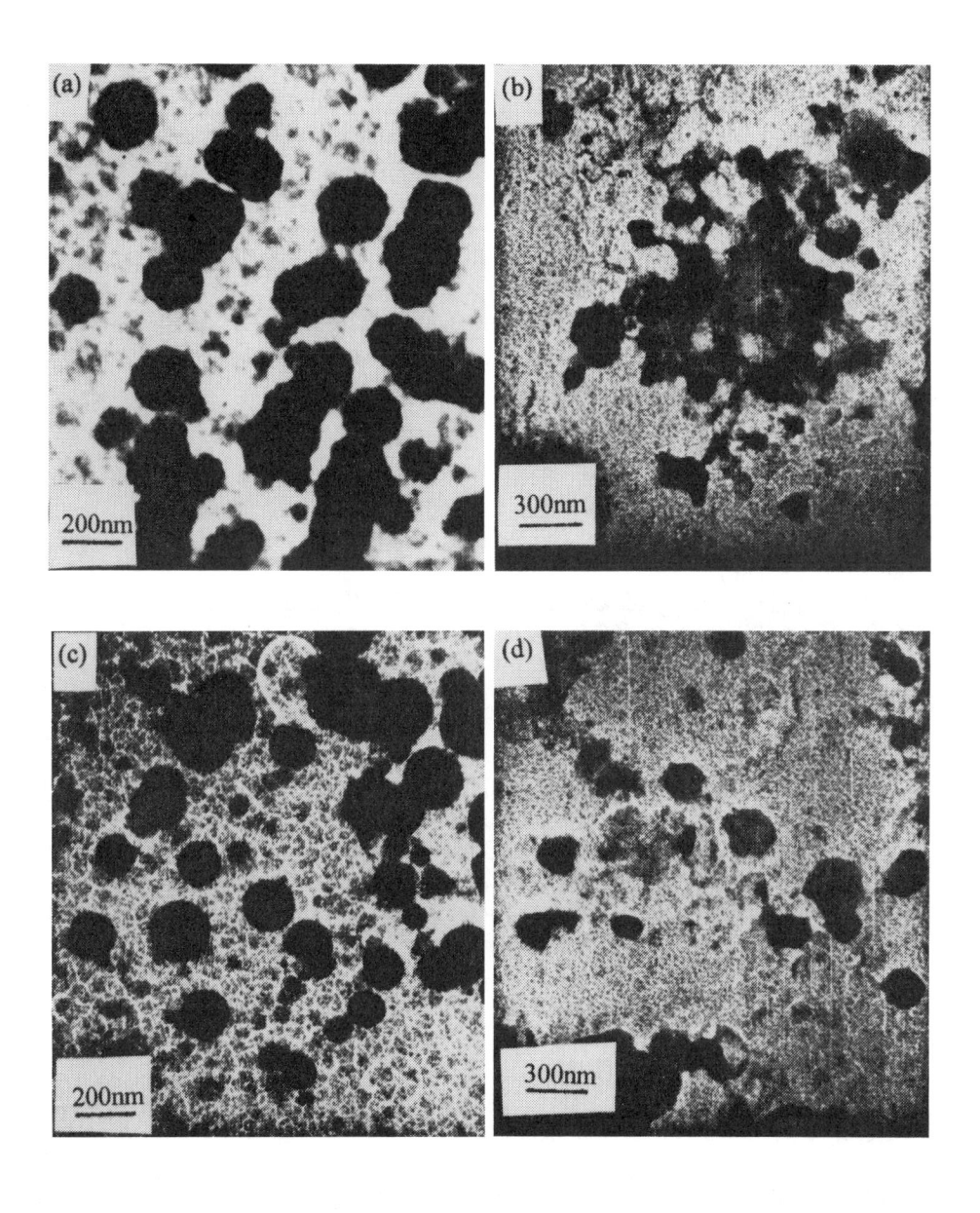

**Fig 2.** TEM micrographs of the samples (a) MZO calcined at 1200°C, (b) EZO calcined at 800°C, (c) SZO calcined at 800°C and (d) CZO calcined at 1300°C.

instead of the conventionally used solvent alcohol. The advantages of this aqueous sol-gel method are as follows:
(i) A series of nano-sized multi-component silicate powders can easily be obtained by this method.
(ii) Here the replacement of metal alkoxides by metal formates and precipitated silica and the use of water as the reaction medium instead of alcohol, which is commonly used as solvent in all-alkoxide sol-gel route, facilitate the reduction in the cost of the product.
(iii) This processing route provides the basis for a technically simple and cost effective method for the preparation of nano-sized multicomponent ceramic powders compared with other conventional methods.

REFERENCES
[1] C.J. Brinker and G.W. Scherer, "Sol-Gel Science: The Physics and Chemistry of Sol-Gel Processing"; Academic Press, New York, 1990.
[2] D.R. Ulrich, "Prospects of Sol-Gel Process," *Journal of Non-Crystalline Solids*, **100** 174-93 (1988).
[3] H. Schmidt, "Chemistry of Material Preparation by the Sol-Gel Process," *Journal of Non-Crystalline Solids*, **100** 51-64 (1988).
[4] L.L. Hench and J.K. West, "The Sol-Gel Process," *Chemical Review*, **90** 33-72 (1990).
[5] A.L. Piere, "Sol-Gel Processing of Ceramic Powders," *Ceramic Bulletin*, **70**[8] 1281-88 (1991).
[6] C.W. Turner, "Sol-Gel- Principles and Applications," *Ceramic Bulletin*, **70**[9] 1487-90 (1991).
[7] R. Roy, "Gel Route to Homogeneous Glass Prepation," *Journal of the American Ceramic Society*, **52**[6] 344 (1969).
[8] R. Roy, "Ceramics by the Solution-Sol-Gel Route," *Science*, **238** 1664-69 (1987).
[9] H.G. Sowman, "Aluminia-Chromia-Metal(IV) Oxide Refractory Fibers Having a Microcrystalline " U.S. Pat.t No. 4125 406 (1978).
[10] S. Sakka and K. Kamiya, "The Sol-Gel Transition in the Hydrolysis of Metal Alkoxides in Reaction to the formation of Glass Fibers," *Journal of Non-Crystalline Solids*, **48** 31-46 (1982).
[11] R. Roy, D.W. Hoffmann and S. Komarneni, "New Sol-Gel Structures for Making Ceramic-Ceramic Composites," *American Ceramic Society Bulletin*, **63**[8] 459 (1984).
[12] R.E. Newham, "Composite Electroceramics," *Journal of Materials Education*, **7**[4] 601-51 (1985).
[13] A. Hardy, G. Gowda., T.J. McMahan, R.E. Riman, W.E.E. Rhine and H.K. Bowen, "Preparation of Oxide Powders"; pp. 407-28 in Ultrastructure Processing

of Advanced Ceramics. Edited by J.D. MacKenzie and D.R. Ulrich, Wiley, New York, 1988.

[14] B.E. Yoldas, "Monolith Glass Formation by Chemical Polymerization," *Journal of Materials Science*, **14** 1843-49 (1979).

[15] J.J. Zelinski, B.D. Fabes and D.R. Uhlmann, "Crystallization Behavior of Sol-Gel Derived Glasses," *Journal of Non-Crystalline Solids*, **82** 307-13 (1986).

[16] U. Selvaraj, S. Komarneni and R. Roy, "Synthesis of Glass-Like Cordierite from Metal Alkoxides and Characterization by $^{27}$Al and $^{29}$Si MASNMR," *Journal of the American Ceramic Society*, **73**[12] 3663-69 (1990).

[17] N.N. Ghosh and P. Pramanik, "Synthesis and characterization of Calcia-Yttria-Alumina-Silica glass-ceramic composition by aqueous sol-gel processing," *British Ceramic Transactions*, **95**[5] 209-11 (1996).

[18] N.N. Ghosh and P. Pramanik, "Aqueous sol-gel synthesis of $SiO_2$-BaO for use in dental composite resins." *British Ceramic Transactions*, **95**[6] 267-70 (1996).

[19] N.N. Ghosh and P. Pramanik, "Aqueous sol-gel synthesis of spodumene and spodumene-zirconia composite powders," *Bulletin of Materials Science*, **20**[6] 247-57 (1996).

[20] N.N. Ghosh and P. Pramanik, "Synthesis of mullite powder using tetraethoxy silane and precipitated silica and aluminium formate as precursors in aqueous medium," *Bulletin of Materials Science*, **20**[6] 283-86 (1996).

[21] N.N. Ghosh and P. Pramanik, "Aqueous sol-gel synthesis of eucryptite and eucryptite-zirconia composite powders," *Materials Science and Engineering B*, **49**[1] 79-83 (1997).

[22] N.N. Ghosh and P. Pramanik, "Aqueous sol-gel synthesis of spodumene and eucryptite powders," *British Ceramic Transactions*, **96**[4] 155-59 (1997).

[23] N.N. Ghosh and P. Pramanik, "Synthesis of mullite and mullite zirconia composite powders using aqueous sol-gel method," *European Journal of Solid State and Inorganic Chemistry*, **34** 905-12 (1997).

[24] J.A. Gadsden, "Infrared Spectra of Minarals and Related Inorganic Compounds"; pp. 189, 193, 229 Butter Worth, London, 1975.

# INNOVATIVE PROCESSES FOR CERAMIC SYNTHESIS USING LIQUID METAL CARBOXYLATES

E. H. Walker Jr., and A.W. Apblett*, Oklahoma State University, Stillwater, OK. 74078-3071.

## ABSTRACT

Liquid carboxylate salts of metals are readily prepared via the reaction of 2-[2-(2-methoxy)ethoxy]ethoxyacetic acid (MEEA) with metal hydroxides, carbonates, and alkoxides. These salts can be used directly in metallo-organic deposition processes for preparation of ceramic films or they may be used to prepare liquid precursors for multi-metallic ceramics by dissolution of metal nitrates or acetates in one of the liquid metal carboxylates. The carboxylates may also be used to prepare ceramics using hybrid MOD/powder processing methods. In a similar approach, liquid ceramic precursors may be prepared by suspension of nano-particulate silica in a liquid metal carboxylate.

## INTRODUCTION

The successful exploitation of ceramics requires both convenient economical methods for their synthesis and techniques for their fabrication into useful morphologies such as coatings, fibers, monoliths, and powders. Meeting these requirements allows the preparation of ceramics for such applications as protective coatings, ceramic membranes, fibers for optical, insulating, and composite materials applications, ceramic honeycombs and powders for catalytic applications, as well as powders for the preparation of ceramic bodies. Chemical routes for the synthesis of ceramics are much more conducive to the development of ceramic for such applications since they provide significantly greater ability to tailor chemical and physical properties than conventional ceramic powder processes using solid state reactions between powder reactants [1].

The recent discovery of a family of metal carboxylates that are liquids at room temperature has provided novel approaches for the preparation of ceramic materials in a variety of morphologies [2-7]. These carboxylates, salts of 2-[2-(2-methoxy)ethoxy]ethoxyacetate, 1, have polyether linkages which undoubtedly contribute to their unusual characteristics by chelating and "solvating" the metal ions. Metal carboxylates have an extensive history for use as preceramic materials and in other applications [8] and the discovery of liquid carboxylate salts has expanded the potential of this class of compounds. Discussed herein are several example of new processes for ceramic synthesis based on these precursors

$$H_3C-O-CH_2-CH_2-O-CH_2-CH_2-O-CH_2-CH_2-\overset{\overset{O}{\|}}{C}-O^-$$

---

To the extent authorized under the laws of the United States of America, all copyright interests in this publication are the property of The American Ceramic Society. Any duplication, reproduction, or republication of this publication or any part thereof, without the express written consent of The American Ceramic Society or fee paid to the Copyright Clearance Center, is prohibited.

## EXPERIMENTAL

The various MEEA salts were prepared according to previously reported procedures [2-7]. All metal salts and carboxylic acids were commercial products and were used as supplied. Water was deionized and distilled before use. All reagents were commercial products and were used without further purification with the exception of 2-[2-(2-methoxyethoxy)ethoxy] acetic acid, MEEAH which was dried over activated 3A molecular sieves for ca. 24 hours. Water was distilled and deionized in a Modulab UF/UV Polishing apparatus before use. Thermogravimetric studies were performed using 20-30 mg samples under a 100 ml/minute flow of dry air in a Seiko TG/DTA 220 instrument or a TA Instruments Hi-Res TGA 2950 Thermogravimetric Analyzer. The temperatures were ramped from 25 °C to 1025 °C at a rate of 2 °C per minute or from 25 °C to 650 °C at a rate of 5 °C/min. Bulk pyrolyses at various temperatures were performed in ambient air in a temperature-programmable muffle furnace using 1-2 g samples, a temperature ramp of 5 °C/minute and a hold time of 6-12 hours. X-ray powder diffraction patterns were obtained using copper $K_\alpha$ radiation on a Scintag XDS 2000 diffractometer equipped with an automated sample changer and a high resolution solid state detector. Jade, a search/match software package, was used in the identification of XRD spectra.

## RESULTS AND DISCUSSION

Typically, metal carboxylates are synthesized by reaction of the carboxylic acid with a metal carbonate or hydroxide or by a metathesis reaction between an ammonium or alkali metal salt of the carboxylate and a metal chloride or nitrate. The latter method is inappropriate for synthesis of MEEA salts because of the extreme difficulty of removing salts such as ammonium or sodium chloride or nitrate from the liquid metal carboxylate. Not only will the metal carboxylate dissolve these salts but it will also solubilize them in non-aqueous solvents. Therefore, the synthesis of MEEA salts is best achieved by different means such as neutralization of metal hydroxides or carbonates by HMEEA in aqueous solution or by reaction of metal alkoxides with HMEEA. The latter reaction is usually accompanied by a side reaction that produces the ethyl ester of 2-[2-(2-methoxyethoxy)ethoxy]acetic acid and a metal hydroxo species. However, the hydroxides do not affect the liquid nature of the MEEA salts so that the products are quite acceptable for ceramic processing.

One technique for the use of these liquid precursors in the preparation of ceramic thin films is metallo-organic deposition, MOD, a non-vacuum, solution-based method of depositing thin films [9,10]. In the MOD process, a suitable metallo-organic precursor dissolved in an appropriate solvent is coated on a substrate by spin-coating, screen printing, or spray- or dip-coating. The soft metallo-organic film is then pyrolyzed in air, oxygen, nitrogen or other suitable atmosphere to convert the precursors to their constituent elements, oxides, or other compounds. Shrinkage generally occurs only in the vertical dimension so conformal coverage of a substrate may be realized. Metal carboxylates with long slightly-branched alkyl chains (e.g. 2-ethylhexanoate or neodecanoate) are often used as precursors for ceramic oxides since they are usually air-stable, soluble in organic solvents, and decompose readily to the metal oxides. MOD processes for the generation of many oxide-based materials have already been developed: e.g.

indium tin oxide [11], SnO$_x$ [12], YBa$_2$Cu$_3$O$_7$ [13] and ZrO$_2$ [14]. Metal salts of 2-[2-(2-methoxy)ethoxy]ethoxyacetate are very attractive MOD precursors since they are hydrolytically-stable and are usually liquids. Thus, they can be used in a variety of MOD processes (Figure 1) without resort to environmentally-damaging organic solvents. Instead, films can be prepared by dip- or spin-coating the neat precursors or their aqueous solutions. For example, Al(MEEA)$_2$OH, a viscous liquid that is readily prepared by reaction of aluminum ethoxide with 2-[2-(2-methoxy)ethoxy]ethoxyacetic acid (HMEEA) has been processed into films by spin-coating the neat liquid on warm (80°C) silicon wafer substrates or by thinning the carboxylate with sufficient ethanol or water to allow spin-coating at room temperature. Both methods yield a thin film of the metal carboxylate on the substrate that converts to a γ-alumina film by pyrolysis at 800°C. Alternatively, protective alumina films can be prepared on carbon-carbon composites and carbon fibers by dip-coating and pyrolyzing at 800°C under a nitrogen atmosphere.

The utility of these liquid metal carboxylates is markedly enhanced by the fact that they are excellent solvents for other metal salts so that precursor solutions for multi-metallic ceramic materials are readily prepared. For example, a homogeneous precursor solution for nickel ferrite was synthesized by dissolution of Ni(NO$_3$)$_2$ in the liquid iron carboxylate, Fe$_3$O(MEEA)$_7$·5.5(H$_2$O)[6]. This precursor solution yielded a metastable, amorphous NiFe$_2$O$_4$ phase at 300°C demonstrating that these precursors are also useful for generating metastable phases. In this case, the amorphous oxide underwent an exothermic crystallization at 374°C to yield the trevorite phase of NiFe$_2$O$_4$. Similarly, barium titanate has also been synthesized using a liquid metal carboxylate precursor that was prepared by dissolving barium acetate in Ti(MEEA)$_4$ [7] and spinel and yttrium aluminum garnet were synthesized at 800°C from liquid precursors derived from dissolution of magnesium or yttrium nitrate in Al(MEEA)$_2$OH [5, 14].

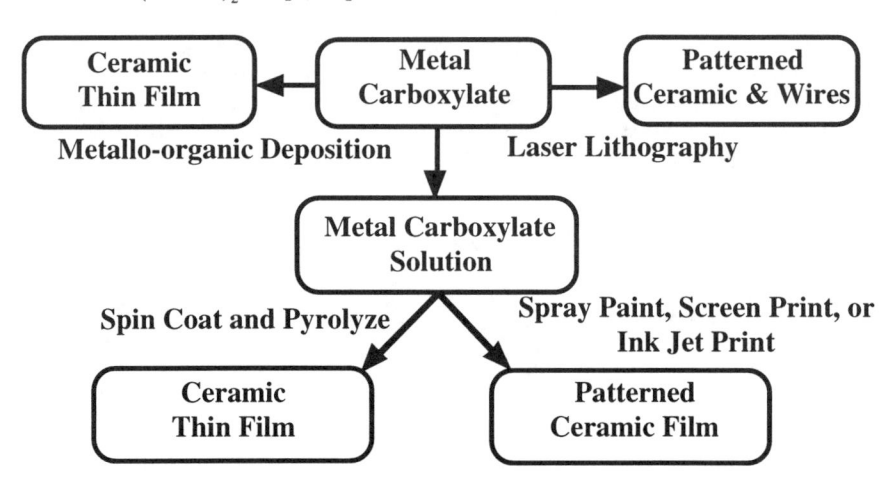

Figure 1. Metallo-Organic Deposition Processes Using Liquid Metal Carboxylates

The application MOD processing in the preparation of integrated circuits or sensors, requires patterning of the ceramic film. This is readily achieved at several different points in the MOD process: during the deposition step, after deposition and prior to pyrolysis, during pyrolysis or after pyrolysis. MOD films have been patterned during deposition by screen printing, and ink jet printing but a variety of other methods such as spraying through a stencil and off-set printing are also applicable [9]. Solutions of traditional MOD precursors often do not have sufficient viscosity for screen printing but the high viscosity of the liquid metal carboxylates make them very suitable precursors for solventless screen printing. Another very useful method of patterning MOD films is patterning during the pyrolysis step by using a laser, electron beam , or an ion beam as a localized heat source [9]. In this type of patterning, the use of the liquid metal carboxylate precursors provides the advantage that thicker features may be grown because the precursor can flow into the area that is being patterned. Therefore, rastering the beam across the desired feature after a slight delay for the precursor to reflood that area will allow thicker films to be developed. The preparation of copper wires by laser pyrolysis of $Cu(MEEA)_2$ in this manner has already been achieved.

Besides metallo-organic deposition, the liquid metal carboxylates may also be used to prepare ceramics by hybrid methods that combine the strengths of chemical processing methods with those of conventional powder processing (Figure 2). Depending on the liquid metal carboxylate this modified powder processing method can produce a variety of ceramics. For example, the metal carboxylate can supply a sintering aid for the ceramic powder. An example of this is the use of a $Y(MEEA)_3$ [3] to coat silicon nitride with $Y_2O_3$ by milling the liquid precursor with the silicon nitride powder, pressing the mixture into a disk with a uniaxial press using 10 tonne pressure, and then firing at 600°C in air followed by sintering at 1200°C. This approach yields a green body with poor strength but the sintering step yields a ceramic with near-theoretical density. Thus, the liquid yttrium carboxylate has potential as a dual binder/sintering aid for production of dense, well-sintered silicon nitride ceramics. However, it would be more effective if a method to polymerize the precursor is developed so that stronger green bodies may be prepared. Currently, we are investigating ammonia infiltration {leading to hydrolysis of $Y(MEEA)_3$} as one method of effecting this polymerization .

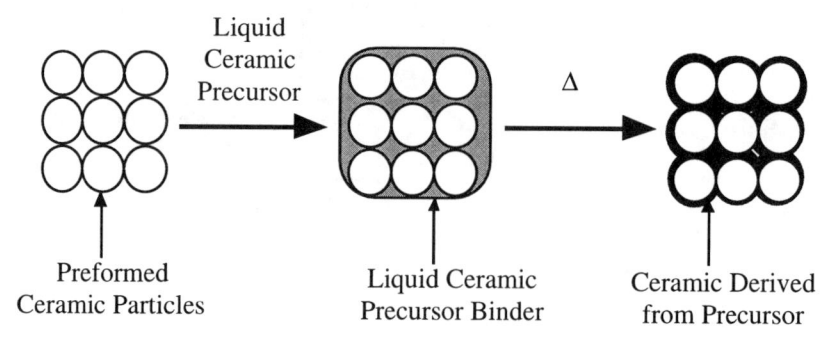

Figure 2. Modified Powder Processing Using Liquid Metal Carboxylates

A second type modified powder processing method occurs when a metal carboxylate is used to deposit a metal oxide of the same composition as the ceramic powder into the pore space of the unfired ceramic. This method has significant potential for the synthesis of high density ceramics with fewer and smaller void spaces. When a liquid precursor for a particular ceramic is mixed with a ceramic powder of the same composition and the mixture was pressed into a body, the void space between ceramic particles becomes filled with liquid precursor rather than air. Subsequent pyrolysis of the liquid deposits the ceramic in the void space, partially filling it. Thus, the "brown body" before sintering would be denser that that achievable by simply pressing with a normal binding aid. This increased density as well as increased necking between the ceramic particles would be expected to lead to improved sintering behavior. Using a similar method, Barron *et al.* have demonstrated that dense alumina ceramic may be prepared from an alumina powder/aluminum carboxylate mixture [16]. We are using this approach to prepare optically-transparent high-density ceramic scintillation elements for medical detectors. The desired ceramic has the composition $Y_{1.34}Gd_{0.60}Eu_{0.06}O_3$ [17] and the liquid precursor can be economically prepared by dissolving the requisite amount of gadolinium and europium nitrates into $Y(MEEA)_3$. Mixing this precursor with ceramic powder by grinding, pressing into a body (using a uniaxial press and 10 tonne pressure), and then firing at 1800°C yields optically-clear ceramic bodies.

The liquid precursor/ceramic powder combination for preparation of a ceramic can provide significant advantages for processing of the ceramic into various shapes including wires. Since the liquid lubricates the powder, the mixture has plasticity and can be molded into a variety of shapes. For example, we have prepared superconducting disks and wires using the approach outlined in Figure 3.

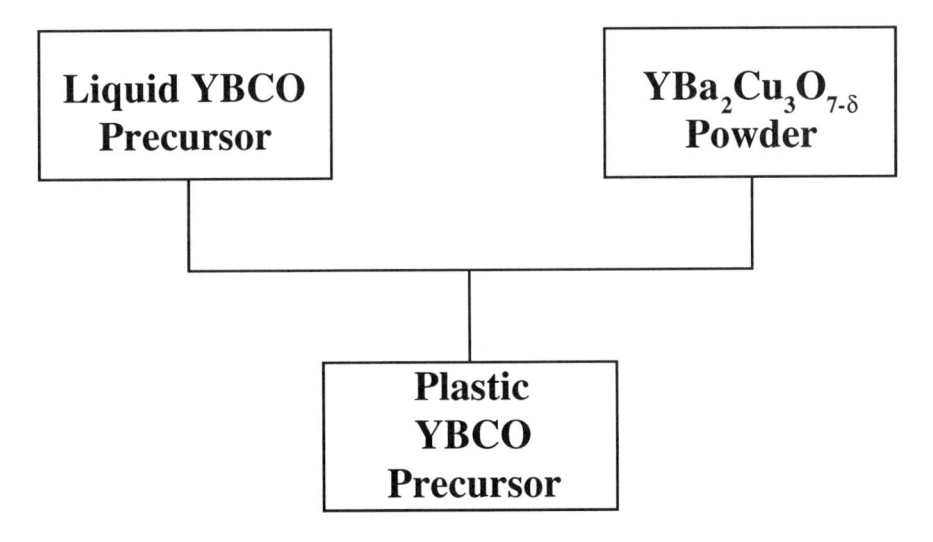

Figure 3. Procedure for Preparation of a Moldable Precursor for YBCO

The liquid precursor was prepared by mixing Y(MEEA)$_3$, Ba(MEEA)$_2$, and Cu(NO$_3$)$_2$ together in the correct ratio. Only 10 percent by weight of this precursor is necessary to confer plasticity on the YBCO powder so that, overall, the ceramic yield upon firing is fairly high.

Another modified powder process results from using the liquid metal carboxylate to supply a second metal oxide reactant as a homogeneous coating on another metal oxide. This approach provides a precursor mixture that has homogeneity that is intermediate between that of pure chemical routes and conventional powder processing (Figure 4). Since, the ions have less distance to diffuse, the solid state reactions are complete at lower temperatures. For example the preparation of spinel, MgAl$_2$O$_4$, by a modified powder process was found to occur quite readily at low temperature. Chromatographic γ-alumina was mixed well with Mg(MEEA)$_2$·2H$_2$O by stirring the reactants together in an alumina crucible. The XRD pattern of this mixture corresponded to spinel after firing at temperatures as low as 700°C [7].

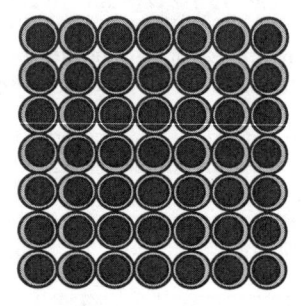

Mixture of Metal Oxide Particles      Metal Oxide-Coated Metal Oxide Particles

Figure 4. Comparison of Conventional Powder Processing and Modified Powder Processing for Reactions Between Two Metal Oxides

Appropriate choice of reactants in this type of processing allows the preparation of ceramics with unusual morphologies. For example, the solid state reaction between nickel oxide and iron oxide to yield nickel ferrite, NiFe$_2$O$_4$ occurs by initial formation of an interfacial layer of nickel ferrite. Subsequent reaction involve diffusion of Fe$_2$O$_3$ through the NiF$_2$O$_4$ to the nickel oxide (Figure 5). Thus, when Fe$_2$O$_3$ particles, coated with nickel oxide by mixing with Ni(MEEA)$_2$ and then heating to 500°C are subsequently sintered at 800°C, the particles expand into collections of extremely small ceramic particles that resemble "broccoli florets" (Figure 6) [6]. The powders produced in this fashion are very porous and it is possible that such reactions could be harnessed for the preparation of ceramic foams for insulation and catalysis applications. The agglomerates are readily broken up by application of an ultrasonic probe so it is possible to make very fine powders for the production os sintered ferrite bodies for transformer cores. Additionally, the ability to make fine-grained ferrites and perhaps other spinels without extensive grinding should be very applicable to production of metal oxide pigments.

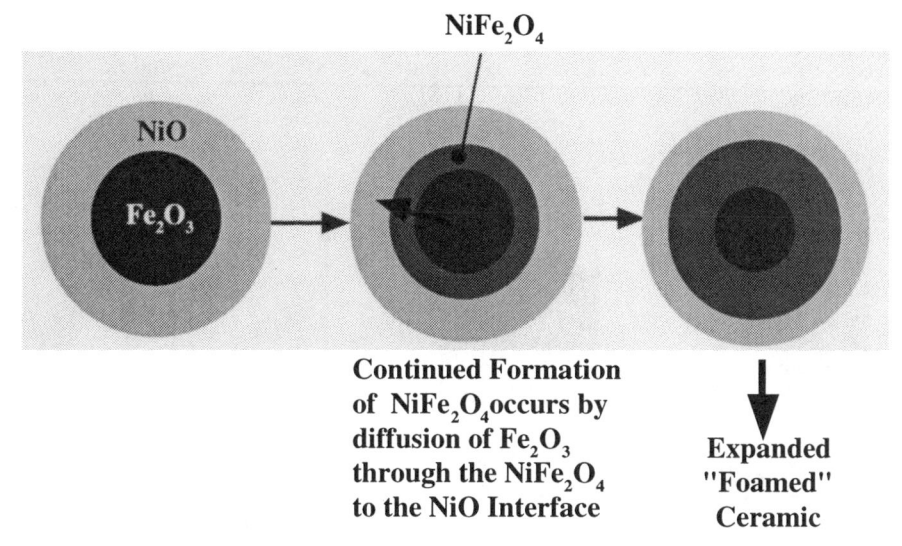

Figure 5. Mechanism of Ferrite Formation by Nickel Oxide Coated Ferric Oxide Powders

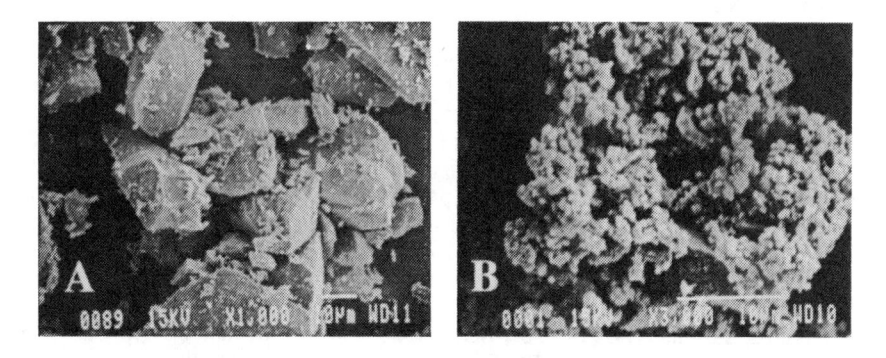

Figure 6. SEM Micrographs of Nickel Oxide Coated Ferric Oxide Powders (A) Before and (B) After sintering at 800°C.

The success of the Mg(MEEA)$_2$/Al$_2$O$_3$ precursor for synthesis of spinel suggests that even more remarkable results could be realized by use of a colloidal or nano-sized oxide phase rather than larger particles. Therefore, a zircon, ZrSiO$_4$, precursor was prepared by dissolving ZrO(MEEA)$_2$ in a 30 weight percent aqueous solution of a colloidal silica (LUDOX® AM-30). After removal of water in vacuum at room temperature, a perfectly clear liquid was obtained that showed no tendency to precipitate the silica phase. The liquid could be converted to a slightly sticky, moldable solid by pyrolysis in air at 150°C. A transparent, glassy, amorphous oxide phase resulted from heating the precursor to 500°C. This glass remained non-crystalline until heated above 1000°C when it crystallized to zircon. Similar results were observed for Al(MEEA)$_2$OH/colloidal silica mixtures (Figure 7) from which glassy nanocomposite materials could be prepared at 800°C while crystalline aluminosilicates were obtained at 1200°C [18].

**30% Aqueous SiO$_2$ Nanoparticles**

Add Al(OH)(MEEA)$_2$    Dry    Liquid Precursor

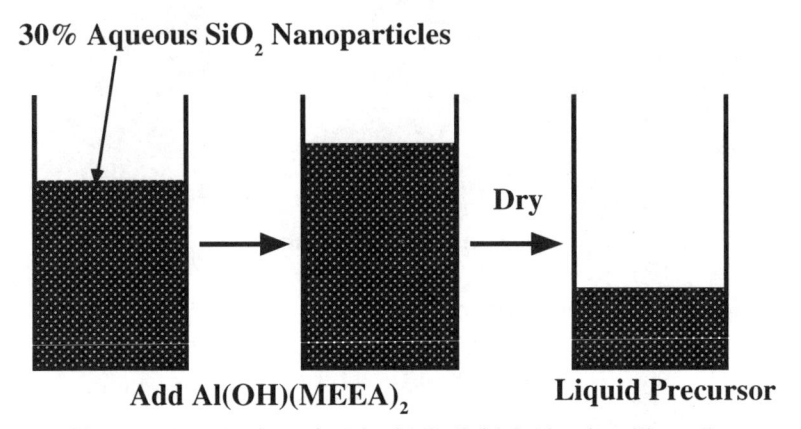

Figure 7. Preparation of a Liquid Colloidal Aluminosilicate Precursor

CONCLUSIONS

The metal salts of 2-[2-(2-methoxy)ethoxy]ethoxyacetate because of their liquid nature and good solvent properties have significant applications in the preparation of oxide materials. There is considerable flexibility in the preparation of precursors for multimetallic oxides since the liquid metal carboxylates may be mixed in any ratio or one or more metal can be supplied as a nitrate or acetate salts or as nano-particulate oxides without losing the fluidity of the precursor. The precursors may be converted to ceramic films by painting or dip- or spin-coating the neat precursors or on appropriate substrates followed by firing at temperatures ranging from 500°C to 1200°C depending on the target ceramic and the desired crystallinity. The liquid metal carboxylates have also been demonstrated to be quite useful for the preparation of bimetallic ceramics via a variety of modified powder processes. Aside from the synthetic potential of such reactions, they may also be applicable for mechanistic investigations of solid state metal oxide reactions and the preparation of supported catalysts and foamed ceramics. Finally, the blending of nano-technology with metallo-organic deposition has opened up numerous exciting possibilities for the preparation of metal oxides with a high ceramic yield as well as nano-composite materials.

ACKNOWLEDGEMENTS

We thank the Louisiana Board of Regents for support of this research through Grant # LEQSF(1996-99)-RD-B-08.

REFERENCES
[1] D. Segal, "Chemical Synthesis of Ceramic Materials" *J. Mater. Chem.*, **7**, 1297-1305 (1997).
[2] M. P. Pechini, "Ceramic Dielectric Materials", U.S. Patent **3,330,697**, 1967.
[3] A. W. Apblett, J. C. Long, E. H. Walker, M. D. Johnston, K. J. Schmidt, and L. N. Yarwood, "Metal Organic Precursors for Yttria" *Phosph. Sulf. Silicon Rel. Elements*, **93-94**, 481-481 (1994).
[4] A. W. Apblett, S. M. Cannon, G. D. Georgieva, J. C. Long, M. I. Raygoza-Maceda, and L. E. Reinhardt, "Polymeric Precursors for Yttria." *Mat. Res. Soc. Symp. Proc*, **346**, 679-683 (1995).
[5] A. W. Apblett, L. E. Reinhardt, and E. H. Walker, "Novel Liquid Precursors for Spinel and Alumina Ceramic Coatings" *Proceedings of Unified International Technical Conference on Refractories*, **Vol III**, 1503-1507 (1997).
[6] A. W. Apblett, M. L. Breen, and E. H. Walker, "Synthesis of Nickel Ferrite Using Liquid Metal Carboxylates" *Chem. Mater.*, **10**, 1265-1269, 1998.
[7] A. W. Apblett, G. D. Georgieva, L. E. Reinhardt, and E. H. Walker, "Precursors for Aqueous and Liquid-Based Processing of Ferroelectric Thin Films"; pp. 95-105 in *Synthesis and Characterization of Advanced Materials*, , Edited by M. Serio, D. M. Gruen, and M. Ripudaman, American Chemical Society, Washington, D.C, 1998.
[8] R.C. Mehrotra and R. Bohra, "*Metal Carboxylates*", Academic Press, New York, 1983.
[9] J. V. Mantese, A. L. Micheli, A. H. Hamdi, and R. W. Vest, "Metal Organic Deposition" *M.R.S. Bull., (XIV),* , 1173 (1989).
[10] R. W. Vest, "Electronic Thin Films from Metallo-Organic Precursors"; pp. 303-347 in *Ceramics Films and Coatings*, Edited by J. B. Wachtman and R. A. Haber, Noyes Publications, Park Ridge, N.J., 1993.
[11] J. J. Xu, A. S. Shaikh, and R. W. Vest, "Indium Tin Oxide Films from Metallo-Organic Precursors" *Thin Solid Films,* **161**, 273-280 (1988).
[12] T. Maruyama and K. Kitamura, "Fluorine-Doped Tin Oxide Thin Films Prepared by Thermal Decomposition of Metallic Complex Salts" *Jpn. J. Appl. Phys.*, **28**, L312-L313 (1989).
[13] A. H. Hamdi, J. V. Mantese, A. L. Micheli, R. C. O. Laugal, D. F. Dungan, Z. H. Zhang, and K. R. Padmanabhan, "Formation of Thin-Film High Tc Superconductors by Metalorganic Deposition" *Appl. Phys. Lett.*, **51**, 2152-2154 (1987).
[14] V. Hebert, C. His, J. Guille, S. Vilminot, and T. L. Wen, "Preparation and Characterization of Precursor of $Y_2O_3$ Stabilized $ZrO_2$ by Metal Organic Compounds" *J. Mater. Sci.*, **26**, 5184 (1991).
[15] A. W. Apblett, L. E. Reinhardt, and E. H. Walker, "Liquid Metal Carboxylate Precursors for Yttrium Aluminum Garnet" *Proceedings of Unified International Technical Conference on Refractories,* , 1525-1529 (1997).
[16] R. L. Callender, C. J. Harlan, N. M. Shapiro, C. D. Jones, D. L. Callahan, M. R. Wiesner, D. B. MacQueen, R. L. Cook, and A. R. Barron, "Aqueous Synthesis of Water-Soluble Alumoxanes: Environmentally Benign

Precursors to Alumina and Aluminum-Based Ceramics" *Chem. Mater.,* **9**, 2418-2433 (1997).

[17] C. Greskovich, D. A. Cusano, and F. A. DiBianca, "Rare Earth-Doped Yttria-Gadolinia Ceramic Scintillators", U. S. Patent **4421671**, Dec. 20, 1983, 1983.

[18] A. W. Apblett, M. L. Breen, and E. H. Walker, "The Application of Liquid Metal Carboxylates to the Preparation of Aluminum-Containing Ceramics" *Comments on Inorganic Chemistry,* in press (1998).

# LANTHANUMHEXALUMINATE AS INTERPHASE MATERIAL IN OXIDE-FIBER-REINFORCED OXIDE-MATRIX-COMPOSITES

B. Saruhan, L. Mayer, H. Schneider
German Aerospace Center (DLR)-Cologne
Institute for Materials Research
D-51147, Köln, Germany

## ABSTRACT

Oxide based fiber-reinforced composites are attractive materials in applications where long-term oxidation and high temperature stability are of primary importance. In such materials strong interfacial reactions are unavoidable due to the high diffusion rates in oxides. An interphase material which provides crack deflection or crack bridging and thereby fiber pull-out, needs to be developed. Lanthanumhexaluminate ($LaAl_{11}O_{18}$) is hereby an interesting interphase material due to its easy cleavage and planar-aligned crystals. The sol-gel process is an effective way to coat fiber yarns with a thin (< 500 nm) layer. Sols can be prepared from both organic and inorganic precursors. The formation of $LaAlO_3$ as an intermediate phase causes $LaAl_{11}O_{18}$ to form at higher temperatures (> 1200°C), resulting in possible difficulties in composite processing. In this study, sols were prepared using various combinations of $La(NO_3)_3$, $Al(NO_3)_3$, Al-sec-butylate, Lanthanum oxalate and Lanthanum-2,4-pentandionate. Subsequent gelation, drying and crystallization behaviour were observed. The formation of $LaAl_{11}O_{18}$ could be suppressed by changing the starting materials. Organic precursors resulted in the formation of $LaAl_{11}O_{18}$ at significantly lower temperatures.

## INTRODUCTION

Oxide based fiber composites become attractive materials for those applications where the high temperature and long-term oxidation resistance are desired. Load transfer from matrix to fiber relies on the presence of a suitable interface material, as well as providing a weaker bonding also leading to crack deflection and to a high work of fracture. Many concepts are developed to meet this requirement. These may be classified as fugitive coatings (1), weakly bonding oxides (2), porous oxides (3), refractory metals (4) to oxides with easy cleavage crystallographic planes (5). C (graphite) and BN are due to their columnar structures present a perfect match for such systems, however they display very poor oxidation resistance (6). Similarly SiC which has been studied as an interface material is not suitable due to lack of long-term oxidation resistance under cyclic conditions. Some oxides, especially if used in the mullite or aluminosilicate matrix-systems, are limited, since they are prone to reaction and display relatively high diffusion and growth rates at elevated temperatures. In alumina composite systems, oxides such as $ZrO_2$(7), $LaPO_4$ (8,9) are applied successfully. However, dense $ZrO_2$, due to its high thermal expansion coefficient and $LaPO_4$, due its reaction with silicates are not suitable for mullite/mullite-composites. La-hexaluminate ($LaAl_{11}O_{18}$) with its magnetoplumbite crystal structure contains easy-cleaving crystallographic planes (10-13). Magnetoplumbite structure consists of spinel blocks which are separated with conductive interspinel layers of $MAlO_3$. Its structure is determined by the charge and the radius of the cation in the mirror plane.

Previous studies concerning the sol-gel chemistry of this material report that formation of $LaAl_{11}O_{18}$ always follows the formation of an intermediate perovskite type of phase, $LaAlO_3$. Thus, occurrence of $LaAl_{11}O_{18}$ can only be achieved only at elevated temperatures (> 1400°C) (13).

Coating of polycrystalline multifilament fibers through sol-gel-route has been previously attempted (14). In these studies, mainly applications related to single oxide systems such as alumina, titania and zirconia have been reported. However, despite the simplicity of these systems, sol-gel coating displayed some problems, such as ineffective wetting of each single fiber in a yarn and bridging of fibers through the coating layer (15-16). A continuous coating apparatus which was developed by Hay et. al. delivers only for thick filaments successful results (17).

Continuous coating of multifilament, thin, polycrystalline fibers is a challenging goal and requires use of optimized sols and methods. This paper presents a comprehensive study on sol-gel synthesis of $LaAl_{11}O_{18}$ at relatively lower temperatures (≤ 1200°C) without the formation of an intermediate phase by employing various sols based on inorganic and organic precursors. Moreover, a method is described which demonstrates successful coating of fiber yarns continuously (for instance, Nextel 720, mullite/alumina-fibers) by means of liquid precursors:

## EXPERIMENTAL METHODS:

Starting materials La- and Al-nitrates; La-oxalate, La-2,4-pentadionate, La-acetate, Al-sec-butoxide have been employed.

## Preparation of sols

*Sample A : Al(NO$_3$)$_3$ and La(NO$_3$)$_3$*

Solutions of La(NO$_3$)$_3$.6H$_2$O (c:0.24 mol/l) and Al(NO$_3$)$_3$.9H$_2$O (c:2.66 mol/L) in ethanol were prepared separately. Solution of Al(NO$_3$)$_3$.9H$_2$O was stabilized with addition of 15 mol. % 2,4-Pentadionate (hereafter referred as AcAc), calculated from Al-content. The amount of AcAc was experimentally determined. The aluminum solution has been co-mixed with ethanol solution of La(NO$_3$)$_3$.6H$_2$O.

*Sample B : Al(NO$_3$)$_3$ and Lanthanum oxalate*

Solution of starting material Al(NO3)3.9H$_2$O (c:2.66 mol/L) in ethanol was prepared and stabilized with addition of 15 mol. % AcAc, calculated from Al-content. 0.12 mol Lanthanum oxalate was dissolved in this sol. Dissolution was carried out under slow heating. At about 76°C and after cooling to RT, a clear sol was obtained.

*Sample C : Aluminum-sec-Butoxide and La-2,4-Pentanedionate*

This preparation has been carried out under inert gas atmosphere, until the gelation began. Firstly, the solution of starting material, Al-sec-butoxide was prepared in 2-propanol (c: 4 mol/l) separately. This solution was stabilized by addition of 38 mol. % AcAc. La-2,4-pentanedionate was dissolved in this sol under intensive stirring.

*Sample D : Aluminum-sec-Butoxide and Lanthanum acetate*

This sol was also entirely prepared under argon atmosphere, until gel formation has started. Al-sec-butoxide was mixed with 2-propanol (c: 4 mol/L) separately and stabilized with addition of 38 mol. % AcAc, calculated from Al-content. La-acetate was dissolved in this sol under intensive stirring. The suspension was peptizied by addition of 60 ml HNO$_3$ to 1l sol. The pH was set to 6-7 and this resulted in a clear sol.

## Gelation and Drying

After completion of sol-homogenization, sols B to D were exposed to air at 60°C to obtain the gel. Sol A was sensitive to air contact, therefore its gelation was carried out under vacuum at RT. Evaporation of solvent was carried out overnight under vacuum mostly at RT or alternatively between 40-60°C in a drying oven. Heat-treatment of sols containing nitrates in air at a temperature higher than 40°C resulted in salt melting and lead to reaction between $NO_3^-$ and crystal water of $Al(NO_3)_3$ to form $HNO_3$. In turn, with all sols, it was observed that vacuum drying up to 200°C yields uniform coating of fibers without causing any vigorous reaction. Extension of the vacuum drying up to temperatures above 200°C (for instance at 400°C) leads to pyrolysis of acetylacetonate and eventually formation of graphite like products which can be recognized from their dark colour in the coating.

**Continuous coating of fibers**

Coating of fiber tows has been carried out in a continuous coating apparatus which was built in our laboratories. Primarily fiber tows passed through a tube furnace at 900°C with a speed of 75 cm/min to remove the sizing. After burn-out of the sizing, fiber tows, consisting of 400 single filaments of 10 µm diameter were immersed in a vessel filled with the coating sol. The vessel contains four sets of rollers, having different functions; such as spreading the fibers, holding the fibers at a certain length in the vessel, etc. Fiber tows which leave the receptacle vertically were set in contact with air and subsequently let gel-formation and curing processes take place. Curing of the gelled coating occurred in a vertical furnace at temperatures approximately 40°C, supported by IR-beam drying and later under vacuum for 24 hours. Solvent evaporation was carried out 60 hours in vacuum furnace up to 200°C as the coated fiber tows were wind on a spool. This process yields a uniform and continuous coating of fibers.

**Characterization of gels and coatings**

Gels which produced simultaneous to fiber coating were characterized by means of X-ray diffractometer and differential scanning calorimetry (NETZSCH-DSC 404). All DSC-measurements were carried out between RT and 1400°C with a constant heating rate of 10K/min and the data collected by computer with respect to a sapphire reference material, considering baseline corrections.

XRD-studies were carried out with a computer-controlled powder diffractometer (SIEMENS D 5000) using Ni-filtered Cu $K\alpha$ radiation. Diffraction patterns were recorded in the $2\Theta$ range 10° to 80°, in step scan mode (3s/0.02°, 2 $\Theta$).

Scanning electron microscopy investigations were performed with a PHILLIPS 525M scanning electron microscope (SEM) on fracture surfaces of coated fibers.

# RESULTS
## Differential Scanning Calorimetry

DSC-measurements of sample A, shown in Fig. 1 displays a large exothermic signal between 250°C and 400°C, having a peak climate at 350°C and a shoulder at about 400°C. Moreover a small exothermic peak at 913°C and a broad exothermic signal between 1106°C and 1320°C can be observed, indicating a sluggish reaction. Sample B show three exothermic signals; a broad and high intensity signal, having a peak maxima at about 350°C, a broad and low intensity signal at about 906°C and a broader signal between 1050° and 1300°C (Fig. 1). DSC-curve of this sample resembles those obtained from inorganic starting materials, given with Sample A. DCS-spectra of sample C in Fig.1 exhibits two exothermic occurrences; a broad and high intensity signal, centering at 410°C and showing two shoulders at 280°C and 520°C and a sharp signal at 933°C. An exothermic decrease of the curve at about 1350°C is also observed. Sample D has two sharp and one small, low intensity signals. First sharp signal appears at 316°C, second at 928°C . The low intensity small signal occurs at 1206°C (Fig. 1a).

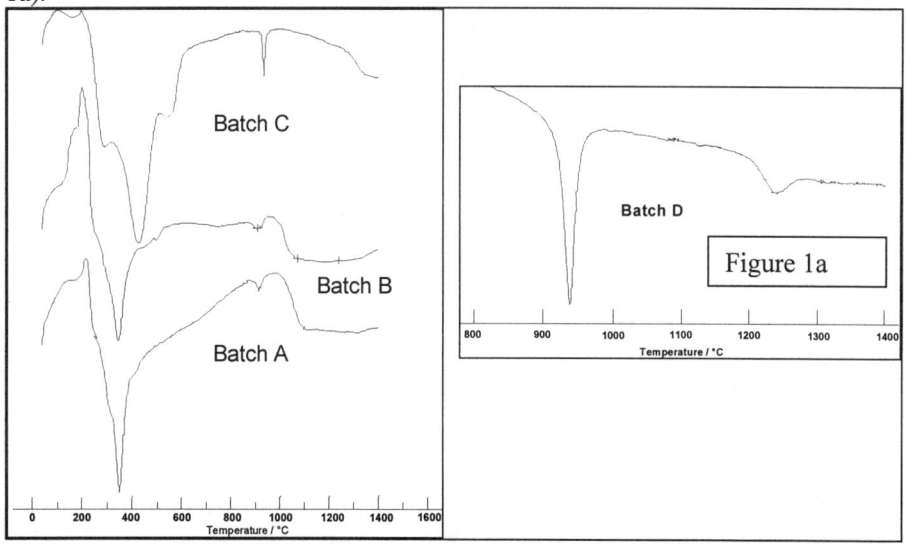

Figure 1: DSC-Curves of Samples A to C shown in the temperature range between RT and 1400°C and of Sample D (Fig. 1a) shown in the range between 800° and 1200°C.

## X-Ray Diffraction Analysis

XRD measurements of the samples, heat-treated at 900°, 1000°, 1100° and

1200°C show that the precursors were amorphous up to 900°C above which formation of a weakly crystalline γ-Al$_2$O$_3$ occurs. LaAlO$_3$ (La-monoaluminate with perovskite structure) crystallizes as the main crystalline phase above 1000°C, while La-hexaluminate (LaAl$_{11}$O$_{18}$) forms as secondary phase (Fig. 2a, 2b, 2c).

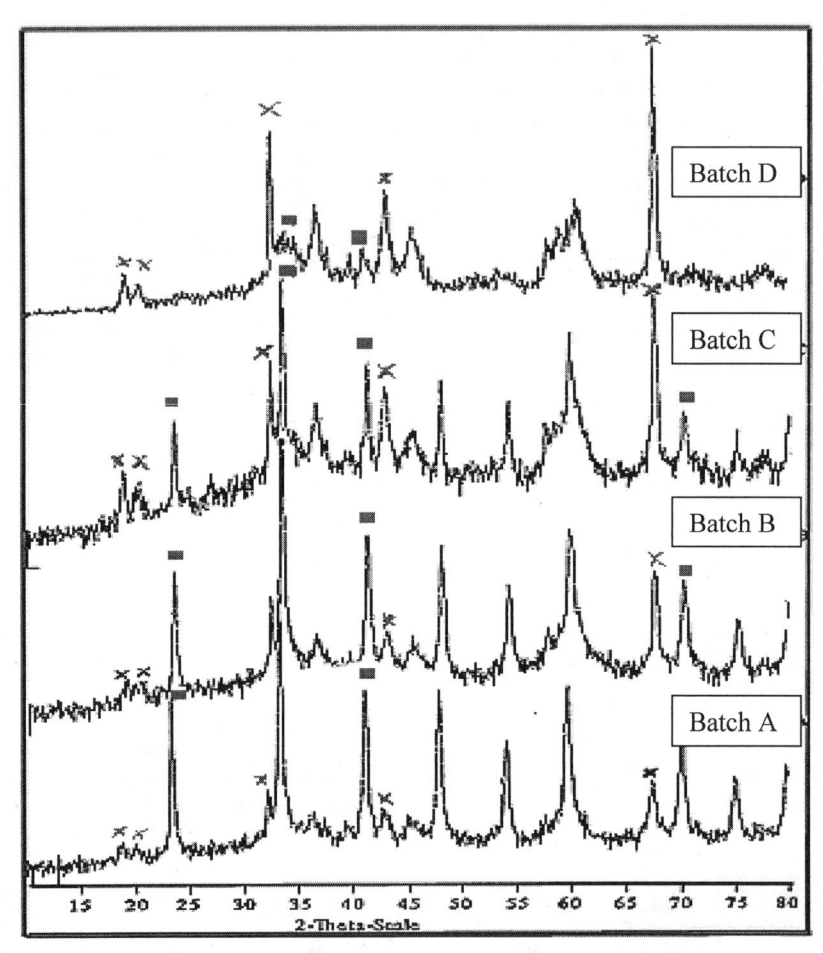

Figure 2-XRD-Curves of Samples A to D after calcining at 1200° for one hour
(■) LaAlO$_3$    (X) LaAl$_{11}$O$_{18}$

The intensities of LaAl$_{11}$O$_{18}$ phase increase from Batch A to C. Batch D shows only LaAl$_{11}$O$_{18}$ as the crystalline phase at 1200°C.

**Microstructural Observations**

SEM-observations on the coated fibers show that the quality of the coating layers depended mostly on drying process and secondarily on the characteristics of the sols. Figure 3 demonstrates 300-400 nm thick continuous and homogeneous layers on Nextel 720 fibers. The thickness of layers varied along the fiber due to flow of the sol previous to gelation. A sudden gelation behavior should be promoted in order to avoid this. It was observed that infrequently coatings contain cracks which did not necessarily destroy the coating layer, however a partial stripping of the layers was unavoidable during sample preparation for the scanning electron microscopy investigations.

Formation of cracks in such systems is not detrimental, however peeling off the coating may cause undesirable results, since at such points strong bonding with matrix may occur and the fiber and consequently composite fails catastrophically.

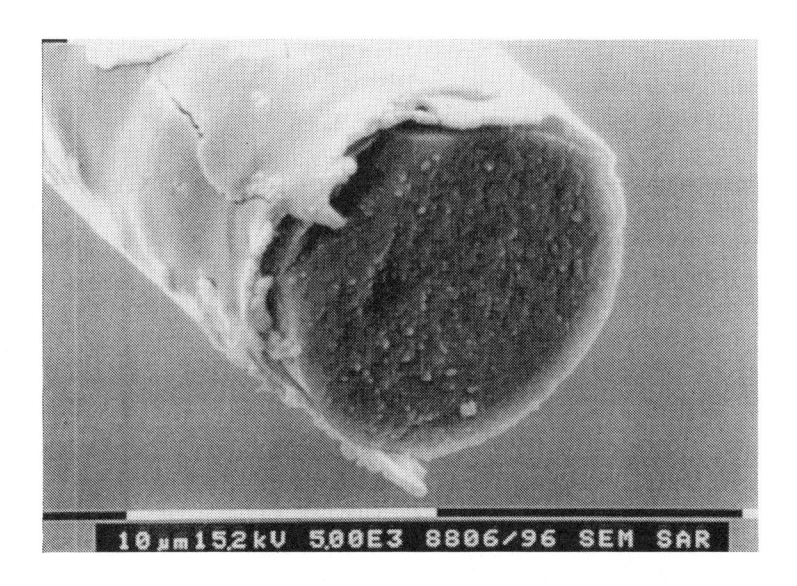

Figure 3- SEM-micrograph of Lanthanumhexaluminate coating, indicating that a thickness of 300-500nm is achievable.

**DISCUSSION**

Multifilament fibers are coated by a revised sol-gel route which involves two

steps:
- precursors synthesis by modifying the starting materials to improve $LaAl_{11}O_{18}$ crystallization conditions

- continuous fiber coating procedure by controlled gelation and drying

Precursor synthesis

One of the main points in sol-gel-processing is the control of condensation reaction. This can be realized by introducing chelating ligands such as 2,4-Pentandionate. The ligand regulates the condensation reaction in its function as a cover group. The ratio of chelating ligands to metal cations in a complex regulates the degree of the condensation reaction. Control of these processes is an essential need for proceeding successful continuous coating of fibers. Gel formation should occur after complete wetting of the fibers is achieved. It is important that the heterogeneous spreading of the sol is avoided. Addition of AcAc helps to control of hydrolysis and wetting conditions, however, it also means a larger amount of organical residue needs to be removed without destroying the coated layer. Thus, a suitable sequence at defining the process steps should be considered to achieve a successful coating. Polar solutions of $Al(NO_3)_3$ and $La(NO_3)_3$ contains trivalent cations, $Al^{3+}$ or $La^{3+}$ and $(NO_3)^-$ anions, as well as ionized water molecules, coming from the crystal water of nitrates. Both cations (M = $Al^{3+}$, $La^{3+}$) are coordinated by six water-molecules and the complexes can be hydrolyzed. On addition of AcAc, a part of trivalent cations will be surrounded with AcAc-complexes, thus stabilizing them against hydrolysis and condensation. As the remaining cations condense and form Al-O-Al- networks. Each $Al(NO_3)_3$ molecule can be surrounded with one AcAc-molecule, while Al-sec-butoxide condenses and thus four condensed molecule are surrounded by one AcAc-molecule (Figure 5). Hence in samples C and D, highly condensed aluminum domains, in other words, an homogeneous compositional range near to stiochiometry ($11Al_2O_3.La_2O_3$) is expected to be achieved. Subsequently the peptization step applies in sample D helps to dissolve the nanodispersed precipitates which may not be observed visually. DSC-curves deliver evidence for these assumptions: There are mainly three signal positions in each sample, occurring at temperatures lower than 400°C, at about 900°C and above about 1100°C. All peaks carry exothermic characteristics. Removal of organic and inorganic specious takes place up to 400°C. Decomposition of AcAc to yield formation of CO and $CO_2$ is responsible for the exothermic signal, occurring between 200-400°C. Moreover, depending on the starting material, either elimination of nitrates, butylates and/or Al-AcAc and La-AcAc may also take place at this temperature range. At the sample D, the sharp 316°C signal demonstrates that only decomposition of AcAc at that temperature occurs and butylate groups leaves the system as buthanol, during low temperature tempering (< 200°C), prior to DSC-measurement.

X-Ray diffraction of 1000°C calcined powders indicates that the exothermic DSC-signal, occurring at about 900°C is responsible for formation of $\gamma$-$Al_2O_3$. Since $\gamma$-$Al_2O_3$ is the only crystalline phase observed in all batches at 900°C, signals which occur at about 900°C seem to be responsible for its crystallization. Intensity of X-ray patterns of $\gamma$-$Al_2O_3$ increases at the sample D in which only $LaAl_{11}O_{18}$ formation was observed, demonstrating that this signal also displays homogeneity of the chemical short-range in the precursors. As a matter of fact, it is likely that this signal behaves similarly to that for mullite precursors where the 980°C signal was related sometime to formation of $\gamma$-$Al_2O_3$, but, mostly with high intensity to formation of mullite (18). Typically, it was reported that $\gamma$-$Al_2O_3$ contains always silicon and some cases up to 18 mol.% and crystallization of mullite occurs as the stiochiometric composition was reached (19). It is suspected that in the case of lanthanum hexaluminate precursors, formation and presence of $\gamma$-$Al_2O_3$ plays a similar role and might contain lanthanum. In the cases where the intermediate formation of $LaAlO_3$ was observed, the deposited amount of lanthanum in $\gamma$-$Al_2O_3$ must be very high, reaching to 1/1-ratio. Therefore, remaining amorphous phase at 900°C in such cases will be alumina-rich and formation of $LaAl_{11}O_{18}$ is diffusion rate-controlled. Above 1100°C, the formation of binary phases differs from sample A to D. All samples on calcination at 1200°C yield XRD patterns of both, $LaAlO_3$ and $LaAl_{11}O_{18}$. It was indicated by Vaidya et al. (11) that intermediate formation of perovskite structured $LaAlO_3$ prior to $LaAl_{11}O_{18}$ at relatively lower temperatures (< 1200°C) is unavoidable. It is also reported that even the compositions richer in $Al_2O_3$ (1:15) forms primarily $LaAlO_3$. Moreover, it was reported that the formation of $LaAl_{11}O_{18}$ is due to a reaction between $\gamma$-$Al_2O_3$ and $LaAlO_3$ (5). $\gamma$-phase disappears as lanthanum aluminates crystallize above 1100°C. Presumably, the formation of $LaAl_{11}O_{18}$ occurs by diffusion rate controlled process. This may explain the sluggish behavior of this formation observed by DSC-spectra. Sample B behaves very similar to Sample A displaying a similar DSC-spectra. Addition of an alkoxide starting material as the lanthanum source do not alter the condensation and polymerization processes, observed in all-inorganic sample A. Formation of $LaAlO_3$ might begin at about 1072°C, at a somewhat lower temperature compared to samples A and C (1106°C). However, $LaAlO_3$ remains relatively stable at temperatures above 1400°C and do not show much changes in phase sequences.

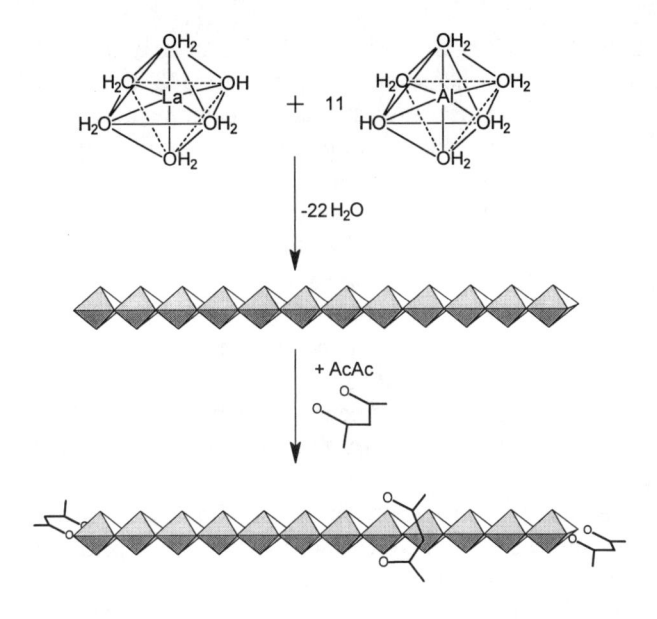

Figure 5 Condensation of hydrolyzed monomers in Batch D

XRD-study shows that the conversion progresses further at 1400°C, but is far away from completion and requires higher temperatures to reach fully conversion. Sample C contains similar phase combination to sample B at 1200°C, however, LaAlO$_3$ decreases significantly at 1400°C as the content of LaAl$_{11}$O$_{18}$ increases. In this case, formation of LaAl$_{11}$O$_{18}$ occurs as a result of degradation of LaAlO$_3$. We assume that excess Al$_2$O$_3$ exists as amorphous phase which causes LaAlO$_3$ to degrade or to react. In sample D, simultaneous removal of organic residues at low temperatures leads to a well-ordered structural formation. Hence, LaAl$_{11}$O$_{18}$ forms directly at about 1200°C. The DSC-signal at 1206°C ought be resulted from this formation. Improvement of short-range order during polymerization leads to a stiochiometric bonding of aluminum and lanthanum-species already during removal of organic substances. γ-Al$_2$O$_3$ forms at about 900°C as the first crystalline phase. Between 900° to 1100°C, the concentration of γ-Al$_2$O$_3$ increases, previous to formation of LaAl$_{11}$O$_{18}$ (Figure 6). Our assumption is that at this stage a gradual incorporation of lanthanum may take place in a similar manner as it is observed in mullite precursors and this leads to formation of LaAl$_{11}$O$_{18}$, without going through the formation of intermediate phase LaAlO$_3$. By combined processing of the sol through peptization and controlled hydrolysis, a better gel formation and a better coating of fibers can be achieved.

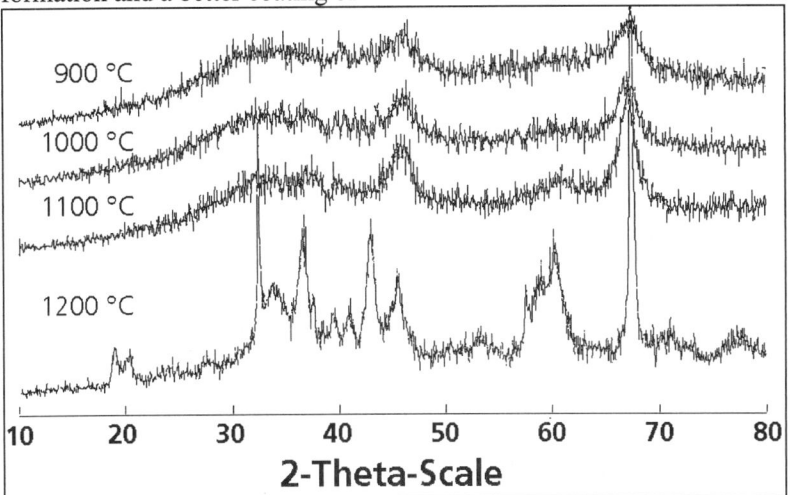

Figure 6- XRD-Data of sample D at the temperature range between 900° and 1200°C.

## CONCLUSIONS

Coating of fibers with $LaAl_{11}O_{18}$ has been achieved by using various starting materials and 2,4-Pentandionate as a condensation and polymerization controlling ligand. Inorganic starting materials such as Lanthanum nitrate and Aluminum nitrate results in formation of $LaAlO_3$ as an intermediate phase. Formation of $LaAl_{11}O_{18}$ as a pure phase is depended on a sluggish decomposition of this phase. In this case of organic starting materials, the formation of $LaAlO_3$ is eliminated and $LaAl_{11}O_{18}$ forms directly at about 1200°C.

Drying process is an important step which should be carried out under controlled conditions to obtain continuous coatings.

## REFERENCES

1- T. Mah and K.A. Keller, T.A. Parthasarathy and J. Guth, Fugitive Interface Coating in Oxide-Oxide Composites, Ceram. Eng. Sci. Proc. **12** (9-10) 1802-15 (1991).
2- P.E.D. Morgan and D.B. Marshall, Functional Interfaces for Oxide/Oxide Composites, Mater. Sci. Eng. **162**, 15-25 (1993).
3- J.B. Davis, J.P.A. Löfvander, A.G. Evans, E. Bischoff and M.L. Emiliani, Fiber Coating Concepts for Brittle Matrix Composites, J. Am. Ceram. Soc., **76** (5) 1249-57 (1993).
4- R.S. Hay, T. Mah and C. Cooke, Molybdenum-Palladium Fiber-Matrix Interlayers for Ceramic Composites, Ceram.Eng.Sci.Proc., **15** (5) 760-68 (1994).

5- M. K. Cinibulk and R.S. Hay, Textured Magnetoblumbite Fiber-Matrix Interphase Derived from Sol-Gel Fiber Coatings, J. Am. Ceram. Soc., **79** (5) 1233-46 (1996).
6- K. K. Chawla, Z. R. Xu, J. S. Ha, M. Schmücker, H. Schneider, Effect of BN coating an the strength of a mullite type fiber, App. Comp. Mater., 4 (1997) 263.
7- E. Mouchon and Ph. Colomban, Oxide Ceramic Matrix/Oxide Fibre Woven Fabric Composites exhibiting dissipative Fracture Bahaviour, Composites, **26**, 175-82 (1995).
8- P. E. D. Morgan and D. B. Marshall and R. M. Housley, High Temperature Stability of Monazite-Alumina Composites, J. Mat. Sci. Eng., ...(1994).
9- S. M. Johnson, Y. Blum, C. Kazanawo, H.-J. Wu, J. R. Porter, P. E. D. Morgan, D. B. Marshall and D. Wilson, Processing and Properties of an Oxide/Oxide Composite, in: Key Engineering Materials,Trans Tech Pub., Vols. 127-131, pp.231-238 (1997).
10- N. Iyi, S. Takekawa and S. Kimura, Crystal chemistry of hexaluminates:ß-alumina and magnetoplumbite structures, J. Solid State Chemistry, **83**, 8-9 (1989).
11- K.J. Vaidya, C.Y. Wang, M. DeGraef and F.F. Lange, Heterepitaxy of rare-earth hexaaluminates on sapphire, J. Mater. Res., Vol.9(2) 410-419 (1994).
12-M. Muira, H. Hongoh, T. Yogo, S. Hirano and T. Fujii, Formation of plate-like lanthanum-ß-aluminate crystal in Ce-TZP matrix, J. Mater. Sci., 29, 262-68 (1994).
13- M.K. Cinibulk, Synthesis and characterization of sol-gel derived lanthanum hexaluminate powders and films, J. Mater. Res., Vol. 10(1) 71-76 (1995).
14- R.S. Hay and E.E. Hermes, Sol-Gel Coatings on continuous ceramic fibers, Ceram. Eng. Sci. Proc., 11 (9-10) 1526-1538 (1990).
15- S. B. Desu, R.O. Claus, R. Raheem and K.A. Murphy, High Temperature Sapphire Optical Sensor Fiber Coatings, SPIE, Vol. 1307, Electro-Optical Materials for Switches, Coatings, Sensor Optics and Detectors, 2-9 (1990).
16- D. B. Gundel, P. J. Taylor, F. E. Wawner, Fabrication of thin oxide coatings on ceramic fibers by a sol-gel technique, J. Mater. Sci. **29**, 1795-1800 (1994).
17- O. Yamagushi, K. Sugiura, A. Mitsui and K. Shimizu, New Compound in the system $La_2O_3$-$Al_2O_3$, J. Am. Ceram. Soc., **68** (2) C-44-C-45 (1985).
18- H: Schneider, B. Saruhan, D. Voll, L. Mervin and A. Sebald, Mullite precursor Phases, J. of Europ. Ceram. Soc. **11** (1993) 87-94.
19- H. Schneider, D. Voll, B. Saruhan, M. Schmücker, Constitution of the γ-$Al_2O_3$ Phase in Chemically Produced Mullite Precursors, J. of Europ. Ceram. Soc. **13** (1994) 441-448.

# INTRODUCTION OF SINTERING AID TO SILICON NITRIDE SYSTEMS VIA COLLOIDAL PROCESSING

A.C. Orlando* and R.A. McCauley
Malcolm G. McLaren Center for Ceramic Research
Rutgers, The State University of New Jersey
Piscataway NJ 08854 USA

## ABSTRACT

Two commercially available silicon nitride powders were coated with various sintering aids by means of colloidal processing. The interaction of the powders and the surfactants is discussed, along with the mechanism that describes the process. TEM and FESEM studies are used in order to observe the absorption of the sintering aid as it coats the silicon nitride powders. Samples of these coated powders were prepared in order to show the advantage of using these coated powders over using conventionally prepared powder for the production of near theoretical density ceramic parts. Colloidal processing is shown to be superior to conventional mechanical mixing of ceramic powders for the controlled, homogeneous incorporation of sintering aids into silicon nitride systems. It is concluded that this colloidal processing technique can be expanded to other systems where the incorporation of small amounts of secondary phases can have profound effects on the final properties and characteristics of ceramic materials.

## INTRODUCTION

The processing of novel ceramic materials often relies on the incorporation of minor ingredients as secondary phases. These secondary phases can act as sintering aids, dispersants, or microstructure enhancers, among other things. Conventional ceramic processing, such as ball milling and co-precipitation very often produce variations in density and distribution of these secondary phases. These variations can produce microstructural defects and non-uniform property behavior in the fired ceramic part. Therefore, it is necessary to find a method of incorporating these materials in order to produce a homogeneous uniform distribution within the major phase of interest.

Conventional ceramic processing often leads to inhomogeneities in the distribution of additives within a ceramic slip. Ball milling is one of the most common techniques for mixing two or more ceramic powders. This technique relies upon physical contact between the constituents to produce an intimate mixture. Often the size of the particles and the particle size distribution can be controlled through this type of mixing, but rarely affords uniformity or reproducibility. The process can also add impurities, due to the media used, and requires subsequent processing in order to eliminate agglomerates or cast ceramic slips. Additions of secondary phases through more advanced techniques, such as precipitation reactions generally afford few improvements over the conventional techniques. Here, more homogeneous additions can be made, but control of the distribution of precipitates can be difficult. Precipitation reactions also rely heavily upon the ability to intimately mix the constituents. Furthermore, these types of advanced mixing techniques often have detrimental effects associated with them such as the very high shrinkage rates that accompany the precipitates, which confounds the situation.[1] Clearly, it is necessary to derive a method of incorporating secondary phases into ceramic systems in a controlled homogeneous fashion.

The amount of secondary phases needed is another important topic of interest when considering which technique to use. The properties of many ceramic systems are often dictated by the properties of the secondary phases, which may often be impurities. When secondary phases are desired for one purpose, they may have adverse effects on other properties of the material. Hence, limitations of a ceramic material are not necessarily the limitations of the bulk phase, but often the properties are limited by the properties of the minor constituents. Here, limitations can be described by judging the ceramic system, including all phases, as a whole. An example of this is the high temperature properties of silicon nitride formed by liquid phase sintering. The inherent high temperature properties of $Si_3N_4$ powders are often compromised by the sintering aids used to promote liquid formation and densification at the sintering temperature. The secondary phases react with the constituent, the $Si_3N_4$, at elevated temperature to form a eutectic liquid which upon cooling remains within the material as an amorphous grain boundary phase.[2] Subsequent use of the $Si_3N_4$ material at elevated temperatures above the eutectic (generally greater than 1000°C) results in softening of the remaining amorphous phases and a resultant decrease in high temperature properties.[3-4] Conventional ceramic processing in these systems typically results in $Si_3N_4$ materials containing as much as 20 wt% oxides as sintering aids to produce full densification.[5] Generally the larger the amount of oxide the greater the softening due to increased amounts of residual amorphous material. The colloidal process for adding sintering aid described above has been able to produce fully dense $Si_3N_4$ with as little as 2 wt% oxide added. This decrease is due to the ability to uniformly distribute the sintering aid among the $Si_3N_4$ grains.

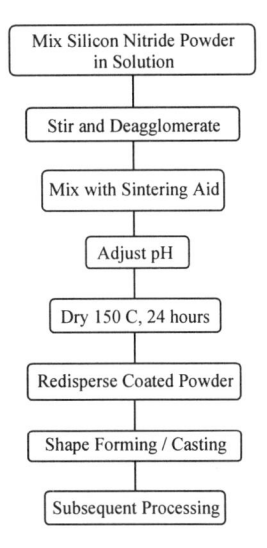

```
┌─────────────────────────┐
│ Mix Silicon Nitride Powder │
│      in Solution         │
└─────────────────────────┘
              │
┌─────────────────────────┐
│  Stir and Deagglomerate │
└─────────────────────────┘
              │
┌─────────────────────────┐
│   Mix with Sintering Aid │
└─────────────────────────┘
              │
┌─────────────────────────┐
│       Adjust pH         │
└─────────────────────────┘
              │
┌─────────────────────────┐
│   Dry 150 C, 24 hours   │
└─────────────────────────┘
              │
┌─────────────────────────┐
│ Redisperse Coated Powder │
└─────────────────────────┘
              │
┌─────────────────────────┐
│  Shape Forming / Casting │
└─────────────────────────┘
              │
┌─────────────────────────┐
│  Subsequent Processing  │
└─────────────────────────┘
```

Figure 1. Processing Flowchart

Attempts have been made by other researchers to improve the homogeneous additions of sintering aids to $Si_3N_4$ systems by various techniques. These techniques have included micro-encapsulation, and sol-gel processing. M. Kulig et al.[6] showed that homogeneous coatings of sintering aid onto $Si_3N_4$ powders was possible via sol-gel processing of nano-sized sintering aid particles. A.K. Garg et al.[7] microencapsulated silicon nitride with yttria and yttria-aliumina precursors, and E. Liden et al.[8] incorporated yttria particles on $Si_3N_4$ by electrostatic adsorption. Many of these techniques are impractical industrially as they are often very difficult to perform or they are very expensive.

The purpose of this research was to devise a method to control the distribution of secondary phases within any primary ceramic system. The systems previously studied were silicon nitride systems with various additives as sintering aids. It was found to be possible to promote complete (near theoretical) densification with a small amount of sintering, aid provided that the sintering aid was uniformly and homogeneously distributed on the silicon nitride powder. The uniform incorporation of sintering aid was assumed when all $Si_3N_4$ grains were in complete contact with the liquid formed from the sintering aids at elevated temperature. This situation facilitated the solution-precipitation mechanism of sintering, prominent in the sintering of covalent ceramics, by decreasing diffusion distances between grains. The final microstructure of sintered bodies consisted of a uniform thin grain boundary phase surrounding every grain, with only a few small pockets of residual glass. By controlling the distribution of additives within the green ceramic body it was possible to tailor the properties of the final material by avoiding any microstructural variation, due to inhomogeneous incorporation of the secondary phase.

EXPERIMENTAL PROCEDURE

The addition of surfactants by the colloidal processing was carried out in aqueous and non-aqueous solutions. A processing flowchart for the procedure is given as Figure 1. The first step involves disbursement of the primary phase in a solution, typically aqueous solutions are chosen. Next, the secondary phase is added and the pH of the solution is adjusted to allow interaction between the phases. Typically, the major phase is dispersed in a solution such that the surface groups of the powder can act as a reagent allowing bonding between the phases. The solution is then milled, the powder is dried and then casting and/or subsequent processing can be done. Subsequent bonding can include the incorporation of other materials coated upon the first. As a result of the process a uniform distribution is obtained for which chemical bonding between the species has resulted.

In this research two silicon nitride powders were coated with different sintering aids. The powders were UBE-E10 silicon nitride and Stark LC9 silicon nitride. One sintering aid system was comprised of yttria and alumina containing alkoxides in the ratio of 2:1, with a total oxide content of 2 wt%. The second system consisted of a yttria based organic resin. The resin consists of yttria ions associated with a weak acid in the ration of 0.1 gm of yttria per 1.0 gm of powder. Again a total of 2 wt% oxide was added to the silicon nitride powder.

Several techniques can be employed to obtain information about the bonding and the uniformity of the coating that is created. Typically, Transmission Electron Microscopy (TEM) is used to observe the surface of the major phases. Usually, TEM can be accompanied by Energy Dispersive Spectroscopy (EDS) to determine the chemical composition of the observed species. More recently, high resolution Field Emission Scanning Electron Microscopy (FESEM) has been employed to observe the three dimensional nature of the coated and the uncoated powders. These techniques can be combined to give one complete information about the surface coverage of the coating upon the major phase.

Other more common ceramic characterization techniques can be employed to get a sense of the completeness of the coating on the ceramic powder. Typically, particle size measurements, and surface area measurements can be made. These parameters are generally varied as one material is coated with a second material. Also, the measurement of the electro-phoretic mobility of the particles can be done. These measurements allow calculation of the iso-electric

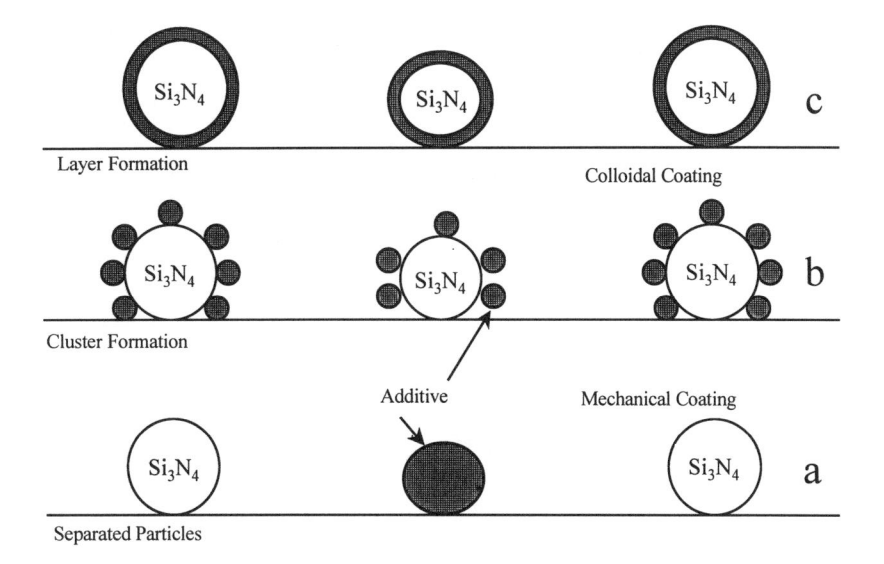

Figure 2. Schematic representation of the adsorption of sintering aid onto silicon nitride particles by various techniques.

points (pH$_{iep}$) of the constituent powders. As one material is coated with another, one should observe a permanent shift in the primary powder's pH$_{iep}$.

RESULTS AND DISCUSSION

Ball milling and other conventional mixing techniques seldom promote a uniform distribution of secondary phases. Typically, the conventional techniques for powder mixing lead to the formation of systems characterized by separated particles (see Figure 2). This situation results in mixing that is of the scale of the particle sizes involved, and is related to the particle size distributions of all phases involved, as well as their respective volume fractions. (Fig. 2a) This is not an acceptable method of mixing materials that require very intimate mixing of the systems involved. Typically, to obtain scales of mixing that are on the order of nanometers, one needs to begin with starting powders that are orders of magnitude smaller. These types of powders are generally quite expensive, and difficult to obtain, if at all. Therefore, a technique of mixing that produces homogeneous uniform distributions of secondary phases needs be devised. The colloidal process of adding secondary phases often leads to an increased uniformity and a homogeneous distribution of secondary phases.(Fig. 2 b & c) Here, the formation of clusters or layers of additives on the primary phase particles can result. The difference between the two situations being related to the

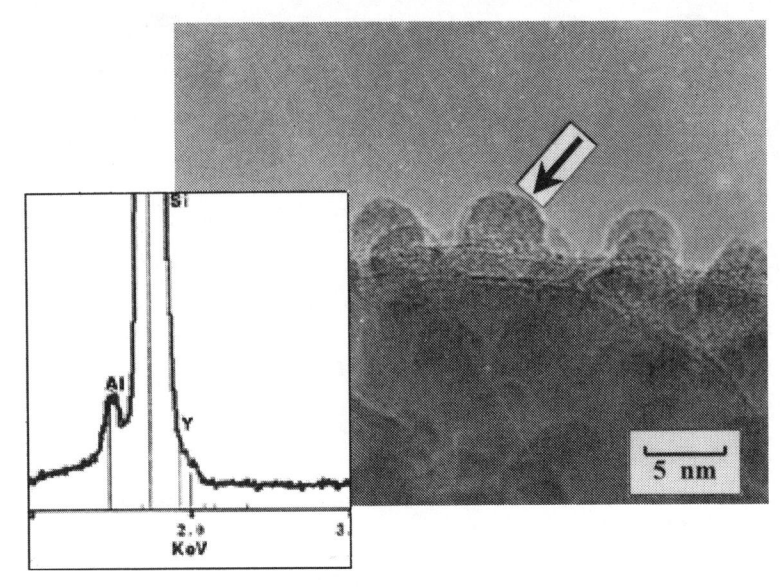

Figure 3. Transmission electron microscope (TEM) image of coated silicon nitride powder. Inset is Energy Dispersive Spectroscopy (EDS) spectra of indicated region showing the presence of yttria and alumina in cluster formation on the surface of the $Si_3N_4$ powder.

exact surface chemistry and included surface species of these primary phase particles.

Coating of primary powders with the desired additives by colloidal processing has been successfully proven to produce uniform homogeneous dispersions. Liquid mediums are employed to allow deagglomeration of the starting powders in order to expose the greatest amount off surface area. Changes in the processing conditions change the surface character of the powders, so these need to be strictly controlled. Also, to have control over the coating, an irreversible interaction between the powder surface and the additive needs be established. This interaction can also provide stability against fluctuations that may occur during the processing. Persson, et al.[9] concluded that weak interactions are reversible and do not allow control over the degree of coating or the uniformity of the coating.

Other researchers have studied the ability to homogeneously incorporate sintering aids into silicon nitride systems via colloidal processing. The $pH_{iep}$ of the $Si_3N_4$ were found to be permanently changed by controlled addition of sintering aid, indicating complete and homogeneous distribution of the sintering

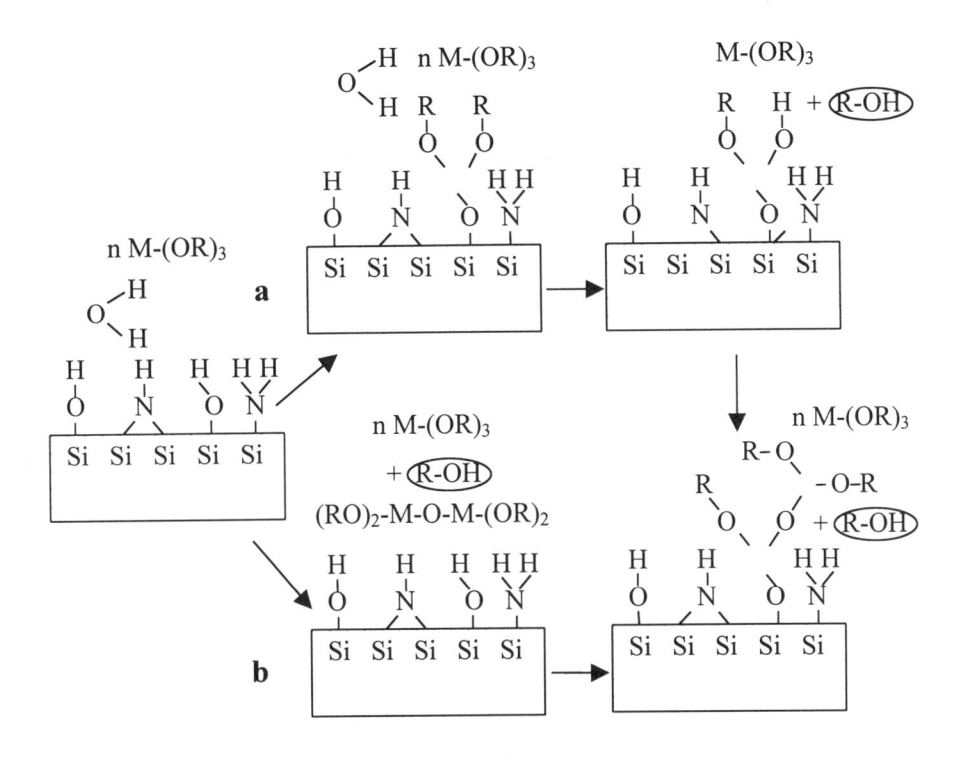

Figure 4. Proposed mechanism for the colloidal coating process. Route (a) represents direct reaction, and route (b) involves hydrolysis that results in dimer formation. Both routes produce chemisorbed coatings on the silicon nitride surface.

aid on the $Si_3N_4$ particles. This distribution was found to be a direct result of controlled pH in an aqueous medium. These studies concluded that uniform incorporation of sintering aids onto $Si_3N_4$ was indeed possible in a reproducible manner, once certain powder characteristics were fully understood.[10] The systems were found to be stable during subsequent processing and they were also found to be susceptible to multiple coatings of various constituents, two very important advances over the conventional ceramic techniques for mixing of powders.

Organometallic complexes with the desired metal cations were incorporated into $Si_3N_4$ systems via colloidal processing. The additives did not have fully coordinated metal ions and could thus interact with polar surface groups, producing very stable chemical bonds. The bonding was produced as the

Table I. BET and Iso-electric Point Measurements of Constituents

| Powder (pH$_{iep}$) | As-Recieved | Coating | As-Coated |
|---|---|---|---|
| UBE-E10 | 6.0 | 9.0 | 8.5 |
| Stark | 8.0 | 5.0 | 5.7 |

| Powder (m$^2$/g) | As-Recieved | | As-Coated |
|---|---|---|---|
| UBE-E10 | 9.3 | | 11.2 |
| Stark | 17.6 | | 18.0 |

pH of the aqueous medium was altered so that interaction between the charged surface of the additive could interact with the powders surface sites. The major advantage to this process was the incorporation of sintering aids, uniformly and in reduced amounts to Si$_3$N$_4$ systems. Other advantages included surface charge modifications which resulted in ease of handling and the ability to produce multiple layered coatings onto the powder surface. The latter being almost impossible using other techniques.

Figure 3, is a TEM image of a Si$_3$N$_4$ Powder (UBE-E10) coated with yttria- and alumina-*tri*-isopropoxides, for sintering aids. The formation of clusters of secondary phases on the Si$_3$N$_4$ particles can clearly be seen. The inset is an EDS spectra of the indicated region, showing the composition to be that of Al and Y. The cluster like formation is due to the relatively high number of silanol groups associated with the surface of this particular Si$_3$N$_4$.

The mechanism of the attachment of the sintering aid species to the silicon nitride has been proposed by Wang. Figure 4 illustrates the two possible cases. The first mechanism (Fig. 4a) involves direct reaction of the M – (OR)$_3$ groups of the surfactant with the surface silanol groups associated with the Si$_3$N$_4$ surface. This reaction produces Si – O – M bonds, and is characterized by chemical adsorption followed by polymerization and hydrolysis. The second mechanism (Fig. 4b) involves an initial hydrolysis of the M – (OR)$_3$ molecules, consumption of residual moisture, and finally the formation of dimers. This process is then followed by subsequent adsorption of the dimers onto the surface silanol sites of the Si$_3$N$_4$ particles. Both of these processes result in a coating that is chemisorbed onto the primary particles.

Figure 5. Field Emission Scanning Electron Microscope (FESEM) image of fracture surface of a sintered silicon nitride sample. Coating can be seen in light contrast to the silicon nitride grains.

The change in surface chemistry of these silicon nitride powders was measured using acoustophoresis and BET measurements. Table I summarizes the results of these tests. The $pH_{iep}$ for the as-received powders along with the measurements for the coated powders is given. The iso-electric points of the powders were observed to be permanently and irreversibly shifted as they were coated with the sintering aids. The $pH_{iep}$ of the UBE-E10 powder shifted from a value of ~6.0 in the as-received state, to a value of ~8.5 when coated with the alumina and yttria containing organo-metallic sintering aids. These sintering aids were determined to have a $pH_{iep}$ value between 9.0 and 10.9. The Stark powder was coated with the yttria containing resin that had an initial $pH_{iep}$ around the value of 5.0. Here, the $pH_{iep}$ of the silicon nitride was found to shift from ~8.0 to ~5.7 when coated with this materials. These observed shifts in $pH_{iep}$, is a direct result of the coating process. Here, the surface of the silicon nitride powders was modified to behave like the appropriate sintering aid. The adsorption of the sintering aid onto the powder can be seen as a change in surface character to that of the sintering aid instead of acting like a surface dominated by silanol species.

The change in surface area as measured by a single point BET method was in agreement with the above findings. An increase in surface area for both powders was observed. This is consistent with the formation of a cluster type structure as indicated by Figure 2. These surface area increases further support the chemisorption of the sintering aid onto the surface of the silicon nitride particles. This cluster formation can be observed by examining the TEM photos of the $Si_3N_4$ coated powders, shown in Figure 3. Thus the results clearly indicate the ability of the colloidal technique to produce a uniform distribution of sintering aid onto the silicon nitride.

The UBE silicon nitride powder was chosen for further densification studies. The small amounts of sintering aid added to this system were found to be sufficient to produce near theoretical density samples. FESEM images of the resultant microstructure of fracture surfaces is given as Figure 5. The hexagonal nature of the $\beta$-$Si_3N_4$ grains can be observed in dark contrast to the higher atomic number Al and Y containing thin grain boundary film. The resultant room temperature MOR of the samples was found comparable to that of other researchers utilizing larger amounts of sintering aid. The necessity of adding larger amounts of sintering aid to the $Si_3N_4$ samples was found to be problematic when high temperature applications are desirable. Cinibulk[12] found that samples containing greater the 10 wt% sintering aid lost 71% of the room temperature strength when tested at 1300°C. The samples used in this research retained nearly 60% of their room temperature strength when tested at the same temperature! This result was found to be a direct result of the ability to add very small amounts of sintering aid to the material and still promote full densification.

The introduction of sintering aid to silicon nitride powders via colloidal processing is superior to other incorporation mechanisms. This process affords the ability to reduce the total amount of sintering aid required to produce full density. In turn, this reduced quantity of secondary phase promotes better properties at room and elevated temperatures. Since, the grain boundary phases have the predominant effect on the overall properties of any ceramic it can be expected that the ability to thoroughly control the distribution of this phase will be beneficial to any material.

CONCLUSION

A colloidal solution of silicon nitride was used to incorporate sintering aids onto the surface of this powder. The incorporation was observed, both by TEM and FESEM, to be uniform over the surface of the silicon nitride. A sintering study was performed, and the results indicated that materials formed via this colloidal processing route had superior high temperature properties than other silicon nitride materials. Therefore, it is concluded that this processing technique is superior for the controlled incorporation of secondary phases into ceramic systems that require only small amounts of these to tailor their specific properties.

ACKNOWLEDGEMENTS

The authors would like to thank the Center for Ceramics Research and the New Jersey Commission of Science and Technology for financial support of this work.

REFERENCES

[1] T.M. Shaw, B.A. Pethica, "Preparation and Sintering of Homogeneous Silicon Nitride Green Compacts," *J. Am. Cer. Soc.*, **69** 88 (1986).
[2] M.H. Lewis and R.J. Lumby, "Nitrogen Ceramics: Liquid Phase Sintering," *Powder Metall.*, **26** 73-81 (1983).
[3] F.F. Lange, "High Temperature Strength Behavior of Hot-Pressed $Si_3N_4$: Evidence for Subcritical Crack Growth," *J. Am. Cer. Soc.*, **57** [2] 84-87 (1974).
[4] R. Kossowsky, D.G. Miller, and S.D. Diaz, "Tensile and Strengths of Hot-Pressed $Si_3N_4$," *J. Mater. Sci.*, **10** [6] 983-987 (1975).
[5] M.J. Hoffman, "High-Temperature Properties of $Si_3N_4$ Ceramics," *MRS Bull.* Feb. (1995).
[6] M. Kulig, W. Oroschin, and P. Griel, "Sol-Gel Coating of Silicon Nitride with Mg-Al Oxide Sintering Aid," *J. Euro. Ceram. Soc.*, **5** 209-217 (1989).
[7] A.K. Garg and L.C. DeJonghe, "Microencapsulation of Silicon Nitride Particles with Yttria and Yttria-Alumina Precursors," *J. Mater. Res.*, **5** [1] 136-142 (1990).
[8] E. Liden, E. Bergstroem, M. Persson, and R. Carlsson, "Surface Modification and dispersion of Silicon Nitride and Silicon Carbide Powders," *J. Am. Cer. Soc.*, **7** [6] 361-368 (1991).
[9] R. Dejong, "Incorporation of Additives into Silicon nitride by Colloidal Processing of Metal Organics in an Aqueous Medium," Ph. D. Thesis. Rutgers University. New Brunswick, NJ (1990). Ann Arbor No. AA19123266.
[10] P.N. Joshi and R.A. McCauley, "Metal Organic Surfactants as Sintering Aids for Silicon Nitride in Aqueous Medium," *J. Am. Cer. Soc.*, **77** [11] 2926-2934 (1994).
[11] H.M. Wang, "Metal-Organic Surfactants as Sintering Aids for Silicon Nitride in Non-Aqueous Media," Ph.D. Thesis. Rutgers University. New Brunswick, NJ (1995). AA19537656.
[12] M.K. Cinibulk and G. Thomas, "Grain Boundary Crystallization and Strength of Silicon Nitride with a YSiAlON Glass," *J. Am. Cer. Soc.*, **73** [6] 1606-1612 (1990).

# Polymer Processing

# SYNTHESIS OF SiC-Si$_3$N$_4$ COMPOSITES FROM A POLYMERIC PRECURSOR

Xujin Bao[a], Mohan J. Edirisinghe[a], Gerard F. Fernando[b] and Michael J. Folkes[b]
[a]Institute of Polymer Technology and Materials Engineering, Loughborough University, Loughborough, Leicestershire LE11 3TU, UK.
[b]Department of Materials Engineering, Brunel University, Uxbridge, Middlesex UB8 3PH, UK.

## ABSTRACT

In this paper, we describe a new approach for the synthesis of SiC-Si$_3$N$_4$ composites using one polymeric precursor. For this purpose, a co-polysilane-cyclodisilazane precursor was synthesized and characterized using Fourier transform-infrared spectroscopy. The pyrolytic yield of this precursor is higher than that of the homopolymers due to a high level of cross-links formed during pyrolysis. X-ray diffraction studies of the crystallized product at 1650°C confirmed the presence of both SiC and Si$_3$N$_4$.

## INTRODUCTION

Considerable interest has been generated in recent years in the pyrolysis of silicon-containing polymeric precursors to produce non-oxide ceramic materials such as silicon carbide (SiC), silicon nitride (Si$_3$N$_4$), silicon carbonitride, silicon oxynitride and their composites [1-7]. The main advantage of precursor route is the processability and low fabrication temperatures of the polymers. Moreover, control of polymer structure, molecular weight and curing mechanism allow the optimization of the processing, the ceramic yield and the final microstructures [8-10].

In the present work, a new co-polysilane-cyclodisilazane precursor with reactive groups (Si-H, N-H, CH$_2$=CH) was synthesized and used as a single precursor for the synthesis of a Si$_3$N$_4$-SiC composite.

## EXPERIMENTAL DETAILS

Polymer synthesis

Polymethylvinylsilane (PMV) and polycyclodisilazane (PCSZ) were synthesized using the reactions shown in **Scheme 1**. PMV and PCSZ are precursors for SiC and $Si_3N_4$, respectively.

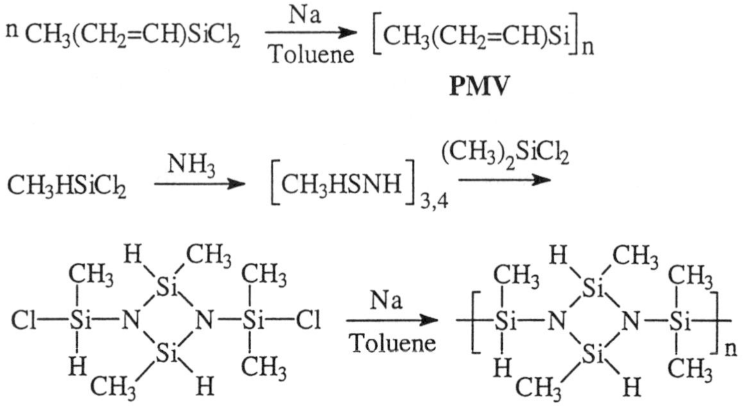

$$n\ CH_3(CH_2=CH)SiCl_2 \xrightarrow[\text{Toluene}]{\text{Na}} [CH_3(CH_2=CH)Si]_n$$
**PMV**

$$CH_3HSiCl_2 \xrightarrow{NH_3} [CH_3HSNH]_{3,4} \xrightarrow{(CH_3)_2SiCl_2}$$

[cyclodisilazane dichloride structure] $\xrightarrow[\text{Toluene}]{\text{Na}}$ [PCSZ structure]

**PCSZ**

**Scheme 1**

A polymethylvinylsilane-cyclodisilazane co-polymer (PMV-CSZ) was prepared as shown in reaction **Scheme 2** and used as the precursor for the $SiC$-$Si_3N_4$ composite.

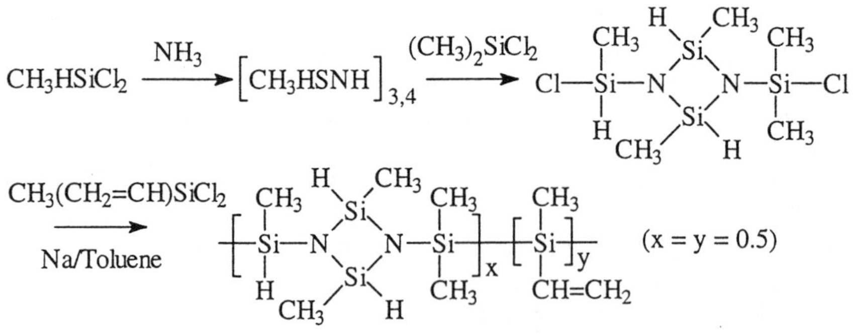

$$CH_3HSiCl_2 \xrightarrow{NH_3} [CH_3HSNH]_{3,4} \xrightarrow{(CH_3)_2SiCl_2}$$ [cyclodisilazane dichloride structure]

$$CH_3(CH_2=CH)SiCl_2 \xrightarrow{\text{Na/Toluene}}$$ [PMV-CSZ copolymer structure] $(x = y = 0.5)$

**PMV-CSZ**

**Scheme 2**

Pyrolysis and crystallization

The precursors were heated from the ambient temperature to 900°C in flowing nitrogen (flow rate of 0.5 cm$^3$ min$^{-1}$) at 10°C min$^{-1}$ using a Perkin-Elmer TGA 7 thermal balance. These samples were also heated in a tube furnace in flowing nitrogen (flow rate approximately 250 cm$^3$ min$^{-1}$) from the ambient temperature to 1650°C at 3°C min$^{-1}$ followed by soaking at this temperature for 3 hours before cooling to the ambient temperature at 3°C min$^{-1}$. A crystallized ceramic product is thus obtained.

Characterization

Fourier transform-infrared (FT-IR) spectra of the PMV, PCSZ and the co-polymer PMV-CSZ were obtained using a Nicolet 710 FT-IR spectrometer in the wavenumber range of 4000-400 cm$^{-1}$.

X-ray diffraction (XRD) was carried out on each residue after heating to 1650°C. Samples for X-ray diffractometry were ground using a boron carbide mortar and pestle. A small amount of industrial methylated spirit (IMS) was added during grinding and the paste produced was placed on a single crystal silicon plate. The IMS was allowed to evaporate before XRD analysis was conducted. A Philips X-ray diffractometer was used with the silicon plate attached to a 20mm diameter stainless steel stub. Ni filtered Cu$K_\alpha$ radiation of wavelength 0.15406nm and a voltage and current of 35kV and 20mA were used, respectively. The scan range was from 10° to 90° with a step size of 0.021° and the scan speed was 0.02° s$^{-1}$.

## RESULTS AND DISCUSSION

Polymer characteristics

The polycyclodisilazane (**Figure 1a**) exhibits characteristic C-H stretching between 3100 and 2700 cm$^{-1}$. Methyl group stretching is observed at 2978 cm$^{-1}$. Additional peaks at 1414 and 1262 cm$^{-1}$ are characteristic of the asymmetric and symmetric bending modes of CH$_3$ bonded to silicon, respectively. A very strong characteristic peak at 2139 cm$^{-1}$ associated with the Si-H stretching. Two peaks at 1156 cm$^{-1}$ and 908 cm$^{-1}$ are associated with the cyclic Si$_4$N$_2$ skeleton. A characteristic band at 3390 cm$^{-1}$ is assigned to the stretching of N-H. The FT-IR spectrum of PMV (**Figure 1b**) shows methyl group stretching at 2966 cm$^{-1}$. Typical C-H vibrations associated with the vinyl group at 3065 cm$^{-1}$ and a characteristic peak at 1589 cm$^{-1}$ represents C = C stretching. A small band at 2144 cm$^{-1}$ is representative of Si-H stretching and shows that there is some silicon-hydrogen bond formation during polymerization. The bands at low wavenumbers of 697 cm$^{-1}$ for Si-C stretching and 464 cm$^{-1}$ for Si-Si, are typical. All the characteristic bands of these two precursors are observed in the spectrum of the copolymer PMV-CSZ (**Figure 1c**) [11-14].

Pyrolytic conversion

The pyrolytic yields of PMV, PCSZ and PMV-CSZ are shown in **Figure 2**. It can be seen that pyrolytic yield of the co-polymer is higher than that of both PMV and PCSZ. This is probably due to increased cross-linking in the co-polymer at the reactive sites of MV (Si-H, CH$_2$=CH) and CSZ (S-H, N-H). It is well known that cross-linking increases the pyrolytic yield [e.g. 15,16]. Possible reactions which could bring about this are illustrated in reaction **Scheme 3** given later in the text [7,14,15].

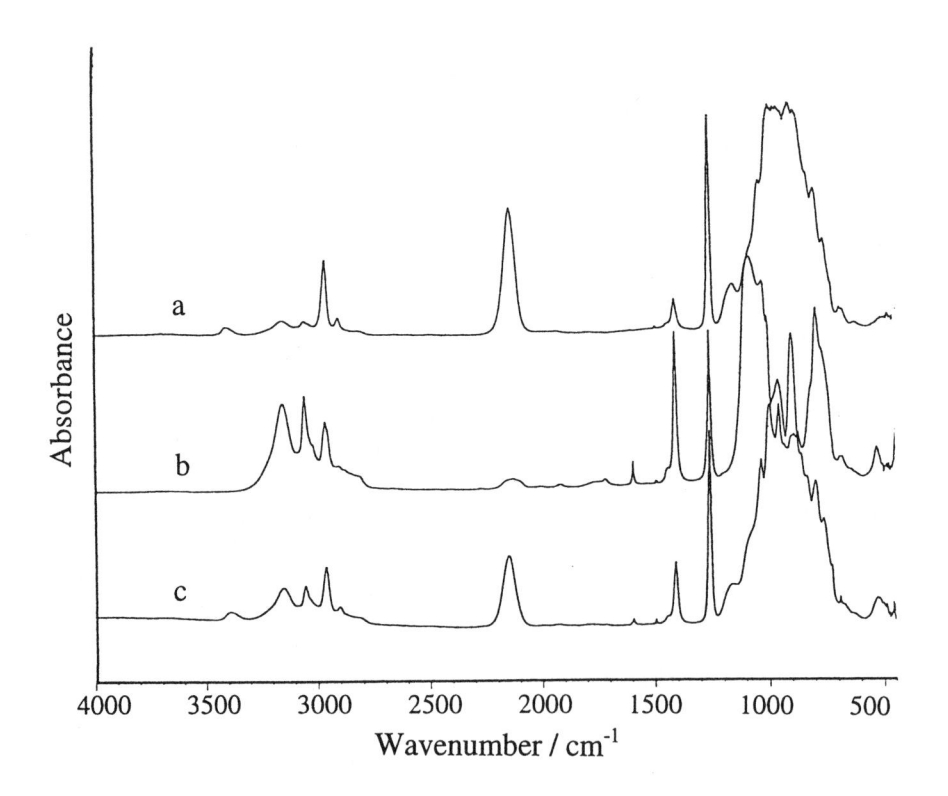

**Figure 1.** FT-IR spectra of (a) PCSZ (b) PMV and (c)PMV-CSZ

## Transamination

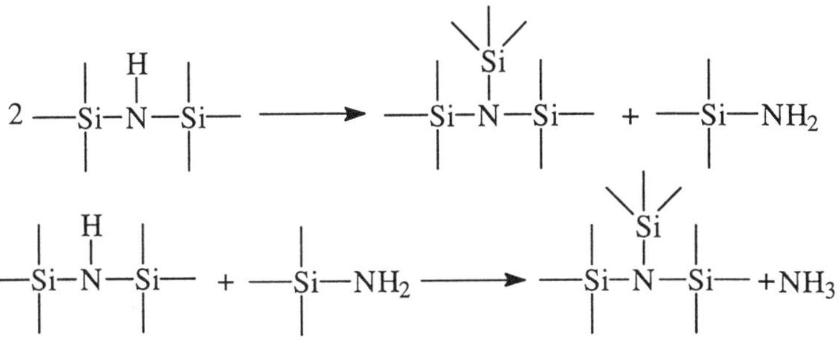

$$2 \ {-}Si{-}N(H){-}Si{-} \longrightarrow {-}Si{-}N(Si{\diagdown}){-}Si{-} + {-}Si{-}NH_2$$

$$-Si{-}N(H){-}Si{-} + {-}Si{-}NH_2 \longrightarrow {-}Si{-}N(Si{\diagdown}){-}Si{-} + NH_3$$

## Dehydrogenation

$$-Si{-}H + {-}Si{-}N(H){-}Si{-} \longrightarrow {-}Si{-}N(Si{\diagdown}){-}Si{-} + H_2$$

$$2\ {-}Si{-}H \longrightarrow {-}Si{-}Si{-}$$

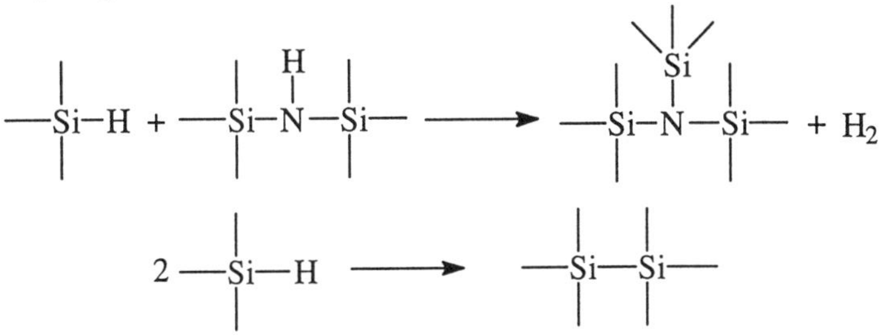

## Polymerization and addition

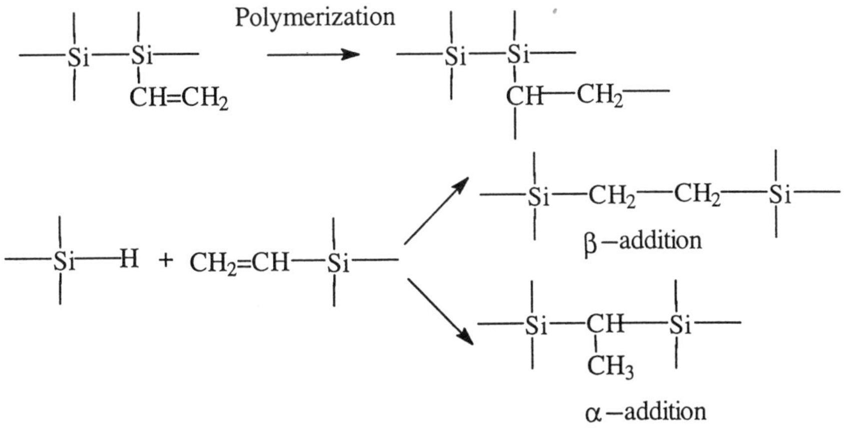

$$-Si{-}Si(CH{=}CH_2){-} \xrightarrow{\text{Polymerization}} {-}Si{-}Si(CH{-}CH_2{-}){-}$$

$$-Si{-}H + CH_2{=}CH{-}Si{-} \longrightarrow \begin{cases} {-}Si{-}CH_2{-}CH_2{-}Si{-} \quad \beta\text{-addition} \\ {-}Si{-}CH(CH_3){-}Si{-} \quad \alpha\text{-addition} \end{cases}$$

**Scheme 3**

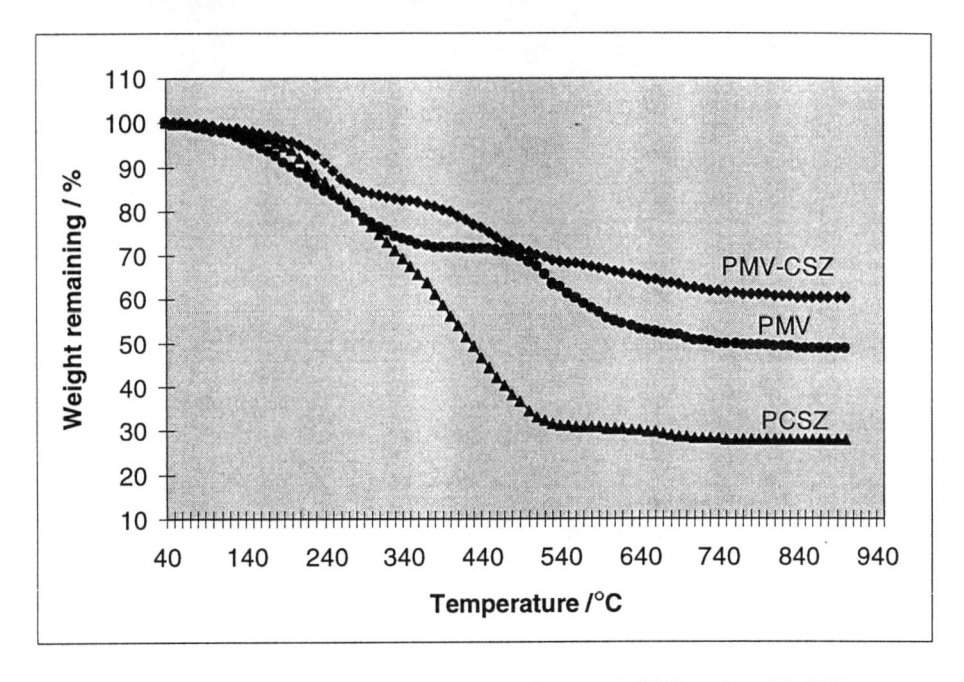

**Figure 2.** Thermogravimetric traces of PMV, PCSZ and PMV-CSZ

SiC-Si₃N₄ composite

Three peaks are observed in **Figure 3a** for the crystallized pyrolytic residue of PMV at $2\theta = 36°$, $61°$, and $72°$, which corresponds to the (111), (220), and (311) planes of $\beta$-SiC, respectively [14]. In **Figure 3b** for the crystallized pyrolytic residue of PCSZ, there are more than 20 peaks, which can be assigned to $\alpha$-Si₃N₄ [17]. However, there are three small peaks at $2\theta = 36°$, $61°$, and $72°$ observed as well. This could be due to the residual carbon reacting with the Si₃N₄ according to the reaction given below [18].

$$Si_3N_4 + 3C \xrightarrow{>1450°C} 3SiC + 2N_2$$

It can be seen clearly in **Figure 3c** that all the characteristic peaks of both $\beta$-SiC and $\alpha$-Si₃N₄ are observed after pyrolyzing and crystallizing the co-polymer, i.e. a SiC-Si₃N₄ composite was synthesized. More importantly, this new appraoch makes it possible to control the final microstructure of the SiC-Si₃N₄ composite by tailoring the composition of the polymeric precursor.

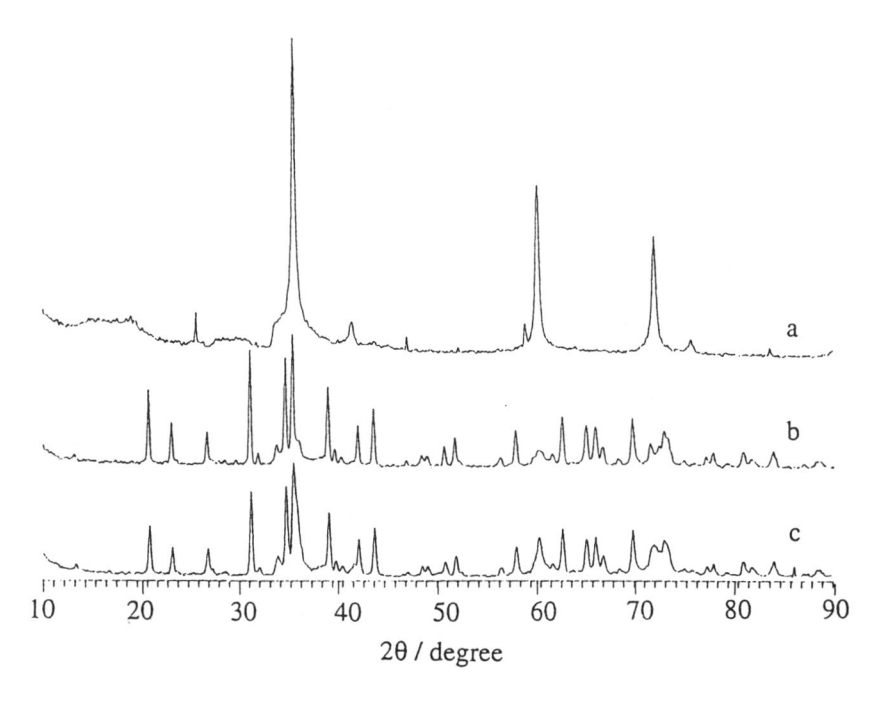

**Figure 3**. XRD patterns of (a) PMV, (b) PCSZ and (c) the copolymer (PMV-CSZ) after heating to 1650°C.

## CONCLUSIONS

A new co-polysilane-cyclodisilazane was synthesized and pyrolyzed in nitrogen. Thermogravimetric analysis showed that pyrolytic yield of the co-polymer was higher than that of individual polysilane and polycyclodisilazane probably because of more cross-links formed in the co-polymer during the pyrolysis. X-ray diffraction showed that the pyrolytic residue of the co-polymer after crystallization by heating to 1650°C was a composite of SiC and $Si_3N_4$.

## ACKNOWLEDGMENTS

The authors wish to thank the Soft Solid Initiative of EPSRC for funding this project (EPSRC Grant No. GR/L21976) and Dr. S. Woodisse for her technical assistance.

# REFERENCES

[1] S. Yajima, J. Hayashi, M. Omori and K. Okamura, "Development of a silicon carbide fiber with high tensile strength", *Nature*, **261** 683-685 (1976).

[2] D. Seyferth, and G. Wiseman, "High yield synthesis of $Si_3N_4$/SiC ceramic materials by pyrolysis of a novel polyorganosilazane", *J. Am. Ceram. Soc.*, **14** C132-133 (1984).

[3] R.W. Rice, "Ceramics from polymer pyrolysis, opportunities and needs. A material perspective", *Am. Ceram. Soc. Bull.*, **62** 889-892 (1983).

[4] K.J. Wynne and R.W. Rice, "Ceramics via polymer pyrolysis", *Ann. Rev. Mater. Sci.*, **14** 297-334 (1984).

[5] R. Riedel, M. Seher, J. Mayer and D.V. Szabo, "Polymer-derived Si-based bulk ceramics, Part I: preparation, processing and properties", *J. Euro. Ceram. Soc.*, **15** 703-715 (1995).

[6] M. Monthioux and O. Delverdier, "Thermal behaviour of (organosilicon)polymer-derived ceramics. V: main facts and trends", *J. Euro. Ceram. Soc.*, **16** 721-737 (1996).

[7] M. Birot, J-P. Pillot and J. Dunogues, "Comprehensive chemistry of polycarbosilanes, polysilazanes and polycarbosilazanes as precursors of ceramics", *Chem. Rev.*, **95** 1442-1477 (1995).

[8] M. Peucket, T. Vaahs and M. Bruck, "Ceramics from organometallic polymers", *Adv. Mater.*, **2** 398-404 (1990).

[9] D. Bahloul, M. Pereira, P. Goursat, N.S. Choong Kwet Yive and R. Corriu, "Preparation of silicon carbonitrides from an organosilicon polymer: thermal decomposition of the cross-linked polysilazanes", *J. Am. Ceram. Soc.*, **76** 1156-1162 (1993).

[10] X. Bao, M.J. Edirisinghe, G.F. Fernando and M.J. Folkes, "Precursors for silicon carbide synthesized from dichloromethylsilane derivatives", *J. Euro. Ceram. Soc.*, in press.

[11] D. Bahloul, M. Pereira and C. Gerardin, "Pyrolysis chemistry of polysilazane precursors to silicon carbonitride Part 1. Thermal degradation of the polymers", *J. Mater. Chem.*, **7** 109-116 (1997).

[12] D.J. Carlsson, J.D. Cooney, S. Gauthier and D.J. Worsfold, "Pyrolysis of silicon-backbone polymers to silicon carbide", *J. Am. Ceram. Soc.*, **73** 237-241 (1990).

[13] L.W. Breed, R.L. Elliott and J.C. Wiley, "The preparation and properties of N,N'-bis(chlorodimethylsilyl)tetramethylcyclodisilazane and its derivatives", *J. Organometal. Chem.*, **24** 315-325 (1970).

[14] W.R. Schmidt, L.V. Interrante, R.H. Doremus, T.K. Trout, P.S. Marchetti and G.E. Maciel, "Ammonia-induced pyrolytic conversion of a vinylic polysilane to silicon nitride", *Chem. Mater.*, **3** 257-67 (1991).

[15] N.S. Choong Kwet Yive, R.J.P. Corriu, D. Leclercq, P.H. Mutin and A. Vioux, "Polymer precursors: thermal cross-linking and pyrolysis of oligosilazane model compounds", *Chem. Mater.*, **4** 141-146(1992).

[16] C.L. Schilling Jr., "Polymeric routes to silicon carbide", *British Polym. J.*, **18** 335-358 (1986)

[17] G-Y. Yu, M.J. Edirisinghe, D.S. Finch, B. Ralph and J. Parrick, "Synthesis of α-silicon nitride powder from a polymeric precursor", *J. Euro. Ceram. Soc.*, **15** 581-590 (1995).

[18] Y.D. Blum, K.B. Schwartz and R.M. Laine, "Preceramic polymer pyrolysis, Part 1 Pyrolytic properties of polysilazanes", *J. Mater. Sci.*, **24** 1707-1718 (1989).

# EFFECT OF POLYMER ARCHITECTURE ON THE FORMATION OF Si-O-C GLASSES

Sandra Dire', Marco Oliver and Gian Domenico Sorarù
Dipartimento di Ingegneria dei Materiali, Università di Trento, Via Mesiano 77, 38050 Trento, Italy.

## ABSTRACT

Si-O-C glasses can be prepared either by direct pyrolysis of polysiloxanes and the pyrolytic conversion of sol-gel derived polymers obtained from monomeric precursors. The different arrrangement of the silicon units in networks having the same chemical composition can affect the pyrolysis chemistry and the ceramic features. In this paper siloxane networks built up with the same silicon units, i.e. $MeSiO_3$ (T) and $MeHSiO_2$ ($D^H$), have been prepared by the sol gel approach starting either from a mixture of methyltriethoxysilane and dimethylethoxysilane and a linear polymethylhydrosiloxane (PMHS). FTIR, DSC and $^{29}Si$ MAS NMR results point out that the two synthesis approaches lead to a different preceramic polymer architecture. This fact influences the pyrolytic conversion of the starting materials and strongly affects the high temperature stability of the derived SiOC glasses, as proved by the TG studies.

## INTRODUCTION

Silicon oxycarbide glasses for high temperature application such as matrices for CMCs are currently prepared from hybrid siloxane networks through a controlled pyrolysis process at 800-1000°C. Mechanical properties[1] and high temperature stability in oxidizing or inert atmospheres[2] are strongly related to the chemical composition of the SiOC glass and specifically to the number of C atoms bonded to Si atoms in the oxycarbide network and to the amount of a secondary free carbon phase. The composition of the starting siloxane network has a significant effect in ruling out the composition of the related SiOC glass[3, 4]. Indeed, it has been shown that siloxanes networks containing Si-Me and Si-H groups form SiOC glasses with a high level of carbon directly bonded to silicon *via* Si-C bonds and with a low amount of a secondary graphite phase[5]. On the other hand, the influence of the molecular arrangement of the various silicon units in the siloxane network, i.e. the "polymer architecture", has not been assessed yet.

In this paper hybrid siloxane networks containing Si-Me and Si-H groups have been prepared following two different approaches: (i) a conventional sol-gel process through hydrolysis-condensation reactions of methyltriethoxysilane, MeSi(OEt)$_3$, and methyldiethoxysilane, MeHSi(OEt)$_2$, and (ii) crosslinking, *via* Si-O-Si bridges, a linear polymethylhydrosiloxane, PMHS, (MeHSiO)$_n$. The two synthesis routes should lead to hybrid networks built by the same silicon units, i.e. trifunctional MeSiO$_3$ (T) and difunctional MeHSiO$_2$ (D$^H$), but with a different structural arrangement and allow to investigate the influence of the polymer architecture on the formation of SiOC glasses. Moreover, the use of PMHS as starting material for the preparation of oxycarbide glasses offers several advantages, compared to the molecular precursors, in the processing of CMCs *via* liquid infiltration and pyrolysis: indeed PMHS is one of the cheapest silicones on the market and it has a high solid content that allows to reduce the number of infiltration/pyrolysis steps that are necessary to obtain a dense component[6].

## EXPERIMENTAL
### Samples Preparation
#### *A) Synthesis of T/D$^H$ gels*

A mixture of methyltriethoxysilane (T), absolute ethanol and water (pH =2 for HCl) in 1: 1: 3 molar ratio was prehydrolyzed at 25 °C. After 15 minutes, different amounts of methyldiethoxysilane (D$^H$) were added to the clear solution in order to obtain T/D$^H$ molar ratios varying from 90/10 to 30/70. The solution was diluted to obtain an EtOH/Si=2 molar ratio. The stoichiometric water to react with the dimethylethoxysilane was added to complete the hydrolysis. The clear solutions were reacted for 30 minutes before casting in open tubes. The transparent and monolithic gels obtained after 10 days were dried at 60 °C for 6 days.

*B) Synthesis of PMHS-derived gels*

In the literature the following synthesis approaches have already been reported for the crosslinking of PMHS to preparare a useful ceramic precursor: (i) the dehydrocoupling reaction catalyzed by transition metal catalysts[7], (ii) the redistribution reactions catalyzed by $Cp_2TiMe_2$[8] and (iii) the hydrosilylation reaction with difunctional hydrocarbon chains[9]. In this work, with the aim of developing a sol-gel compatible approach, PMHS was crosslinked using a controlled hydrolysis reaction catalyzed with KF. A solution of polymethylhydrosiloxane (ABCR, MW=2270) in tetrahydrofurane (1:120 molar ratio) was reacted with water containing 0.05 moles/l of KF. The average number of Si-H groups in a PMHS chain is 35. Consequently, different hydrolysis molar ratios $r = H_2O/$ PMHS were used, ranging from r= 35 to r= 8.75. The mixtures were kept at different temperatures and reacted for different times before casting in open vessels. The clear sols lead, after several hours, to transparent gels. The gels were then dried at 60°C for 6 days.

The gel samples prepared according to A and B procedures were ground to fine powders (<100 μm) and pyrolyzed, in a silica tube furnace, up to 1000 °C at 5 °C/min under Ar flow (100 ml/min) and kept at 1000 °C for one hour. The samples heat-treated at 1000 °C were further heated, in a graphite furnace, to 1400°C and 1500°C at 10 °C/min under Ar and held at the final temperature for one hour.

Characterization Techniques

The FTIR spectra were recorded in transmission mode in the 4000-400 cm$^{-1}$ interval on a Nicolet 5 DXC spectrophotometer. 64 scans were collected on KBr pellets with 2 cm$^{-1}$ resolution. The Si-Me moieties give rise to an absorption peak at ca 1273 cm$^{-1}$ in the MeSiO$_3$ (T) units or at 1263 cm$^{-1}$ in the MeHSiO$_2$ (D$^H$) units (See Figure 1). This difference allows to get a quantitative estimation of the amount of the two silicon sites through simulation and deconvolution of the experimental spectra.

The $^{29}$Si MAS NMR experiments were performed on a Bruker spectrometer at 59.627 MHz with 2.5 µs pulse width and 60 s delay between pulses. The experimental spectra were simulated to get a quantitative estimation of the various silicon sites.

Differential Scanning Calorimetry (DSC) measurements were carried out on a Mettler TC 10A instrument. 50 mg samples were heated at 10 °C/min from -150 to 100 °C and scans were repeated to verify trend reproducibility.

The thermogravimetric analyses were done on a STA 409 -Netzsch thermobalance with 10 °C/min heating rate and 100 cc/min Ar flow.

## RESULTS AND DISCUSSION
### Characterization of T/D$^H$ Gels

FTIR spectroscopy of the gels prepared from the co-hydrolysis of T/D$^H$ mixtures shows a progressive change of the spectral features as a function of the composition. The spectra of the 90/10, 60/40 and 30/70 T/D$^H$ gels are presented in Figure 1. The absorption peak at 2172 cm$^{-1}$, due to the Si-H stretching vibration, decreases in intensity according to the concentration of the D$^H$ units. Two overlapped Si-CH$_3$ stretching vibrations are present with different intensity at 1275 and 1263 cm$^{-1}$; they are attributed to the Si-CH$_3$ bond in T and D$^H$ units, respectively[10]. The spectra show absorptions in the 1100-1050 cm$^{-1}$ range due to the Si-O stretching vibrations. The position and intensity of these peaks depend on the composition; the two main absorption peaks are present at 1106 and 1042 cm$^{-1}$ in the T/D$^H$ 30/70 sample and shift to 1125 and 1035 cm$^{-1}$ for the T/D$^H$ 90/10 composition.

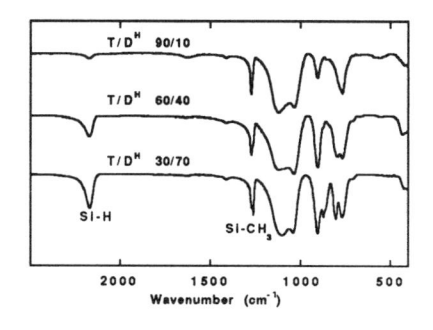

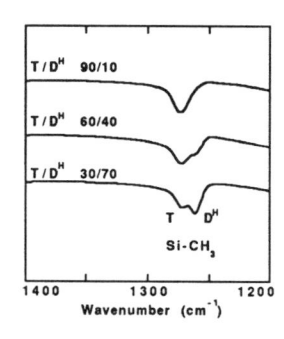

Figure 1. Evolution of the FTIR spectra as a function of T/D$^H$ gel composition.

The signals in the range 905-770 cm$^{-1}$, related to Si-H, Si-C and Si-O bonds vibrations are also characterized by different overlapping of the T and D$^H$ units contributions. Almost fully condensed T/D$^H$ networks appear from FTIR results since the signals due to residual Si-OR or Si-OH terminal groups are not detected.

The structural changes in the T/D$^H$ gel networks as a function of the composition, highlighted by the FTIR spectra, result in different DSC patterns. The DSC curves show an inflection point at the glass transition temperature (T$_g$). A progressive shift of T$_g$ towards high temperatures is observed with increasing the amount of T units. Moreover, the thermal effect associated to the glass transition becomes less defined for T/D$^H$ gels rich in trifunctional MeSiO$_3$ units.

Table 1.- Glass transition temperature, Tg, as determined from DSC analysis, for the T/D$^H$ gels.

| T/D$^H$ 30/70 | T/D$^H$ 50/50 | T/D$^H$ 60/40 | T/D$^H$ 80/20 | T/D$^H$ 90/10 |
|---|---|---|---|---|
| -98 °C | -56 °C | -25 °C | 35 °C | 39 °C |

Chracterization of PMHS-derived Gels

Different gels have been prepared by KF catalyzed hydrolysis of a commercial Me$_3$SiO terminated, linear polymethylhydrosiloxane through the partial conversion of the MeHSiO$_2$ (D$^H$) silicon units, forming the polymer backbone, into MeSiO$_3$ (T) units. Our results show that the advancement of the hydrolysis reaction can be controlled, up to a certain extent, by selecting the synthesis parameters: indeed, varying the hydrolysis ratio (r= 35, 8.75), the temperature (40-60 °C) and the reaction time (30 and 60 min) leads to siloxane networks with a different content of T unit ranging from ≈60 to ≈80%, as calculated from the FTIR and $^{29}$Si MAS NMR spectra. Figure 2 shows the FTIR spectra of three selected PMHS-derived gels (F61, F63 and F45), having a T unit content of 60, 70 and 80% respectively, that will be compared with the T/D$^H$ 60/40, 70/30 and 80/20 gels. The typical absorption peaks of MeSiO$_3$ and MeHSiO$_2$ silicon units are present in the spectra. The signals related to Si-CH$_3$, Si-O and Si-H bonds do not show any significative change with respect to those detected in the case of gels prepared from methyltriethoxysilane and methyldiethoxysilane. Thus, with respect to the individual silicon units forming the siloxane network, the FTIR spectra suggest a

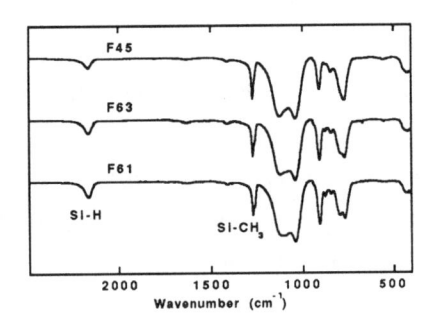

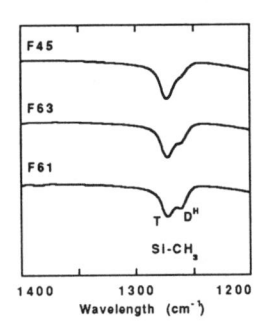

Figure 2. FTIR spectra evolution for the F61, F63 and F45 PMHS-derived gels

substantial equivalence between the PMHS-derived gels and the corresponding T/D$^H$ gel compositions. However, the DSC curves recorded on the F61, F63 and F45 samples, do not show any detectable glass transition, pointing out a different structural arrangement of the variuos silicon atoms, i.e. a different polymer architecture, compared with the alkoxide-derived gels.

Figure 3 reports two typical $^{29}$Si MAS NMR spectra recorded on the T/D$^H$ gels and on the PMHS-derived networks. In both case the spectra are dominated by components at ca -37 and ca -66 ppm due to the presence of MeHSiO$_2$ and MeSiO$_3$

silicon units respectively. In the spectra of the PMHS-derived gel is also evident a peak at 9.3 ppm due to the terminal Me$_3$SiO units of the starting polymer. On the other hand, in the spectrum of the alkoxide-derived gel is present a component at -56.8 that can be attributed to trifunctional MeSiO$_{2/2}$OX, (X= H or OEt) with two bridging and one terminal oxygen atoms. The polymer-derived network seems fully condensed compared with the T/D$^H$ gels in which a certain amount of terminal Si-O bonds are still present. Finally, the linewidth of the peaks due to MeHSiO$_2$ and MeSiO$_3$ silicon sites is always larger in the PMHS-derived samples compared with the T/D$^H$ gels, particularly for the T units (Table 2).

Table 2. Relevant $^{29}$Si MAS NMR data for T/D$^H$ gels and three selected PMHS-derived gels.

| Sample | MeSiO$_3$ | MeSiO$_3$ | | MeHSiO$_2$ | |
|---|---|---|---|---|---|
| | (%) | (ppm) | FWHL | (ppm) | FWHL |
| T/D$^H$ 50/50 | 50 | -66.1 | 2.1 | -37 | 1.3 |
| T/D$^H$ 70/30 | 70 | -65.8 | 3.4 | -36.9 | 2.6 |
| T/D$^H$ 90/10 | 90 | -65.4 | 4.5 | -36.7 | 3.6 |
| F43 | 74 | -66.4 | 4.5 | -36.5 | 3.3 |
| F58 | 76 | -66.4 | 4.7 | -37,3 | 3.3 |
| F45 | 79 | -66.5 | 4.8 | -37.2 | 3.1 |

This evidence suggests a more disordered architecture of the T and D$^H$ silicon units in the PMHS-derived network compared with the alkoxide-derived gels, in agreement with the difference already pointed out by the DSC analysis.

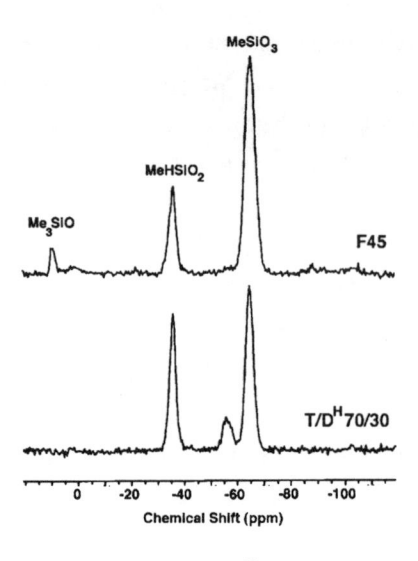

Figure 3. $^{29}$Si MAS NMR spectra of T/D$^H$ 70/30 and F45 gel samples.

Pyrolysis and HT Stability Study

The pyrolytic conversion from the pre-ceramic networks to the SiOC glasses was investigated with DTA/TG analysis. The DTA traces do not show any relevant thermal effect. On the other hand, TG curves provide very interesting information. Figure 4 compares two TGA patterns recorded on samples having the same composition, in terms of T and D$^H$ units, namely 80% T and 20% D$^H$, but with different polymer architecture. The two TG curves reported in Figure 4 are characteristic, in their general trends, of all the alkoxide- and PMHS-derived samples reported in this study.

The pyrolytic conversion of the T/D$^H$ gels is characterized by two well distinct weight loss steps: between 450°C and 550°C, due to the evolution of silanes and between 750 °C and 850 °C, related with the organic/inorganic transformation with evolution of H$_2$ and CH$_4$[11]. The weight changes associated with the two weight loss steps are correlated to the gel compositions, and, as a general trend, increase from T/D$^H$ 90/10 to T/D$^H$ 30/70 samples. Above 1400°C the TG curve shows the onset of a carbothermal decompositon process.

TG curve recorded on the PMHS-derived network is more complicated and show, at least, three weight loss steps, that are less defined compared with the alkoxide-derived gels. The first weight change is observed between 150°C and 550°C, the second one between 550°C and 750°C and the last one between 750°C and 850°C. This last decomposition step is quite similar, either in terms of temperature range and in terms of shape, to the one observed for the T/D$^H$ gel

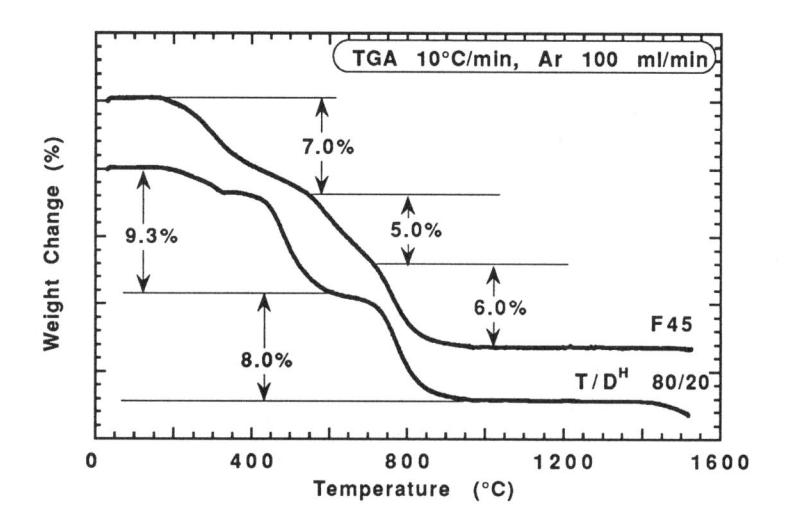

Figure 4. TG curves recorded on the T/DH 80/20 and F45 gel samples.

assigned to the organic/inorganic transformation. On the other hand, the precise description of the pyrolysis chemistry ruling out the two weight loss steps at lower temperature needs more investigation and will be the subject of a future TG/MS study. Nevertheless, these results clearly show that despite the same composition in terms of T and D$^H$ silicon units the two pre-ceramic gels, having a different polymer architecture, display a quite different pyrolytic transformation that could lead to different composition of the SiOC glasses. It is also worth noting that the PMHS-derived network does not show any decomposition process related to the carbothermal reduction. These results prompted us to investigate the high temperature stability of the SiOC glasses derived from the two different precursors. Thus, PMHS-derived samples with a T unit content ranging from 60% to 80% (F61, F63 and F45) have been pyrolyzed at 1000°C and further heated at 1400°C and 1500°C for 1 h, and their weight change recorded For comparison purposes, gel samples with the same composition, i.e. T/D$^H$ 60/40, T/D$^H$ 70/30 and T/D$^H$ 80/20 have been treated with the same pyrolysis cycle. The results are reported in Table 3.

Table 3. Weight loss for selected PMHS- and T/D$^H$- derived SiOC glasses

|        | F61   | T/D$^H$ 60/40 | F63   | T/D$^H$ 70/30 | F45   | T/D$^H$ 80/20 |
|--------|-------|---------------|-------|---------------|-------|---------------|
| 1400°C | 5.8%  | 8.6%          | 3.6%  | 8.7%          | 3.9%  | 5.0%          |
| 1500°C | 38.5% | 47.4%         | 20.7% | 52.0%         | 25.2% | 47.7%         |

From the results of Table 3 it is clear that the PMHS-derived SiOC glasses display a better high temperature stability compared with alkoxide-derived SiOC glasses. Indeed, the weight losses for T/D$^H$ derived glasses are always higher than the PMHS-derived samples. At 1500°C the weight loss for the PMHS-derived SiOC is roughly 50% of the value observed for the T/D$^H$ derived glasses.

CONCLUSIONS

Pre-ceramic networks formed by the same trifunctional (T) and difunctional (D$^H$) silicon units, have been obtained with two synthesis approaches: a conventional sol-gel process starting from methyltriethoxysilane and methyldiethoxysilane; and the partial hydrolysis of a linaer polymethylhydrosiloxane. The FT-IR, $^{29}$MAS NMR and DSC analyses indicate that the two synthesis approaches lead to a different arrangements of the silicon units in the gel network: i.e. to a different polymer architecture. The TGA studies show that this difference turns out into a different pyrolysis chemistry and ultimately into a strong difference in the high temperature stability of the derived SiOC glasses.

## ACKNOWLEDGEMENTS

MURST 40% and Programma Galileo are kindly acknowledged for the financial support. The authors are grateful to C. Gavazza for the DSC measurements. Dr. R. Simonutti, University of Milan, is greatly acknowledged for the MAS NMR spectra collection.

## REFERENCES

1 G.D.Sorarù, E. Dallapiccola & G. D'Andrea, "Mechanical Characterization of Sol-Gel-derived Silicon Oxycarbide Glasses", *J.Am. Ceram. Soc.,* **79** 2074-80 (1996).
2 F. I. Hurwitz, P. Heimann, S. C. Farmer & D. M. Hembree Jr., " Characterization of the Pyrolytic conversion of polysilsesquioxanes to silicon oxycarbides", *J. Mater. Sci.,* **28** 6622-30 (1993).
3 R. J. P. Corriu, D. Leclerq, P. H. Mutin & A. Vioux, "$^{29}$Si nuclear magnetic resonance study of the structuyre of silicon oxycarbide glasses derived from organosilicon precursors", *J. Mater. Sci.,* **30** (1995) 2313-2318.
4 A. M. Wilson, G. Zank, K. Eguchi, W. Xing, B. Yates & J. R. Dahn, "Polysiloxane Pyrolysis", *Chem. Mater.* **9** 1601-0 (1997).
5 G. D. Soraru, G. D'Andrea, R. Campostrini, F. Babonneau & G. Mariotto, "Structural Characterization and High Temperature Behavior of Silicon Oxycarbide Glasses Prepared from Sol-Gel Precursors Containing Si-H Bonds" *J. Am. Ceram. Soc.* **78** 379-87 (1995).
6 P. Carri & G. D. Sorarù, "Sol-Gel Processing of Continuous Fiber Reinforced Composites by the Liquid Infiltration and Pyrolysis (**Lip**) Method", in: "Innovative Processing and Synthesysis of Ceramics, Glasses and Composites", Edited by N. P. Bansal, K. V. Logan & J. P. Singh (Ceramic Transactions Vol. 85, 1997, p.405-416).
7 H. J. Wu, Y. D. Blum, S. M. Johnson, C. Kanazawa, J. R. Porter & D. M. Wilson, "Preceramic Polymer Application-processing and Modification by Chemical Means", *Mat. Res. Soc. Symp. Proc.,* Vol. 435, 431-36 (1996).
8. R. M. Laine, J. A. Rahn, K. A. Youngdahl, F. Babonneau, M. L. Hoppe, Z.-F. Zhang & J. F. Harrod, "Synthesis and High-Temperature Chemistry of Methylsilsesquioxane Polymer Produced by Titanium-Catalyzed Redistribution of Methylhydridooligo-and -polysiloxanes", *Chem. Mater.,* **2** 464-472 (1990).
9 R. Kalfat, F. Babonneau, N. Gharbi & H. Zarrouk, "$^{29}$Si MAS NMR investigation of the pyrolysis process of cross-linked polysiloxanes prepared from polymethylhydrosiloxane", *J. Mater. Chem.,* **6** 1673-78 (1996).
10 P. J. Launer, "Infrared analysis of organosilicon compounds: spectra-structure correlations in silicon compounds, Petrarch Systems Inc. 1984, pp. 69-72.
11 R. Campostrini, G. D'Andrea, G. Carturan, R. Ceccato, & G. D. Sorarù, "Pyrolysis Study of Methyl-Substituted Si-H Containing Gels as Precursors for Oxycarbide Glasses, by Combined Thermogravimetric, Gas Cromatographic and Mass Spectrometric Analysis", *J. Mater. Chem.* **6** (1996) 585-594.

# POLYMER DERIVED CERAMIC HARD MATERIALS BY ADDITION OF TUNGSTEN

Keon-Taek Kang, Deug-Joong Kim
Sung Kyun Kwan University, Department of Material Engineering, Suwon, Korea

Annette Kaindl, Peter Greil
University of Erlangen-Nuernberg, Department of Material Science, Erlangen, Germany

## ABSTRACT

The formation, microstructure and properties of novel ceramic composite materials by active filler controlled polymer pyrolysis was investigated. In the presence of active filler particles such as transition metals, bulk components of various geometry can be fabricated from siliconorganic polymer. During pyrolytic decomposition of the polymer, the filler particles reacted with gaseous and solid decomposition products resulting in a volume expansion of the filler phase. This volume expansion can be used to compensate for the polymer shrinkage. Polymethylsiloxane filled with W is particularly interest because of the formation of ceramic bonded hard materials (WC-W$_2$C-SiOC) for wear resistant applications. High metal-filled polymer suspensions were prepared and their conversion to ceramic composites by annealing in N$_2$ or CH$_4$ atmosphere at 1000~1600 ℃ were studied. Dimensional change, porosity and phase distribution (filler network) was analyzed and correlated to the resulting material properties. Microcrystalline composites with the filler reaction products embedded in a silicon oxycarbide glass matrix were formed. Depending on the pyrolysis conditions, ceramics with a density up to 95 TD%, a hardness of 7~9.5 Gpa, a fracture toughness of 5~6.8 MPam$^{1/2}$, and a flexural strength of 380~470 MPa were obtained.

## INTRODUCTION

The manufacturing of ceramic materials from the pyrolysis of Si-containing polymers has been attained increasing interest because of the application of plastic shaping technologies and low manufacturing temperature.[1-3] Silicon containing polymer precursor is transformed into a oxycarbide ceramic by pyrolysis. However, the tremendous shrinkage and density increase during the pyrolysis limit the formation of bulk ceramics. Recently, near net shape ceramic with low shrinkage and porosity was successfully performed by the addition of transition metals as active filler materials[4].

The volume increase during reaction of the active filler with the solid and gaseous decomposition products compensates the polymer shrinkage. The microstructure of the resulting ceramic composites consists of carbide as well as silicide hard phases embedded in the silicon oxycarbide matrix. The varying the preceramic polymer, filling materials and atmospheric condition allows to synthesize tailored ceramic composites with minimum shrinkage and high hardness. In the present work, new synthesizing route of anti-abrasive hard materials was developed by using W as the active filler dispersed in polymethylsiloxane. The microstructure formation and properties of novel ceramic hard materials derived from polymer/active filler mixtures were investigated.

EXPERIMENTAL PROCEDURES

Preceramic precursor used in this experiment was a polymethylsiloxane, $[(CH_3)SiO_{1.5}]_n$, (NH2100, Chemische Werke Nuenchritz, Germany). Tungsten with a grain size 1.08 $\mu m$ (Korea Tungsten Co, Korea) was used as an active filler. Polymer (60 vol%) and active filler (40 vol%) were mixed with acetone and homogenized with mechanical and ultrasonic stirring. Evaporating the solvent in a rotavapor finally yielded a powder, which was ground in a mortar. The mixture was then formed to tablets at 30 MPa and 230 ℃ using a thermal press.

The pyrolysis was performed in an electrically heated tube furnace under 0.1 MPa of $N_2$ or $CH_4$ atmosphere. The typical heating cycle involved heating up to 550 ℃ at 5 ℃/min rate, holding at this temperature for 4 hr, a second ramp at 1~2 ℃/min up to final temperature of 1000~1500 ℃ with a 4 hr hold and final cooling at 5 ℃/min.

Densities of the ceramic products were measured by a helium picnometer. Theoretical densities were estimated on samples after milling. Open and closed porosities were characterized by dimensional measuring and additional calculation from the result of helium picnometer. Phases were identified by an X-ray diffraction using Cu $K_\alpha$ radiation. Microstructures were examined by SEM. Hardness and fracture toughness were measured using a Vickers indentation fracture method under an indentation load 98 N. Fracture strength was obtained by the 3-point bending test with the span length 16 mm.

RESULTS AND DISCUSSION

Figure 1 shows the variation of porosities with the pyrolysis temperature in $N_2$. Decomposition of the polymethylsiloxane started above 400℃ and was almost completed at approximately 800℃.[5] During the pyrolytic decomposition release of gaseous reaction products such as methane occurs, accompanied with the formation of open porosities in the material. The decrease of this open porosity above 800℃ can be explained by the increasing shrinkage of the polymer derived silicon oxycarbide matrix.[5] Above 800℃ an amorphous silicon oxycarbide glass was formed and rearrangement of inorganic particles took place, which resulted in shrinkage and further formation of closed pores at 1400℃. With increasing pyrolysis temperature up to 1500℃, a total porosity of 4.5% was obtained with open porosity to be less than 0.3%.

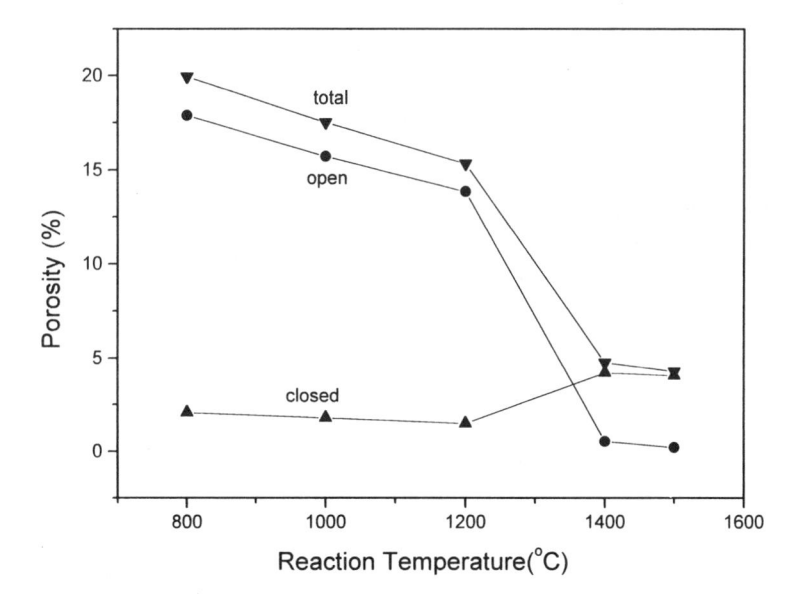

Fig. 1   Porosity with various reaction temperature under nitrogen atmosphere

Figure 2 summarize the carburization of W as active filler during the pyrolysis under $N_2$ or $CH_4$ atmosphere, which was determined by XRD. The formation of tungsten carbide, WC and $W_2C$, occurs at pyrolysis temperature of 1000°C. Above 1400°C, $W_2C$ becomes a major phase indicating the reverse eutectoid reaction, W + WC = $W_2C$, has occured. Tungsten is in equilibrium with WC at low temperature and $W_2C$ at high temperature. The temperature of the eutectoid reaction has been reported as 1575K.[6] In methane atmosphere the equilibrium moves to WC - $W_2C$ two phase region. The residual W reacts with $CH_4$, yielding WC and $W_2C$ carbide phases. Figure 3 shows the microstructures of ceramic composites pyrolyzed at 1500°C in $N_2$ or $CH_4$ atmosphere. The microstructures consisted of unreacted W and carbide particles, which were embedded in a silicon oxycarbide matrix. Faceted carbide grains were predominant in the microstructure of ceramic composite from methane atmosphere.

The mechanical properties are given in Table 1. The mechanical properties were strongly influenced by the phase changes and porosities developed during pyrolysis. Due to the conversion of a polymer into a ceramic matrix and increased carburization of W to WC and $W_2C$, a maximum hardness of 9.5 MPa was obtained after pyrolysis at 1500°C in $CH_4$. Fracture strength of 380~476 MPa and fracture toughness of 5~6.8 $MPam^{1/2}$ were obtained.

Table 1. Mechanical Properties of specimens reacted at different temperature and atmosphere

| Samples \ Property | Fracture strength (MPa) | Fracture toughness ($MPam^{1/2}$) | HV (GPa) |
|---|---|---|---|
| 1400($N_2$) | 476 | 6.83 | 7.3 |
| 1500($N_2$) | 381 | 6.10 | 8.8 |
| 1500($CH_4$) | 435 | 5.04 | 9.5 |

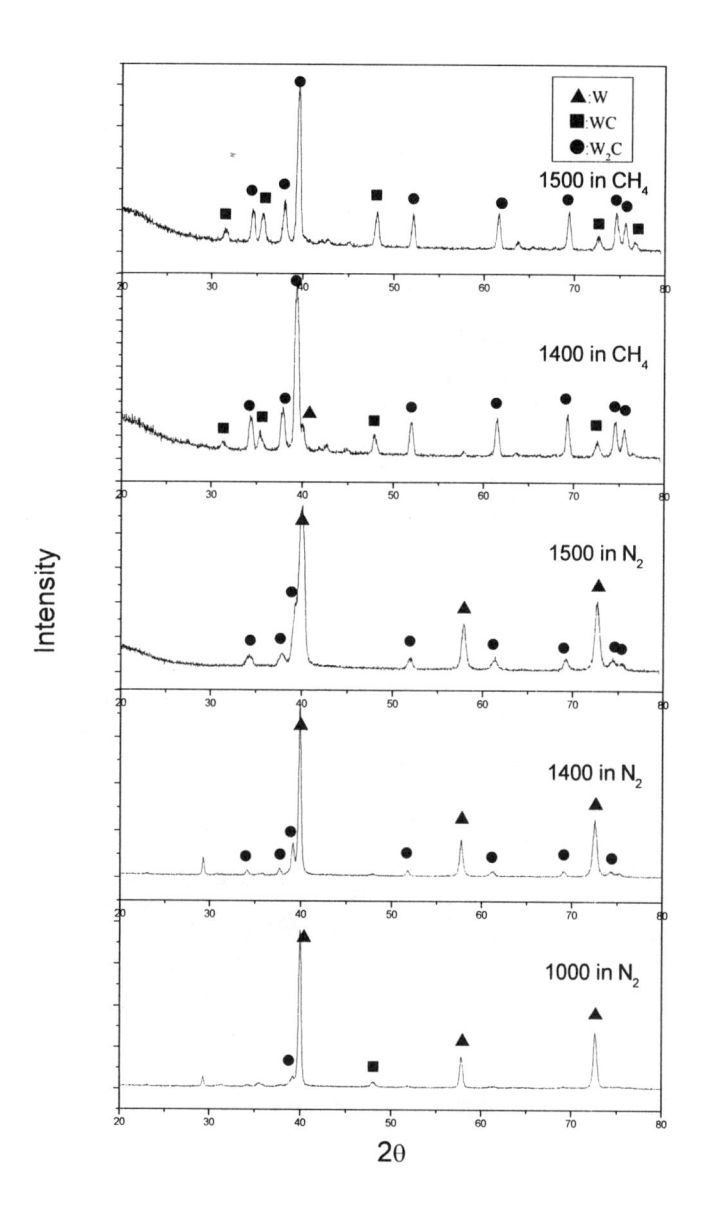

Fig. 2 XRD-plots of specimens reacted at different temperatures and atmospheres

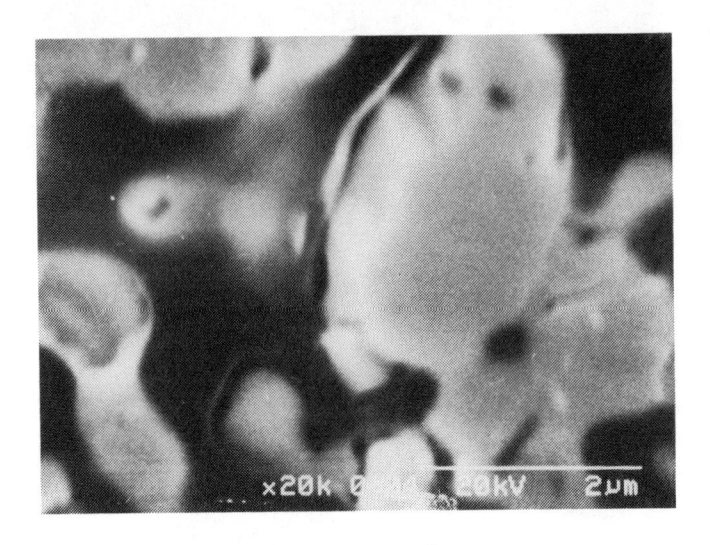

Nitrogen atmosphere

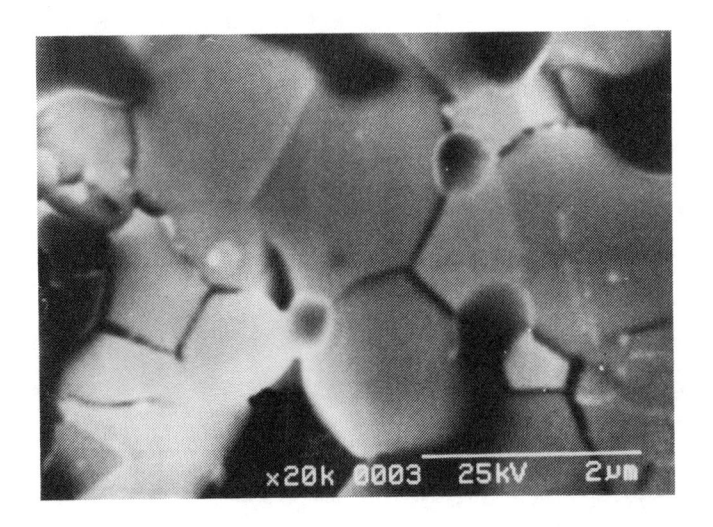

Methane atmosphere

Fig. 3 Microstructure of specimens reacted at 1500 ℃ under different atmosphere

CONCLUSION

Ceramic composites with less than 5% porosity were obtained through the incorporation of reactive W powders in preceramic polymethylsiloxane with subsequent pyrolysis. Viscous silicon oxycarbide glasses from polymer and volume expansion upon reaction of fillers offered a high potential for shrinkage and porosity reduction. Microcrystalline composites with the carbide reaction product embedded in a silicon oxycarbide glass matrix were formed. Mechanical properties were strongly affected by the microstructural evolution related with pores and reacted phases on pyrolysis condition. Ceramic hard materials with densities more than 95TD%, hardness of 7.3~9.5 GPa, fracture strength of 380~476 MPa and fracture toughness of 5~6.8 MPam$^{1/2}$ were obtained from Polymethylsiloxane/W mixture after pyrolysis.

ACKNOWLEDGEMENT

The Korea-German Cooperative Science Program supported this project. Korea Science & Engineering Foundation (KOSEF) and Deutsch Forschung Gesellschaften (DFG) are gratefully acknowledged.

REFERENCE

1. R. W. Rice, Ceramics from Polymer Pyrolysis, Bull. Am. Ceram. Soc., 62 889~892 (1983)
2. K. J. Wynne, Ceramics via Polymer Pyrolysis, Ann. Rev. Mat. Sci., 14 297~334 (1984)
3. M. Peuckert, T. Vaahs, M. Brueck, Ceramics from Organmetallic Polymer, Adv. Mat., 2,398~404 (1990)
4. P. Greil, Active Filler Controlled Pyrolysis of Preceramic Polymer (AFCOP), Am.Ceram. Soc., 78, 835~848 (1995)
5. T. Erny, M. Seibold, O. Jarchow, P. Greil, Microstructure Developement of Oxycarbide Composites during Active-Filler-Controlled Polymer Pyrolysis, J. Am.Ceram. Soc., 76, 207~213 (1993)
6. D. K. Gupta, L. L. Seigle, Free Energies of Formation of WC and W$_2$C, and the Thermodynamic Properties of Carbon in Solid Tungsten, Met. Trans. 6A, 1939~1943 (1975)

# Shock Synthesis

# SHOCK SYNTHESIS OF MoSi2 - SiCp/SiCw COMPOSITES
# FROM MECHANICAL ALLOYING PRETREATED PRECURSORS

Tatsuhiko Aizawa
Department of Metallurgy, University of Tokyo
7-3-1 Hongo, Bunkyo-ku, Tokyo 113-8656, Japan

Naresh N. Thadhani
Department of Materials Science and Engineering, Georgia Institute of Technology
Atlanta, GA 30332-0245

## ABSTRACT

The shock reactive synthesis is one of the most promising methodologies to yield non-traditional materials. Use of the pretreated powders by the mechanical alloying enables us to make shock reactive synthesis with full reactivity. Furthermore, various intermediate phases can be synthesized during the shock induced reaction. Mo-Si system is employed here to investigate the effect of pretreatment on the synthesis of $MoSi_2$. Of great importance is that amorphous $MoSi_2$ can be synthesized as an intermediate phase. Both the SiC particles and whiskers are mixed with the mechanical alloying pretreated powders for shock expriment to discuss the fast quenching effect on the shock induced reaction process. $MoSi_2$ - SiCp/SiCw composites are successfully synthesized into full-dense billets. Morphological difference between SiC particle and whisker reflects on the local interfacial reaction; no reactions are seen between the SiC particle and the synthesized $MoSi_2$ matrix, while the tertiary phase was formed on the interface of thin, long SiC whisker. Through close microstructure observation, mass mixing mechanism during shock loading is discussed.

## INTRODUCTION

The shock reactive synthesis is one of the most effective tools to yield non-equilibrium phase materials even for refractory metal aluminides and silicides through the solid-state chemical reaction. As had been pointed out by Thdhani [1], this shock synthesis can be classified into two categories: 1] Shock assisted synthesis, and 2] Shock induced synthesis. In the former, high temperature and pressure are necessary even to make partial reactions from the blended powder mixture to the targeting aluminides and silicides. In the shock assisted synthesis [2] of molybdenum di-silicide, its liquid phase must be existing to sustain the reaction like a typical self-heating synthesis [3]. In the latter, application of high pressure pulse to materials should work as a trigger to ignite the reaction pass to synthesize the intermetallic compounds. As had been stated in Refs. [4,5], the powder morphology in the starting materials affected the shock reactivity.

One of authors [6-10] has proposed the shock reactive synthesis from the pretreated powders by using the mechanical milling and grinding. Its advantageous features can be summarized in what follows: 1] Full shock reactivity into targeting aluminides and silicides can be attained irrespective of the shock loading conditions once the applied shock pressure exceeds over the critical value, and 2] Intermediate phase of a product should be present to have been synthesized from the pretreated elements and to react into a regular compound of aluminides and silicides. The above feature 1] implies that the necessary shock conditions to ignite reactions should be reduced when starting from the MA pretreated powders. Use of lower flyer velocities in the shock reactive synthesis leads to successful recovery of dense, reacted materials without cracks or defects. Owing to the feature 2], the shock induced reaction mechanism can be experimentally described to understand various theoretical models. Horie [11] proposed a new model to explain the transient behavior of powders under shock compression, where wide deviation of Hugoniot for individual powder from the equilibrium Hugoniot drives mass mixing and reaction during shock loading. Furthermore, use of intermediate phase materials enables us to synthesize new type of materials. Hence, various experimental trials are still necessary to propel this shock reactive synthesis by making full use of these two features.

In the present study, one-stage light gas gun is used for shock reactive synthesis of $MoSi_2$ - SiC composites from the pretreated precursors. Although the maximum flyer velocity is limited by 1.0 km/s in using the gas gun, use of the pretreated powders enables us to make full reactions even in the lowered shock conditions; the crack-free bulk composites can be synthesized from the mixes of the pretreated Mo-Si powders with SiC particles and whiskers. It is of importance that $MoSi_2$ - SiCp/SiCw composites can be synthesized even when low thermal transients are experienced during shock loading due to the quenching effect. With increasing the volume fraction of solids, the specific volume change is further reduced than the porous medium, so that both the transient and the residual temperatures are significantly reduced. Irrespective of the SiC volume fraction, the shock reactivity remains the same as seen in the system without SiCp or SiCw. This fact must be a good proof demonstrating that the shock induced reaction from the MA precursor should be free from thermal transients and intrinsic to shock wave compression behavior. Through precise microstructure observation, the morphological effects on the shock reactivity is also discussed between $MoSi_2$ - SiCp and $MoSi_2$ - SiCw.

**EXPERIMETAL PROCEDURE**

Mo-Si system was employed to experimentally demonstrate the possibility to synthesize the SiC reinforced $MoSi_2$ from the mixture of SiC particle or whiskers into the MA pretreated powder mixes of Mo and Si. Mo powder (average size of 40 μm) and Si powder (average size of 10 μm) were employed as the starting powders. Their purity has three to four nines. These elemental powders were ball milled for 24 h and, poured into PSZ (Partially Stabilized Zirconia) vial with PSZ balls. As had been stated in Refs. (12,13), PSZ vial and balls are preferable for mechanical alloying of transient metal elements and silicon since contamination from vial and balls can be suppressed to the minimum level. If a significant amount of contamination were noticed, there were little possibility for the shock synthesis to be harmed by this contamination. Spex-type milling machine (SPEX-8000) was used for mechanical alloying from the blended powders to the refined elemental mixture.

The grain refinement can be controlled by the milling time in this pretreatment. All through this powder preparation, the powder handling took place under inert gas atmosphere. As listed in Figure 1, these pretreated powders were mixed and further blended with SiC particles and whiskers into mixes. The average diameter of SiC particle (SiCp) and SiC whisker (SiCw) was selected to be 2 - 3 μm. Their volume fraction was also controlled to be 20 % and 50 %.

These powder mixes were uniaxially pressed into a cylindrical billet with the diameter of 10 mm and the thickness of 4 mm. For the pretreated powder billet without SiCp and SiCw, its relative density can be controlled to be about 80 %. For the pretreated powder billet with SiCp, its relative density ranges from 75 to about 80 %; while, it becomes 70 to 75 % for the billets with SiCw. These samples were capsuled into a stainless steel sample holder with the mild steel momentum traps. Three samples can be recovered at the same shock condition at one time; the effect of sample constituents on the shock reactivity can be directly evaluated by comparison of XRD profiles and microstructure observation. Table I lists the shock experimental conditions.

As had been indicated for Ti-Si system [14,15], the pretreatment effect on the shock reactivity has close relationship with the milling time or the refinement process during mechanical alloying. Variation of XRD profiles with increasing the milling time is depicted in Figure 2. Refs. (12,13)

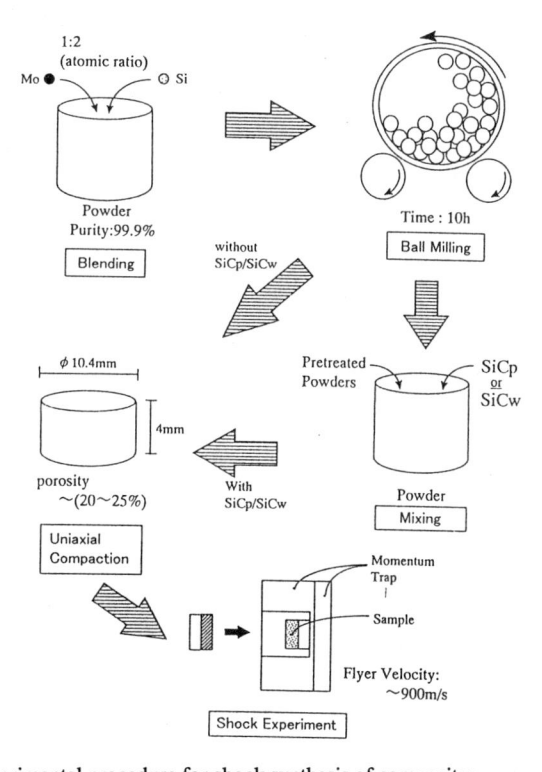

Figure 1: Experimental procedure for shock synthesis of composites.

also stated that mechanically induced self-heating synthesis takes place after a critical milling time even in the mechanical alloying; if Si grain size were refined into this critical size (which was estimated to be 9 nm or less than), the regularization reaction from elemental refined particles into di-silicide could be ignited and self-sustained by the formation energy of di-silicide even at the ambient temperature. In the present study, the milling time was selected to be constant by 10 h when the measured grain size of Si becomes 20 to 30 nm. At this stage of mechanical alloying, Si peak was noticeably reduced and broadened, but both peaks for Mo and Si were still distinguished in Figure 2.

Table I: List of samples to be used for the shock reactive synthesis by the gas gun.

| Sample Number | Morphology | Volume fraction | Flyer velocity |
|---|---|---|---|
| #1 | None | None | 980 m/s |
| #2 | particle | 20 % | 980 m/s |
| #3 | particle | 50 % | 980 m/s |
| #4 | None | None | 915 m/s |
| #5 | whisker | 20 % | 915 m/s |
| #6 | whisker | 50 % | 915 m/s |

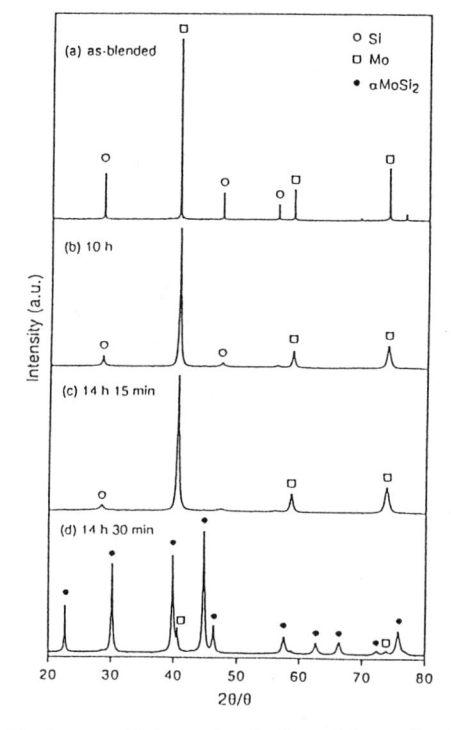

Figure 2: XRD profile change with increasing the time of the mechanical alloying.

**EXPERIMETAL RESULTS**

The shock reactivity from the MA pretreated powders is investigated with and without the reinforcing SiC particles and whiskers. The volume fraction of SiC particles and whiskers are varied by f = 20 % and 50 % in order to describe the quenching effect on the shock reactivity, especially the thermal transients during shock loading.

Shock Synthesis of $MoSi_2$ from MA Pretreated Precursor

Ref. (5) stated residual Mo and Si were still left in the recovered sample together with $MoSi_2$ when the blended powder mixture with Mo-2Si was shot even by increasing the flyer velocity; in the shock experiments by lower shock conditions, little or no reactions took place from the blended powders. Even when the fly velocity (V) was controlled to be about 900 m/s, the pretreated powder green compact can be fully reacted into $MoSi_2$ and densified. Furthermore, this pretreated MA precursors were fully reacted irrespectively of the flyer velocity for 0.9 km/s < V < 1.5 km/s. This assures that the shock induced solid-state reaction from the premixed elemental mixture into $MoSi_2$ should become indifferent to the applied shock pressure once the ignition condition was satisfied in the shock reaction mechanism. The point to be noted in the microstructure observation is that the densified part at the center of a sample is composed of round-shaped $MoSi_2$ crystal with non-equilibrium phase. Since its chemical composition pointwise analyzed by EDX varies form $Mo_{60}Si_{40}$ to $Mo_{35}Si_{65}$, this phase can be thought to be amorphous.

Axisymmetric calculation of pressure distribution was made by using the hydrodynamic code. The shock pressure distributes in the inside of a sample: for the sample with the initial porosity of 25 %, the center part of the sample experiences 25 - 30 GPa, while P = 20 GPa at either end of the sample. This difference reflects on the densification during shock compression.

Shock Synthesis of $MoSi_2$-SiCp Composites from MA Pretreated Precursor - SiCp Mixes

As shown in Figure 3, the pretreated powders of Mo-Si were fully reacted into $MoSi_2$, and mixed and consolidated with SiCp without any interfacial reactions in the macroscopic sense. Even increasing the volume fraction of SiCp, the starting green compact was successfully synthesized and densified into $MoSi_2$ - SiCp composite. Since all XRD profiles correspond to a mixture of $MoSi_2$ and SiCp, the shock reactivity from Mo + Si to $MoSi_2$ matrix might be indifferent to the volume fraction of SiCp. In practice, however, increase of SiCp volume fraction reflects on the handling of a recovered sample: since the recovered sample including SiCp by f = 50 % was cold just after shock experiment, it can be manually handled. Figure 4 shows a typical microstructure of this composite for f = 20 %. Although the randomly mixed SiCp particles in the starting materials seem to be a little agglomerated, all SiCp particles were completely embedded into $MoSi_2$ matrix. EDX analysis was made to investigate the occurrence of local reaction between SiCp and matrix phase and the formation of amorphous phase in the matrix. As depicted in Figure 4, no interfacial reactions were seen althrough the observation. In addition, no amorphous phase could be found in the matrix. This is partially because the initial porosity differs by 5 to 10 % from the pretreated Mo-Si powder compact without SiC particles; local pressure is apt to be increased with increasing the porosity of the sample.

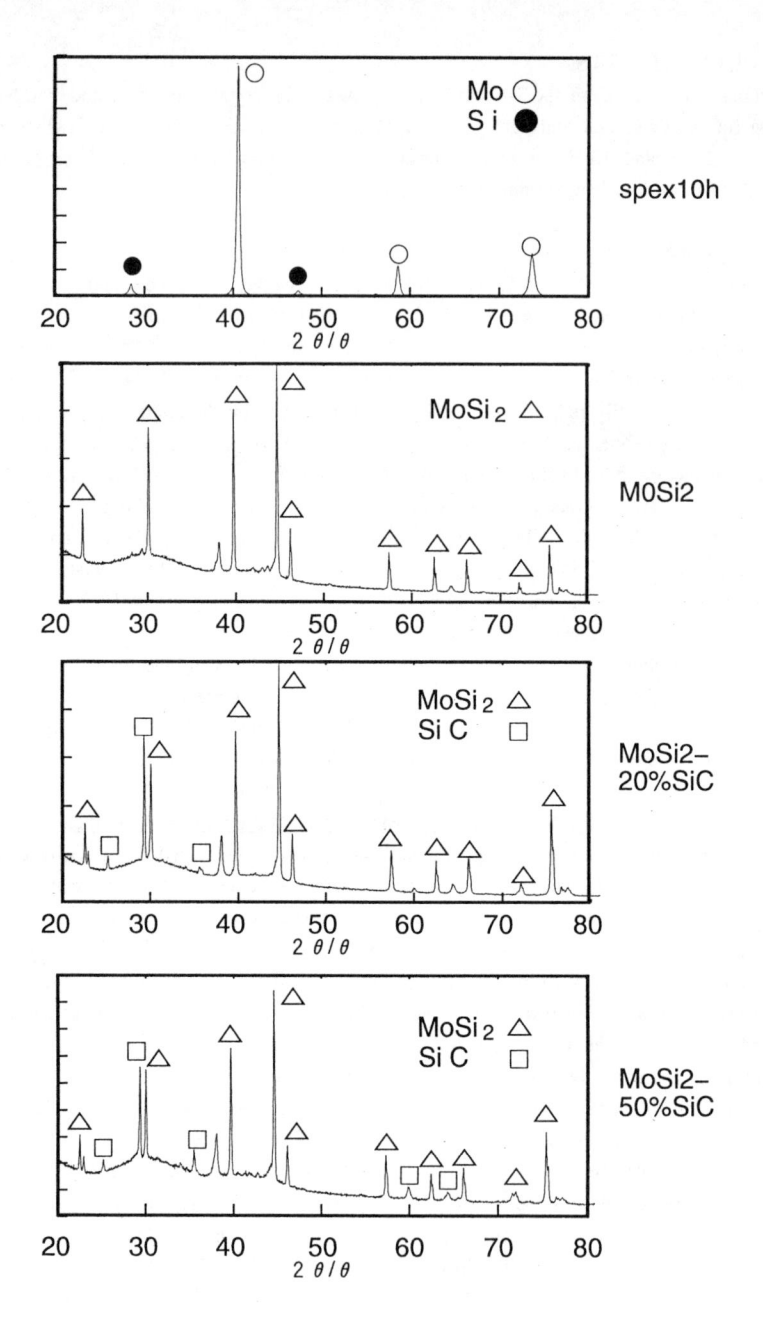

Figure 3: Comparison of XRD profiles among the samples with and without SiCp.

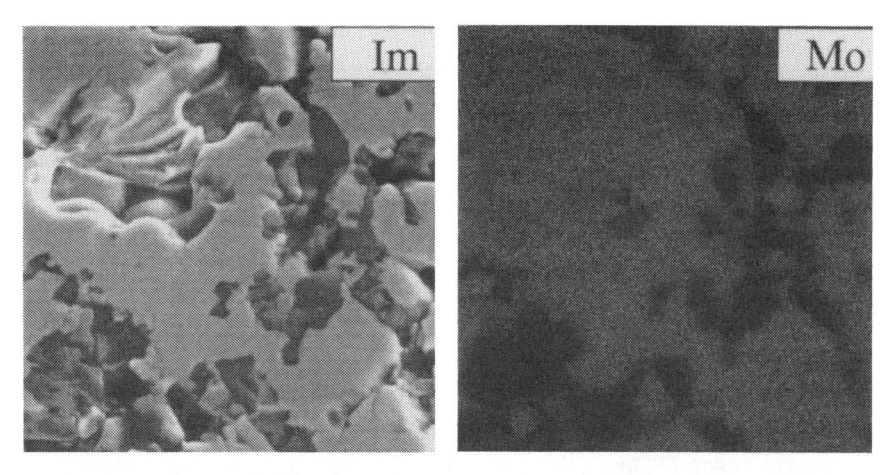

Figure 4: Microstructure of the recovered sample for the volume fraction of SiC particles by f =20 % with results of Mo X-ray dot map.

Shock Synthesis of MoSi$_2$-SiCw Composites from MA Pretreated Precursor - SiCw Mixes

Although the flyer velocity was reduced by 65 m/s from the shock experiment to synthesize MoSi$_2$ - SiCp, the shock reactivity can be thought to be mainly indifferent to the morphological change of a round particle to a whisker. In fact, as compared between Figure 3 and Figure 5, the XRD profiles both for MoSi$_2$ - SiCp and MoSi$_2$ - SiCw became nearly the same. Only, the peak intensity for SiCp and SiCw increases with increasing their volume fraction in the starting materials.

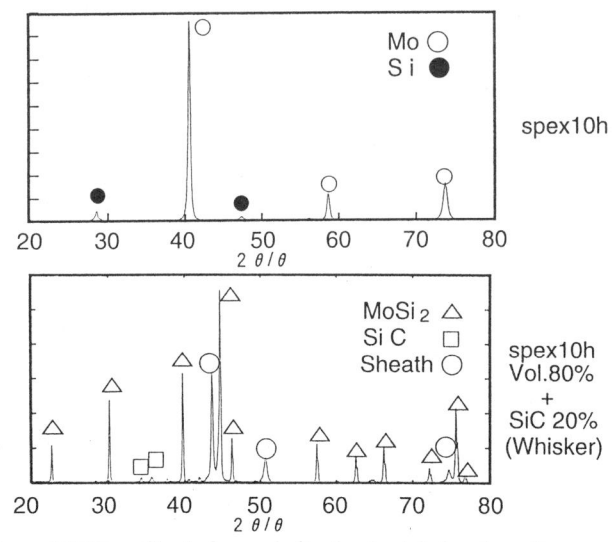

Figure 5: Comparison of XRD profiles before and after the shock induced reaction.

Innovative Processing/Synthesis: Ceramics, Glasses, Composites II

Figure 6 shows the microstructure of MoSi$_2$ - SiCw for f = 20 % at its center part and the right-side end. As stated before, the initial porosity ranges from 30 to 25 % in the average since the green compact including SiCw cannot be easily uniaxilly pressed to the targeting density. The flyer velocity was selected to be lowered nearly to 900 m/s. Hence, significant pores were still left both in the center and the end parts. Different from the microstructure for MoSi$_2$ + SiCp, the same amorphous zones as had been reported in Ref. 5) were also formed in the MoSi$_2$ matrix. It must be of more interest that SiC whiskers randomly distributed in the powder mixes at the initial stage should be regularly aligned during shock loading. In the microstructure for the center part in Figure 6, no SiC whiskers were included in the field. While, SiC whiskers were forced to be shaped as a line at the right-side end part.

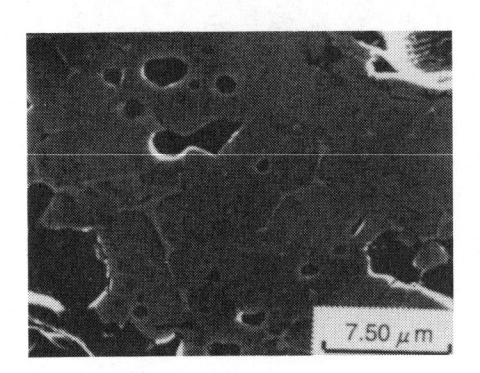

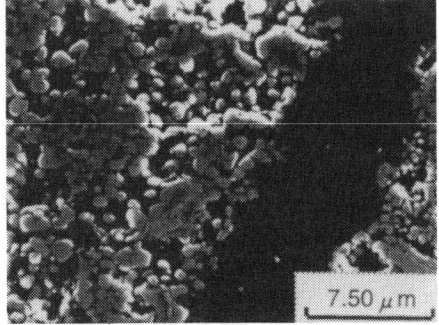

Figure 6: Microstructure of the recovered sample for the volume fraction of SiC whiskers f=20 %.

**DISCUSSION**

In the shock reactive synthesis from the pretreated MA precursor, the non-equilibrium phase materials as well as the round-shaped MoSi$_2$ grains were formed. Since the chemical composition of Si detected by EDX ranges from 40 at% to 70 at% in this non-equilibrium phase, this synthesized phase must be amorphous and intermediate phase material to be further synthesized into MoSi$_2$. Through the systematic study by varying the flyer velocity, the volume fraction of this amorphous phase significantly changes itself. As depicted in Figure 7, the volume fraction (Va) of this amorphous phase drastically increases with decreasing the flyer velocity less than 1.0 km/s. This implies that there should be a critical state where the starting mixture can be fully reacted into amorphous phase of MoSi$_2$. As shown in Figure 6, formation of this amorphous phase materials in MoSi$_2$ matrix was also detected in using the gas gun for shock experiment. Although further studies are still necessary to investigate the formation mechanism of this intermediate phase, the ignition condition must be described as the primary process in the shock induced reaction, which leads to the regularization process into the formation of intermetallic compounds.

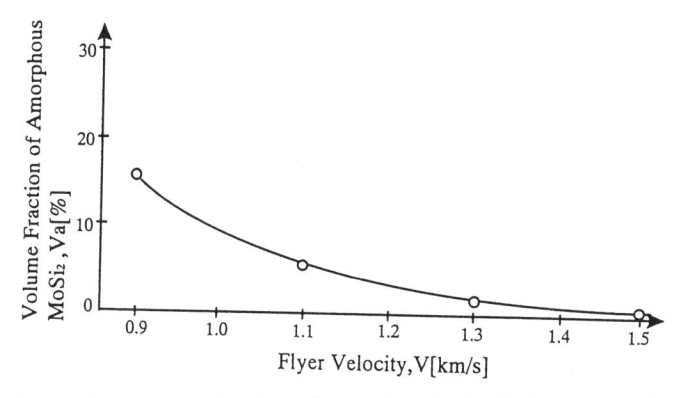

Figure 7: Variation of the volume fraction of amorphous $MoSi_2$ in the recovered sample with decreasing the flyer velocity.

In general, inclusion of solid, dense matter into porous material significantly reduces the thermal transient by decreasing the change of specific volume in shock compression, and, the residual temperature is also lowered because of far better thermal conductivity for solid matter. The temperature transient was calculated by assuming the mixture rule for the Hugoniot for $MoSi_2$ + SiC particle mixes without consideration of formation energy. The residual temperature was noticeably reduced, but the reactivity should be indifferent to inclusion of SiC particles and whiskers in the shock experiment. This assures that the shock reactivity should be free from the thermal transients during shock loading. Hence, the shock induced reaction can be distinctly defined by the solid-state reaction, having little or nothing to do with any thermal transients during shock loading.

Comparing the XRD profiles between samples with SiC particles and whiskers, no difference was distinguished, but looking into the details in microstructure, the morphology effect on the shock reactivity can be found. Figure 8 shows the microstructure both for $MoSi_2$ - SiCp and $MoSi_2$ - SiCw for f = 50 %. No interfacial reactions took place between $MoSi_2$ phase and SiCp; no new products were seen in the microstructure except for $MoSi_2$ and SiCp. The EDX result reveals that the tertiary phase in Mo-Si-C should be synthesized at the vicinity of SiCw fragments. Since little or no reactions into those tertiary phases took place on the interfaces between large SiCw and $MoSi_2$ matrix, this reaction was only activated at the site of long, thin SiCw fibers. This morphological effect on shock reactivity might be thought as a mechano-chemical reaction during shock loading.

As theoretically proposed by Horie [11], the driving mechanism to shock induced reaction lies in the transient Hugoniot of particle individuals converging to the equilibrium Hugoniot during shock loading. If each powder particle had sufficient reactivity into compound, it could be synthesized only by application of shock waves into materials. Hence, the pretreatment by mechanical alloying must be necessary to activate the grain boundary energy between Mo and Si by refining the grain size and straining the grain boundaries. At the critical shock condition to synthesize the intermediate, amorphous $MoSi_2$ phase before the regularization reaction, the activated grains might be subjected to intense shear deformation and compaction by transient Hugoniot, resulting in the random structure even in the atomic scale, or, the amorphous phase formation. This mechanical interaction between constituent particles and shock wave propagation might be often frozen in the microstructure of the recovered sample. As shown in Figure 9, the SiC whiskers which were randomly mixed into the pretreated powders at the initial stage, were forced to be shaped lamellar. This alignment must be attributed to shear deformation of MA precursor and stacking of SiC whiskers; this lamellar structure formation in normal to the shock wave front might be a proof of intense flow and deformation in materials during the passage of shock waves.

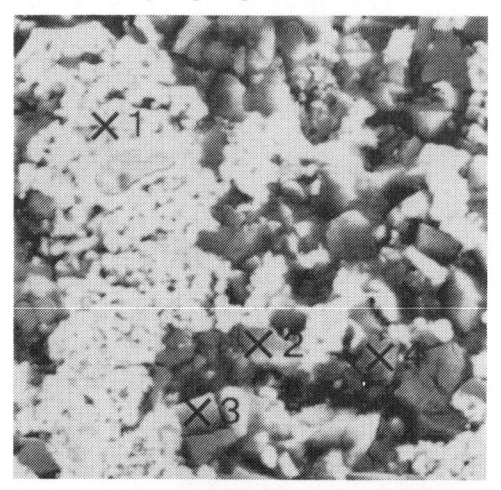

| Sample Point | Si (at %) | Mo (at %) | C (at %) |
|---|---|---|---|
| 1 | 66.47 | 33.53 | 0 |
| 2 | 40.92 | 0 | 59.08 |
| 3 | 40.74 | 0 | 59.26 |
| 4 | 63.26 | 0 | 36.74 |

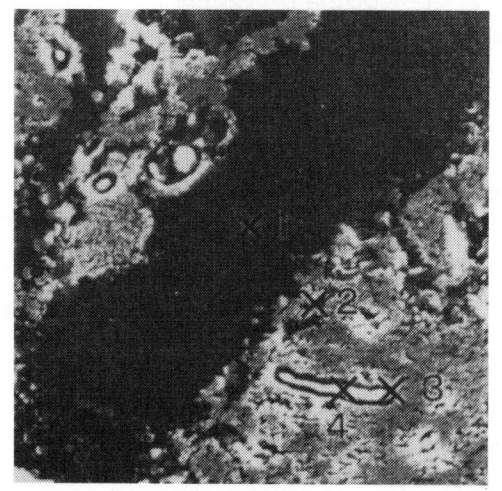

| Sample Point | Si (at %) | Mo (at %) | C (at %) |
|---|---|---|---|
| 1 | 45.42 | 0.30 | 54.29 |
| 2 | 50.41 | 25.75 | 23.84 |
| 3 | 43.37 | 22.94 | 33.69 |
| 4 | 10.14 | 43.54 | 46.31 |
| 5 | 42.82 | 0.24 | 56.94 |
| 6 | 66.39 | 33.61 | 0 |

Figure 8: Comparison of EDX results between $MoSi_2$ - SiCp and $MoSi_2$ - SiCw for the volume fraction of 50 %: a) $MoSi_2$ - SiCp composite for f = 50 %, and b) $MoSi_2$ - SiCw for f = 50 %.

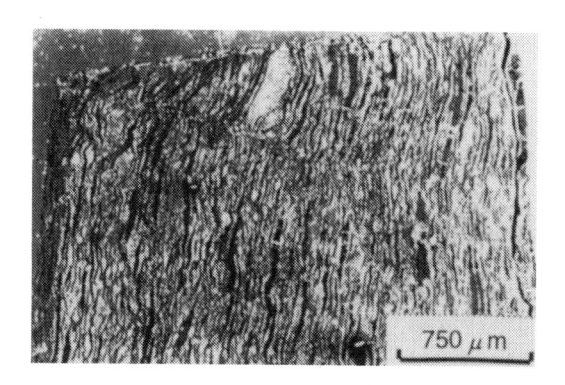

Figure 9: Intrinsic lamellar structure formation to the shock induced reaction of MoSi$_2$ - SiCw composites.

## CONCLUSION

Since full reactions never take place when starting from the blended elemental powder mixes, it must be difficult or nearly impossible to yield ceramic particle or whisker reinforced di-silicide materials. Use of the pretreated powders by the mechanical alloying enables us not only to make shock induced synthesis into di-silicide with full reactivity but also to yield particle or whisker reinforced di-silicide under the lower shock conditions.

The gas gun can be available to synthesize these new composites without cracks or defects when starting from the mixes of reinforcing ceramic particles or whiskers with the pretreated powders. By high volume loading of SiCp or SiCw into the pretreated powders, the thermal transients were quenched and the residual temperature was significantly reduced. Indifference of shock reactivity to this thermal transients must be a proof that demonstrates that the shock induced reaction should be pressure-oriented mechano-chemical behavior.

The shock induced reaction process also consists of two steps in other transition metal - silicon systems; however, the intermediate phase to be synthesized is strongly dependent on each system. The present experimental methodology is effective to describe the shock reaction mechanism.

## ACKNOWLEDGEMENTS

Authors would like to express their gratitude for Mr. P.J.Counihan and Mr. Ken-ichi Ichige for many helps in experiments. One of authors would like to express also his thanks to Prof. Y. Syono at the Institute of Materials Research, Tohoku University for his suggestion. The present work was performed as a part of international collaboration between the Georgia Institute of Technology and the University of Tokyo.

**REFERENCES**
[1] N. N. Thadhani, et al., "Shock Induced and Shock-Assisted Solid-State Chemical Reactions in Powder Mixtures," J. Applied Physics 76 (1994) 2129-2138.

[2] S. Batsanov, F.D.S. Marquis and M.A. Meyers, "Shock Induced Synthesis of Silicides," Metallurgical and Materials Application of Shock Wave and High-Strain-Rate Phenomena. (1995) 715-722.

[3] A.G. Merzhanov, "Self-Propagating High-Temperature Synthesis: Twenty Years of Search and Findings," Combustion and Plasma Synthesis of High Temperature Materials (Eds. Z. Munir and J. Holt). VCH (1990) 1-53.

[4] N.N. Thadhani, et al., "Shock-Induced Chemical Reactions in Titanium-Silicon Powder Mixtures of Different Morphologies: Time-Resolved Pressure Measurements and Materials Analysis," J. Applied Physics. 82 (1997) 1113-1128.

[5] T. Aizawa, "Shock Reative Synthesis of Refractory Metal Aluminides and Silicides," Ceramic Engineering and Science Proceedings. 18 (1997) 573-580.

[6] T. Aizawa, Y. Asakawa and J. Kihara, "Shock Reactive Synthesis of Refractory Metal Aluminides from a Mechanically Alloyed Precursor," Annales de Chimie. 20 (1995) 181-196.

[7] T. Aizawa, "Shock Reactive Synthesis of Refractory Metal Aluminides from Pretreated Materials," Proc. 3rd NIRIM Int Symposium on Advanced Materials. (1996) 45-50.

[8] N.N Thadhani and T. Aizawa, "Materials Issues in Shock Compression Induced Chemical Reactions in Porous Solids," Ch. 5 in High Pressure Shock Compression of Solids. (Ed. Y. Horie) IV (1997).

[9] T. Aizawa, B.K. Yen and Y. Syono, "Shock-Induced Reaction Mechanism to Synthesize Refractory Metal Silicides," in Shock Compression of Condensed Matter-1997 (1998) (in press).

[10] T. Aizawa, "Micro-Physical Modeling of Shock Induced Reactions in Solid Synthesis of Silicides," Proc. International Seminar on Vapor Explosions and Explosive Eruptions (AMIGO-IMI-Sendai) (1997).

[11] Y. Horie and K. Yano, "A New Computer Code: DM2," Abstracts of Int. Workshop on Industrial Applications of Explosion, Shock-Wave and High Pressure Phenomena. (1997, Oct. Kumamoto, Japan) O5.

[12] B.K. Yen, T. Aizawa and J. Kihara, "Synthesis and Formation Mechanism of Molybdenum Silicides by Mechanical Alloying," Mat. Sci. Eng. A220 (1996) 8-14.

[13] T. Aizawa, B.K. Yen and J. Kihara, "Mechanical Alloying of Mo-Si System," J. Graduate School and Faculty of Engineering, University of Tokyo. XLII [4] (1996) 501-519.

[14] T. Aizawa, N. Sakakibara, B.K. Yen and Y. Syono, "Shock Reactive Synthesis of (Mo, Ti) Silicides from Pretreated Powders," J. American Ceramic Society. (1998) (in press).

[15] T. Aizawa, "Shock Induced Reaction to Synthesize Refractory Metal Silicides," Proc. 5th NIRIM Int Symposium on Advanced Materials. (1998) 189-192.

# Mechanical Alloying

# MECHANOCHEMICAL SYNTHESIS OF REFINED Ag- AND Zn-COMPOSITE POWDERS STARTING FROM OXIDES

S.D.De la Torre[*,a,b], K.N.Ishihara[a], P.H.Shingu[a], D.Rios-Jara[b] and H.Miyamoto[c].

[a] *Department of Energy Science & Eng. Kyoto University. Sakyo-Ku, Kyoto 606 Japan.*
[b] *Centro De Inv. en Materiales Avanzados, CIMAV. S.C., 120, CP.31109. Chihuahua, Mexico.*
[c] *Technology Res. Inst. of Osaka Prefecture. 2-7-1Ayumino, Izumi, Osaka 594-1157, Japan.*

## ABSTRACT

De-oxidation of $Ag_2O$ and $ZnO$ powders has been accomplished by the mechanochemical process (MCP) using Si and Fe as reducing agents. Exception is the reduction reaction of ZnO with Fe since it involves a positive enthalpy, and so even after 300 h milling its reduction does not take place. A solid solution of nanosized $Fe_2O_3$ into an amorphous-like ZnO matrix is developed instead. In contrast to reduced products prepared by conventional thermal decomposition of oxides, a number of advantages are conferred into the MCP-synthesized $Ag-SiO_2$, $Ag-Fe_2O_3$, $Zn-SiO_2$ and $ZnO-Fe_2O_3$ submicrometre composite products, such as a homogeneous dispersion of active oxides into the reduced metal matrix and nanometre tailored particles.

## INTRODUCTION

Metal matrix composites (MMCs) have been of interest for several decades. Several methods have been developed to incorporate large volume fractions (up to ~50%) of oxide, carbide, boride, and other particles into metal matrices. Dispersion-strengthened alloys were the earliest MMCs and they contained several volume % particles, usually oxides. The mechanical alloying (MA) technology has been used to produce strengthened MMCs with oxides dispersion and controlled fine microstructures [1-3]. Researchers have studied, for example, the structures and properties of Al-MO (where M is Cu, $Fe^{+2}$, Si and Mg) powders prepared by MA [4] and $Al-M_2O_5$ (M is V, Nb, and Ta) [5]. Derived from the MA concept, the mechanochemical (MCP) process fulfills the formation of complex precursor oxides for advanced materials technology applications [6-9]. Ball milling also has been carried out to induce polymorphism of $Fe_2O_3$ and its reaction with NiO and ZnO [10]. Schaffer and McCormick [11], reported direct synthesis of Cu starting from its oxide using various reducing elements, such as: Al, Ca, Ti, Mg, Mn, Fe and Ni. In practice, an essential factor governing the solid synthesis, via ignition propagation of a combustion front [12-13] is the high exothermic heat of a given reaction. This work aims the synthesis of Ag- and Zn-matrix composites having a homogeneously dispersed oxide of high melting point and small crystal size, and to establish microstructural differences of those materials produced by conventional heating and the MCP route. The powders microstructural formation is emphasized.

---

To the extent authorized under the laws of the United States of America, all copyright interests in this publication are the property of The American Ceramic Society. Any duplication, reproduction, or republication of this publication or any part thereof, without the express written consent of The American Ceramic Society or fee paid to the Copyright Clearance Center, is prohibited.

**Table I**. The Oxides Reduction Reactions Studied

| Reaction No. | △H | △G | MP (°C) | structure* |
|---|---|---|---|---|
| 1) Ag$_2$O + 1/2 Si → 2 Ag + 1/2 SiO$_2$ | - 424 | - 417 | 1710 | polymorph |
| 2) 3 Ag$_2$O + 2 Fe → 6 Ag + Fe$_2$O$_3$ | - 730 | - 708 | 840-980 | cubic |
| 3) 2 ZnO + Si → 2 Zn + SiO$_2$ | - 214 | - 220 | 1800 | hcp |
| 4) 3 ZnO + 2 Fe → 3 Zn + Fe$_2$O$_3$ | + 221 | + 213 | 1457 | hcp |

Where: △H and △G are respectively enthalpy and standard free energy change for the reaction at room temperature and they are given in kJ/mol. MP means melting point of the underlined oxides. * stands for the structure of underlined oxide at room temperature.

## EXPERIMENTAL

The studied reactions are listed in Table I. Powder mixtures of Ag$_2$O (purity 98% up, -300mesh), Si (99.99%, -10μm), and Fe (99.9%, -300 mesh) were appropriately mixed and poured into low energy (90 rpm) stainless containers (12 cm in diameter ϕ), filled up with 4 kg of stainless balls (9 mm ϕ). A 50% stoichiometric excess of reducer agent was set for the reactions. ZnO powder was 99.90% pure under -300 mesh. The ball to powder weight ratio was 80:1. The milling process was performed under Ar gas atmosphere. All reactions were carried out without lubricant addition. Exception is reaction No.3, which was the only one analyzed by addition of 0 and 3ml of anhydrous toluene (99.5% C$_6$H$_5$CH$_3$). The MCP'ed powders were examined with a scanning electron microscope SEM, supplied with an electron probe micro-analyzer EPMA (Hitachi X-650 microscope, Japan). Thermal analysis of powders was carried out by using a Perkin-Elmer DSC-7, USA calorimeter, operated at a heating rate of 20°C/min. X-ray powder diffractometry XRD was performed using a Rigaku RAD-B apparatus.

## RESULTS AND DISCUSSION

Apart from the innovative electronic (piezoelectric) applications that are expected from Ag- and Zn- MCP-prepared composites, the constituent elements for reactions 1, 2 and 3 were chosen because of their refractory character (high MP) and the large free energy change (see table I) associated with each reaction, which ensures a large thermodynamic driving force. In contrast, reaction 4 would only proceed if enough energy were supplied into the reactants. We challenged its synthesis via the MCP process. Results of the studied reactions are presented and discussed in the following subsections.

### *Reactions 1 and 2*

Fig.1 shows the XRD patterns obtained from the original Ag$_2$O + Si mixture for reaction 1 and those resulted following several hours milling. After 24h, all the peaks intensity significantly reduced. This is both due to a decrease on the powders crystallite size and to the onset of the solid state reduction reaction SSRR, as reflected from the pure Ag peaks emerging in the 24h pattern. SEM observations on powder at this milling stage revealed refined Si in closer contact with broken Ag$_2$O particles. After 48h milling, (111) and (200) pure Ag peaks are perfectly identified, where the real identity of the (200) Ag$_2$O peak corresponds to (111) Ag. Further milling fulfills the reduction reaction, showing wider and enlarged Ag peaks. Since Si is chemically in excess, its peaks after 110h MA are still visible, whereby small traces (<5µm) of non-reacted Si particles are seen in Fig.2. SiO$_2$ particles are smaller and well distributed in the ~25nm sized Ag matrix. As inferred from the larger heat content and in contrast to reaction 1, Fe (in react.2) has a greater ability to reduce Ag from its oxide. The reaction rate progress occurred when using Si and Fe as reducing agents is plotted vs. the milling time in Fig.3. It was estimated using the ratio Ag$_2$O/Ag, from their main peaks area. Reaction 2 was respectively started and completed at about 4 and 10h milling. In its turn, reaction 1 took place more slowly, so that 50h were needed at least.

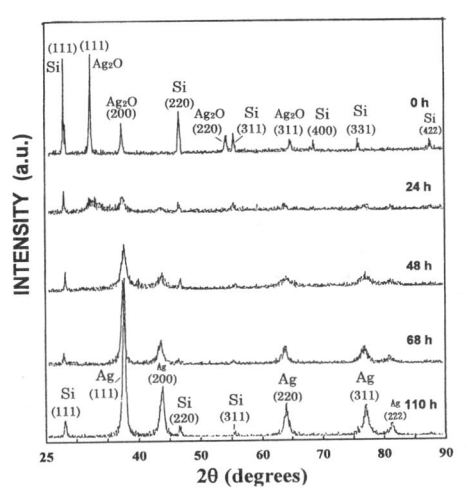

Fig.1 XRD patterns displayed from react.1, Ag$_2$O + Si, as a function of milling time.

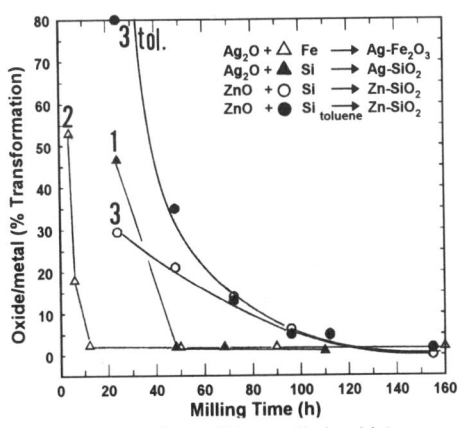

Fig.3 A comparison of the studied oxide's reduction rate by using Fe and Si as reducers.

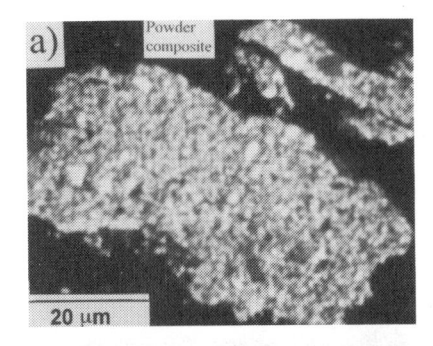

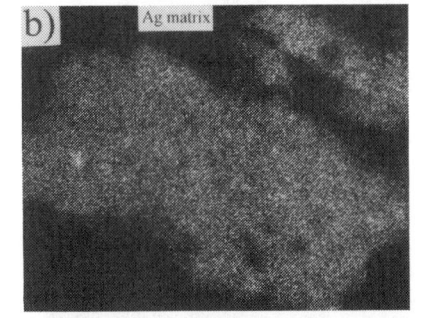

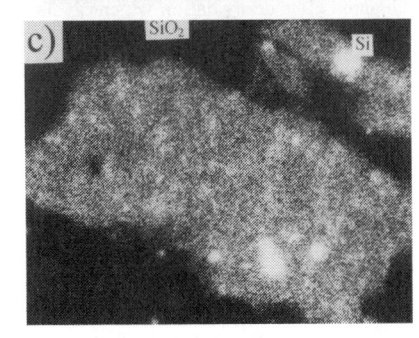

Fig.2. SEM-EPMA images of Ag$_2$O+Si, MA for 110 h (a) sec.image, (b) Ag-kα, (c) Si-kα.

React.2 is thermodynamically favored, whereby Fe is a better electron donor than Si to reduce Ag from its oxide. It should be noted that even if Si were absent in reaction 1, by simple heating Ag$_2$O powder one might destabilize the oxide to have the Ag reduced. Thermodynamically, that is possible since that makes $\Delta G < 0$. However, important micrometre/nanometre microstructural differences between the particle sizes of the powder product prepared by conventional heating and the MCP route will markedly influence the final composites mechanical and electronic properties.

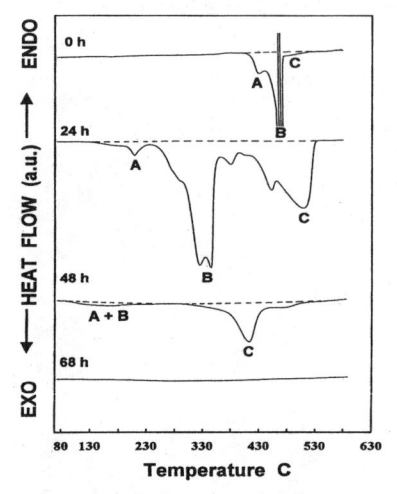

Fig.4 DSC traces registered from react.1 as a function of the milling time (refer to text for explanation).

Fig.5 XRD patterns of the original Ag$_2$O + Si mix of powder before and after being ball treated, and analyzed by DSC at 600°C.

Fig.4 shows normalized DSC traces obtained from reaction 1 for different milling time. When the original powder mixture (0 h) was heated up to 600°C, its DSC trace disclosed a series of irregular exothermic peaks placed between 400 and 500°C. Its resulted XRD pattern is presented in Fig.5b. Looking at the peaks width of such pattern, i.e., following its Bragg peaks obtained after de-oxidation we can induce that the resultant particle size is comparable to that of the starting powder (Fig.5a). In addition, for this sample (Fig.5b) some unknown peaks are diffracted at around 80deg., which are attributed to secondary reactions. Once the reaction between Ag$_2$O and Si has reached 24h milling (Fig.4) the exothermic peaks temperature range became broader as compared to the as-mixed sample. Separate heating runs were performed for the 24h specimen at 250, 370 and 600°C, each followed by XRD. The X-ray pattern corresponding to 250°C showed light increase on all the Ag peaks intensities, as well as on the (111) Ag$_2$O peak intensity. Therefore, it is believed that the first exothermic peak, pointed out by letter A is in connection with the outset of reaction 1. Peak B is formed because of the intrinsic heat evolved from the reaction to reduce Ag. At this point, Ag peaks were developed almost completely, but again difficult to index Bragg peaks appeared. Peak C is both in relation with these secondary reactions annihilation and grain growth of the SiO$_2$ embedded into the Ag matrix. After 48h milling the intermediate reactions (peak C in Fig.4) almost disappeared. Fig.3 also shows that Ag$_2$O+Si powders after 48h milling were almost fully reacted, hence its corresponding DSC trace revealed a much broader and short exothermic peak, indicated in Fig.4 by letters A+B.

The peak C, at this milling time is associated with grain growth and recrystallization of Ag and SiO$_2$. On further milling, neither exo nor endothermic peaks are evolved when react.1 was in the stage of 68h because at this time the SSRR has already been completed.

Assuming that the A peak temperature in Fig.4 corresponds to the ignition temperature for developing reaction 1 ($T_{i-1}$) and B peak temperature is the apparent reduction temperature ($T_{r-1}$) they have been plotted vs. the milling time in Fig.6, together with data of reac.2. The longer milling time, the smaller values of those. Both temperatures became smaller since the mechanical stress and strain supplied into the reactants is gradually stored and enlarged, activating them as to accomplish the SSRR. That is, the driving force needed to reduce the original oxides is supplied parallel to the MCP time and so the heat content coming from the DSC apparatus was substantially each time less needed from specimens long time ball-treated. In contrast, free-stress containing and much more larger Ag crystals were obtained when reduced by the heating route. Full conversion into refined Ag having homogeneously dispersed SiO$_2$ particles is better obtained by the MCP route. This can be verified by comparing the Ag peaks width in the as-mixed sample+600°C (Fig.5b) and its MCP'ed counterpart (Fig.5c). In the latter case, the volume fraction of Ag parallel increased with the relative intensity of the Ag peaks. Another difference is that in the case of ball treated material no secondary compounds are left in the products. In general, few hours reacted powder mixtures demanded larger energy (heat) for Ag$_2$O to decay.

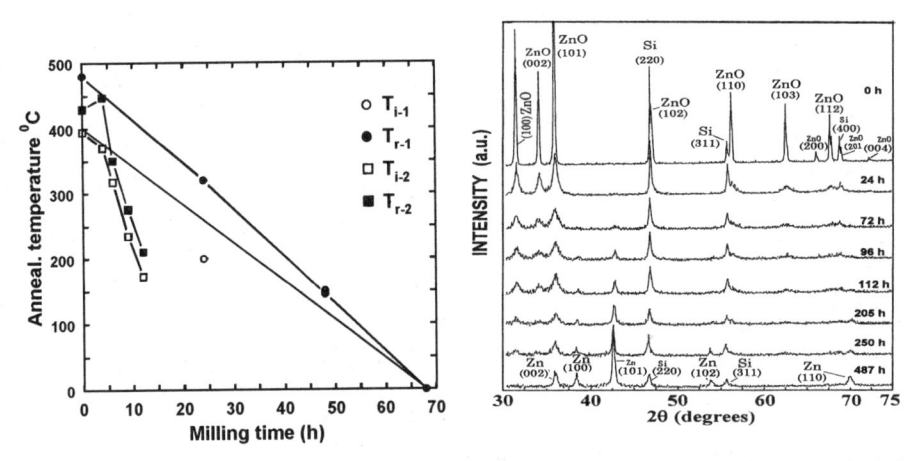

Fig.6 Ignition ($T_i$) and apparent reduction ($T_r$) temperatures for reactions 1 and 2 vs milling.

Fig.7 XRD patterns displayed from react.3, ZnO + Si, vs milling. 3 ml of lubricant (toluene) added.

### Reaction 3, with and without addition of lubricant

Apart from attempting the reduction of Zn from its oxide at room temperature by using Si (in 50% stoichiometric excess) and Fe (10% excess, in react.4) as reducing agents, there was interest in understanding how the usage of lubricant influences the reactions rate. Most of the next results are to establish the main differences found out. The XRD patterns in Fig.7 show the gradual interaction between ZnO and Si when toluene was used as the process control agent. The reaction was essentially completed after ~150 h milling. The effect of using lubricant is seen in Fig.3. When toluene is absent the same reactants interact faster. Roughly, half of the milling time is required to accomplish the reduction reaction. Larger Bragg peaks intensities resulted from dry-treated samples. At 70 h milling, the lubricant scarce and therefore the ZnO/Zn conversion rate becomes nearly equal in both cases. A very similar microstructural evolution was observed from reaction 3 performed with no lubricant addition. As occurred in the case of reactions 1 and 2, the SSRR took place parallel to the diminution of the reducers particle size. Once reduced and after long milling time, refined Zn particles became harder and brittle, eventually randomly dispersing with $SiO_2$. From EPMA analysis performed on these specimens it was evident that apart from the presence of non-reacted excess Si particles (<10μm), $SiO_2$ and much finer Zn particles became in the nanometric scale [14]. Fig.8 shows a normalized DSC curve recorded from the ZnO + Si mixture, MCP'ed for 370h without lubricant addition. Inserted XRD patterns are respectively those obtained after the original powder was heated by DSC up to 0 and 600°C, and 370h MCP'ed powder heated to 0, 250, 370, and 600°C. The large endothermic peak which appeared at 419.5°C matches good with the Zn melting temperature. Under 300°C there is an overlapping of at least two exothermic peaks. To further interpret them it was convenient to perform several XRD examinations along this DSC trace. For comparison, a reference pattern composed of the dealing reactants mixture was plotted in the upper most part of all collected XRD patterns.

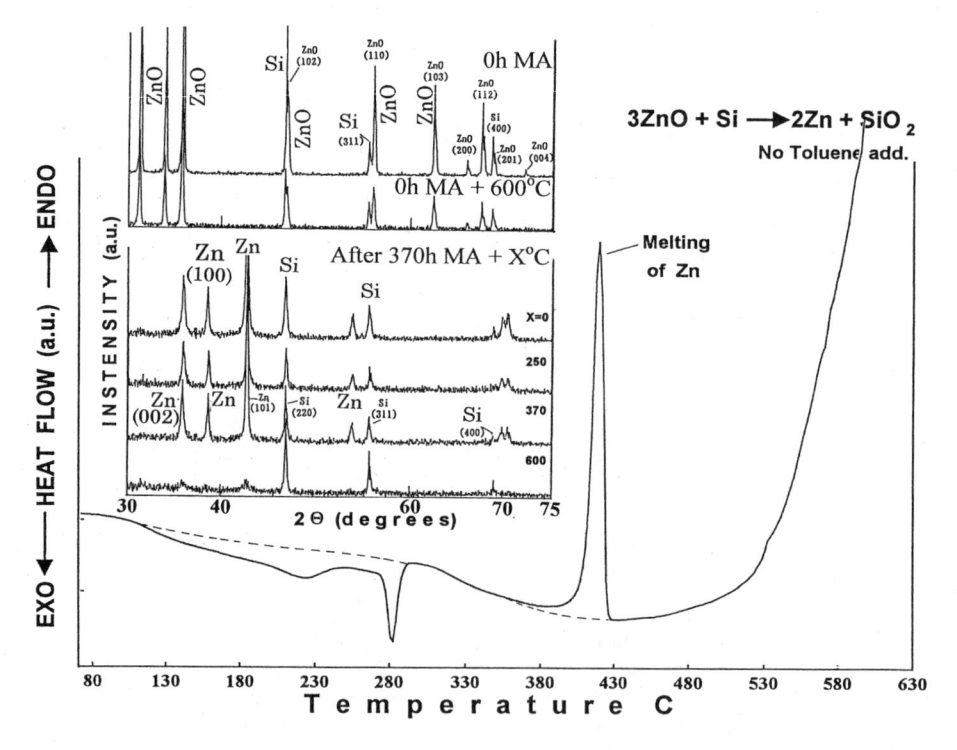

Fig.8 DSC trace of 370 h ball milled ZnO + Si powders (no toluene added). Inserted XRD patterns are those obtained after the powders mixture was milled for 0 and 370h. Each pattern corresponds to a different point in the DSC curve.

Examining the second pattern (from top to the bottom), no structural change was evident indicating that elemental ZnO can not be reduced directly by increasing its temperature to 600°C. Once the 370 h mechanically activated powder was heated to 250°C, its DSC trace displayed a wide and large exothermic area [14]. However, since its corresponding XRD pattern did not reveal other change rather than just a little more narrowed peaks, one can associate the wide DSC area to a strain relaxation phenomenon. The sharp exothermic peak displayed at about 280°C is connected with relaxation and recrystallization of Zn, Si and Si$_2$O (undetected) particles. The XRD pattern placed at the bottom confirms melting of Zn. Si peaks are basically its only feature.

## *Reaction 4*

The MCP-interaction occurred between ZnO and Fe powders can be followed in Fig.9 by analyzing the XRD patterns evolution. Contrary to react.3 (ZnO and Si), even though 300 h milling were set in this case, it is evident that the positive enthalpy of the reaction, intrinsically associated with the interacting materials, could not be overcame by our low energy milling set conditions. That is, the free energy of formation of this reaction system resulted difficult to surmount because $\triangle G_{ZnO} < \triangle G_{Fe2O3}$. Instead of reducing Zn from its oxide with Fe, ZnO was gradually structure-disordered, while Fe peaks shortened and shifted from their original positions towards lower angles. For example, the (110) Fe peak moved from 44.40 to approx. 43.69 degrees. The X-ray pattern recorded after 300h milling is typical of a partially amorphized solid solution. Although the particle size is too small to enable precise structure identification by X-ray diffraction, considering that only two weak Fe$_2$O$_3$ peaks; (13·4) and (22·6) were detected at around 35deg.,at present we speculate on the fact that to some extent, nanometre tailored Fe oxidized into Fe$_2$O$_3$ forming metastable ZnO-Fe$_2$O$_3$ powder composites.

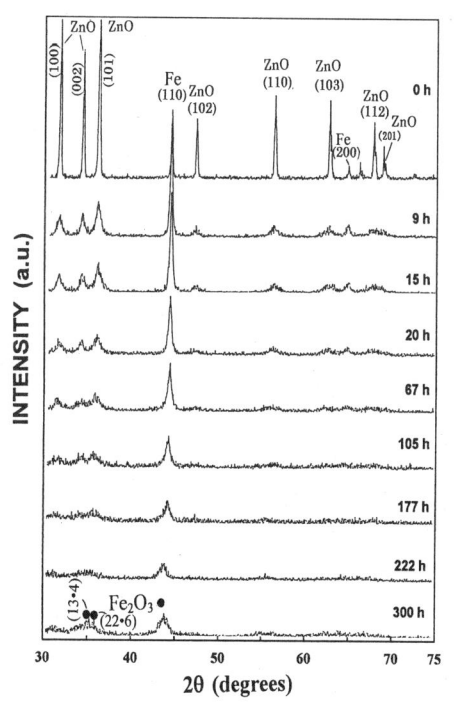

Fig.9 XRD patterns displayed from ZnO+Fe→ZnO-Fe$_2$O$_3$ vs MA time.

## CONCLUSIONS

The MCP process, as performed on the studied MMCs afforded the energy required to start chemical interaction between the set reactants, whose equilibrium enthalpy of formation is negative. The mechanical energy supplied into the powders initiated thermodynamically favored exothermic reactions, leading to the SSRR development. In contrast to the free-strain, heterogeneous and large grain size features exhibited from the powder products obtained by thermal decomposition of elemental oxides, the MCP-synthesis of nanometre sized and homogeneous Ag-SiO$_2$ and Ag-Fe$_2$O$_3$ composite powders was successfully accomplished, starting from Ag$_2$O by using pure Si and Fe powders as reducers. The same is true in the Zn reduction case. Processing of ZnO and Si resulted in the formation of submicrometre refined Zn-SiO$_2$ composites. In the latter case and in contrast to dry MCP, wet processing decreased the reaction rate to half of the milling time. The solid solubility of nanosized Fe$_2$O$_3$ particles and structural-disordered (partially amorphized) ZnO particles was obtained.

# REFERENCES

1. J.S.Benjamin, "Dispersion Strengthened Superalloys by Mechanical Alloying" *Metal.Trans.* **1** (1970) 2943.
2. J.S.Benjamin, "Fundamentals of MA", *Mater.Sci.Forum,* **88-90** (1992) 1.
3. J.J.de Barbadillo and G.D.Smith, "Recent Developments and Challenges in the Application of Mechanically Alloyed, Oxide Dispersion Strengthened Alloys", *Mater. Sci. Forum,* **88-90** (1992) 167.
4. D.G. Kim, J.Kaneko and M. Sugamata, "Structures and Properties of Mechanically Alloyed Aluminum-Metal Oxide Powders and Their P/M Materials", *J. Japan Inst. Metals* **57** 6 (1993) 679.
5. D.G. Kim, J.Kaneko and M. Sugamata, "Structures and Properties of Mechanically Alloyed Aluminum-Transition Metal Oxide (V$_2$O$_5$, Nb$_2$O$_5$, Ta$_2$O$_5$) Powders and Their P/M Materials", *J. Japan Inst. Metals* **57** 11 (1993) 1325.
6. M.Senna, "Incipient Chem. Interaction between Fine Particles under Mechanical Stress - A Feasibility of Producing Advanced Materials via Mechano-Chemical Routes", *Solid State Ionics,* **63-65** (1993) 3.
7. P.H. Shingu, (ed.), "Mechanical Alloying". *Mater. Sci. Forum,* **88-90**. Trans. Tech. Publications, Zürich (1992).
8. E.M. Gutman (ed.), "Mechanochemistry of Solid Surfaces", World Sci. Pub. (1994).
9. T.Tsuzuki, J.Ding and P.G.McCormick, "Ultrafine Powder Synthesized by Mechanochemical Processing", *in Proc. of the Int. Symp. on Designing, Processing and Prop. of Advanced Eng. Materials, ISAEM-97.* Toyohashi, Japan. 1997.
10. T.Kosmac and T.H. Courtney, "Milling and Mechanical Alloying of Inorganic Nonmetallics" *J. Mater. Res.* **7** 6 (1992) 1519.
11. G.B. Schaffer and P.G. McCormick, "The Direct Synthesis of Metals and Alloys by Mechanical Alloying", *Materials Sci. Forum,* **88-90** (1992) 779.
12. Z.A.Munir, "Synthesis of High Temperature Materials by Self-Propagating Combustion Methods", *American Ceramic Bulletin* **67** 2 (1988) 342.
13. Z.A.Munir and J.B.Holt (ed.), "Combustion and Plasma Synthesis of High-Temp. Materials", VCH Publishers (1990).
14. S.D.De la Torre, PhD Thesis. Dept. of Energy Sci. & Eng., Kyoto University, 1995.

# CERAMIC OXIDE (MeO$_2$) SOLID SOLUTIONS OBTAINED BY MECHANICAL ALLOYING

Federica Bondioli, Marcello Romagnoli, Luisa Barbieri and Tiziano Manfredini
Department of Chemistry, University of Modena
Via Campi 183, 41100 Modena (Italy)

## ABSTRACT
Cerium and zirconium oxide powders (MeO$_2$) doped with 10 mol% of praseodymium were prepared by mechanosynthesis in a high energy ball mill to investigate whether mechanical alloying is possible in these ceramic systems industrially used as inorganic pigments. The structure variation of the powders was followed by X-ray diffraction (XRD) and scanning electron microscopy (SEM) and to confirm the effective formation of the solid solution, leaching tests were performed. The results prove that mechanosynthesis occurs in both ceramic systems.

## INTRODUCTION
High energy ball milling (mechanosynthesis) has become a very used processing method for producing and synthetising intermetallic compounds, extended solid solutions, amorphous alloys and nanocrystalline materials [1,2]. However it was mainly applied to alloys formed from pure metals or intermetallic compounds; recently it was showed that mechanical alloying could also be used for ceramic materials. Chen et al. [3] and Chen and Yang [4] first reported the mechanical alloying of ceramics in the ZrO$_2$-CeO$_2$ system.

The work described in this paper has to be regarded as an initial study about the synthesis of ceramic pigments by mechanosynthesis. Colour is usually introduced into ceramic glazes or bodies by dispersing a coloured crystalline phase insoluble in the matrix. The crystalline phase, commonly called a pigment, imparts its colour to the matrix. Pigments for the production of coloured

---

To the extent authorized under the laws of the United States of America, all copyright interests in this publication are the property of The American Ceramic Society. Any duplication, reproduction, or republication of this publication or any part thereof, without the express written consent of The American Ceramic Society or fee paid to the Copyright Clearance Center, is prohibited.

traditional glazed and unglazed tiles must show thermal and chemical stability at high temperature (1200-1250°C) and must be inert to the chemical action of the molten glaze (frits or sintering aids). There are only a limited number of pigments suitable for use in ceramic industry because only few materials can withstand this high temperature and corrosive environment. This drove to an increase in the search of new pigments or new synthetic routes for the old ones.

In ceramic industry pigments were traditionally obtained by calcination of raw materials [5]. These heat treatments were carried out at high temperature, adding mineralizers (fluorides) to lower the synthesis temperatures. The possibility to obtain them at room temperatures and without mineralizers is of great interest: avoiding high calcination temperatures can cause drastic reduction in toxic gas emission and minimises high temperature technological investments.

The two systems (ceria and zirconia powders containing 10 mol% praseodymium) were chosen because of their thermal and chemical properties and (from an industrial point of view) because of the interesting red and yellow tonalities of colour that they can develop. Studies regarding these two systems are few, and the existing ones [6] involve only traditional or co-precipitation synthesis of the praseodymium doped ceria system, while, even if zirconia is well known in ceramic, there are no publications about the praseodymium doped ziconia system.

In this study the effect of milling time and specific surface area of starting materials on the solid solution formation were analysed to determine optimal conditions under which pigments can be obtained.

EXPERIMENTAL

High purity $MeO_2$ (Me = Ce, Zr) and praseodymium oxide ($Pr_6O_{11}$, Carlo Erba) powders were mixed in order to obtain the stoichiometric composition of $Me_{0.9}Pr_{0.1}O_2$. In order to determine the effect of the physical properties of the starting materials on samples obtained, only for the system zirconia praseodymia, three types of zirconia were chosen: monoclinic zirconia with low (15 m$^2$\g, A) and high (100 m$^2$\g, B) specific surface area, and an yttria stabilised zirconia (C).

The powders were charged in a cylindrical agate vial of 40 mm in diameter together with agate balls of determined diameter (12 mm for the small type, 14 mm for the big type as milling media). Mechanical alloying was performed in a Fritsch "Pulverisette 7" planetary ball mill using a rotation speed of 200 rpm keeping constant the ball-to-powder weight ratio at 20:1. Milling was interrupted every hour for 30 minutes to prevent the powder's temperature from rising too high. At predetermined intervals of time, depending on the system involved (1, 5, 10, 12, 15 and 20 hrs for ceria and 5, 10, 15 and 20 hrs for zirconia) samples of milled powders were removed from the mortar and subjected to X-ray diffraction analysis.

All the X-ray diffraction experiments were made with a Philips PW 3710 diffractometer, using CuKα radiation, acquiring data from 20° to 60° 2θ, stepsize 0.02°, 5 sec acquisition time. The morphology and microstructure of the various samples were examined by scanning electron microscopy (SEM, Philips XL40) with EDS analysis (EDAX PV9900).

To confirm the effective formation of the solid solution, leaching tests in boiling solutions of concentrated (37 wt%) hydrochloric acid were performed. The Pr content of the solutions was determined by ICP spectroscopy (Variant, mod. Liberty 200).

RESULT AND DISCUSSION
Ceria-Pr System

*Solid Solution Formation*: Figure 1 shows the X-ray diffraction patterns of the samples milled for 0, 10, 15 and 20 hrs, respectively, in 20-40° 2θ range.

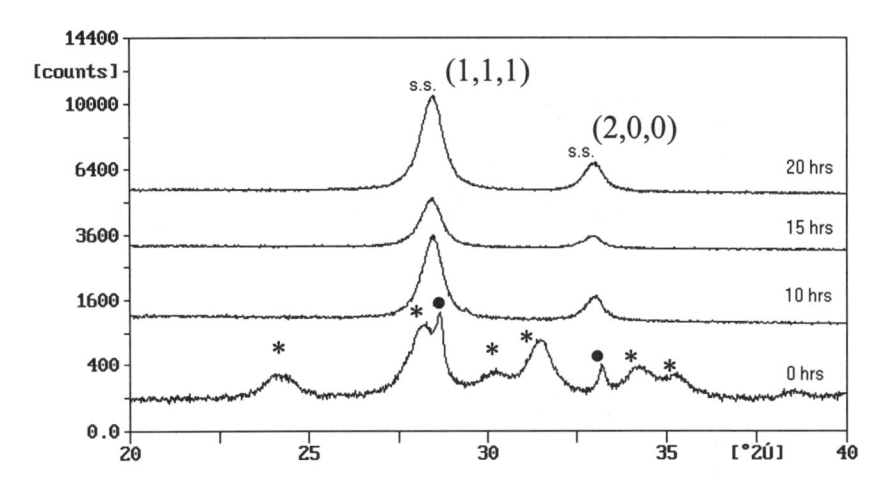

Figure 1 - XRD patterns of $Ce_{0.9}Pr_{0.1}O_2$ mixed powders after milling for different time.
\* = $CeO_2$, • = $Pr_6O_{11}$, s.s. = $Ce_{0.9}Pr_{0.1}O_2$

From figure 1 it can be seen that the ceria before milling exists in a fluorite (f.c.c.) structure. The position of the (1,1,1) peak showed that the formation of the solid solution $Ce_{0.9}Pr_{0.1}O_2$ occurred for milling time higher than 10 hrs. The solid solution formation is accompanied by the disappearance of the praseodymium oxide peaks.

This trend is confirmed by SEM and EDS analysis. In figure 2a and b are reported the SEM and EDS analysis of the sample obtained after 15 hrs milling. Chemically homogeneous powders of $Ce_{0.9}Pr_{0.1}O_2$ solid solution were obtained in the form of agglomerates of about 1 μm diameter, consisting of particles with lower average size. EDS analysis excluded the presence of Si coming from milling media even for milling time higher than 20 hrs.

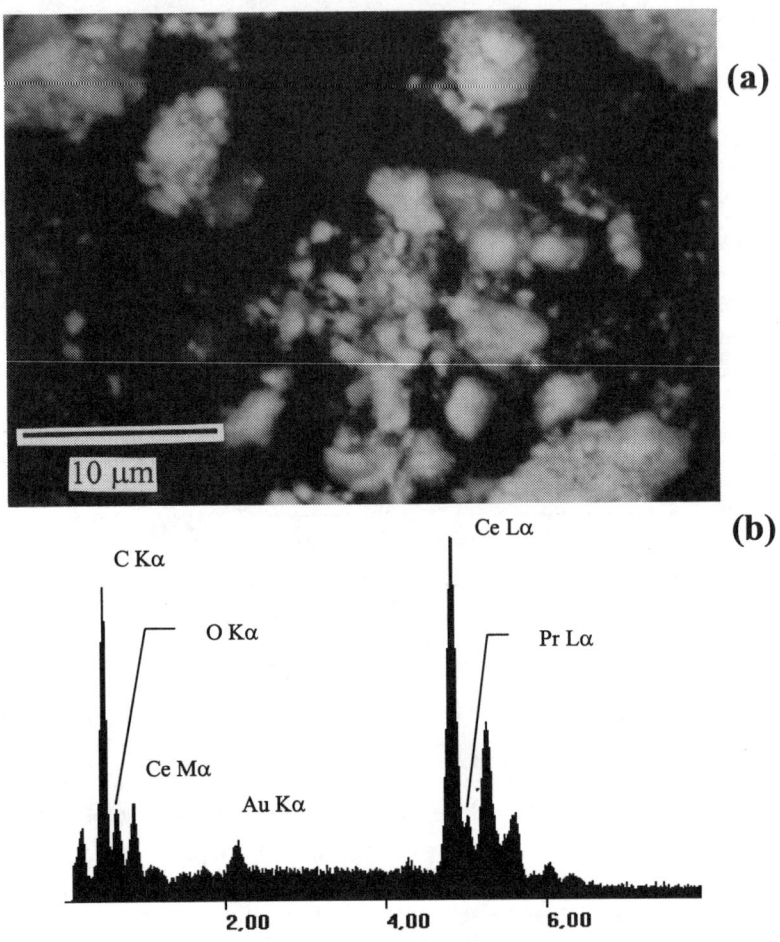

Figure 2 - SEM (a) and EDS (b) analysis of powders after 15 hrs milling.

Leaching tests in boiling solutions of concentrated (37 wt%) hydrochloric acid may be considered suitable to evaluate the effective diffusion of praseodymium in the ceria lattice and the stability of the so-obtained pigment.

This selective chemical attack permits the determination of the amount of unreacted praseodymium oxide. Infact free praseodymium oxide ($Pr_6O_{11}$, $Pr_2O_3$ and $PrO_2$) is easily solubilized in hydrochloric acid to form soluble praseodymium chloride. Contrariwise, ceria is a very low soluble compound being stable in many strong inorganic acids even in concentrated solutions.

Tests, carried out on the powdered samples, demonstrated a lower solubility of the praseodymium in samples obtained for milling time higher than 15 hrs (table I).

Table I - Release (%) of praseodymium on the total amount present in the pigment as a function of milling time.

| Milling Time (hrs) | Release of Pr (%) |
|---|---|
| 0 | 83.9 |
| 5 | 74.2 |
| 10 | 42.4 |
| 15 | 28.9 |
| 20 | 6.1 |

Zirconia-Pr System

*Solid Solution Formation*: In figure 3 are reported the XRD patterns of zirconia with low (a) and high (b) specific surface area and that of the stabilised zirconia (c).

From the figure 3a, it can be seen that starting zirconia is monoclinic with peaks overlapping those of praseodymium oxide. After 10 hrs of milling, the sharp crystalline lines of the starting powders have disappeared and a relative broad diffuse pattern shows up [7]. This amorphization is a clear indication that the crystal structures of the starting materials have been modified by high energy ball milling. After 15 hrs, a series of new lines appears attributable to the t-$ZrO_2$ phase, while those of m-$ZrO_2$ remain. Peaks position and the absence of praseodymium oxide peaks indicated that this phase is a t-$ZrO_2$ solid solution. On further milling, the diffraction intensity of the new lines increases gradually and that of the m-$ZrO_2$ decreases, simultaneously. After 20 hrs milling all the diffraction peaks of the m-$ZrO_2$ phase have completely disappeared.

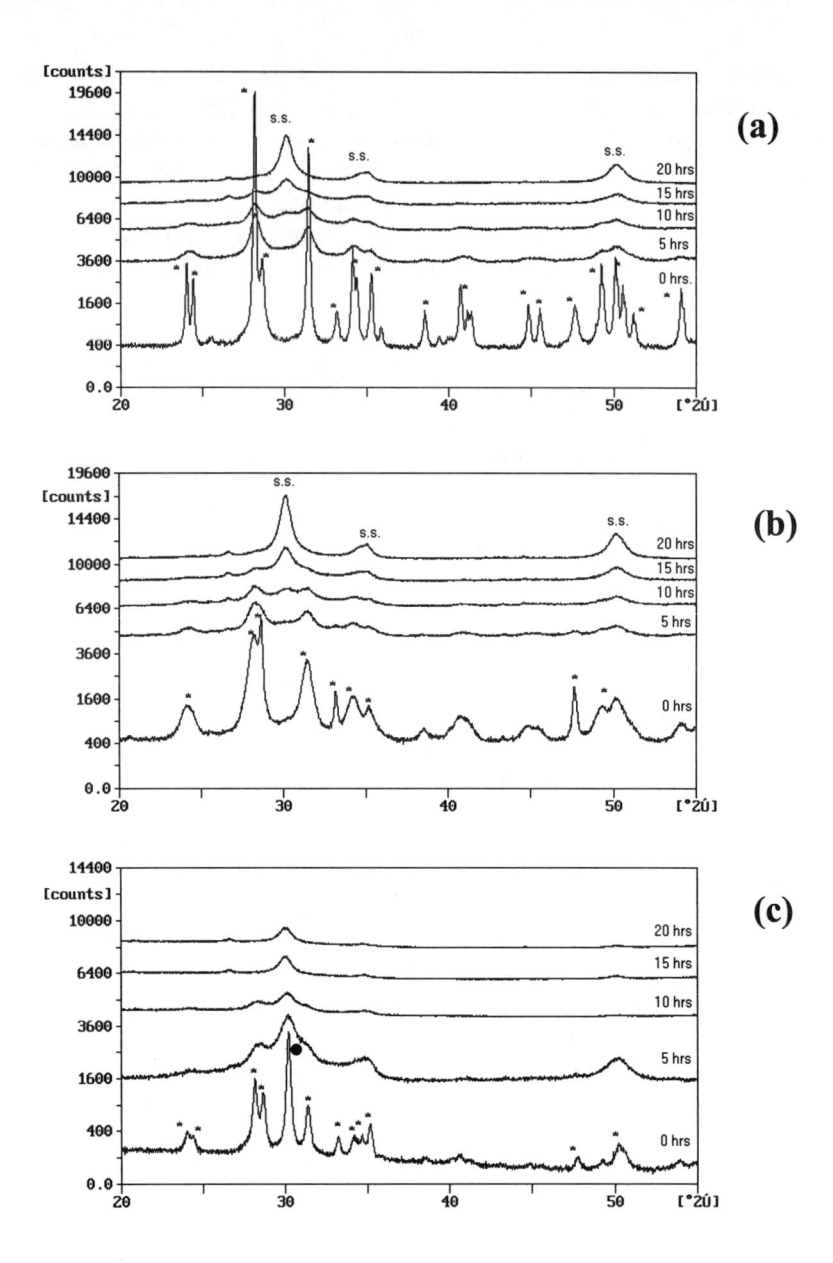

Figure 3 - XRD pattern of $Zr_{0.9}Pr_{0.1}O_2$ mixed powders after milling for different time. (a) low, (b) high specific surface area zirconia and (c) stabilised zirconia.

* = m-ZrO2, ● = t-ZrO2, s.s. = t-$Zr_{0.9}Pr_{0.1}O_2$

The complete transformation of monoclinic structure to tetragonal was observed also by Tonjec et al. [9] in opposition to the work of J.E. Bailey et al. [10] that showed the possibility to transform monoclinic zirconia to the tetragonal modification, but limited to approximately 45% of the material even after very long milling.

In the figure 3b the XRD patterns of the B zirconia type are reported. As previously shown, even in this case m-$ZrO_2$ completely transformed to t-$ZrO_2$ phase. The solid solution develops for milling time higher than 15 hrs.

Figure 3c shows that in this case the starting zirconia is composed of two structures, m- and t-$ZrO_2$. It can be seen that the presence in the starting zirconia of tetragonal structure does not affect the 15 hrs milling time required in the two previous cases to obtain the $(Pr,Zr)O_2$ solid solution.

Leaching tests on the powdered samples demonstrated, as shown in table II, a lower solubility of praseodymium in samples obtained with unstabilized zirconia for 20 hrs milling time. In particular with zirconia with lower specific surface area the Pr release is small even for shorter milling time.

Table II - Release (%) of praseodymium on the total amount present in the pigment as a function of milling time.

| Milling Time (hrs) | Release of Pr (%) | | |
|---|---|---|---|
| | Zirconia A | Zirconia B | Zirconia C |
| 0 | 81.4 | 93.6 | 100 |
| 5 | 16.4 | 77.9 | 65.5 |
| 10 | 8.5 | 52.6 | 39.8 |
| 15 | 5.6 | 8.2 | 10.8 |
| 20 | 5.7 | 5.0 | 11.1 |

The behaviour of partially stabilised zirconia can be explained considering that the tetragonal yttria stabilised phase present in the starting material does not take part in the solid solution formation. The higher release measured in this case leads to think that the final material could be composed by a mixture of the yttria-stabilised t-$ZrO_2$, remained unaltered during the milling process, and of the praseodymia doped t-$ZrO_2$ solid solution, originated by the monoclinic structure transformation.

To better understand the process, the powders obtained starting from zirconia with lower specific surface area were characterised by SEM and EDS analysis. In figure 4 are reported the SEM micrographs of the samples obtained after 10 (a) and 20 (b) hrs milling and the EDS analysis of this last sample (c). After 10 hrs milling the powders, chemically homogeneous, were obtained in the form of agglomerates of about 5 μm. SEM micrographs showed that an increase in milling time leads to reagglomeration of powders. The 20 hrs milling powders are contaminated by $SiO_2$, coming from the agate container and balls (Fig. 4c). The impurity level being low it may not affect the phase transition process.

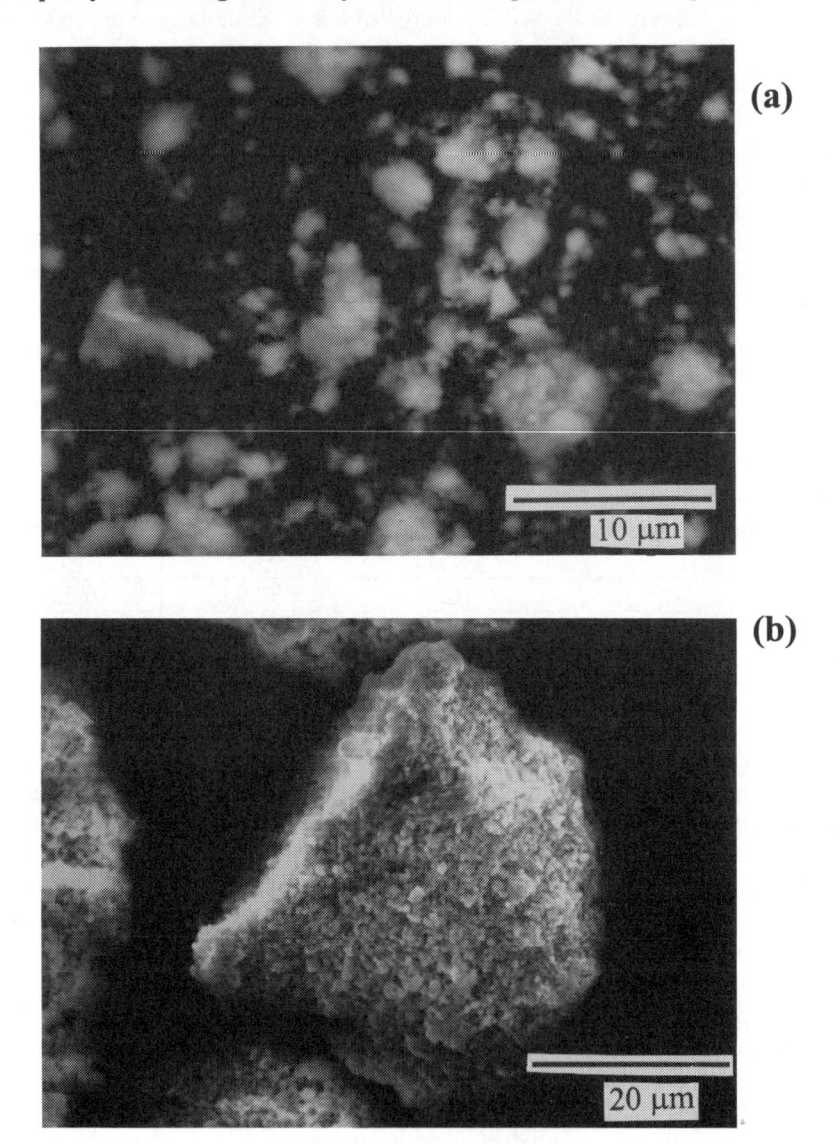

(a)

(b)

**(c)**

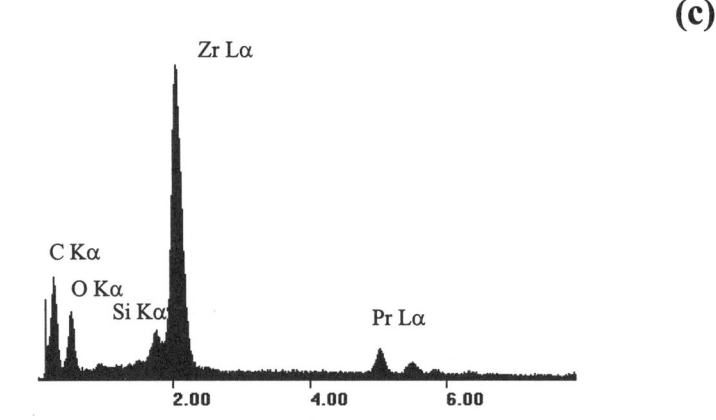

```
         Zr Lα


C Kα
  O Kα
    Si Kα              Pr Lα

       2.00   4.00   6.00
```

Figure 4 - (a) SEM micrograph of 10 hrs milling powder. SEM (b) and EDS (c) of the 20 hrs milling sample.

## CONCLUSION

By mechanical alloying we succeeded to synthetize praseodymium doped $CeO_2$ and $ZrO_2$ solid solutions starting from pure zirconia or ceria and praseodymium oxide ($Pr_6O_{11}$).

The $Ce_{0.9}Pr_{0.1}O_2$ solid solution, in form of agglomerates of about 1 µm diameter consisting of particles with lower average size, was obtained for milling times higher than 15 hrs. The experiments carried out with zirconia demonstrated that the solid solution t-$Zr_{0.9}Pr_{0.1}O_2$ could be obtained, starting from a monoclinic structure, for milling time higher than 15 hrs. The better results (inferior release of praseodymium), were obtained with monoclinic zirconia with low specific surface area (15 m$^2$/g).

This study confirms the experimental result of Y.L. Chen *et al.* [3,4,8] that mechanosynthesis of oxide ceramics is possible and offers new possibilities for the ceramic materials research.

## ACKNOWLEDGEMENT

We are very grateful to Paolo Veronesi for performing the milling experiments.

REFERENCE

[1] W. Weeber and H. Bakker, "Amorphisation by Ball Milling. A Review", *Physica B,* **153**, 93 (1988).

[2] C.C. Koch, "The Synthesis of Non-equilibrium Structures by Ball Milling", *Material Science Forum*, **88-90,** 243 (1992).

[3] Y.L. Chen and D.Z. Yang, "Formation of Supersaturated Solid Solution in $ZrO_2$-$CeO_2$ System Induced by Mechanical Alloying", *Scripta Metallurgica et Materialia,* **29**, 1349 (1993).

[4] Y.L. Chen, M. Qi, J.S. Wu, D.H. Wang and D.Z. Yang, "Mechanical Alloying Process of the Zirconia-8 mol% Yttria Ceramic Powder", *Applied Physic Letter,* **65** (3), 303 (1994).

[5] R.A. Eppler, "Selecting Ceramic Pigments", *Ceramic Bulletin*, **66**, 11, 1600 (1987).

[6] R. Olazcuaga, G. Le Polles, A. El Kira, G. Le Flem and P. Maestro, "Optical Properties of $Ce_{1-x}Pr_xO_2$ Powders and Their Applications to the Coloring of Ceramics", *Journal of Solid State Chemistry*, **71**, 570 (1987).

[7] Y.L. Chen, M. Qi and D.Z. Yang, "Mechanical Alloying of Ceramics in Zirconia-Ceria System", *Materials Science and Engineering*, **A183**, L9 (1994).

[8] A.M. Tonjec and A. Tonjec, "Zirconia Solid Solutions $ZrO_2$-$Y_2O_3$ (CoO or $Fe_2O_3$) Obtained by Mechanical Alloying", *Materials Science Forum*, **225-227**, 497 (1996).

[9] J.L. Bailey, D. Lewis, Z. Librant, and L.J. Porter, *Journal of the British Ceramic Society*, **71**, 25 (1991)

**Films/Coatings**

# FORMATION OF SELF-ASSEMBLED MONOLAYERS AND CERAMIC FILMS ON SEMICONDUCTOR AND OXIDE SUBSTRATES

U. Sampathkumaran, M. R. De Guire, and A. H. Heuer, Case Western Reserve University, Cleveland, OH, USA;
T. Niesen, J. Bill, and F. Aldinger, Max-Planck-Institut für Metallforschung, Stuttgart, Germany.

## ABSTRACT

Organic self-assembled monolayers (SAMs) have been shown to be effective surfaces for promoting the formation of thin oxide films onto (100) oriented Si from aqueous solutions at low temperatures. This paper reports deposition of SAMs for the first time on other substrates including silicon carbide, silicon-germanium, and sapphire, and of zirconia thin films on SAMs on silica glass and sapphire. Also reported are investigations of the topographies of the substrates, the SAMs, and the oxide films using atomic force microscopy (AFM).

## INTRODUCTION

The increasing trend towards greater integration of electronic circuits with sensing and output devices is the driving factor to explore alternative routes for thin oxide film deposition on a variety of semiconductor substrates. Organic self-assembled monolayers (SAMs) have been shown to be effective surfaces for promoting the formation of thin (typically < 100 nm) oxide films from aqueous solutions at low temperatures. All our earlier studies have been carried out on silicon (100) substrates and the conditions for optimum SAM properties and subsequent oxide film formation have been established. Thin oxide films of titania ($TiO_2$) [1-5], zirconia ($ZrO_2$) [6], yttria ($Y_2O_3$) [7], yttria-stabilized zirconia $Y_2O_3$-$ZrO_2$ [6], magnetite ($Fe_3O_4$) [8], and tin oxide ($SnO_2$) [9] have been synthesized using SAMs as a template for ceramic deposition on the silicon surface. Several groups have studied the role of synthetic organic surfaces having particular surface chemistries or topographies in the deposition of inorganic materials from liquid solutions [10-12].

In this study, the deposition of SAMs is reported for the first time on 1) polycrystalline silicon carbide thin films on silicon, 2) silicon-germanium thin films on silicon, and 3) sapphire substrates. The deposition of zirconia thin films on SAMs on silica glass and sapphire substrates is also reported. The first step in the process was to deposit SAM coatings, having properties equivalent to those of similar coatings on silicon, on these new substrates. The characterization of the surfaces of the SAMs, and of the oxide films deposited on these surfaces, is discussed.

The trichlorosilane bonding group (-SiCl$_3$) of the surfactant (*i.e.*, the molecules that bond to the substrate to form the SAM), in the presence of trace moisture, readily attaches to substrates with a surface oxide that possesses an -OH character in ambient humidity. This results in the formation of siloxane bonds between the organic molecules and the oxide surface. Additionally, siloxane cross-linkages are formed between the organic molecules themselves. The trichlorosilane-based SAMs formed in this way are particularly rugged; they withstand fairly aggressive environmental conditions, including strong acids at temperatures up to 100°C and mild bases at temperatures up to 80°C.

The deposition of SAMs on Si has been described elsewhere [13]. The sulfonic acid (-SO$_3$H) functionality for the surface end group has been used extensively for oxide film depositions in our previous work. Other surface functionalities that have been studied include thioacetate (-SCOCH$_3$) and methyl (-CH$_3$) groups. This paper reports the deposition of SAMs on the new substrates, and their characterization by wettability, ellipsometry, and x-ray photoelectron spectroscopy (XPS) studies. Atomic force microscopy (AFM) studies of silicon substrates, of SAM coatings on silicon, and of oxide thin films on SAMs are also reported.

EXPERIMENTAL PROCEDURE

Substrates

The substrates used in this study were p-type (B-doped) single-crystal (100) silicon wafers polished on one side (Silicon Sense, Nashua, NH, USA); SiO$_2$ glass (Herasil; Heraeus GmbH); c-axis sapphire (single-crystal α-Al$_2$O$_3$) (Kristallhandel Kelpin GmbH, Karlsruhe, Germany); polycrystalline thin films (400 nm thick) of cubic (3C) silicon carbide (SiC) deposited via CVD on silicon; and CVD-grown Si$_{0.9}$Ge$_{0.1}$ thin films (800 nm thick) on silicon. All substrates were approximately 10-15 mm square × 0.3-0.5 mm in size.

Prior to any characterization or further surface treatments, all substrates were degreased manually with acetone, ethanol, and methylene chloride or chloroform in sequence, using a fresh low-lint tissue for each wafer and solvent; followed by rinsing in methylene chloride and blowing dry with nitrogen.

Chemical Treatments

The deposition of siloxy-anchored SAMs requires a uniformly wetting oxide layer on the substrate. Previous work has established that immersion in "piranha" solution (7 ml concentrated H$_2$SO$_4$ (95-97%) plus 3 ml chilled aqueous 30% H$_2$O$_2$ solution, slowly stirred) for 20 min at 80°C yields a suitable hydrophilic amorphous oxide layer on silicon. Variations in temperature and time of exposure to piranha solution, as described in the results section, were used to obtain similar surfaces on the new substrates.

For Si-Ge, two alternative chemical treatments were studied: "1:2:3" etchant (a solution of HF:H$_2$O$_2$:CH$_3$COOH in 1:2:3 volume ratio [14]) at room temperature for 15-30 s; and buffered oxide etch (BOE) (HF:NH$_4$F = 1:8), commonly used to strip the native oxide on silicon at room temperature for 30 s. Both treatments were followed by rinsing twice in de-ionized water and drying under flowing nitrogen.

Deposition of SAMs

Two kinds of surfactants were used in this work: octadecyltrichlorosilane ($Cl_3Si$-$(CH_2)_{17}$-$CH_3$) (OTS), a commercially available (Aldrich) surfactant that yields SAMs with a methyl (-$CH_3$, hydrophobic) termination; and 1-thioacetato-16-(trichlorosilyl)hexadecane ($Cl_3Si$-$(CH_2)_{16}$-$SCOCH_3$) (TA). The surfactant was dissolved (1% by volume) in dicyclohexyl. The SAMs were deposited by immersing the substrates in this solution for 5 h at room temperature in a glove bag, under inert atmosphere. The synthesis of the TA surfactant and deposition conditions for SAM coatings have been described in more detail elsewhere [7,13].

TA SAMs were converted to sulfonate (-$SO_3H$) surfaces using an *in situ* transformation, by two different methods [15]. The first was wet oxidation using a saturated solution of oxone ($2KHSO_5 \cdot KHSO_4 \cdot K_2SO_4$, potassium peroxomonosulfate). The second was by photo-oxidation (used here on Si-Ge) in a slow oxygen purge using a Mineralight UV lamp Model UV-254 and wavelength of 254 nm.

Characterization of Substrates and SAMs

The contact angles with water of surfaces at various stages of preparation were measured using a Rame-Hart or Krüss goniometer. A PHI 5400 X-ray photoelectron spectrometer (XPS) was used to determine the elemental compositions of as-received, chemically treated, and SAM-coated surfaces. All XPS peaks were referenced to the carbon 1s peak at 284.5 eV. A variable-angle spectroscopic ellipsometer (J. A. Woollam Co.) was used to measure the thickness of the oxide layers and SAMs.

Atomic Force Microscopy (AFM)

Before AFM examination, all specimens were cleaned ultrasonically in acetone and dried in air. A Topometrix 2000 Explorer AFM was used to obtain the images reported here; similar images were obtained using a Digital Instruments Nanoscope III. All scans were taken at room temperature, in air. Imaging was performed using pyramidally shaped silicon nitride tips (4 μm base, 4 μm height, aspect ratio ca. 1:1, radius < 50 nm) on silicon nitride cantilevers (spring constant = 0.032 N/m). The most reproducible images were obtained using non-contact mode for the Si wafers, and contact mode for the the SAMs and oxide thin films. A large-area scanner was used to examine regions ranging in size from 10 μm × 10 μm to 100 μm × 100 μm, and a small-area scanner for regions ranging in size from 0.5 μm × 0.5 μm to 10 μm × 10 μm. Scan rates were scaled to the image size, so that 1-1.5 lines were scanned per second.

RESULTS AND DISCUSSION

Effect of Substrate Preparation on SAM Deposition

The values of contact angle with water indicate the general hydrophobicity (if high) or hydrophilicity (if low) of the surface, a measure of the surface's wettability by aqueous solutions. Another important indicator of a well-packed, uniform SAM surface is low hysteresis, *i.e.*, a small difference between advancing and receding contact angles.

*Silica glass and sapphire:* Table I shows that, as for silicon, a 20-min piranha treatment at 80°C was sufficient to obtain good wetting oxide surfaces on silica glass and sapphire substrates. A low contact angle with water at this stage indicates that the substrate is adequately hydrolyzed for subsequent SAM attachment. For thioacetate SAMs on silica glass, the contact angles were comparable to those on silicon, while the contact angles were low for sapphire. Though sapphire had a reasonably hydrophilic surface after the piranha treatment, the packing of the TA-SAM was not optimal, as indicated by its low contact angles and large hysteresis. Poor packing of a SAM can result from excessive substrate roughness (among other things), but the roughness of the piranha-treated sapphire surfaces has not yet been characterized to confirm or disprove this possibility.

Table I: Contact angles of water on various substrates after piranha treatment (80°C, 20 min) and after subsequent thioacetate (TA) SAM deposition. (adv.: advancing; rec.: receding.)

| Substrate | piranha oxidation °, adv. | °, rec. | TA SAM °, adv | °, rec. | -SO$_3$H SAM °, adv | °, rec. |
|---|---|---|---|---|---|---|
| Si | 16 ± 2 | 11 ± 1 | 72 ± 1 | 68 ± 1 | 24 ± 4 | < 10 |
| SiO$_2$ glass | fully wetting | | 75 ± 2 | 67 ± 2 | 34 ± 2 | < 10 |
| Sapphire | 25 ± 2 | < 10 | 56 ± 2 | 32 ± 3 | fully wetting | |

Subsequent oxone oxidation of the thioacetate coatings to sulfonate coatings on the glass and sapphire substrates gave wetting surfaces (Table I).

Using the same conditions as for deposition on SAMs on Si [6], zirconia films were deposited on sulfonate-functionalized sapphire substrates (4 immersions, 1 h each) and on both sides (top and bottom simultaneously) of sulfonate-functionalized silica glass (24 h). Figure 1 shows the XPS spectra of zirconia thin films simultaneously deposited on sulfonate SAMs on silicon and on both sides of silica glass. The XPS spectra are indistinguishable from one another, indicating identical chemistries of the films' surfaces in all three cases. Furthermore, the absence of silicon signals from the underlying substrates indicates complete coverage of the substrate and a lack of pinholes in the zirconia films. TEM characterization of these films is currently in progress.

*SiC on Si:* Compared to Si, glass, or sapphire, the SiC film required a longer exposure (90 min vs. 20 min) in piranha solution to achieve a fully hydrolyzed surface. Whereas the as-grown SiC surface was strongly hydrophobic, piranha treatment made it hydrophilic (from 100° adv. to 20° adv., Table II). XPS confirmed that the as-grown surface consisted mainly of SiC. In addition, XPS revealed trace amounts of surface hydroxyl groups and adventitious carbon. After piranha treatment, only trace carbon and the oxide (SiO$_2$) were present on the surface (in the ultra-high vacuum of the XPS chamber).

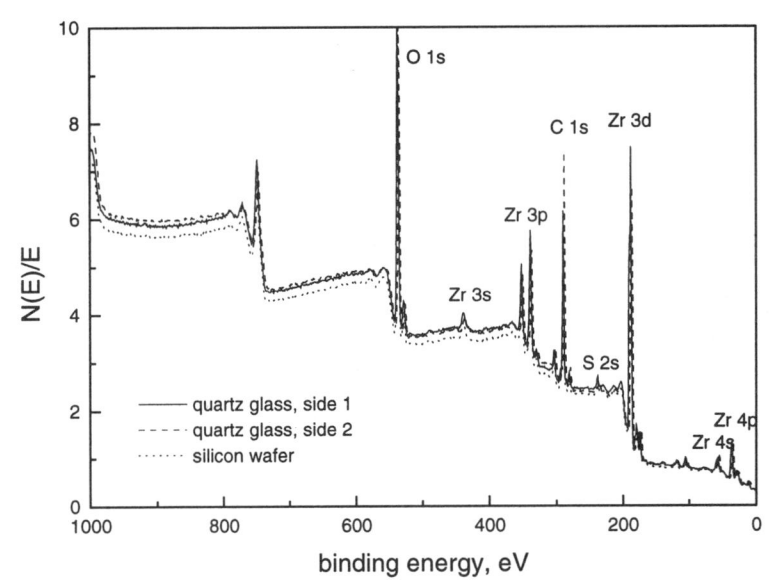

Figure 1: XPS spectra of zirconia thin films deposited simultaneously on sulfonate-functionalized SAM coatings on silicon and both sides of silica glass.

Table II: Contact angle and XPS data on piranha-etched 3C-SiC/Si (400 nm)

| Surface | Contact angles °, adv. | °, rec. | XPS |
|---|---|---|---|
| As-received | 100 ± 5 | 68 ± 6 | SiC, C, OH |
| Piranha etched | 20 ± 2 | < 5 | $SiO_2$, C |
| OTS SAM | 108 ± 2 | 100 ± 2 | $SiO_2$, C |
| TA SAM | 73 ± 1 | 65 ± 2 | CO, C, $SiO_2$, S |

SAMs deposited on the piranha-treated SiC exhibited wetting behavior comparable to similar coatings on Si, with a slightly larger hysteresis (*e.g.*, 8° for OTS on SiC (Table II), compared to 5° for OTS on silicon (Table III)). The OTS-coated surface showed the presence of hydrocarbon, while the TA coating exhibited sulfur in the divalent state and carbon (-$CH_3$, C=O in -$SCOCH_3$) as expected.

A similar SiC film with a thermally grown oxide layer 12.5 nm in thickness was also studied. This oxide was more hydrophilic than the as-grown SiC surface. However, SAMs on the thermal oxide showed greater hysteresis in wetting angles, indicating poorer packing compared to SAMs on the piranha-oxidized surface.

*SiGe on Si:* As-received $Si_{1-x}Ge_x$, like silicon, has a native oxide layer present on the surface. In the present specimens, its thickness was 2.5-3 nm, and it was significantly enriched in Ge: Ge/(Si+Ge) = 0.4, vs. the film composition of x = 0.1.

It is known that a wet etch of $Si_{1-x}Ge_x$ can cause a pileup of Ge below the "chemical" oxide, since Ge is preferentially dissolved in favor of forming $SiO_x$ [16]. Accordingly, piranha treatment at 80°C for 20 min yielded an oxide surface with a reduced Ge content (x = 0.04), higher wetting angles, and very large hysteresis. When the immersion time was reduced to 5 min and temperature was reduced to 25°C, the wetting nature of the Si-Ge surface was comparable to that of silicon (Table III), but the oxide layer was too thick (5-7 nm). The large hysteresis (18°) for OTS coatings deposited after piranha treatment indicated that poor packing of the SAM coating might have resulted from a rough surface oxide layer. The piranha treatment was apparently too aggressive for this substrate, even with a shorter exposure and a lower temperature.

Table III: Effect of various treatments on the surfaces of substrates and on the deposition of SAMs on Si and $Si_xGe_{1-x}$.

| Surface | Silicon °, adv. | Silicon °, rec. | thickness (nm) | $Si_xGe_{1-x}$ (x=0.9) °, adv. | $Si_xGe_{1-x}$ °, rec. | thickness (nm) |
|---|---|---|---|---|---|---|
| as-received | 35 ± 2 | 25 ± 5 | 2.8 | 67 ± 5 | 50±10 | 2.9 |
| piranha-oxidized* | 16 ± 2 | 11 ± 1 | 1.8 | 25 ± 2 | < 10 | 5-7 |
| BOE[†] | — | — | — | 18 ± 2 | 12 ± 1 | 1.7 |
| -CH₃ | 105 ± 1 | 100 ± 1 | 2.5 | 103 ± 2[†] | 96 ± 2[†] | 2.6 |
|  | — | — | — | 108 ±1* | 90 ± 2* |  |
| -SCOCH₃ | 72 ± 1 | 67 ± 1 | 2.6 | 71 ± 1[†] | 67 ± 1[†] | 2.6 |
| -SO₃H (Oxone) | 24 ± 4 | <10 | 2.8 | 35 ± 2[†] | < 10[†] | 4.7 |
| -SO₃H (UV) | 26 ± 2 | <10 | 2.9 | 28 ± 2[†] | < 10[†] | 2.8 |

The 1:2:3 etchant yielded oxide layers with thicknesses from 2.8 nm to 4.5 nm, with no obvious dependence on etch time (15 s to 30 s) or hold time (2-3 h) of the solution before etching.

These results indicated a need to strip some of the native oxide from the as-received Si-Ge. Buffered oxide etch reduced the layer thickness from 3.5 nm to 1.7 nm and brought the Si-Ge ratio of the oxide close to that of the semiconductor layer. The subsequent OTS and TA depositions on this surface produced coatings with contact angles comparable to those on silicon surfaces (Table III).

Oxone oxidation, a wet process, did not yield good -SO₃H SAMs on Si-Ge. The advancing contact angle was high and the hysteresis large. Also, the thickness of the SAM was too high (4.7 nm), suggesting a higher degree of surface rough-

---

* For Si: 20 min, 80°C; for Si-Ge: 5 min, 25°C, prior to SAM coating.
[†] BOE: buffered oxide etch (see text) prior to SAM coating.

ness. This is probably a consequence of the solubility of $GeO_2$-rich surfaces in aqueous media.

Photo-oxidation, on the other hand, yielded a more hydrophilic surface (Table III) with thickness of 2.8–3.0 nm, in agreement with the expected value of 2.7-3.0 nm.

## AFM

Figure 2 shows a series of AFM images taken over 20 μm × 20 μm regions, of specimens at various stages of the film deposition process: a) shows an oxidized Si wafer, b) a TA-SAM on an oxidized Si wafer, and c) and d) $ZrO_2$ films on a -$SO_3$H SAM. Figure 3 shows a similar series as Figure 2, but taken with the small-area scanner (2.5 μm × 2.5 μm).

Table IV shows the mean ($R_a$) and rms ($R_{rms}$) roughness values obtained from the areas shown in Figures 2 and 3. For any one type of sample, the roughness tends to decrease with decreasing size of the area examined. This trend was often (though not always) observed as well in other data that are not reported here; it indicates that comparisons to literature data should be made with caution if the sizes of the areas examined are not the same (or are not known).

*AFM of substrates and SAMs:* The roughness values for oxidized silicon ($R_a$<1.0 nm for areas up to 20 μm × 20 μm) are within the range reported in the literature [17,18].

The present roughness values for TA SAMs ($R_a$ = 0.2 nm, $R_{rms}$ ≤ 0.3 nm) are slightly higher than those reported by Calistri-Yeh *et al.* [18] on -$CH_3$-terminated surfaces (0.1-0.15 nm) synthesized under normal lab conditions, and are similar to those reported by Tsukruk and Reneker (0.2-0.5 nm) [19]. Nevertheless, the roughness data support the literature finding (as well as the qualitative impression conveyed by Figures 2a,b and 3a,b) that the surfaces of siloxy-anchored SAMs can be smoother than the silicon substrates underneath.

*AFM of ceramic thin films on SAMs:* Figure 2c shows a 20 μm × 20 μm region of a $ZrO_2$ film that was deposited in four successive 1-h immersions from an aqueous solution of 4 mM zirconium sulfate and 0.4 N hydrochloric acid at 70°C. Figure 3c shows a smaller region (2.5 μm × 2.5 μm) of the same specimen. These surfaces exhibit roughness values (Table IV) ranging from 1 to 4 nm, somewhat rougher than either the SAMs or the oxidized silicon substrates. These values, as well as the point-to-point height variations (essentially all ≤ 10 nm) seen in line scans (not shown), are comparable in magnitude to the sizes of the *t*-$ZrO_2$ crystallites seen in TEM images of $ZrO_2$ films on SAMs [7].

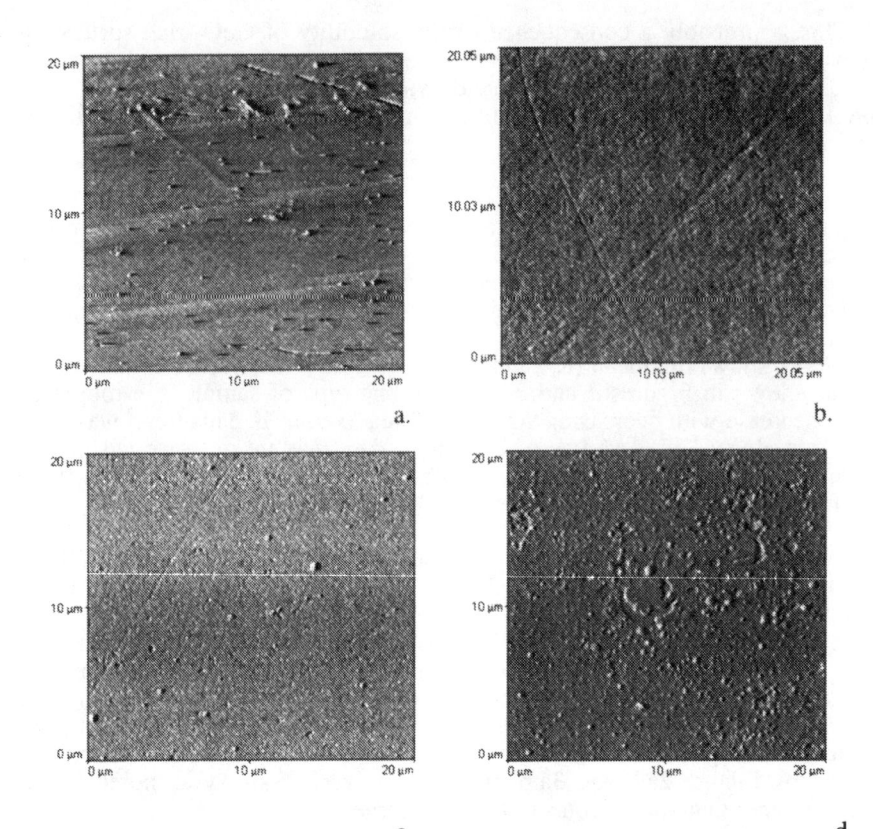

Figure 2. AFM images (20 μm × 20 μm) of: a) the surface of (100) Si after piranha oxidation (non-contact mode), z-axis range: 9.87 nm; b) the thioacetate-functionalized SAM on piranha-oxidized Si (contact mode), z-axis range: 3 nm; c) an as-deposited $ZrO_2$ thin film on sulfonate-functionalized SAM showing homogeneous topography (contact mode), z-axis range: 17 nm; and d) an as-deposited $ZrO_2$ thin film, the topography showing both nanoscale background roughness and micron-scale features (contact mode), z-axis range: 71 nm.

In contrast, Figures 2d (20 μm × 20 μm) and 3d (2.5 μm × 2.5 μm) show regions of another $ZrO_2$ film, deposited in a single 6-h immersion, that show depressions and protrusions up to 40 nm in depth/height and typically 0.5-1 μm in width. This depth probably corresponds to the thickness of the film in this area. The roughness values of these regions (Table IV) were consistently higher (by factors of 2-6) than those of the "smooth" regions of Figures 2c and 3c.

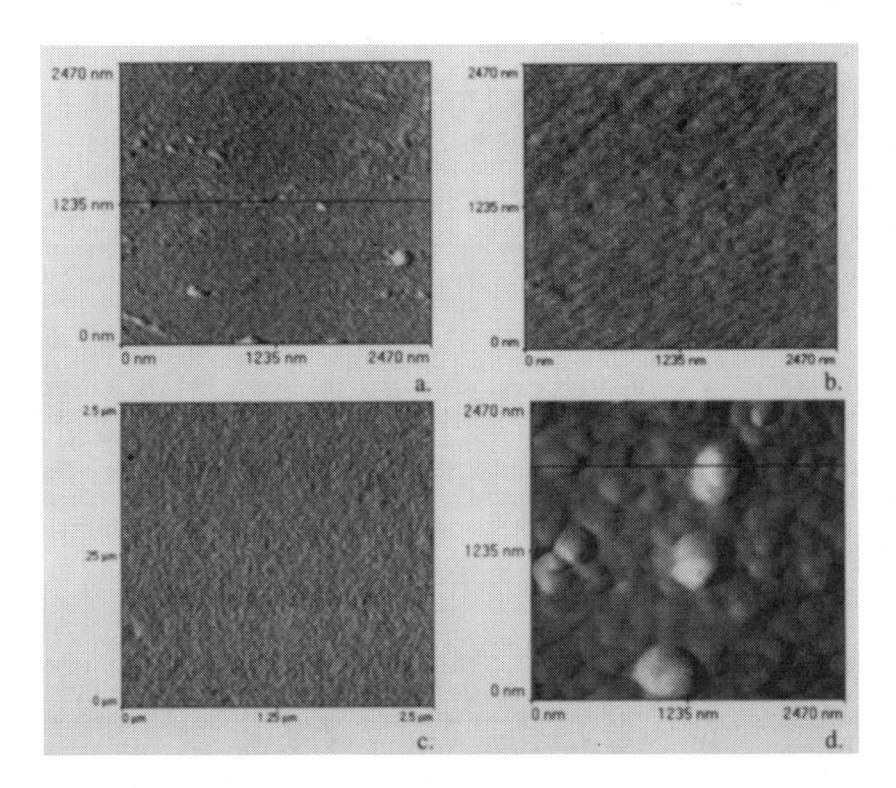

Figure 3. AFM images (2.5 μm × 2.5 μm) of: a) the surface of (100) Si after piranha oxidation (non-contact mode), z-axis range: 10 nm; b) the TA-functionalized SAM on piranha-oxidized Si (contact mode), z-axis range: 1.86 nm; c) an as-deposited $ZrO_2$ thin film on a $SO_3H$-functionalized SAM showing homogeneous topography (contact mode), z-axis range: 10 nm; and d) an as-deposited $ZrO_2$ thin film, the topography showing both nanoscale background roughness and submicron-scale, shallow protrusions (contact mode), z-axis range: 47 nm.

Table IV. AFM Roughness data for the films shown in Figs. 2-3.

| Surface | Area, μm x μm | $R_a$, nm | $R_{rms}$, nm | Fig. No. | Mode |
|---|---|---|---|---|---|
| $SiO_2$ on Si | 20 × 20 | 0.91 | 1.71 | 2a | non-contact |
|  | 2.5 × 2.5 | 0.78 | 1.02 | 3a | non-contact |
| TA SAM | 20 × 20 | 0.22 | 0.33 | 2b | contact |
|  | 2.5 × 2.5 | 0.19 | 0.24 | 3b | contact |
| $ZrO_2$ "smooth" | 20 × 20 | 1.69 | 4.25 | 2c | contact |
|  | 2.5 × 2.5 | 1.21 | 1.57 | 3c | contact |
| $ZrO_2$ "rough" | 20 × 20 | 10.02 | 14.68 | 2d | contact |
|  | 2.5 × 2.5 | 2.77 | 4.03 | 3d | contact |

CONCLUSIONS

Good SAM coatings, with wetting behavior typical of the respective surface functionalities (-CH$_3$, -SCOCH$_3$, and -SO$_3$H), can be deposited on silica glass, (0001) sapphire, and polycrystalline 3C-SiC on silicon, after immersion in piranha solution at 80°C. The optimum immersion time was 20 min for the glass and sapphire, similar to that for (100) Si, while the SiC required 90 min to achieve full hydrolysis of the surface. SAMs could be deposited on silicon-germanium alloys after a buffered oxide etch. Good quality zirconia thin films with complete surface coverage have been deposited on silica glass and sapphire substrates. Further characterization and optimization of these thin films is in progress. Future work will focus on the deposition of ceramic oxide thin films on the multicomponent semiconductor substrates for various applications.

AFM images of the substrates show that thioacetate SAMs can exhibit smoother surfaces than those of bare, hydrolyzed silicon wafers on which they are deposited. AFM images of zirconia films on sulfonated SAMs derived from these TA-SAMs show roughness on two length scales: nano-scale roughness indicative of the sizes of the crystallites observed in the films using TEM, and submicron- to micron-scale features that may be associated with physisorbed agglomerates or with defective regions of the underlying SAMs.

ACKNOWLEDGMENTS

The authors would like to thank Cpt. L. Henry (Hanscom AFB, Bedford MA) for providing the Si-Ge substrates, C. Jacob and C. Zorman (CWRU) for the SiC substrates, and T. Wagner (MPI) for the sapphire substrates; G. Maier (MPI) for AFM measurements; Prof. A. Hiltner (CWRU) for use of the ellipsometer; and M. Wieland (MPI) and W. Jennings (CWRU) for XPS. This work was supported in part by the U.S. Basic Missile Defense Office through an STTR Award to Solid State Scientific Corporation. Part of this work was done while MRD was on sabbatical leave at the Max-Planck-Institut für Metallforschung, Stuttgart, Germany in the research groups of Prof. Fritz Aldinger and Prof. Manfred Rühle, whose interest and support is gratefully acknowledged.

REFERENCES

1. H. Shin, R. J. Collins, M. R. De Guire, A. H. Heuer and C. N. Sukenik, "Synthesis and Characterization of TiO$_2$ Films on Organic Self-Assembled Monolayers: I. Film Formation from Aqueous Solutions," *J. Mater. Res.* **10** [3] 692-8 (1995).
2. *idem*, "Synthesis and Characterization of TiO$_2$ Films on Organic Self-Assembled Monolayers: I. Film Formation via an Organometallic Route," *J. Mater. Res.* **10** [3] 699-703 (1995).
3. S. Baskaran, L. Song, J. Liu, Y. L. Chen and G. L. Graff, "Titanium Oxide Thin Films on Organic Interfaces Through Biomimetic Processing," *J. Am. Ceram. Soc.* **81** [2] 401-408 (1998).

4. D. Huang, Z. D. Xiao, J. H. Gu, N. P. Huang, and C.-W. Yuan, "TiO$_2$ Thin Films Formation on Industrial Glass through Self-Assembly Processing," *Thin Solid Films*, **305** 110-115 (1997).
5. Z. Xiao, J. Gu, D. Huang, Z. Lu and Y. Wei, "The Deposition of TiO$_2$ Thin Films on Self-Assembled Monolayers Studied by X-ray Photoelectron Spectroscopy," *Applied Surface Science*, **125** 85-92 (1998).
6. M. Agarwal, M. R. De Guire and A. H. Heuer, "Synthesis of ZrO$_2$ and Y$_2$O$_3$-Doped ZrO$_2$ Thin Films Using Self-Assembled Monolayers," *J. Am. Ceram. Soc.* **80** [12] 2967-81 (1997).
7. M. Agarwal, M. R. De Guire and A. H. Heuer, "Synthesis of Yttrium Oxide Thin Films with and without the Use of Organic Self-Assembled Monolayers," *Appl. Phys. Lett.* **71** [7] 891-3 (1997).
8. M. Maiti, M.S. thesis, Case Western Reserve University, Cleveland, Ohio (1994).
9. S. Supothina and M. R. De Guire, submitted to *Thin Solid Films*.
10. P. Calvert and S. Mann, "Review: Synthetic and Biological Composites formed by *in-situ* precipitation," *J. Mater. Sci.* **23** 3801-15 (1988).
11. A. H. Heuer, D. J. Fink, V. J. Laraia, J. L. Arias, P. D. Calvert, K. Kendall, G. L. Messing, J. Blackwell, P. C. Reike, D. H. Thompson, A. P. Wheeler, A. Veis, and A. I. Caplan, "Innovative Materials Processing Strategies: A Biomimetic Approach," *Science* **255** 1098-1105 (1992).
12. B. C. Bunker, P. C. Reike, B. J. Tarasevich, A. A. Campbell, G. E. Fryxell, G. L. Graff, L. Song, J. Liu, J. W. Virden, and G. L. McVay, "Ceramic Thin-Film Formation on Functionalized Interfaces Through Biomimetic Processing," *Science* **264** 48-55 (1994).
13. N. Balachander and C. N. Sukenik, "Monolayer Transformation by Nucleophilic Substitution: Applications to the Creation of New Monolayer Assemblies," *Langmuir* **6** 1621-27 (1990).
14. T. K. Carns, M. O. Tanner & K. L. Wang, "Chemical Etching of Si$_{1-x}$Ge$_x$ in HF:H$_2$O$_2$:CH$_3$COOH," *Journal of the Electrochemical Society* **142** [4] 1260-66 (1995).
15. R. J. Collins and C. N. Sukenik, "Sulfonate-Functionalized, Siloxane-Anchored, Self-Assembled Monolayers," *Langmuir* **11** 2322-4 (1995).
16. I. S. Goh, J. F. Zhang, S. Hall, W. Eccleston and K. Werner, "Electrical Properties of Plasma-Grown Oxide on MBE-Grown SiGe," *Semicond. Sci. Technol.* **10** 818-28 (1995).
17. M. Anast, Å. Jamting, J. M. Bell, B. Ben-Nissan, "Surface Morphology Examination of Sol-Gel Deposited TiO$_2$ Films," *Thin Solid Films* **253** 303-307 (1994).

18. M. Calistri-Yeh, E. J. Kramer, R. Sharma, W. Zhao, M. H. Rafailovich, J. Sokolov, and J. D. Brock, "Thermal Stability of Self-Assembled Monolayers from Alkylchlorosilanes," *Langmuir* **12** 2747-2755 (1996).
19. V. V. Tsukruk and D. H. Reneker, "Scanning Probe Microscopy of Organic and Polymeric Films: From Self-Assembled Monolayers to Composite Multilayers," *Polymer* **36** 1791-1808 (1995).

# SYNTHESIS OF $\beta$-SiC CONVERSION COATING LAYER BY CHEMICAL VAPOR REACTION PROCESS

Young-Hoon Yun[*], Sung-Churl Choi

Dept. of Inorg. Mat. Eng., Hanyang univ.
17, Haengdang, Sungdong, Seoul, Korea

The preparation of SiC conversion coating layers using chemical vapor reaction (CVR) process and the resultant cross-sectional microstructure have been studied. In this process, the SiO gas was liberated from the reaction between silica powder and graphite substrate, and the surface region of CBY and HK-1 graphite substrates was converted into the SiC coating layer through the infiltration of SiO reactant gas at 1750℃ and 1850℃. The Conversion layers on substrates showed a $\beta$-SiC(3C) structure in XRD patterns. The transformation of graphite (layer structure) into SiC (cubic structure) was achieved through the interaction between SiO gas and coke crystalline planes. In the cross-sectional morphology, the CBY specimens showed a 200 thick SiC conversion layer and an interlayer between SiC layer and substrate, and the HK-1 specimens showed a 150 thick SiC conversion layer and an interface. The microstructural features of specimens were controlled by the density and pore size distribution of substrates, hence, it was found that these properties of substrates influenced on the infiltration of SiO gas and the conversion behavior. On the other hand, the oxidation of graphite substrates could be retarded by the presence of CVR-SiC coating layers in the oxidation condition to air at 1000℃.

## 1. Introduction

Graphite materials and carbon/carbon composites have been used as a high-temperature structural material because of its thermal shock resistance and mechanical properties. To improve the oxidation resistance of high performance carbon materials, SiC has been commonly used as a coating material to prevent direct oxygen attack on the carbon material.[1-3] SiC coating has attracted attention as a protective coating because of its relatively low CTE value and excellent oxidation resistance at elevated temperature. The most important role of a protective SiC coating on graphite is to prevent oxygen from penetrating through coating layer to graphite.

With increasing demands of refractory materials, such as Si-wafer pot, and nozzle material for space shuttle, the SiC coated graphites or C/C composites have been tried to substitute the sintered bodies.[4-6] Generally, the CVD method has been applied to prepare the SiC coating on graphite. But, fresh reactant gases in fabricating a thick protective coating have to permeate to a deep level against the out-flowing streams of product gases. It is possible to build up conveniently a thick SiC coating layer on substrate by the CVR process which is based on the carbothermal reduction,[7-9] because the SiO reactant gas has a high vapor pressure on substrate during the conversion process. On the other hand, the reactions in silica-carbon system are dependent upon the composition of starting materials, as well as the reaction temperature. Thus, it is necessary to control the disposition of substrate and staring materials in reaction chamber to suppress the consumption of SiC coating layer in manufacturing process.

This work is aimed of preparing the SiC coating layer by the chemical vapor reaction using graphite substrates and silica powder as starting materials. The chemical conversion mechanism, and the interface or boundary between coating layer and substrate in the cross-sectional specimens are also investigated. This study describes the dependence of graphite characteristics upon the cross-sectional microstructure and XRD patterns of coating layers. The oxidation resistance of SiC-coated graphites was also evaluated.

## 2. Experimental Procedure

The graphite substrates used for this study were CBY(density: 1.66g/cm$^3$, specific surface area: 9.132m$^2$/g, UCAR co.), HK-1(density: 1.85g/cm$^3$, specific surface area 8.427m$^2$/g, Seung-lim Carbon, co.). Fig. 1 shows the surface morphology of CBY and HK-1 graphite substrates. Fig. 2 shows the pore size distribution of CBY and HK-1 graphites, which was measured by mercury porosimeter. Graphite substrates were cut into 3mm×5mm× 7mm pieces, and were ground with SiC abrasive papers. Silica powder(average particle size: 17μm, Junsei, Chemical Co. Japan) and graphite pieces were put into the graphite crucible coated with BN-slurry (Advanced ceramics co., USA), and then, heated at 1750℃ and 1850℃ in flowing nitrogen gas.

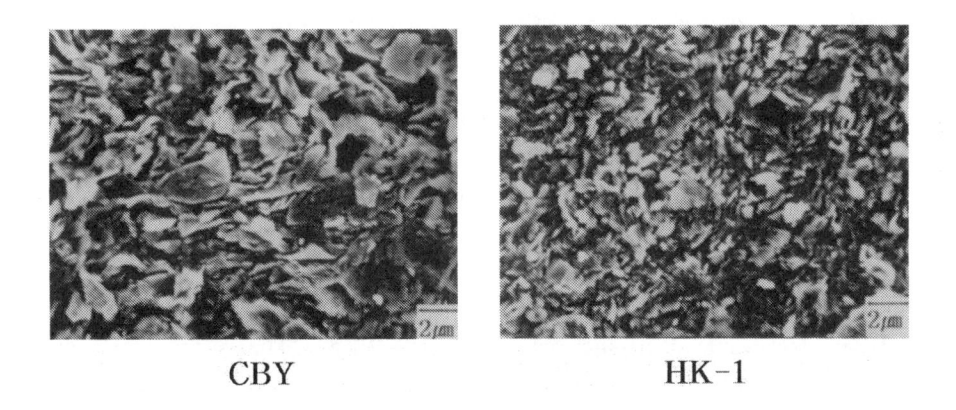

Fig. 1. Surface morphology of CBY and HK-1 graphite substrates.

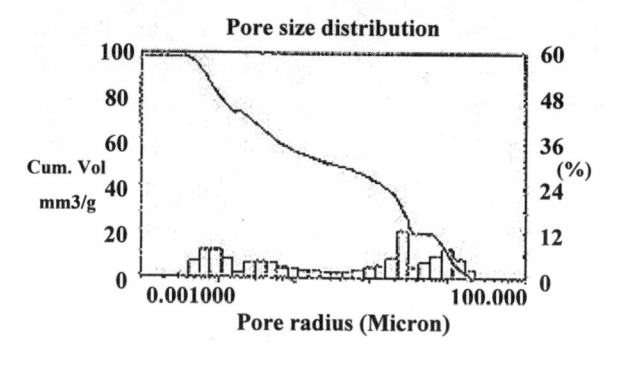

(a) CBY graphite substrate

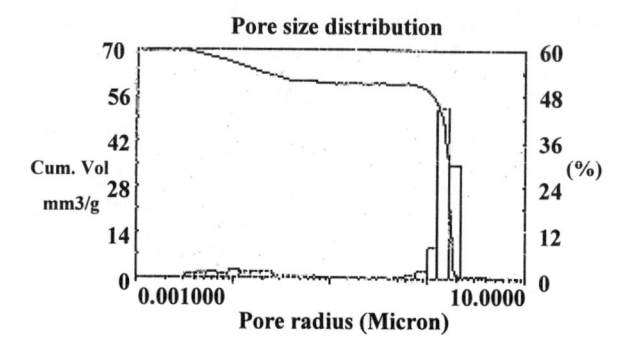

(b) HK-1 graphite substrate

Fig. 2. Pore size distribution of CBY and HK-1 graphite substrates.

The surface of coating layer was analyzed by X-ray diffractometer (XRD, Ni-filter, Cu $K_\alpha$-radiation), and the cross-sectional morphology of specimens was observed with scanning electron microscopy (SEM). The chemical composition of interface or boundary of the cross-sectional specimens was analyzed by wave dispersive spectroscopy (WDS). The substrates coated with SiC coating layer and the unconverted samples were isothermally oxidized to air at 1000°C during 10min, 30min, 1hr, 2hrs, 3hrs, and then, these specimens were cooled rapidly to room temperature. The weight loss of specimens was measured.

## 3. Results & Discussion

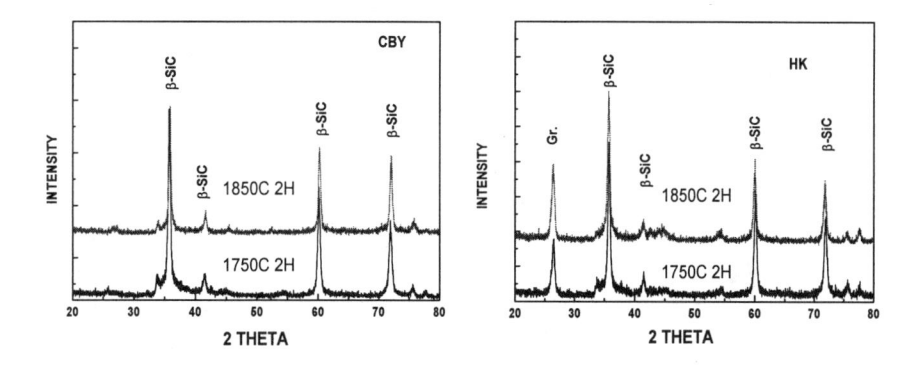

Fig. 3. XRD patterns of surface of SiC coating layer produced on CBY and HK-1 substrates at 1750℃ 2hrs and 1850℃ 2hrs.

Fig. 3 shows the XRD patterns of surface of coating layers. The XRD patterns of SiC-coated CBY specimens showed the presence of $\beta$-SiC (3C type, zinc blende structure), but that of coated HK-1 specimens contained the unconverted graphite peak as well as the SiC peaks. Therefore, it was recognized that the conversion reaction of CBY-graphite having lower density and larger pore size as compared to HK-1 substrate was completed easily at 1750℃ and 1850℃. These results indicate that the XRD patterns of SiC coated substrates are influenced by the substrate properties such as pore size distribution and density.

On the other hand, on the transformation of layer structure into cubic structure, it was supposed that the graphite substrate surface is converted by the interaction between SiO gas and coke crystalline planes(or atomic sites) which have larger reactivity than others. Because the graphite materials at elevated temperature show an appreciable C-axis expansion behavior so that the prismatic planes have a large reactivity. Also, such process will include the decomposition of SiO molecules and the insertion of Si into graphite. Thus, some of carbon atoms will be substituted by silicon atoms. As a result, the crystalline units of graphite are converted into the tetrahedron units which constitute the $\beta$-SiC structure.

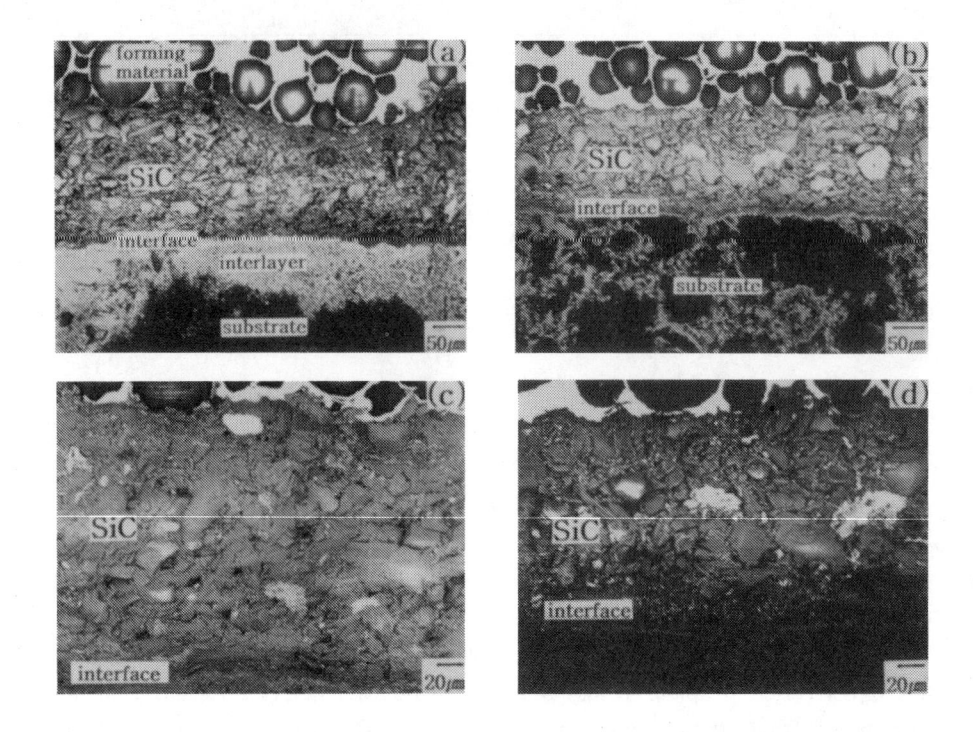

Fig. 4. Back scattered image of cross-sectional SiC conversion layer prepared on CBY and HK-1 substrates.
(a), (c): SiC coating layer & interface(interlayer) in coated CBY specimen.
(b), (d): SiC coating layer and interface in coated HK-1 specimen.

Fig. 4 shows the cross-sectional micrographs of coated CBY and HK-1 specimens manufactured at 1850℃, 2hrs. The cross-sectional morphology shows the SiC coating layer and the interlayer or boundary between outer layer and substrate, as shown in Fig. 4(a) and (b). The coated CBY specimen of Fig. 4(a) contains the interlayer showing average SiC/C ratio of 1:3. However, the coated HK-1 specimen in Fig. 4(b) showed a definite boundary showing the SiC/C ratio from 1:3 to 1:4. Based upon these results, it was thought that the interlayer or boundary structure of specimens was controlled by the nucleation rate of SiC phase on coke particles, and the connection & growth rate of grains during the CVR process. In consequence, it seemed that the formation of interlayer on CBY graphite was induced by high nucleation rate according to the low density, and the substrate pore size distribution. On the other hand, as shown in Fig. 2, the CBY graphite substrate showed a continuous pore size distribution in the pore size range from 0.001μm to 10.0μm, and the HK-1 substrate showed a discrete distribution in the range of 0.001μm to 1.0μm. Especially, it was proved that the pore size distribution required for the infiltration of SiO gas and the conversion behavior was in the range of 1.0 to 10.0μm. As a result, it was concluded that the pore size distribution was an important factor to determine the interfacial or boundary structure and the shape or size of grains in SiC coating layer.

The SiC conversion layers in Fig. 4 (c), (d) represent the morphology of connected and grown grains. As shown in Fig. 4 (c), (d), the conversion coating layer on HK-1 specimen reveals rather a coarse SiC grains in comparison to those of CBY specimen. The distinctive feature of the HK-1 specimen is the unevenness in grain size of conversion layer. It indicates that the HK-1 graphite compared to CBY graphite is composed of smaller coke particles and has relatively dense microstructure as shown in the surface morphology of Fig. 1. Also, the morphology of conversion layers in the CBY and HK-1 specimens can be attributed to the pore size distribution of substrates shown in Fig. 2, because the infiltration of SiO into substrate surface and the conversion behavior are strongly influenced by pore size distribution. As a result, it can be presumed that the infiltration of SiO and the connection & growth in the HK-1 graphite showing smaller pore size and higher density than that of CBY graphite show a heterogeneous behavior.

$$SiO_2 + C \rightarrow SiO + CO \quad (1)$$
$$SiO + 2C \rightarrow SiC + CO \quad (2)$$

In conclusion, the formation of SiC coating layer is composed of the creation of SiO vapor from reduction of silica, and the reaction of SiO and graphite as described in reaction (1) and (2). In this process, the SiO vapor is infiltrated into substrate surface, and SiC phase is formed on the coke particles (nucleation step). The reaction between SiO and coke particles is carried out continuously, thus, the coke particles are converted into SiC particles. Also, the converted particles will be connected and grown, finally, the graphite surface is converted into the SiC coating layer. However, as the reactions proceed, the partial pressure of CO gas is increased, which is dependent on the density or pore size distribution of substrates. Thus, the temperature dependences of substrate properties in conversion process described above will be increased. The reaction rate for SiC coating layer is decreased gradually according to the variation of equilibrium $P_{CO}/P_{SiO}$ for $2C+SiO \rightarrow SiC+CO$ reaction.[10-12]

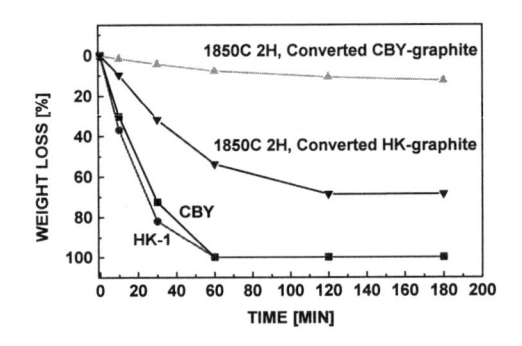

Fig. 5. Weight loss of unconverted graphite and converted graphite in the isothermal oxidation (1000°C).

Fig. 5 shows the weight loss of SiC coated specimens and unconverted substrates. In the weight loss of specimens at 1000°C, the bare graphites showed a complete oxidation after 60 min, and the SiC coated specimens did not show further an oxidation, after 120 min, the variation of weight loss with oxidation time shows a constant behavior. Therefore, it was thought that the weight loss of coated specimens could be constrained by the SiC coating layer. Also, the oxide layer created at the beginning of oxidation and the interfacial structure between substrate and coating layer would have influenced on oxidation behavior. The residual carbon and coarse grains in coating layer of HK-1 specimens increased the weight loss of specimens. In conclusion, the oxidation resistance of coated specimens was enhanced as compared with the unconverted substrates, and the coated CBY specimens showed a strong oxidation resistance.

## 4. Conclusion

In CVR process, the SiO gas was obtained from the reduction of silica powder by graphite substrate surface, and the SiC conversion coating layer was produced on substrate by the reaction between SiO reactant gas and the coke particles of substrate. In XRD patterns, the coating layer of CBY specimens showed the only SiC peak, but, in that of HK-1 specimens, the SiC and graphite peak were detected. On the other hand, the coating layers and the boundary structure between outer layer and substrate in cross-sectional morphology showed the dependences upon pore size distribution and density of substrates. The SiC coated CBY and HK-1 specimens showed resistance to oxidation in air at 1000℃.

### Acknowledgements

This work was supported by the Korea Science and Engineering Foundation(KOSEF) through the Ceramic Processing Research Center(CPRC) at Hanyang University.

## REFERENCES

[1] T. M. Wu, W. C. Wei and S. E. Hsu, "The Effect of Boron Additive on the Oxidation Resistance of SiC-Protected Graphite", Ceramics International, **18**, 167-172 (1992).

[2] Y. G. Jung, S. W. Park, S. C. Choi, "Effect of $CH_4$ and $H_2$ on CVD of SiC and TiC for possible fabrication of SiC/TiC/C FGM", Materials Letters, **30** 339-345 (1997).

[3] J. E. Sheehan, "High-Temperature Coatings on Carbon Fibers and Carbon-Carbon Composites"; pp. 224-253 in Carbon-Carbon Materials and Composites, Edited by J. D. Buckley, D. D. Edie. np.

[4]S. David., "Method of manufacturing hybrid fiber-reinforced composites nozzle materials", U. S. Pat. No. 5525372, June 11, 1996.

[5]S. Nogami, "Development of High Purity Silicon Carbide Block", Proceedings of the 12th Japan-Korea Seminar on ceramics, Tsukbar, Japan, 414-418 (1995).

[6]K. Kurahashi, Y. Mizuuo, "II-15 Silicon Carbide Ceramics of Tokai Konetsu"; pp. 417-425 in SiC Ceramics, Edited by S. Somiya, Y. Inomata, Uchida Rokakuho Publishing Co, Tokyo, 1988.

[7]J. J. Biernacki, G. P. Wotzak, "Stoichiometry of the C+SiO$_2$ Reaction," Journal of American Ceramic Society, **72** [1] 122-129 (1989).

[8]F. Viscomi and L. Himmel, "Kinetic and mechanistic study on the formation of silicon carbide from silica flour and coke breeze", Journal of METALS, June [6] 21-24 (1978).

[9]G. C. T. Wei, "Beta-SiC Powder Produced by Carbothermic Reduction of Silica in a High-Temperature Rotary Furnace", Communications of the American Ceramic Society, **66** [7] C111-C113 (1983).

[10]A. W. Weimer, "Thermochemistry and Kinetics"; pp 94-97 in Carbide, Nitride and Boride Materials Synthesis and Processing, 1st ed. Edited by A. W. Weimer. CHAPMAN & HALL, London, 1997.

[11]P. Kennedy, B. North, "The Production of Fine Silicon Carbide Powder by the Reaction of Gaseous Silicon Monoxide with Particulate Carbon", Proceedings of British Ceramic Society, **33**, No. Fabrication Science 3, 1-15 (1983).

[12]P. D. Miller, J. G. Lee and I. B. Cuttler, "The Reduction of Silica with Carbon and Silicon Carbide", Journal of American Ceramic Society, **62** [3-4] 147-149 (1979).

# ADHESION OF COPPER NITRIDE FILM TO SILICON OXIDE SUBSTRATE

Kwang Ho Kim*, S.O. Chwa, and H.C. Park

Department of Inorganic Materials Engineering, Pusan National University, Pusan 609-735, Korea

Dong. W. Shin

Division of Materials Sci. & Eng., Kyongsang National University, Chinju, Kyongnam 660-701, Korea

## ABSTRACT
Films of copper nitride on a $SiO_2(4000 Å)/Si$ wafer were made by sputtering copper target in a nitrogen plasma state. The adhesion property and its bonding characteristics to $SiO_2$ substrate were investigated for a possible application for copper metallization in devices such as micro-inductor etc. The behavior of copper nitride film with deposition variables was investigated. $Cu_3N$ film was successfully deposited in the substrate temperature range from 100℃ to 200℃ by r.f. reactive sputtering technique. Regardless of the film thickness of copper nitride, it can be used as a bonding promoter between copper film and $SiO_2$ substrate. The fabricated $Cu/Cu_3N/SiO_2$ film showed much improved adhesion property and passed the Scotch-tape test in accordance with ASTM standards. The copper nitride film was, however, degraded as heat-treated above 400℃.

## INTRODUCTION
Copper is one of the promising materials for metallization in micro magnetic devices[1-3] and in ultralarge scale integrated (ULSI) circuits due to its

** This work was financially supported by Korean Ministry of Education
(Advanced Materials: N-97-63)

low resistivity and high electromigration resistance.[4-7] In spite of its merits, the application of copper metallization has been obstructed because copper has poor adhesion to ceramic substrate such as $SiO_2$.[5]

Recently, an interesting material, $Cu_3N$ was synthesized in a nitrogen plasma state by the sputtering technique[8] and was investigated for its possible applications for recording media.[9-11] However, other properties of the copper nitride film are not well-known yet and its application as a bonding promoter for the copper metallization has never been reported. The present work was, therefore, undertaken to investigate the bonding characteristics of copper nitride film to $SiO_2$ substrate and to try a successful copper metallization using the copper nitride film.

**EXPERIMENTAL**

Copper nitride films were prepared on $SiO_2$(4000 Å)/Si wafer by reactive magnetron sputtering technique. Copper target(99.99%, CERAC), 2 inches in diameter and 1/8 inch in thickness was sputtered in the plasma of a gas mixture composed of nitrogen and argon(purity $\geq$ 99.999%), the gas ratio of which was controlled by MFC(mass flow controller). The substrate temperature was controlled in the range from room temperature to 400 ℃. The sputtering process was conducted at a chamber pressure of 10 mtorr under an R.F. power of 30 watt.

Phase and crystallinity of the deposited film was analyzed by an X-ray diffractometer(Rigaku) using Cu$K\alpha$ radiation. Film thickness was measured by a stylus(Tencor, $\alpha$-step) and scanning electron microscope(Hitachi, S-4200). However, the thickness of very thin film was estimated from the deposition time with the measured thickness of the thick film. Adhesion property of copper nitride film to $SiO_2$ substrate was tested in accordance with standards of both D3330 (Method for Peel Adhesion of Pressure-Sensitive Tape of 180 Angle) and D3359-87 (Test Method for Measuring Adhesion by Tape Test) of American Society for Testing Materials(ASTM).[12,13] For the tape test, a half inch wide, semi-transparent and pressure-sensitive tape (3M) was attached on the deposited area of $4 \times 4$ mm$^2$. In order to attach the pressure-sensitive tape to the film with uniform attaching force, the tape was suppressed with a constant pressure of 10 bar using a press instead of pressing by a pencil with an eraser described in ASTM standards. The attached tape was removed by seizing a free end and pulling it off rapidly back on itself at as close to angle of 180° as possible. Figure 1 shows a schematic illustration of our Scotch-tape method for conducting the adhesion test.

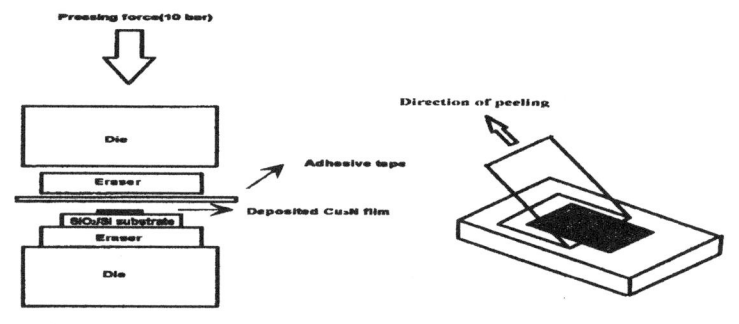

Fig.1. Schematic illustration of our Scotch-tape method for conducting the adhesion test

## RESULTS AND DISCUSSION

Figure 2 shows X-ray diffraction patterns of films deposited at various substrate temperatures. Nitrogen gas ratio in sputtering gas mixture was fixed

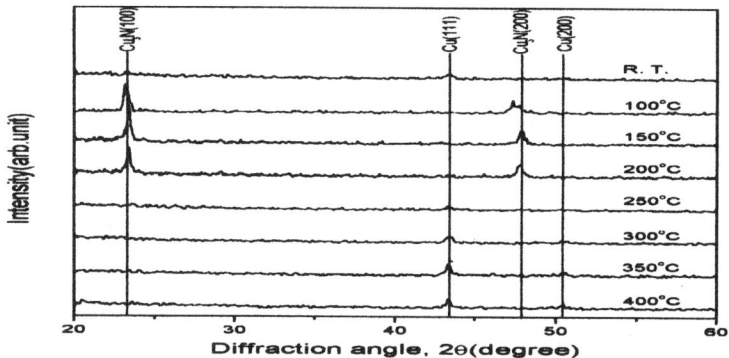

Fig. 2. X-ray diffraction patterns of film deposited at various substrate temperature

at $X_{N_2} = 0.5$. At room temperature metallic Cu peak was seen, however, at substrate temperature in the range of 100-200 ℃ Cu$_3$N films were deposited. Further increase in the substrate temperature above 250 ℃ seem to show Cu peaks again. From our experimental results, it was found that the optimum temperature range to synthesize Cu$_3$N was from 100 ℃ to 200 ℃. At the low substrate temperature of room temperature thermal energy to synthesize copper nitride is considered to be insufficient, whereas at high substrate temperature thermal decomposition of synthesized copper nitride is known to occur. T. Maruyama et al.[11] have reported that at high substrate temperature around 300℃, re-evaporation of atomic nitrogen in the copper nitride film occurred. Figure 3 shows X-ray diffraction patterns of films deposited at various nitrogen gas ratios. The substrate temperature was fixed at 150℃ and

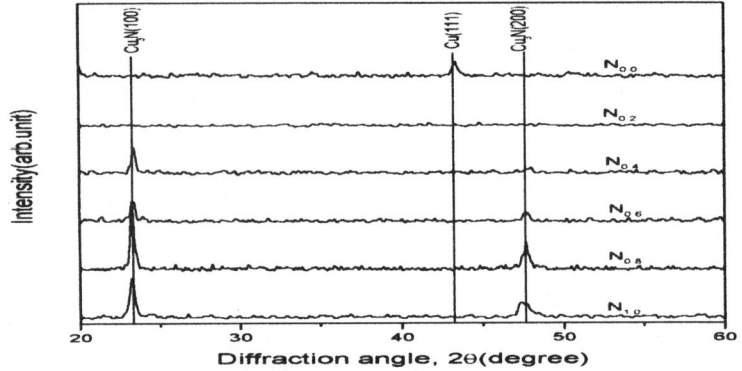

Fig. 3. X-ray diffraction patters of films deposited at various nitrogen gas ratios

nitrogen gas ratio, $X_{N_2}$ was changed from zero to one. At pure argon atmosphere of $X_{N_2}$= 0.0, metallic copper was deposited. At $X_{N_2}$= 0.2, any diffraction peaks corresponding to either metallic copper or copper nitride phase did not appear. At nitrogen gas ratios above $X_{N_2}$= 0.4, however, crystalline copper nitride films with [100] preferred orientation were deposited.

Table I shows the tape test results of films deposited on $SiO_2$ with various substrate temperatures and nitrogen gas ratios. The thickness of

Table I. Tape test results of films deposited on $SiO_2$ with various substrate temperatures and nitrogen gas ratios

| Sample | Result | Sample | Result |
| --- | --- | --- | --- |
| $T_{R.T.}N_{0.5}$ | × | $T_{400}N_{0.5}$ | × |
| $T_{100}N_{0.5}$ | ○ | $T_{150}N_{0.0}$ | × |
| $T_{150}N_{0.5}$ | ○ | $T_{150}N_{0.2}$ | ○ |
| $T_{200}N_{0.5}$ | ○ | $T_{150}N_{0.4}$ | ○ |
| $T_{250}N_{0.5}$ | ○ | $T_{150}N_{0.6}$ | ○ |
| $T_{300}N_{0.5}$ | × | $T_{150}N_{0.8}$ | ○ |
| $T_{350}N_{0.5}$ | × | $T_{150}N_{1.0}$ | ○ |

where, T: substrate temperature(℃), N: nitrogen gas ratio( $X_{N_2}$ ) in sputtering gas, ×: peeled off, ○: passed.

copper nitride films for tape test was in the range of 1800-2500 Å. All Cu₃N films deposited at substrate temperatures in the range of 100-200 ℃ did not peel off from the substrate and passed the tape test, whereas Cu films detached from the substrate by the tape and showed poor adhesion property. Figure 4 shows optical micrographs of the film surface after the tape test. The copper

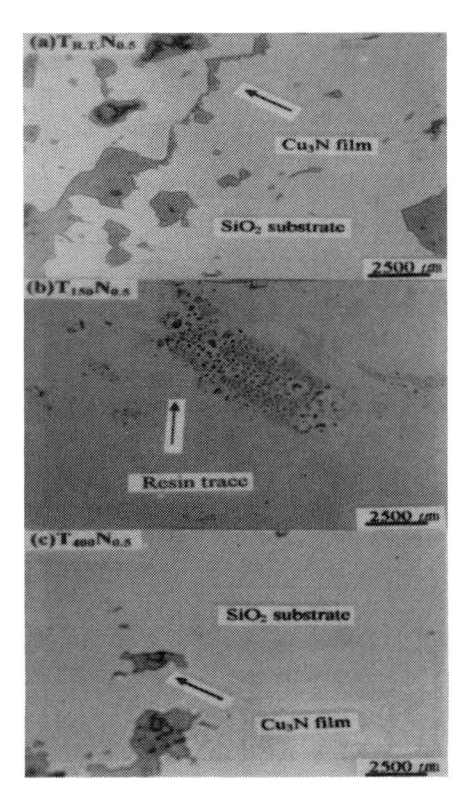

Fig. 4. Optical micrographs of the film surface after tape test
(a) room temperature   (b) 150 ℃   (c) 400 ℃

films deposited both at room temperature and at 400 ℃ peeled off from SiO₂ substrate after the performing tape test. On the other hand, copper nitride film deposited at 150 ℃ remained as it had been and rather left the trace of resin on its surface, detached from the adhesive tape. This reflects the fact that the adhesion force of copper nitride film to SiO₂ substrate at least exceeds maximum adhesion capability of the tape.

To test the adhesion of copper metallization to $SiO_2$ using a copper nitride film, two types of FGM(functional graded materials) structured films were deposited on $SiO_2$ substrate. They were double layer of $Cu/Cu_3N$ film and $Cu/CuN_X$ film, respectively. The schematic drawings showing the variation of nitrogen concentration in their FGM structured film are shown in Fig. 5.

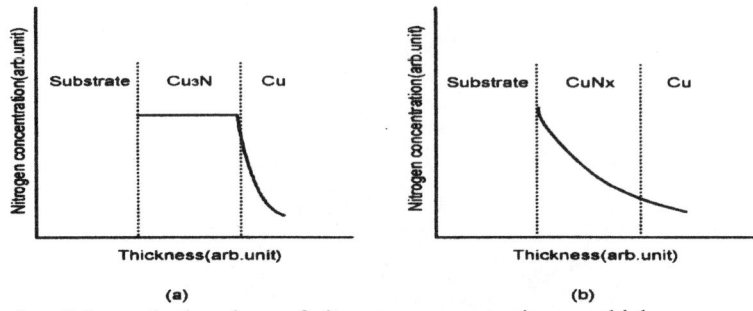

Fig. 5. Schematic drawings of nitrogen concentration vs. thickness
(a) in double layer $Cu/Cu_3N$ structure and (b) in $Cu/CuN_x$ structure.

These FGM structured films were prepared by changing the nitrogen gas ratio of plasma gas mixture during the sputtering process. Table II shows the tape test results of double layer $Cu/Cu_3N$ films and $Cu/CuN_X$ films with variation of interlayer thickness. All FGM structured films using copper

Table II. Tape test results of double layer $Cu/Cu_3N$ films and $Cu/CuNx$ films deposited on $SiO_2$ with various interlayer film thickness.

| Sample | Result | Sample | Result |
| --- | --- | --- | --- |
| $Cu/Cu_3N$(2400 Å, measured) | ○ | $Cu/CuN_x$(2400 Å, measured) | ○ |
| $Cu/Cu_3N$(60 Å, estimated) | ○ | $C/CuN_x$(60 Å, estimated) | ○ |
| $Cu/Cu_3N$(10 Å, estimated) | ○ | | |

where, ×: peeled off, ○: passed

nitride interlayer showed good adhesion properties to $SiO_2$ and passed the tape test regardless of interlayer thickness.

In summary, $Cu_3N$ film was successfully deposited in the substrate temperature range from 100 ℃ to 200 ℃ by R.F. reactive sputtering technique and was found to be a possible bonding material as adhesive interlayer between copper and $SiO_2$. We are investigating the chemical bonding characteristics of the interface between copper nitride film and silicon oxide substrate, and are in the progress of quantifying the adhesion force on various copper nitride films.

## CONCLUSIONS

$Cu_3N$ film was successfully deposited in the substrate temperature range from 100℃ to 200℃ by R.F. reactive sputtering technique. Regardless of the film thickness of copper nitride used as a bonding promoter between copper film and $SiO_2$ substrate, the fabricated $Cu/Cu_3N/SiO_2$ film showed much improved adhesion property and passed the Scotch-tape test in accordance with ASTM standards. The copper nitride film, however, degraded when the substrate was heat-treated above 400℃.

## REFERENCES

[1] C. Patrick Yue, C. Ryu, J. Lau, T. H. Lee, and S. S. Wong, "A Physical Model for Planar Spiral Indicators on Silicon"; pp. 155-158, Proceedings of International Electronic Devices Metting. Edited by IEEE, Sanfrancisco, 1996.

[2] E. Tarvainen, H. Ronkainen, H. Kattelus, T. Riihisaari, and P. Kuivalainen, "1.8GHz Current-Controlled Oscillator IC Implemented by Using Integrated Inductors and 0.8 $\mu$m BiCMOS Technology," *Electronics Letter*, **32(16)** 1465-1467 (1996).

[3] W. B. Kuhn, A. Elshabini-Riad, and F. W. Stephenson, "Centre-Tapped Spiral Inductors for Monolithic Bandpass Filters," *Electronics Letters,* **31(8)** 625-626 (1995).

[4] J. N. Burghartz, D. C. Edelstein, K. A. Jenkins, C. Jenkins, C. Uzoh, E. J. O'Sullivan, K.K. Chan, M. Soyuer, P. Roper, and S. Cordes, "Monolithic Spiral Inductors Fabricated Using a VLSI Cu-Damascene Interconnect Technology and Low-Loss Substrate"; pp 99-102, Proceedings of International Electronic Devices Metting. Edited by IEEE, Sanfrancisco, 1996.

[5] S. P. Muraka, R. J. Gutmann, A. E. Kaloyeros, and W. A. Lanford, "Advanced Multilayer Metallization Schemes with Copper as Interconnection Metal," *Thin Solid Films* **236** 257-266 (1993).

[6] E. S. Choi, S. K. Park, and H. H. Lee, "Growth and Resistivity Behavior of Copper Film by Chemical Vapor Deposotion," *J. Electrochem. Soc.* **143(2)** 624-627 (1996).

[7] C. -K. Hu, K. Y. Lee, K. L. Lee, C. Cabral, Jr., E. G. Colgan, and C. Stanis, "Electromigration Drift Velocity in Al-Alloy and Cu-Alloy Lines," *J. Electrochem. Soc.* **143(3)** 1001-1006 (1996).

[8] S. Terada, H. Tanaka, and K. Kubota, "Heteroepitaxial Growth of $Cu_3N$

Thin Film," *J. of Crystal Growth* **94** 567-568 (1989).

[9]M. Asano, K. Umeda, and A. Tasaki, "Cu3N Thin Film for a New Light Recording Media," *Japn. J. Appl. Phys.* **29(10)** 1985-1986 (1990).

[10]T. Maruyama and T. Morishita, "Copper Nitride and Tin Nitride Thin Films for Write-Once Optical Recording Media," *Appl. Phys. Lett.* **69(7)** 890-891 (1996).

[11]T. Maruyama and T. Morishita, "Copper Nitride Thin Films Prepared by Radio-Frequency Reactive Sputttering," *J. Appl. Phys.* **78(6)** 4104-4107 (1995).

[12]Annual Book of ASTM Standards, Vol 06.01.

[13]Annual Book of ASTM Standards, Vol 15.09.

# SYNTHESIS AND CHARACTERIZATION OF $SiO_2$-BASED POROUS GLASS MEMBRANES FOR $CO_2$ SEPARATION

Jin-Joo Park, Chihiro Kawai, Seiji Nakahata, Masamichi Yamagiwa, Takao Nishioka, Hisao Takeuchi and Akira Yamakawa
Itami Research Laboratories, Sumitomo Electric Industries, Ltd.
1-1-1, Koya-Kita, Itami, Hyogo, 664-0016 Japan

## ABSTRACT

$SiO_2$-based porous glass membrane was synthesized on porous $Si_3N_4$ substrates through the new cake-filtration/melting process. The microstructure and gas separation characteristics of the membrane were investigated. $SiO_2$-$B_2O_3$-$Na_2O$ glass powder suspension in alcohol (the glass composition: $SiO_2$ 72.4mol%, $B_2O_3$ 20.4mol%, $Na_2O$ 7.2mol%) was filtered through a disk-like porous $Si_3N_4$ substrate to form the cake layer thereon. It was heated under vacuum at a temperature of 1473 K for 0.6 ks, then quenched to room temperature in air, and soaked in 3N $HNO_3$ at 363K for 0.9ks to 7.2ks to leach out the soluble components. The obtained membrane had micropores of 0.46nm diameter and a gradient in composition from the surface to core. The permeance of $N_2$ through the membrane from the gas mixture $CO_2$ and $N_2$ ($CO_2:N_2$=1:9 ratio) was larger than that of $CO_2$. The permeability ratio ($P_{CO_2}/P_{N_2}$) decreased from 0.17 to 0.04 with the increase of temperature from room temperature to 373K. From these results, it was concluded that the membrane had a high selectivity of $N_2$ permeation.

To the extent authorized under the laws of the United States of America, all copyright interests in this publication are the property of The American Ceramic Society. Any duplication, reproduction, or republication of this publication or any part thereof, without the express written consent of The American Ceramic Society or fee paid to the Copyright Clearance Center, is prohibited.

## 1. INTRODUCTION

The global warming phenomenon attributed to $CO_2$ is a serious problem for human beings. The membrane separation of $CO_2$ could be one of the promising processes for eliminating $CO_2$ exhaust in the air. Inorganic membranes are advantageous compared with organic membranes for exhaust gas separation especially at high temperature because of their chemical and thermal stability, and high mechanical strength. It has been reported that porous glass membranes such as porous hollow-fiber glass membranes[1], microporous glass membranes[2], molecular-sieve glass membranes[3], Vycor glass membranes[4], silica hollow-fiber membrane[5], xerogel-coated porous glass membrane[6], and glass capillary[7] have the potential for $CO_2$ separation. The objective of this study is to synthesize and characterize new glass membranes for effective $CO_2$ separation. We tried the synthesis of $SiO_2$-based porous glass membranes by cake-filtration, melting and leaching process. To substrates for supporting the glass membranes, the high strength porous $Si_3N_4$ made in Sumitomo Electric Industries was applied because of its excellent mechanical properties[9]. The resulting membranes showed $N_2$ selectivity from the gas mixture of $CO_2$ and $N_2$, in contrast with $CO_2$ selectivity shown by the other membranes[1,3,6,7]. In this paper we present the experimental results of the microstructure analysis and gas permeation measurements of the newly synthesized $SiO_2$-based porous glass membranes.

## 2. EXPERIMENTAL PROCEDURE

### 2.1 Material

$SiO_2$-based porous glass membranes were synthesized by the following process (the schematic illustration is shown in Fig. 1). $SiO_2$-$B_2O_3$-$Na_2O$ glass powder suspension in alcohol (the glass composition: $SiO_2$ 72.4mol%, $B_2O_3$ 20.4mol%, $Na_2O$ 7.2mol%) was filtered through the disk-like (25mm diameter and 1mm thick) porous $Si_3N_4$ substrate (Sumitomo Electric Industries, Ltd., Type-M) to form the cake layer of

200 $\mu$m thickness thereon. It was heated under vacuum at a temperature of 1473K for 0.6 ks, then quenched to room temperature in air to suppress the $B_2O_3$-$Na_2O$ phase separation, and soaked in 3N $HNO_3$ at 363K for 0.9 to 7.2ks to leach the soluble components out of the glass film.

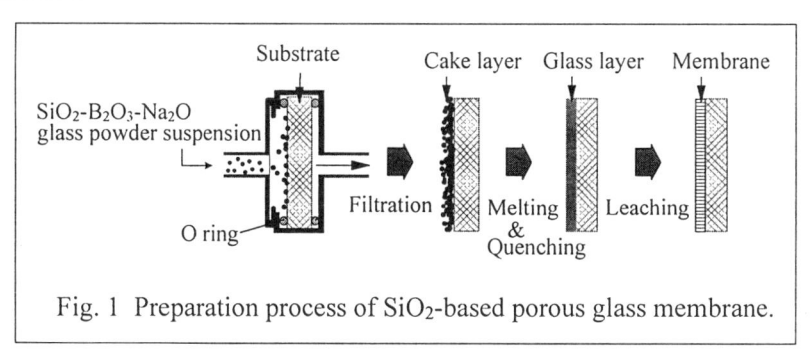

Fig. 1 Preparation process of $SiO_2$-based porous glass membrane.

## 2.2 Analysis of the membrane microstructure

A cross-section of the membrane on the substrate was observed by Scanning Electron Microscopy (SEM). Diameter distribution of the pore size below 1nm in the membrane was measured by the helium adsorption method. Elemental distribution from surface to inside of the membrane was analyzed by the use of sputter depth profiling in conjunction with X-ray Photoelectron Spectroscopy (XPS). Analyzed elements were silicon, boron and sodium. The surface layer was sputtered away at the rate of 0.03nm/s (calibrated with pure $SiO_2$ film) by use of an argon ion gun.

## 2.3 Measurement of $CO_2/N_2$ separation performance

Figure 2 shows a schematic diagram of the test system. The membrane sample with the substrate was attached to the end of the support tube by use of inorganic adhesives. This membrane unit was set into an electric oven with gas inlets and outlets. The gas mixture of $CO_2$ and $N_2$ ($CO_2$:$N_2$=1:9 volume ratio) was fed into the surface of the membrane at the flow rate of $1.7 \cdot 10^{-6} m^3/s$. The permeate through the membrane was swept by helium stream (flow rate: $0.17 \cdot 10^{-6} m^3/s$) and analyzed by gas chromatography. Flow of the permeate was determined with a bubble flow meter. Gas pressure of the feed side and the permeate side was kept at 0.15MPa and 0.1MPa, respectively. Measuring temperature was 293K and 373K, respectively. Permeability ratio was calculated as the ratio of permeance of $CO_2$ to permeance of $N_2$.

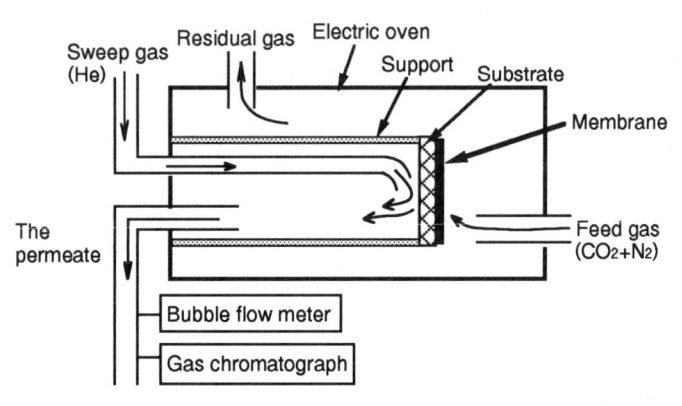

Fig. 2 Schematic diagram of the test system for $CO_2/N_2$ separation performance of the membrane.

## 3. RESULTS AND DISCUSSION

### 3.1 Membrane microstructure

Figure 3 shows the SEM photograph of a cross-section of the $SiO_2$-based porous glass membrane on the porous $Si_3N_4$ substrate for $CO_2$ separation. A part of the membrane infiltrated into the pores of the substrate at the interface, which suggested a high affinity between membrane and substrate. The membrane thickness was approximately 30 $\mu$ m. The micropores for gas separation were formed by elution of $Na_2O$ and $B_2O_3$ components from $SiO_2$-based glass film during $HNO_3$ leaching process as shown in Fig. 4. The elution ratios of the oxides, which were based on the original amount of the

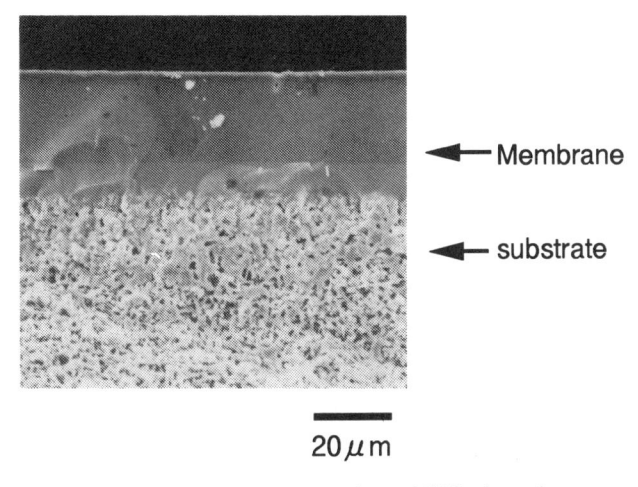

Fig. 3 SEM photograph of a cross-section of SiO$_2$-based porous glass membrane on substrate.

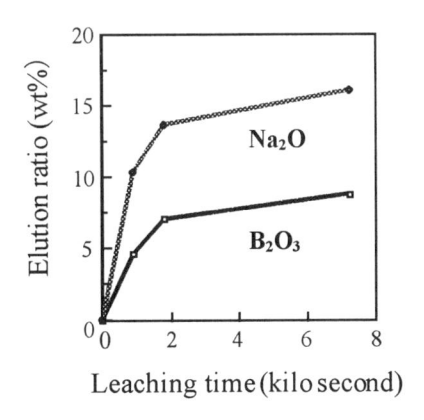

Fig. 4 Elution behavior of Na$_2$O and B$_2$O$_3$ from SiO$_2$-based glass film by HNO$_3$ leaching. Elution ratio is based on the original amount of the corresponding oxide in the glass before leaching.

corresponding oxide in the glass before leaching, increased with leaching time up to 2ks and were nearly saturated after that (the data were not linear on a log-log scale, either). Pore diameter distribution after leaching for 0.9ks (as shown in Fig. 5) indicates that the membrane had micropores of 0.46nm diameter. Figure 6 shows the results of the elemental distribution analysis from surface to core of the membrane after leaching process. Si ratio decreased from 100 to 85mol% and B and Na ratios increased from 0 to 11 and 4mo% respectively with the depth from 0 to 300nm. At 300nm depth, the element ratios were equivalent to that of the bulk. The formation of these gradients seems to be attributed to volatilization and elution of the $Na_2O$ and $B_2O_3$ components out of the glass film during the melting process in vacuum and leaching process in acid.

In summary, it was determined that the $SiO_2$-based porous glass membrane had micropores of 0.46nm diameter and a gradient in composition from the surface to core.

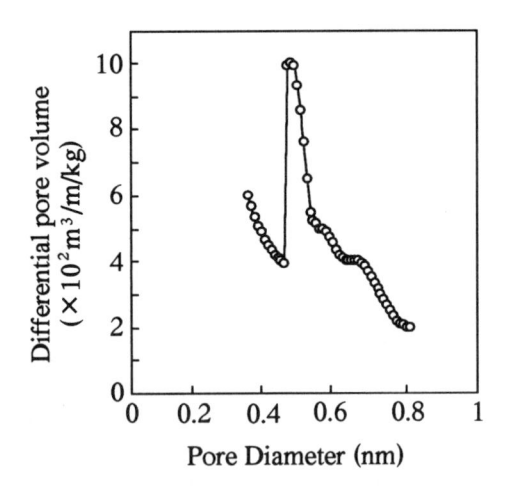

Fig. 5 Pore diameter distribution of $SiO_2$-based porous glass membrane, measured by helium adsorption method.

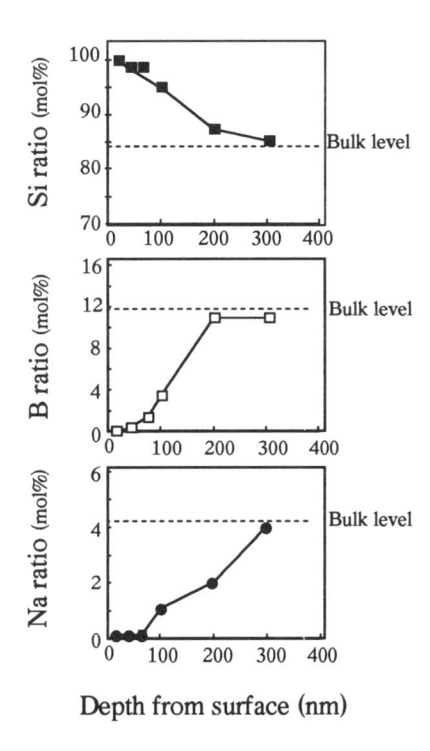

Fig. 6 Elemental distribution analysis of SiO$_2$-based porous glass membrane.

## 3.2 CO$_2$/N$_2$ separation performance

Effects of HNO$_3$ leaching time on the CO$_2$/N$_2$ separation performance at the temperature of 293K are shown in Fig. 7. The permeance of N$_2$ (P$_{N2}$) was larger than that of CO$_2$ (P$_{CO2}$) at every leaching time, and the permeability ratio (P$_{CO2}$/P$_{N2}$) was lowest at 0.9ks. With the increase of leaching time the permeances increased but the permeability ratio was close to 1.0. These results suggest that long time leaching causes the enlargement of the micropores in the membrane. Figure 8 shows the effects of temperature on permeances and permeability ratio. The HNO$_3$ leaching time of the tested membrane was 0.9ks. The permeance of $N_2$ increased and that of $CO_2$ decreased with the increase of temperature. The permeability ratio reached 0.04 at 373K, which indicates that the $SiO_2$-based porous glass membrane has a high selectivity of $N_2$ permeation from $CO_2+N_2$ mixed gas at elevated temperature.

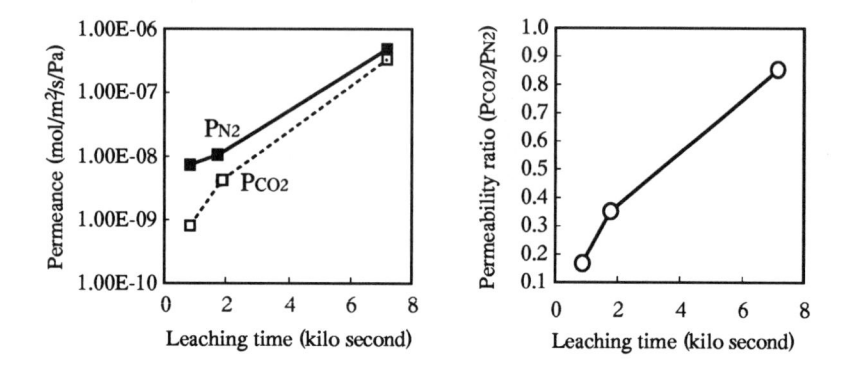

Fig. 7 Effect of $HNO_3$ leaching time on the $CO_2/N_2$ separation performance. Measuring temperature was 293K.

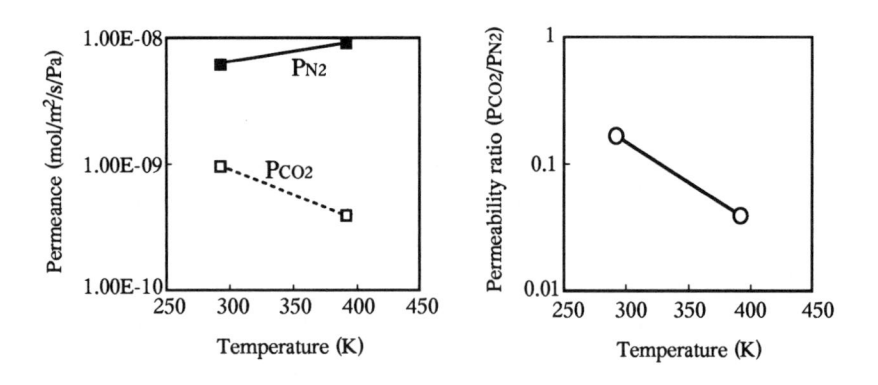

Fig. 8 Effect of temperature on the $CO_2/N_2$ separation performance of $SiO_2$-based porous glass membrane. ($HNO_3$ leaching time was 0.9ks.)

Yazawa et al.[7] reported that the glass capillaries made from $SiO_2$-$Na_2O$-$B_2O_3$ had a high selectivity of $CO_2$ permeation. The raw material composition and acid leaching process of our glass membranes were similar to those of Yazawa's glass capillaries, but our membranes showed reverse selectivity to Yazawa's. Kawahara et al.[8] studied the $N_2$ selectivity of porous $BaTiO_3$, which showed a minimum permeability ratio ($P_{CO_2}/P_{N_2}$) of 0.56. In comparison, the $SiO_2$-based porous glass membranes in our study proved to have much greater potential for $CO_2$ separation by blocking $CO_2$ permeation. The separation mechanism is now being investigated.

## 4. CONCLUSION

We succeeded in the synthesis of $SiO_2$-based porous glass membranes for $CO_2$ separation through the new process. The membranes have micropores of 0.46nm diameter and compositional gradients from the surface to core. The process consists of $SiO_2$-$B_2O_3$-$Na_2O$ powder preparation, cake-filtration, melting, quenching and leaching. The membranes showed extraordinary gas separation properties. The permeability ratios ($P_{CO_2}/P_{N_2}$) were 0.17 at 293K and 0.04 at 373K. These results indicate that the membranes have a high selectivity of $N_2$ permeation and therefore the potential for trapping and thus eliminating $CO_2$ from exhaust gas streams.

## 5. REFERENCES

1) Y.H.Ma, M.Bhandarkar and Y.C.Yang, *Key Eng. Mater.*, 61&62(1991)187-194
2) M.Bhandarkar, A.B.Shelekhin, A.G.Dixon and Y.H.Ma, *J. Membrane Sci.*, 75(1992)221-231.
3) Y.H.Ma, A.G.Dixon and A.B.Shelekhin, *Sep. Technol.*, edited by E.F.Vansant(1994)637-646.
4) Y.H.Ma, S.Pien, Y.She and A.Shelekhin, *Fundam. Adsorpt.*,(1996)553-562.

5) M.H.Hassan, J.D.Way, P.M.Thoen and A.C.Dillon, *J. Membrane Sci.*, 104(1995)27-42.
6) K.Kuraoka, H.Tanaka and T.Yazawa, *J. Mater. Sci. Lett.*, 15(1996)1-3.
7) T.Yazawa, K.Kuraoka, Zu Qun and K.Kushibe, *Membrane Symp.*, 7(1995)117-120.
8) H.Kawahara, T.Ishihara and Y.Takita, *Nippon Kagakukai Shunki Nenkai Kouen Yokoushu 69*, 3C104(1995).
9) C.Kawai and A.Yamakawa, *J. Am. Ceram. Soc.*, 80(1997)2705-708.

## 6. ACKNOWLEDGMENTS

This work was supported by the New Energy and Industrial Technology Development Organization (NEDO) Japan for Research and Development of High Temperature $CO_2$ Separation and the Utilization Technology. We also thank Dr. Inada at the Japan Fine Ceramics Center for his support and helpful discussions on the experiment.

# MECHANICAL PROPERTIES OF PLASMA-SPRAYED $Mo_5Si_3$-MoB-$MoSi_2$ SYSTEM

S.C. Okumus, Ö.Ünal, M.J. Kramer and M. Akinc
Ames Laboratory, Iowa State University, Ames, IA 50010.

## ABSTRACT

Mechanical properties of plasma-sprayed $Mo_5Si_3$-MoB-$MoSi_2$ specimens were studied between room temperature and 1200°C. The comparison of hardness, strength and fracture toughness results from perpendicular and parallel as-deposited specimens, showed that the local anisotropy has little effect on macroscopic properties. However, temperature had a major impact on the deformation and fracture behavior. The specimens exhibited nonlinear behavior at 800°C and the strength increased up to 1000°C, and the crack growth at elevated temperatures was unusually stable. The presence of a significant amount of amorphous phase(s) at the splat boundaries appeared to be the main reason for temperature dependent observations.

## INTRODUCTION

Molybdenum based silicides have attracted considerable attention as high temperature structural materials due to their high melting temperatures, low density, good oxidation and corrosion resistance. However, no single silicide phase appears to have all the properties desired in elevated temperature applications. For instance, $MoSi_2$ has undesirable creep properties although it possesses exceptional oxidation resistance [1]. Similarly, $Mo_5Si_3$ has been shown to have poor oxidation properties while it has very attractive creep resistance [1,2,3]. So, the goal often becomes how to engineer multiphase materials system where one can obtain satisfactory properties from the combination of different phases. The processing difficulties of preparing a single phase also make such an approach attractive. Since plasma spray process was shown to be a viable method to produce dense materials [4], it was used in this study. The goal of this preliminary investigation was to study mechanical properties of plasma-sprayed ($Mo_5Si_3$-MoB-$MoSi_2$) specimens between room temperature and 1200°C through hardness, strength and fracture toughness measurements.

## EXPERIMENTAL

Commercial (Exotherm, Trenton, NJ) powders used in this study had the following elemental composition: 81.5 wt % Mo, 16.2 wt % Si, 1.6 wt % B and 0.7 wt % O, which places the initial composition within the $Mo_5Si_3$-$MoSi_2$-$MoB$ phase field. In order to achieve equilibrium and to remove possible metastable phases, as-received powders were heat-treated at 1800°C for 4 hrs. X-ray diffraction phase analysis on annealed powders showed that the composition was about 83 vol % $Mo_5Si_3$, 14.5 vol % MoB and 2.5 vol % $MoSi_2$. Deposition was made on stainless steel substrates by low-pressure plasma spraying (LPPS) in argon. Deposits were then removed from the substrates to prepare specimens. To eliminate the characteristic layer structure and to increase the overall density, the samples were sintered at 1800°C in argon for 2, 6, 10 hrs. The details of processing and microstructure are discussed in a companion paper [5]. Specimens were machined with respect to the substrate surfaces: parallel and perpendicular. Local anisotropy was studied by indentation while its effect on global properties was investigated by flexure tests using bulk specimens. Indentation tests were carried out at 25°C by a Vickers indenter using 9.8 N load. Flexure tests were conducted to determine both the strength and plane-strain fracture toughness values, $K_{1C}$. The overall specimen dimensions for the strength and toughness specimens were 25x2.5x2 mm and 25x4x3 mm, respectively. The $\alpha_o$ ($=a_o/W$) value in the latter specimens was 0.375. Both strength and fracture tests were carried out by a 4-point bend fixture, where the inner and outer span distances were 10 and 20 mm, respectively. The cross-head speed in all tests was 2 μm/sec. Flexure tests were carried out between 25 and 1200°C in $N_2$. Optical and scanning electron microscopy were utilized to study indentation cracks and the surfaces of test specimens.

## RESULTS
### Hardness

Indentation experiments were carried out to study local anisotropy and its change with heat-treatment time. Figures 1a and 1b show the indentation traces produced on the as-sprayed and 2hr heat-treated specimens, respectively. Note that both specimens were tested in perpendicular orientation, i.e., the splats are end-on. To observe the effect of heat-treatment on crack propagation mode, one of the diagonals of Vickers indent was positioned parallel to splat boundaries, so as to induce preferential cracks in these regions. Since indent diagonals introduce very high local stresses, the weaker regions are expected to delaminate. As expected, the splat boundaries cracked and preferential parallel cracks formed readily on both diagonals. In contrast, heat-treated specimens showed remarkably different crack pattern. As expected, the cracks in Fig. 1b primarily originated from the high stress regions near diagonals without preference. This type of crack pattern is usually observed in isotropic solids where the crack initiation/ propagation is determined by the external loads, rather than the microstructural features. Also note that although both indentations were made by the same load, the trace in Fig. 1b is much smaller than that in Fig. 1a, indicating a significant increase in hardness.

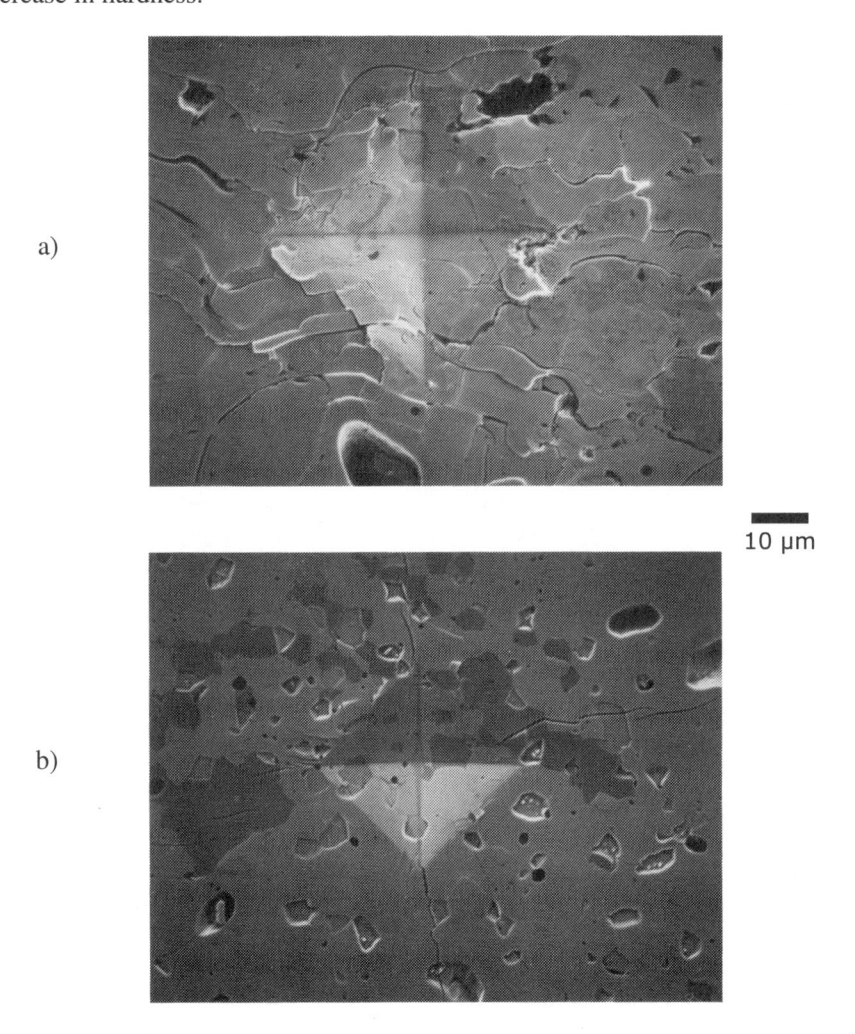

Fig. 1 SEM micrographs of indents on a) as-sprayed and b) 2 hr heat-treated specimens.

Figure 2 shows the effect of heat-treatment on the hardness of plasma sprayed Mo$_5$Si$_3$-MoB-MoSi$_2$ specimens. The hardness values were measured on both the parallel and perpendicular specimens through 30 indentations at each datum point. Considering the scatter in values, it can be concluded from Fig. 2 that hardness shows very little dependence on specimen orientation. As-deposited specimen had the average hardness value of 8.5 GPa. With 2hr heat treatment, hardness increased rapidly to about 12.5 GPa, an increase of about 50%. Further increase in heat-treatment time, however, did not affect the hardness values in either orientation.

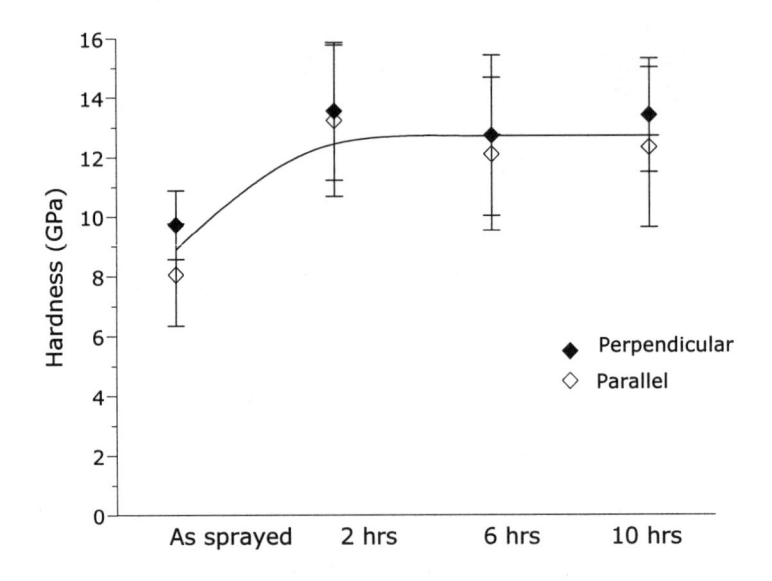

Fig. 2 Variation of hardness with heat-treatment.

Recrystallization and the healing of microcracks in the presence of amorphous phase(s) are believed to be the primary reason for the rapid hardness increase at the beginning. Although there were clear differences in the microstructure of as-sprayed specimens parallel and perpendicular to the spray direction, the agreement in their hardness values in Fig. 2 can be attributed to the magnitude of the indentation load used. As the load level increases, the measured hardness values between the two are expected to converge due to increased indentation area and thus, the improved sampling of the inhomogeneities in the microstructure. It appears that 9.8 N load in this study was high enough to extract bulk values without the significant effect of local anisotropy.

Strength

The comparison of stress/displacement curves obtained from the as-sprayed specimens at 25, 800, 1000 and 1200°C is shown in Fig. 3. The material was linear elastic at room temperature and the failure took place in a brittle manner at the maximum stress level of ~130 MPa. With increasing temperature to 800°C, the strength increased significantly (~45%). Also, the material exhibited some nonlinear behavior at this temperature. The deviation from linearity (elastic limit) started at ~120 MPa and continued up the maximum stress of ~190 MPa, where it failed catastrophically. The nonlinear deformation at 1000°C started at a lower stress level, ~70 MPa.

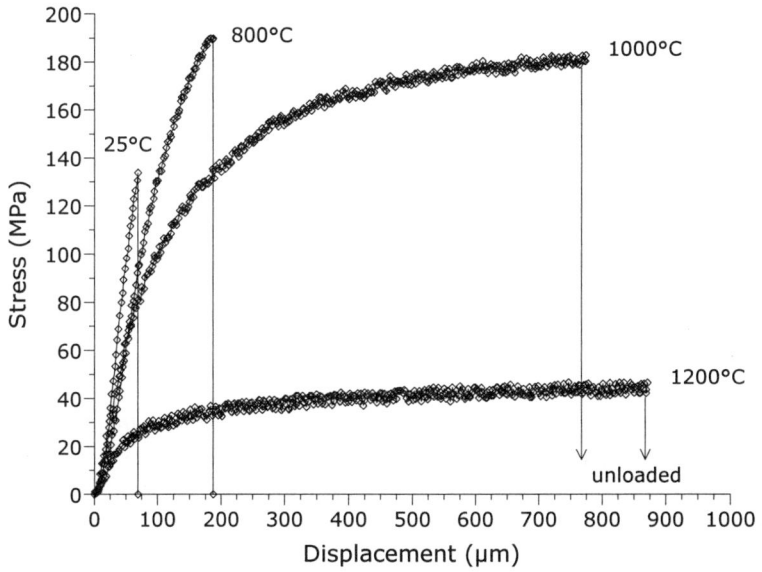

Fig. 3 Stress/displacement curves of as-sprayed specimens.

However, the material still had very high strength, ~180 MPa. An additional test carried out at this temperature confirmed the observations that the increase in strength with temperature up to 1000°C is genuine. Most importantly, the material did not fail at 1000°C in spite of very large deformation, thus it was unloaded at the plateau region. At 1200°C, both the elastic limit and strength dropped significantly, ~17 MPa and ~45 MPa, respectively. As in the case of the 1000°C test, the stress/displacement curve at 1200°C contained a distinct plateau region and the specimen did not fracture. The plateau region in loading curve during the pseudo-dynamic test indicates the equilibrium (steady-state) condition between the mechanisms, which would lead to hardening and softening at the same time.

Thus, the information could be used to predict an approximate creep rate. So, the creep rate at 1000°C and 180 MPa and at 1200°C and 40 MPa was the same, ~$1.6 \times 10^{-4}$ 1/s. These values indicate that the current specimens have poor creep properties.

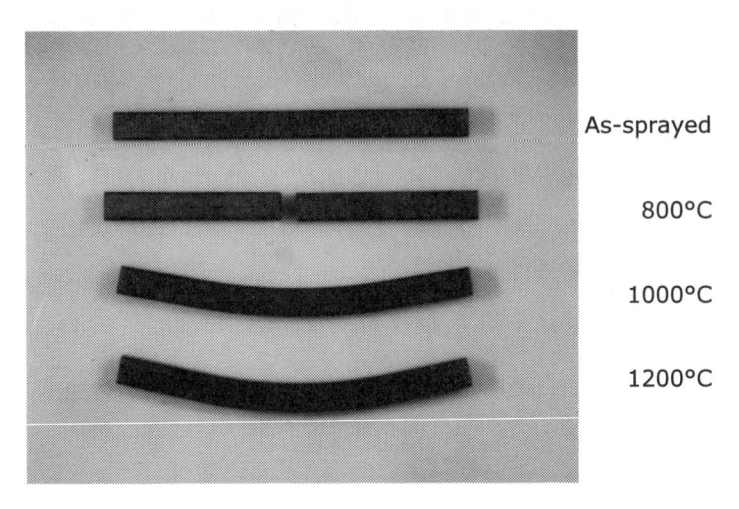

Fig. 4 Deformation of as-sprayed specimens as a function of temperature.

Figure 4 shows the comparison of as-sprayed specimen, and those as-sprayed specimens tested at 800, 1000 and 1200°C. The specimen tested at 25°C was not included since, macroscopically, it was similar to the specimen tested at 800°C. Although the 800°C-specimen exhibited some nonlinear behavior ( Fig.3), no visible deformation on the specimen was detected. Moreover, fractography did not reveal any notable features. Notice, however, that there is significant residual deformation in both specimens tested at 1000°C and 1200°C. The residual strain in these specimens was calculated by measuring the curvature within the inner span where the stress state was pure bending. Calculations showed that the strain in 1000°C and 1200°C specimens were ~4% and ~4.2%, respectively. Obviously, these are significant strain values, which are not typical for brittle materials to sustain, particularly at 1000°C. Thus, there must be other mechanisms in addition to plastic deformation of grains. In fact, SEM investigations showed the accumulation of significant pores and cracks in both materials although the distribution was much more uniform in the 1200°C specimen. The SEM micrograph of tensile cracks on the surface of specimen tested at 1200°C is shown in Fig. 5. The large arrows at outside and the small arrows inside the specimen show the direction of applied stresses and the resulting tensile cracks, respectively. These cracks usually nucleate from both the intrinsic and newly generated pores. On further loading the isolated cracks coalesce, eventually forming macroscopic cracks on planes with maximum stresses. The presence of cracks is a sign for the loss of continuity within the material where the deformation rates of the phases differ greatly. Since the microstructure in this study contained $Mo_5Si_3$, MoB, $MoSi_2$ and, as will be shown later, glassy phase and they have different thermo-mechanical properties, the crack formation was inevitable.

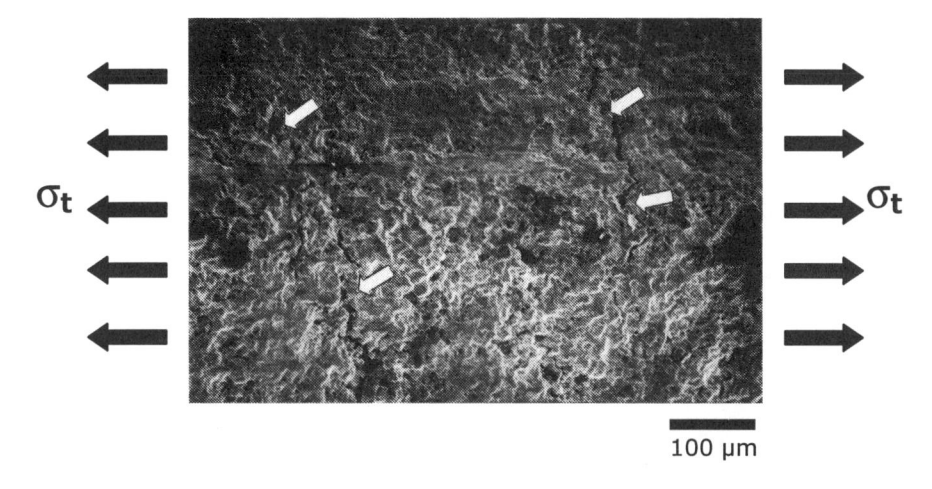

Fig. 5 Tensile cracks on the specimen tested at 1200°C.

The summary of strength and elastic limit values obtained from both the perpendicular and parallel as-sprayed specimens with temperature is shown in Fig. 6. As indicated before, strength was about 30% higher at 800°C and 1000°C. Beyond 1000°C, however, it decreased quickly. Elastic limit on the other hand, showed a continuous drop with temperatures starting at 800°C. These results do not appear to show much dependency on orientation. The figure also includes two additional strength measurements, which correspond to specimens exposed to 1000°C for a short time. After observing the increase in strength with temperature up to 1000°C, it was decided to find out the possible contributions into this observation. First, the specimens were heated to 1000°C for a short period of time to simulate the exposure time of the typical 1000°C test. Then, the exposed specimens were tested in the usual way at room temperature. At room temperature, the strength of unexposed specimens was ~130 MPa. As a result of 1000°C exposure, however, the strength increased to ~150MPa. Although only two specimens were tested, the effect is believed to be genuine since these

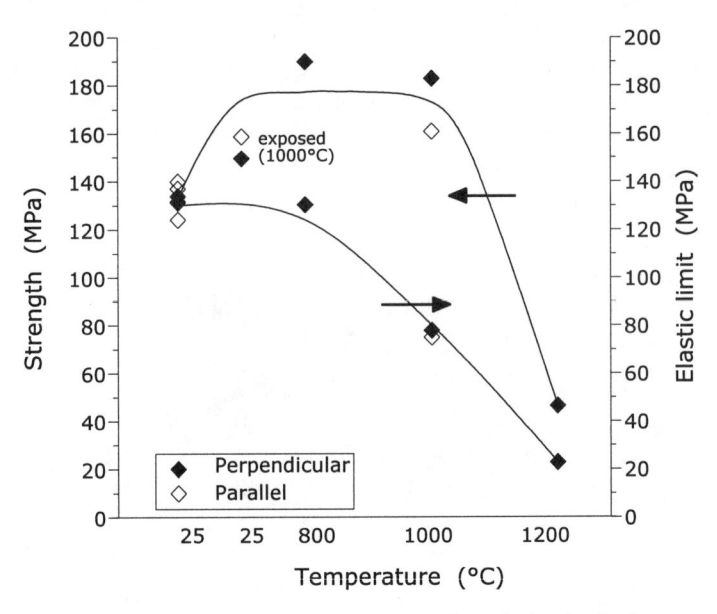

Fig. 6 Effect of temperature on strength and elastic limit.

exposed specimens also showed much higher stiffness values as compared to unexposed specimens.

Fracture Toughness

Fracture toughness values of as-sprayed specimens were measured using chevron-notch specimen configuration since it does not require precracking. Figure 7 shows the comparison of typical load/displacement curves obtained from parallel specimens at room temperature and 1200°C. The origin of the 1200°C-test was displaced to emphasize differences in the fracture behavior. At room temperature, the applied load increased rapidly due to the elastic behavior of specimen. Prior to maximum load, ~50 N, however, the loading curves exhibited deviation from linearity due crack initiation from the sharp notch. Beyond maximum load, the failure was non-catastrophic. These are the typical features which the load/displacement curve of the chevron-notch specimen must contain to obtain a valid fracture toughness value. At 1200°C, however, the fracture behavior of the specimen changed completely. Although the loading curve at 1200°C contained an initial linear region, it was mostly non-linear and the scale was very much different. The maximum load was only ~18 N and the separation of specimen ligament lasted for very large displacements. Such behavior is the manifestation of stable crack growth by the complex deformation mechanisms

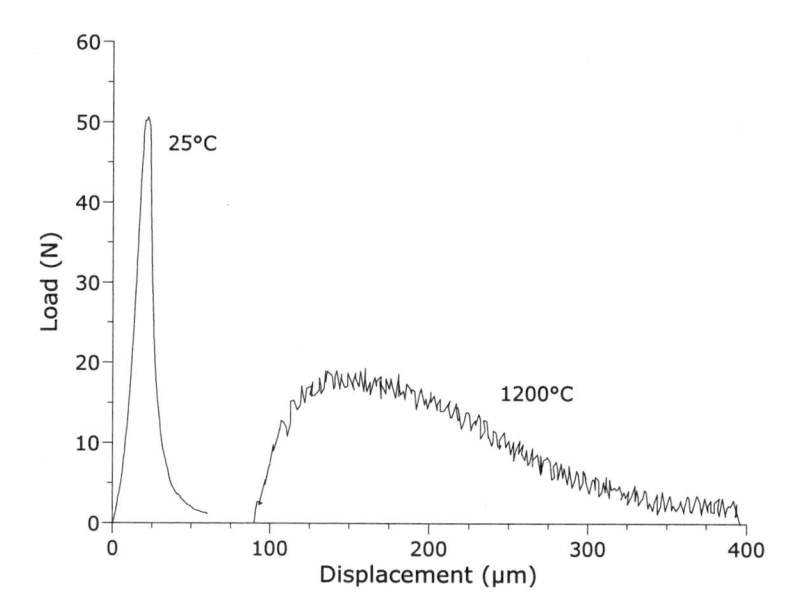

Fig. 7 Load/displacement curve obtained from chevron-notch specimens.

in the chevron region. The plane strain fracture toughness, $K_{1C}$, values were calculated using,

$$K_{1C} = \frac{P_{max}}{B\sqrt{W}} f(a/W) \qquad (1)$$

where $P_{max}$, B, W, and f(a/W) are the maximum load, width, depth and compliance function, respectively [6]. The fracture toughness value at room temperature was 4.1 MPa.m$^{1/2}$ while at 1200°C it was 1.62 MPa.m$^{1/2}$. As will be discussed later, however, the value corresponding to 1200°C test is not valid since the assumption of linear elasticity is violated due to very large deformation. The observations made at 1000°C were also similar to those at 1200°C.

The fracture surfaces of the chevron-notched specimens tested in Fig. 7 are shown in Fig. 8. At 25°C, the fracture surface was relatively planar for porous plasma sprayed specimen, and the fracture took place in both intergranular and transgranular fashion (Fig. 8a). At 1200°C, on the other hand, the specimen surface became rougher and the fracture was intergranular, Fig. 8b. Considering the magnitude of displacement in Fig. 7, it is no surprise that crack path was very tortuous.

a)

100 μm

b)

Fig. 8 Fracture surface of chevron-notch specimens tested at a) 25°C, b) 1200°C.

In the absence of a dominant mechanism, the tortuousity and the expected wake contribution could not account for the large deformations observed since the crack tip stresses are expected to be very large at such large crack openings. Fortunately, the high magnification SEM images on the fracture surface of 1200°C-test specimen provided the evidence for a likely mechanism.

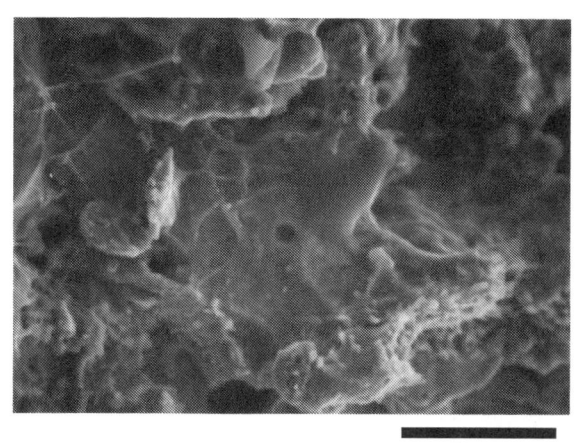

Fig. 9 Glassy phase on the fracture surface of chevron-notch specimen tested at 1200°C.

As can be seen clearly in Fig. 9, the fracture surface is covered with a distinct amorphous phase, which was determined by EDS to be mostly $SiO_2$.

DISCUSSION

Initial tests on the mechanical tests focused on the quantification of anisotropy. Indentation experiments showed that the initial microstructure was directional. With heat-treatment, the indentation crack patterns became symmetrical due to structural homogenization as a result of recrystallization and local sintering. The hardness changed rapidly and reached a plateau region with longer times. Overall, the hardness showed very little dependency on specimen orientation. The present study showed that relatively dense materials can be readily produced by plasma spraying and that the initial anisotropy can be greatly reduced by pressureless sintering.

Brittle-to-ductile transition temperature (BDTT) is influenced strongly by the phases present and their volume fractions. As shown in Fig. 3, the material was linear elastic at 25°C, but exhibited nonlinear behavior at ~800°C. Such deviation in the loading curve is often treated as a sure sign for this transition. However, a caution needs to be exercised. As this study clearly shows, the information from loading curves alone is inconclusive and supporting information from microstructure must be gathered for unequivocal determination. The observance of significant $SiO_2$ on the fracture surface, the fact that $SiO_2$ softens below 1000°C and the knowledge that the primary phase ($Mo_5Si_3$) does not deform at this temperature, suggest that the nonlinear behavior at 800°C is not intrinsic and cannot be an indication for BDTT. Rather, it indicates the start of viscous flow of the amorphous phase at the splat boundaries.

The room temperature strength of 130 MPa is relatively low for this class of materials. For instance, the plasma sprayed $MoSi_2$ was reported to have the strength value of 280 MPa [4]. It should be indicated, however, that the porosity level and the test method strongly effect the measured values. In Ref. 4, the specimens were denser and they used three-point bending test; both of which lead to higher strength values. At temperatures up to 1000°C, strength showed a clear increase. This is a rather unusual observation. Three factors are believed to have contributed to this. They are: (i) densification, (ii) stress relaxation and (iii) reduction in stress intensity due to viscous flow. Although 1000°C is a low temperature for substantial densification, the significant increase in stiffness as a result of the exposure to 1000°C clearly implies that some densification could take place. The stress relaxation can also be a factor since plasma spraying typically induces tensile stresses during rapid cooling. At high temperatures the stress-state tends to reverse itself, leading to higher strength. However, the largest contributor is believed to be the viscous flow of amorphous phase, which shields the crack tip from the externally applied loads. But, large deformation such as seen in this study cannot be induced by densification and stress relaxation. The continuous drop in elastic limit with temperature in Fig. 6 is another sign of the impact of the glassy phase. The sudden drop in strength between 1000 and 1200°C can also be attributed to decreasing viscosity of amorphous phase with temperature. As the glass becomes more fluid it loses its ability to protect the cracks and as a result, the material behaves as if the amorphous phase was not present.

As shown in a companion paper [5], TEM investigations showed that while considerable deformation took place in MoB and $MoSi_2$ grains, dislocation activity was absent in $Mo_5Si_3$ at 1200°C. However, since the volume fractions of MoB and $MoSi_2$ were relatively small and large cracks were observed in deformed specimens, their contributions into the observed permanent deformation were assumed to be negligible. Thus, the primary contribution came from viscoelastic deformation of glassy-phases and inelastic deformation due to porous microstructure. Since the volume percentage of the glassy phase was significant, a judgement on the deformability of major phase ($Mo_5Si_3$), based on this study could not be made. In the presence of weaker glassy phase the stress level cannot reach critical levels to deform $Mo_5Si_3$. Instead, deformation becomes localized mostly into the amorphous phase and pores and cracks appear, as shown in Fig. 5. Therefore, to get information on the plasticity of the primary phase, $Mo_5Si_3$, tests must be conducted with specimens free of glassy phase(s). It should also be indicated that in addition to $SiO_2$, TEM studies revealed a second amorphous phase with the elemental composition of mainly Si and Mo. This Mo-Si phase was present mainly at the thin spat boundaries. It is likely that Mo-Si is a metallic glass formed during spraying process as a result of rapid quenching.

Since large deformation near the crack violates the assumption of linear elasticity upon which the toughness equation was derived, a single fracture parameter, such as $K_{1C}$, cannot describe the fracture behavior of the material. Thus, it is often meaningless to talk about fracture toughness values at elevated temperatures. The single parameter, $K_{1C}$, is valid at low temperatures where the material is linear elastic and exhibits flat R-curve. At elevated temperatures, nonlinear deformation mechanisms become dominant and the materials tend to exhibit R-curve behavior, i.e., their toughness increases with the crack extension. As a result, high temperature properties including the strength and fracture toughness become rate-dependent, which is particularly true for systems containing glassy phases. Therefore, for the given test rate, the $K_C$ value computed from $P_{max}$ at elevated temperatures is only a point on the R-curve. The complete determination of R-curve requires the compliance calibration of test specimen, which can be done either numerically or experimentally. In this study, no attempt has been made to determine the R-curve. In the absence of R-curve, high temperature fracture may be better characterized by the concept of work-of-fracture.

$$W = \frac{1}{2A} \int P du \qquad (2)$$

where, A, P and u are the area, load and displacement, respectively. The calculated W for the chevron-notched specimen tested at 1000 and 1200°C are 484 and 395 $J/m^2$, respectively.

CONCLUSIONS

1. The comparison of hardness, strength and fracture toughness data from perpendicular and parallel as-deposited specimens showed that the local anisotropy has little effect on the macroscopic properties.

2. Temperature had a major impact on the deformation and fracture behavior of $Mo_5Si_3$-MoB-$MoSi_2$ specimens. The specimens exhibited nonlinear behavior at 800°C and the strength increased up to 1000°C. Moreover, at high temperatures the crack growth in fracture specimens was unusually stable.

All these observations were linked to the presence of a significant amount of amorphous phase(s) in the structure.

3. While secondary phases of MoB and $MoSi_2$ deformed plastically, the major phase $Mo_5Si_3$ did not. The overall high temperature properties were dominated by $SiO_2$.

## ACKNOWLEDGEMENTS

The Ames Laboratory is operated by the U.S. Department of Energy (DOE) by Iowa State University under Contract No. W-7405-ENG-82. This work was supported by the Office of Energy Research, Office of Computational and Technology Research, Advanced Energy Projects Division.

## REFERENCES

1. A.K. Vasudevan, and J.J. Petrovic, "A Comparative Overview of $MoSi_2$ Composites", Mater. Sci. Eng., A155, 1 (1992).
2. D.L. Anton, "High Temperature Properties of Refractory Intermetallics", Mat. Res. Soc. Symp. Proc., 213, 733 (1991).
3. M.K. Meyer, M.J. Kramer and M. Akinc, "Compressive Behavior of $Mo_5Si_3$ with the addition of Boron", Intermetallics, 4, 273 (1996).
4. R. Tiwari and H. Herman, "Vacuum Plasma Spraying of $MoSi_2$ and its Components", Mater. Sci. Eng., A155, 95 (1992).
5. M.J. Kramer, S.C. Okumus, O. Ünal, and M. Akinc, "Microstructure of a Plasma-Sprayed Mo-Si-B Alloy", in this volume.
6. D. Munz, R.T. Bubsey, and J.L. Shannon, "Fracture Toughness Determination of $Al_2O_3$ Using Four-Point-Bend Specimens with Straight-Through and Chevron Notches", J. Am. Ceram. Soc., 63, 300 (1980).

# EFFECT OF Si-O-C COATINGS ON THE STRENGTH OF SAPHIKON™ AND NICALON™ FIBERS

J.R. Hellmann, M.P. Petervary, J.M. Priest, and C.G. Pantano
Department of Materials Science and Engineering
The Pennsylvania State University
University Park, PA 16802

## ABSTRACT

The effect of Si-O-C coatings on the tensile strength of Saphikon™ single crystal sapphire filaments and ceramic grade Nicalon™ fibers was examined. A moderate strengthening of Saphikon™ was observed for fibers in the as-coated form, and was attributed to suppression of environmentally-assisted crack growth. Pyrolysis of the coating at 1000°C in argon, and subsequent oxidation at 1000°C for 24 hours in air, yielded no strength degradation relative to the as-received Saphikon™ strength. No strength modification was observed for the Nicalon™ fiber in the as-coated form. However, a 13% strength degradation was observed after coating pyrolysis at 1200°C in argon, and was attributed to microstructural modification and CO evolution in the fiber during coating pyrolysis.

## INTRODUCTION

It has been well established that tailoring of the interfacial strength and elastic compliance between the fiber and matrix is required to achieve desired combinations of fracture toughness, stiffness, and strength in ceramic matrix composites[1-3]. Typically, this has been achieved through the application of coatings to the fibers prior to composite consolidation. To date, carbon- and BN-based coatings have demonstrated the greatest potential as interphases for achieving *graceful failure* in oxide and non-oxide based CMC's. However, their susceptibility to oxidation limits their utility in high temperature industrial and aerospace applications where long term thermochemical and thermomechanical durability is required.

Recent studies have demonstrated that sol-gel derived Si-O-C coatings can yield desirable interfacial shear behavior while exhibiting significantly greater oxidation resistance relative to BN- and carbon-based coatings[5]. Furthermore, through control of the coating precursor chemistry, and subsequent coating pyrolysis conditions, the coatings can be manipulated to adjust cohesive strength, compliance, and thermal stability[6]. This combination of characteristics makes Si-O-C an attractive candidate for interfacial coatings in oxide and non-oxide CMC's for use in high temperature oxidative environments. However, little is known about the thermochemical and thermomechanical compatibility of Si-O-C coatings on ceramic fibers.

---

To the extent authorized under the laws of the United States of America, all copyright interests in this publication are the property of The American Ceramic Society. Any duplication, reproduction, or republication of this publication or any part thereof, without the express written consent of The American Ceramic Society or fee paid to the Copyright Clearance Center, is prohibited.

The purpose of this study is to examine the effect of Si-O-C coatings on the strength of Saphikon™ single crystal sapphire filaments and ceramic grade Nicalon™ fibers.

EXPERIMENTAL PROCEDURE

Unsized Saphikon™ single crystal sapphire fibers (c-axis oriented) with an average diameter of 117 µm, and ceramic grade Nicalon™ silicon carbide fibers with PVA sizing were examined in this study.

Si-O-C Blackglas™ coatings were prepared from Allied Signal's 493A monomeric solution and 493B curative via a polymerization route. Solutions containing 41:1 weight ratio of 493A:493B in iso-octane were prepared, and refluxed at 110°C for 3 hr while stirring. Coatings were applied to cleaned fibers via a drain coating technique. Saphikon™ filaments were cleaned using a two-step oxidizing and complexing treatment with hydrogen peroxide solutions (described in detail elsewhere[7]) to remove organic and metallic ions from the surface of the fibers. Nicalon™ fibers were cleaned oxidatively at 600°C in air for 30 minutes, and then rinsed in methyl ethyl ketone. Coating pyrolysis was performed at 1000°C and 1200°C for 1 hr in UHP argon for the Saphikon™ and Nicalon™ fibers, respectively. Subsequent oxidation treatments were performed at 1000°C in air for 24 hours. The pyrolysis conditions were selected to yield microporous coatings comprised solely of Si-O-C in which all organic constituents have been eliminated. A more detailed description of the role of pyrolysis and subsequent oxidative exposure on the composition, structure, and microporosity of bulk Si-O-C prepared in this manner is presented in reference 8.

Fibers were tested in tension at a strain rate of $4 \times 10^{-3}$ min$^{-1}$. Fractographic analysis was performed using scanning electron microscopy to identify failure sources. The censored strength data was analyzed using a two-parameter Weibull distribution function as described elsewhere[7], and the maximum likelihood estimation of Weibull strength distribution parameters was employed[9].

RESULTS AND DISCUSSION

Figure 1 shows the strength distribution for the Saphikon™ and Nicalon™ fibers in the as-cleaned condition. The Saphikon™ filaments exhibited a characteristic strength of 3.0 GPa, and the Nicalon™ fibers exhibited a characteristic strength of 1.87 GPa, both with Weibull moduli of 7; these results are in good agreement with prior studies in our laboratory[7,10].

Figure 2a illustrates that whereas there was no strength modification observed in the Nicalon™ fibers in the as-coated form ($\sigma_\theta$ = 1.93 GPa,) there was an apparent strengthening of the Saphikon™ filaments after the application of the coating ($\sigma_\theta$ = 3.4 GPa; Figure 2b.)

We speculated that the apparent strengthening in the coated Saphikon™ filaments could be due either to the filling of surface flaws by the low viscosity coating, resulting in crack blunting[11], or due to suppression of environmentally-assisted slow crack growth during testing. It is well established that sapphire is

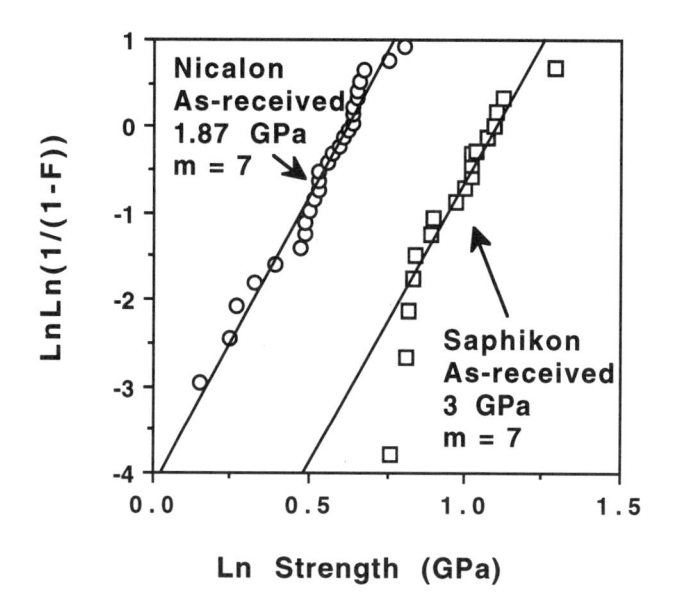

Figure 1. Weibull strength distributions for Saphikon™ and Nicalon™ fibers in the as-received condition

susceptible to stress corrosion; when tested in inert ambients containing no water, higher strengths are obtained due to the absence of environment assisted slow crack growth[12]. Therefore, it is possible that in this study, the unpyrolyzed coating serves as a barrier to moisture migration to the crack tip, thereby resulting in slower crack growth rates and higher measured strengths for the sapphire fiber. This may be due to the hydrophobic character of the carbon groups remaining on the surface of the as-deposited gel coating, thereby inhibiting moisture adsorption and migration through the coating, as demonstrated previously in our laboratory[13-14].

To test this hypothesis, we coated cleaned fibers with silicone vacuum grease to eliminate migration of water vapor to the crack tip during testing, and repeated the strength measurements. Since silicone grease will not bear any appreciable load during testing, nor act as a flaw healing or crack blunting agent at room temperature, any strength enhancement observed relative to the uncoated fibers must be due to suppression of stress corrosion during testing. Figure 3 indicates that the fibers tested under these "inert" conditions (covered with vacuum grease) exhibited significantly higher strengths than in the uncoated form, and virtually identical characteristic strength relative to the Si-O-C coated fibers prior to

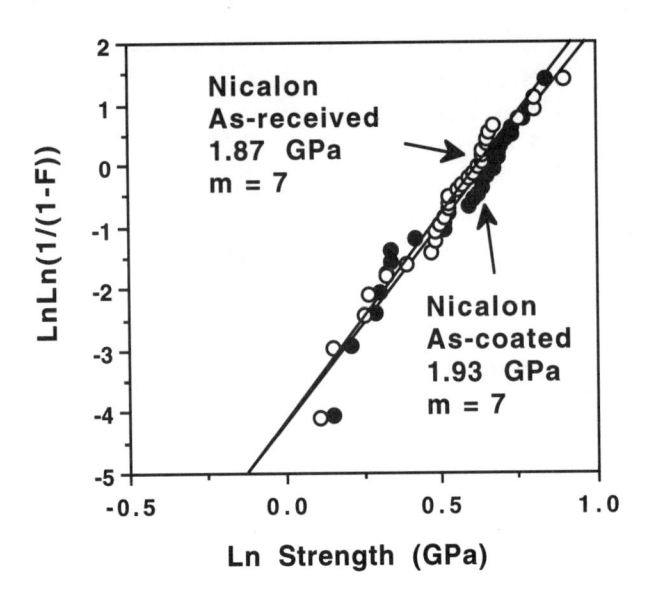

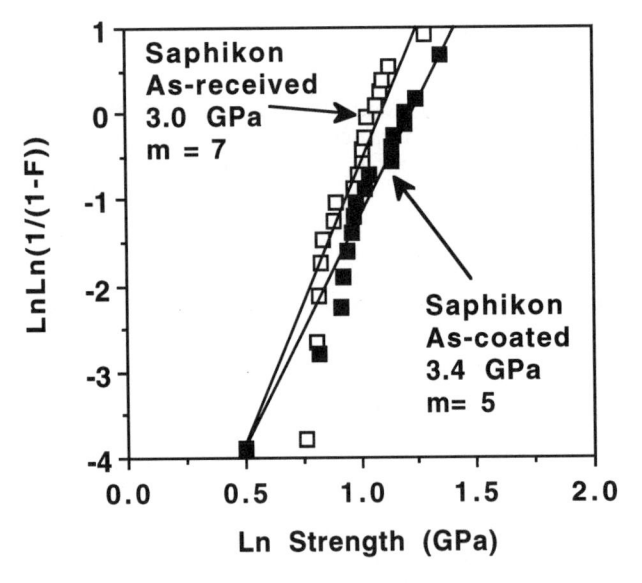

Figure 2. Weibull strength distributions for as coated fibers: a) Nicalon™ and b) Saphikon™

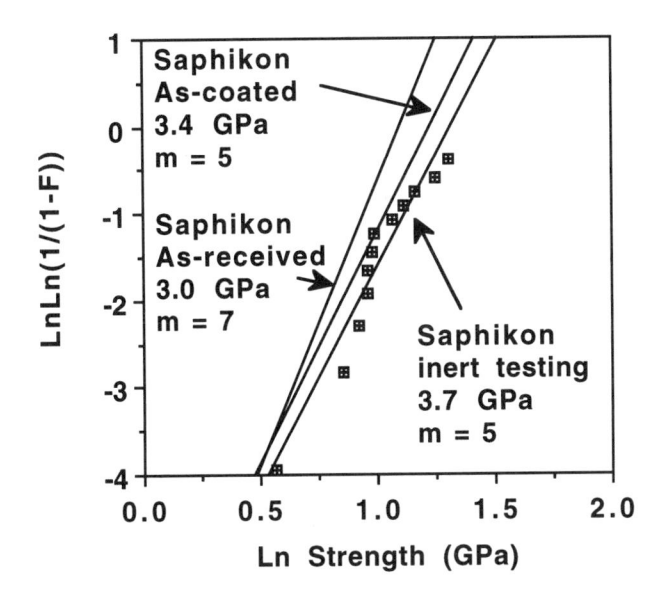

Figure 3. Weibull strength distribution for Saphikon™ filaments tested under inert conditions to suppress stress corrosion, relative to the as-received and as-coated conditions

pyrolysis. This is consistent with the hypothesis that the unpyrolyzed Si-O-C coating suppresses stress corrosion in the Saphikon™ filaments. It is apparent from this experiment that the unpyrolyzed coating may be exploited as a protective sizing to mitigate fiber damage and strength degradation during composite processing.

Figure 4 shows the effect of pyrolyzing the coating on fiber strength for Saphikon™ filaments. No strength degradation was observed for the pyrolyzed, coated Saphikon™ relative to the as-cleaned condition ($\sigma_\theta$ = 3.0 GPa, m=7 for the uncoated fiber; $\sigma_\theta$ = 3.0 GPa, m=8 for the pyrolyzed coated fibers.) However, the pyrolyzation process reduces the efficacy of the coating as a moisture barrier, thereby eliminating the stress corrosion inhibition effect observed for the unpyrolyzed coating. Because the pyrolyzed coatings are quite thin (<0.25 μm,) we believe that the microporosity which further develops during pyrolysis, coupled with loss of the hydrophobic carbon groups during pyrolysis, nullifies the moisture barrier effect[13-15].

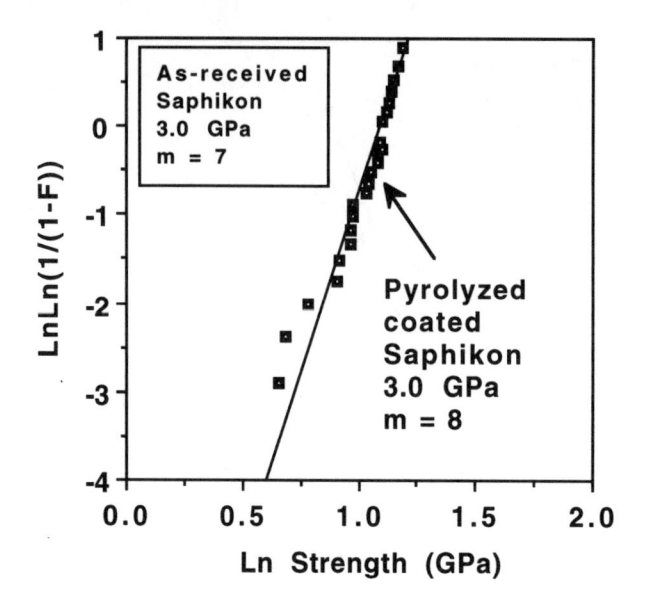

Figure 4. Weibull strength distribution of Saphikon™ filaments after coating pyrolysis at 1000°C in UHP argon for 1 hr.

In contrast, a minor strength degradation (≈13%) was observed after pyrolyzing coated Nicalon™ fibers (Figure 5a.) To assess whether the strength degradation was associated with the coating, or due solely to the heat treatment used for pyrolysis, a group of uncoated fibers was tested after heat treating at 1200°C for 1 hour in UHP argon. Figure 5b reveals that the heat treated fibers, and the pyrolyzed coated fibers possess essentially identical strength distributions, indicating that it is the pyrolyzation heat treatment, and not the coating itself, that yields the strength degradation in Nicalon™ fiber. These results are consistent with prior findings that heat treatment of Nicalon™ fibers at elevated temperature in UHP argon results in decomposition of the amorphous $SiC_xO_y$ phase in the fiber, forming β-SiC and evolving CO gas in the process[10, 16-18]. This microstructural evolution results in the formation of internal voids and strength limiting flaws in the heat treated fiber.

The effect of exposure to high temperature oxidative environments on the strength of the coated, pyrolyzed Saphikon™ fibers is presented in Figure 6. A moderate strength increase was observed for the oxidized, coated fibers relative to the as-received fiber strength ($\sigma_\theta = 3.4$ GPa versus $\sigma_\theta = 3.0$ GPa, respectively.)

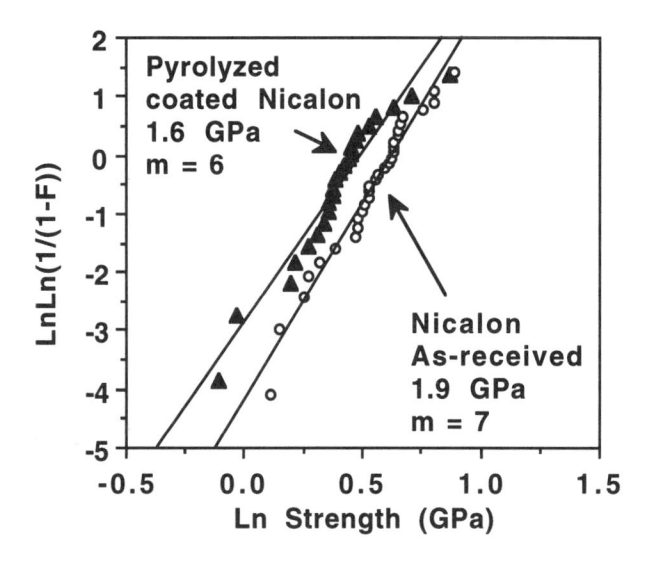

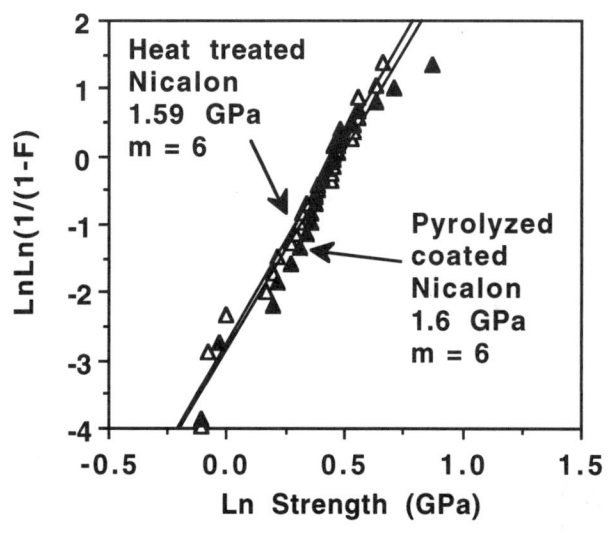

Figure 5. Weibull strength distributions for a) as-received Nicalon versus after coating pyrolysis, and b) As-received Nicalon after pyrolysis heat treatment only.

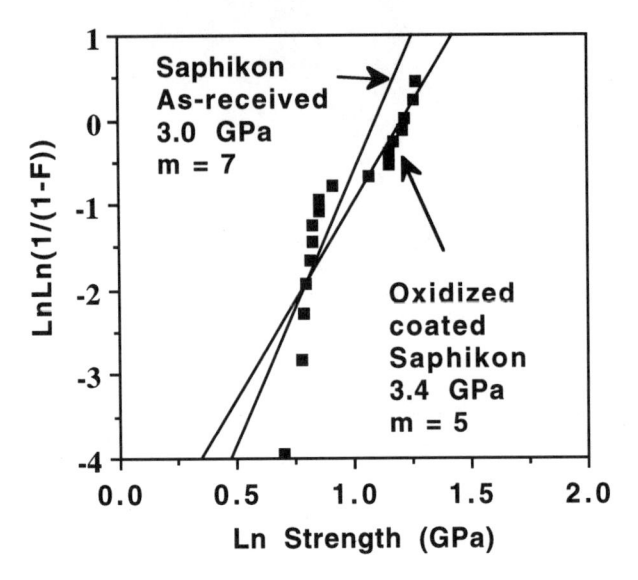

Figure 6. Weibull strength distribution for pyrolyzed, coated Saphikon™ filaments after oxidation at 1000°C for 24 hours in air

Fractography revealed a distinct shift in the sources of failure from combined surface- and volume-flaws in the as received Saphikon™ filaments, to predominantly volume flaws in the oxidized, coated filaments. Tensile tests on uncoated filaments subjected to the oxidation heat treatment revealed a smaller, but significant strength enhancement ($\sigma_\theta$ = 3.1 GPa; m=8) relative to the as-received Saphikon™ filaments ($\sigma_\theta$ = 3.0 GPa; m=7.) This suggests that the strength increase may be due in part to thermally-induced flaw healing at the sapphire fiber surface[19], or to the annihilation of surface flaws due to reaction of the sapphire fiber with the Si-O-C coating at high temperature. Studies are currently underway to more completely interrogate this interpretation.

SUMMARY AND CONCLUSIONS

Sol-gel derived silicon oxycarbide has been proposed as an oxidation resistant interphase for CMC's. In this study, we evaluated the effect of Si-O-C coatings on the strength of two candidate reinforcements for CMC's: Saphikon™ single crystal sapphire filaments and ceramic grade Nicalon™ silicon carbide fibers. In the unpyrolyzed form, the Si-O-C coatings yielded no modification in the Nicalon™ fiber strength distribution. However, a moderate (≈10%) increase in strength was observed for the Saphikon™ filaments in the as-coated form. The strength enhancement was found to be due to the coating serving as a barrier to water vapor migration from the test environment to the crack tip, thereby inhibiting the effects of stress corrosion during testing. This result indicates that the unpyrolyzed coating can also serve as a fiber sizing to mitigate fiber damage and strength degradation during composite processing.

No effect of coating pyrolysis on the strength of Saphikon™ filaments was observed; the as-coated filaments exhibited essentially the same strength distribution as clean, as-received filaments. However, the pyrolyzed coatings no longer inhibit the stress corrosion of the filaments during testing.

Pyrolysis of the coated Nicalon™ fibers resulted in a minor strength degradation relative to clean, as-received fibers. The degradation was subsequently attributed to microstructural evolution occurring in the fiber during the pyrolysis heat treatment, rather than due to the coating itself.

Oxidation of the coating yielded a moderate strength increase in Saphikon™ filaments. The increase was partially attributed to thermally-induced flaw healing; however, the role of fiber/coating interaction during oxidative heat treatment on flaw healing and/or annihilation requires more detailed study.

ACKNOWLEDGMENTS
The authors thank Ms. Mary Strezlecki and Dr. David Shelleman for their assistance in coating synthesis and Weibull analyses, respectively. The authors gratefully acknowledge the support of the U.S. Department of Energy Continuous Fiber Ceramic Composites program, Lockheed Martin Energy Research Corporation (contract #11X-SF881C,) and the NASA HiTEMP Program (grant #NAGW-1381.)

REFERENCES
[1] A.G. Evans and D.B. Marshall, "The Mechanical Behavior of Ceramic Matrix Composites," Acta Met., 37(10)2567-83(1989).
[2] R.J. Kerans, R.S. Hay, N.J. Pagano, and T.A. Parthasarathy, "The Role of the Fiber-Matrix Interface in Ceramic Composites," Bull. Am. Ceram. Soc., 68(2)429-442(1989).
[3] P.D. Jero, R.J. Kerans, and T.A. Parthasarathy, "Effect of Interfacial Roughness on the Frictional Stress Measured Using Pullout Tests," J. Am. Ceram. Soc., 74(11)2793-2801(1993).
[4] J.B. Davis, J.P.A. Lofvander, A.G. Evans, E. Bischoff, and M.L. Emiliani, "Fiber Coating Concepts for Brittle-Matrix Composites," J. Am. Ceram. Soc., 76(5)1249-1257(1993).
[5] T. Paulson, M. Hammond, and C.G. Pantano, "Porous, Microcomposite Interface Coatings for Ceramic Composites," <u>HITEMP Review 1994</u>, Proceedings of the 7th Annual HITEMP Review, NASA CP-10146, (1994).
[6] G.D. Soraru, E. Dallapiccola, and G. D'Andrea, "Mechanical Characterization of Sol-Gel Derived Silicon Oxycarbide Glasses," J. Am. Ceram. Soc., 79(8)2074-2080(1996).

[7] E.R. Trumbauer, J.R. Hellmann, D.L. Shelleman, and D.A. Koss, "Effect of Cleaning and Abrasion-Induced Damage on the Weibull Strength Distribution of Sapphire Fiber," J. Am. Ceram. Soc., 77(8)2017-2024(1994).

[8] M.T. Strzelecki and C.G. Pantano, "Blackglas™ Silicon Oxycarbide as a Porous, Oxidation Resistant Interphase for CFCMCs," submitted to the Journal of the American Ceramic Society.

[9] S.F. Duffy and E.H. Baker, Maximum Likelihood Estimator (MLE) Macro for Excel, Civil Engineering Department, Cleveland State University (1996).

[10] D.J. Pysher, K.C. Goretta, R.S. Hodder, and R.E. Tressler, "Strengths of Ceramic Fibers at Elevated Temperature," J. Am. Ceram. Soc., 72(2)284-288(1989).

[11] Fabes, et al., "Strengthening of Silica Glass by Gel-Derived Coatings," J. Non-Cryst. Solids, 82(1986)349-355.

[12] S.M. Wiederhorn, Fracture of Ceramics, NBS Special Publications-303, May 1969.

[13] A.K. Singh and C.G. Pantano, "Surface Chemistry and Structure of Silicon Oxycarbide Gels and Glasses," J. Sol Gel Sci & Tech., 8(1997)371.

[14] A.K. Singh and C.G. Pantano, "Porous Silicon Oxycarbide Glasses," J.Am. Ceram. Soc., 79(10)2696-2704(1996).

[15] T.M. Chaundry, et al., "Silicon Oxycarbide Coatings on Graphite Fibers, II. Adhesion, Processing, and Interfacial Properties," Mater. Sci. Eng. A, 195(1995)237-249.

[16] R. Bodet, J. Lamon, N. Jia, and R.E. Tressler, "Microstructural Stability and Creep Behavior of S-C-O (Nicalon) Fibers in Carbon Monoxide and Argon Environments," J. Am. Ceram. Soc., 79(10)2673-2686(1996).

[17] N. Jia, R. Bodet, and R.E. Tressler, "Effects of Microstructural Instability on the Creep Behavior of Si-C-O (Nicalon) Fibers in Argon," J. Am. Ceram. Soc., 76(12)3051-3060(1993).

[18] G.S. Bibbo, P.M. Benson, and C.G. Pantano, "Effect of Carbon Monoxide Partial Pressure on the High-Temperature Decomposition of Nicalon Fibre," J. Mater. Sci., 26(1991)5075-5080.

[19] F.P. Mallinder and B.A. Proctor, "The Strength of Flame Polished Sapphire Crystals," Phil. Mag., 13(121)197-207(1966)

**Electronic Ceramics**

# SYNTHESIS, MICROSTRUCTURE AND MAGNETIC PROPERTIES OF SINTERED Mn-Zn FERRITES FROM HYDROTHERMAL POWDERS

Ralph Lucke[a], Ernst Schlegel[b] and Rolf Strienitz[b]
[a]Siemens Matsushita Components GmbH & Co. KG
Ferrites Division
P.O. Box 80 17 09
Munich, D-81617
[b]Institute of Silicate Engineering
Techn. University Bergakademie Freiberg
Agricolastr. 17
Freiberg, D-09599
Germany

## ABSTRACT

Soft magnetic materials of the system manganese zinc ferrite have been made from hydrothermal nanosized powders. The conditions for the synthesis of the powders were optimized in terms of qualitative and economical targets. The achieved magnetic properties are influenced by the microstructure of the ferrite material. The magnetic material properties as well as the ferrite processing conditions are dependent on the conditions of hydrothermal powder synthesis. The initial permeability of highest permeability materials could be improved by 12 % while sintering under 50 K lower temperature. The power losses of high frequency power transformer materials were reduced by 10 - 20 %. The simplified powder synthesis is compared with the conventional mixed oxide processing.

## INTRODUCTION

Soft magnetic manganese-zinc-ferrites are widely used in telecommunications and related applications as well as for power transformers. The ferrite cores are conventional produced by a mixed oxide route. This is a multi-stage process that influences, with all steps, the nature of the product and its magnetic properties. Basically the properties are influenced by raw materials and subsequent powder preparation[1, 2]. The ferrite powder preparation includes calcination in order to remove volatile ingredients and to adjust the sintering behavior. The sintering activity is dependent on the grain size of the powder and finally on the milling parameters. Highly sophisticated ferrite materials can be achieved by complete control of all powder preparation steps and the subsequent processing. The formulation of a target for the magnetic properties includes the question for their physical limits: Highest permeabilities are limited by the level of energy necessary for the magnetic reversal. That energy consists of crystal anisotropy energy, magnetostriction energy and stray field energy. All these energies should be low and are influenced by the composition. Additionally the composition-dependent magnetization influences the permeability. For highest permeability ferrites a compromise has to be found between an adequate temperature and frequency dependence, practicable magnetization behavior and powder-dependent sintering conditions. The mentioned magnetic properties are strongly influenced by the microstructure of the ferrite material[3].

The requirement for low power losses in ferrite transformers can be met among others by lowering the core losses. There are three components, each with a different frequency-dependent contribution to the core losses[4,5]:

- hysteresis losses
- eddy current losses
- residual losses (domain wall resonance and damping)

The hysteresis losses are composition- and microstructural-dependent and can be influenced similarly as the permeability. The eddy current losses are a result of non-desired Ohmic losses within the microstructure. They can be reduced by an increased electric resistivity of the ferrite grains as well as the grain boundaries and by a fine grained microstructure[6,7]. Because of the strong frequency dependence, suitable measures for lowering the eddy current losses have to be chosen very carefully, especially for high frequency applications. The residual losses are directly linked to the resonant frequency of the ferrite material and can be lowered by well-dispersed additives[8].

Obviously the microstructure is a central feature determining the properties. The processing of well-defined powders of homogeneous composition provides good chances for designing the microstructure.

The formation of the desired phases is dependent on the homogeneity of the mixture. However, solid-state reactions during powder preparation are limited by the diffusion within crystal lattice. Alternative preparation routes through the dissociated state of the appropriate metal ions could be suitable for more homogeneous powders. Following powder properties are desired:

- constant chemical composition
- homogeneity down to the atomic structures
- high chemical purity
- constant grain shape and grain size
- high sintering activity

The origin of the hydrothermal synthesis method is the genesis of minerals in geological periods. It is based on the solubility of raw materials at elevated temperatures. In order to get a significant solubility the temperature of the water-powder mixture often has to exceed the boiling point. This can only occur at increased pressures according to the water vapor pressure. The reactor used for this process is the autoclave. It can be used as a batch reactor or as a continuous reactor. The combination of several autoclaves leads to comparably low energy consumption because of possible steam exchange between them[9].

All processing steps have to meet economic targets. Therefore the selection of the raw materials in consideration of their processability and their cost is important. The powder preparation under wet conditions with raw materials available for a large scale production should be the basis for the process. The hydrothermal powder preparation route is one of those processes. It offers at low temperatures, a direct route to oxidic powders without any hydroxide content, avoiding the calcination step required for sol-gel processing. Consequently it should be suitable for an economic achievement of a very homogeneous ferrite powder[10, 11].

The hydrothermal powder preparation provides the chance to simplify the ferrite powder processing (Fig. 1).

| Conventional powder processing | Hydrothermal powder processing |
|---|---|
| Weighing of raw materials | Weighing of raw materials |
| Mixing / Pelletizing | |
| Calcination | Hydrothermal synthesis |
| Milling | Desagglomeration |
| Granulation | Granulation |

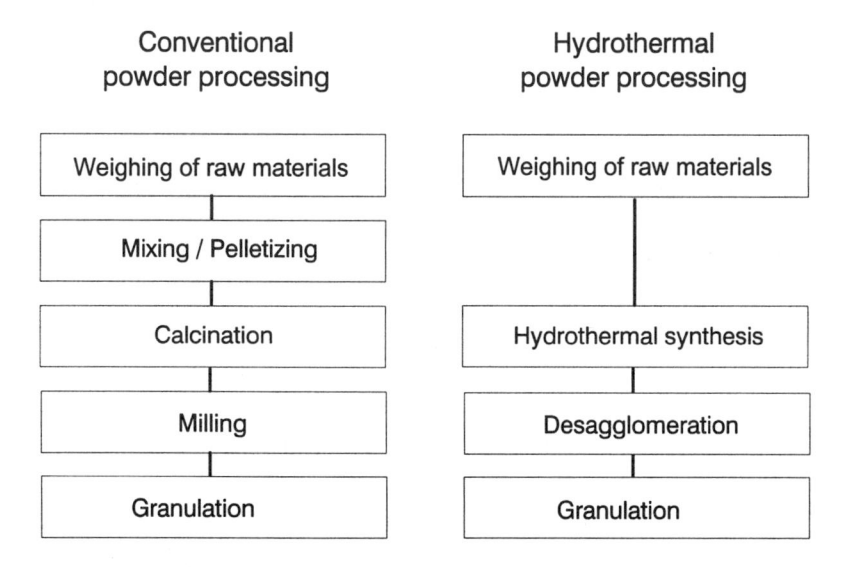

**Figure 1:** Comparison of technological steps for conventional and hydrothermal ferrite powder processing

The hydrothermal preparation of ferrite powders is used with following targets:
- preparation of powders with low composition gradients, with extremely high purity and with homogeneous distribution of additives
- preparation of powders with well-defined grain sizes and grain shapes
- preparation of powders with high sintering activity
- preparation of powders under comparable or lower cost than the equivalent mixed oxide processing

Consequently the target of the available literature dealing with the chemical ferrite powder preparation[12, 13] can be extended to the magnetic properties of products made of them.

The problems when dealing with very fine grained ferrite powders have to be investigated under large scale production.

EXPERIMENTAL PROCEDURE

The raw materials used for the ferrite powder synthesis were hematite (99.45 % $Fe_2O_3$, $d_{50}$ = 2.1 µm), hausmannite (98.55 % $Mn_3O_4$, $d_{50}$ = 1.3 µm) and zincite (99.56 % ZnO, $d_{50}$ = 0.6 µm). Stoichiometric amounts of the appropriate oxides were filled together with deionized water into the autoclave (4 kg solid, 7 kg water, autoclave volume 10 dm³, stainless steel). The hydrothermal synthesis were conducted at temperatures from 200 °C to 300 °C under an equilibrium water pressure. The time of synthesis was varied from 7 to 72 h. After hydrothermal treatment, the autoclave was cooled and the product was desagglomerated in an agitation mill.

The mixed oxide process was carried out under manufacturing conditions. Starting with the same raw materials, the calcination in a rotary kiln requires flowable agglomerates. The current mixing / pelletizing process was used for this purpose. After calcination an intensive milling process is necessary in order to produce a reactive ferrite powder. The granulation was carried out for both powder routes by appropriate binder additions and subsequent built-up granulation.

The TGA was performed using a Shimadzu TGA 50 with 10 K/min heating rate in air. The shrinkage behavior was measured with 5 K/min in air using a Linseis dilatometer L 75/20. The XRD-measurements were performed using a Philips-X´Pert, Cu-Kα-radiation. The particle size distribution was analyzed using the method of laser diffraction / scattering (LS230 - Coulter). The grain size distribution of the sintered microstructures were measured by evaluating SEM images (Type 6400, JEOL, Japan) using the system "Analysis" from Soft Imaging Systems, Muenster, Germany. All magnetic properties were investigated at ferrite toroids and compared with conventional prepared toroids of the same size. The measuring conditions correspond to the desired reliability values of the production. The magnetic properties were measured by HP4194A Impedance Analyzer (permeability, hysteresis material constant), Tektronix TDS744A (flux density, coercive field strength) and Clarke-Hess Model 258, New York (power losses).

RESULTS AND DISCUSSION
Powder preparation

One thermodynamically stable spinel phase is Franklinite ((Zn, Mn)$Fe_2O_4$). It is formed by conventional calcination as well as by hydrothermal reaction.
Commercial hematite contains considerable amounts of chlorine (1500 ppm) which causes a weak acid reaction of the water suspension (pH app. 6). The chlorine cannot be removed during the hydrothermal process. This seriously affects the composition of the synthesized ferrite powder and leads to a zinc rich franklinite. The higher the temperature the more manganese oxide can react to ferrite. That behavior is strongly dependent on the valency of the manganese oxide. $Mn_3O_4$ has a better hydrothermal reactivity then $MnO_2$. The content of residual $Mn_xO_y$ after the hydrothermal treatment influences the reoxidation behavior as well as the shrinkage behavior of the pressed powder in the temperature range of 400 - 800 °C (Fig. 2 and 3).

For practical evaluations it is necessary to know the internal pressure and the residence time of the ferrite batch within the autoclave. The internal pressure is dependent on the reaction temperature corresponding to the water-vapor-pressure.
No hydrothermal reaction is detectable for conditions below 150 °C. Only at higher temperatures franklinite and residual oxide are obtained (s. table 1). There were no hydroxides present; therefore no calcination step is necessary. The solubility of the different oxides in water is strongly temperature dependent which affects the composition of the franklinite formed. The higher the temperature, the higher the insertion of manganese oxide into the franklinite lattice. Considering the changing manganese content, a slight shifting of the typical x-ray franklinite peaks in the order of 0.05 ° become understandable. That can cause decreasing franklinite contents measured at increasing temperatures.

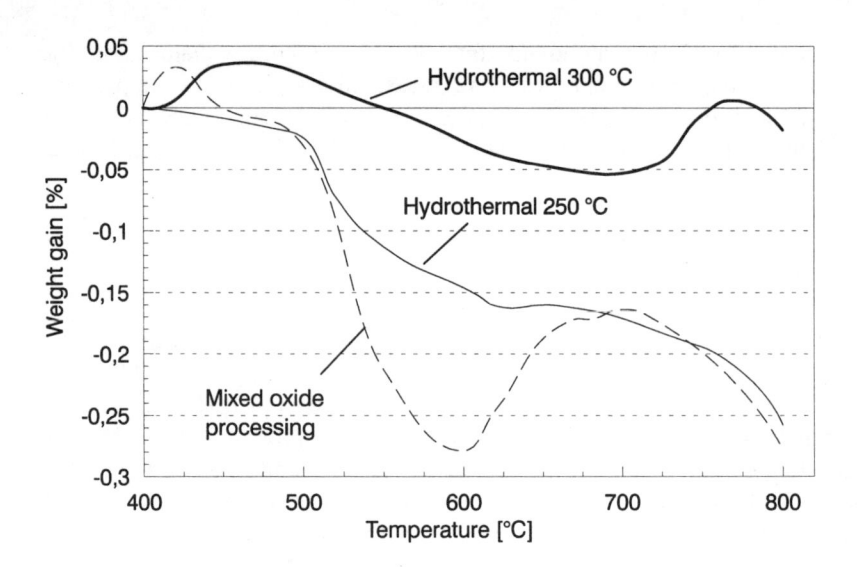

**Figure 2:** TGA of hydrothermal and conventional powders

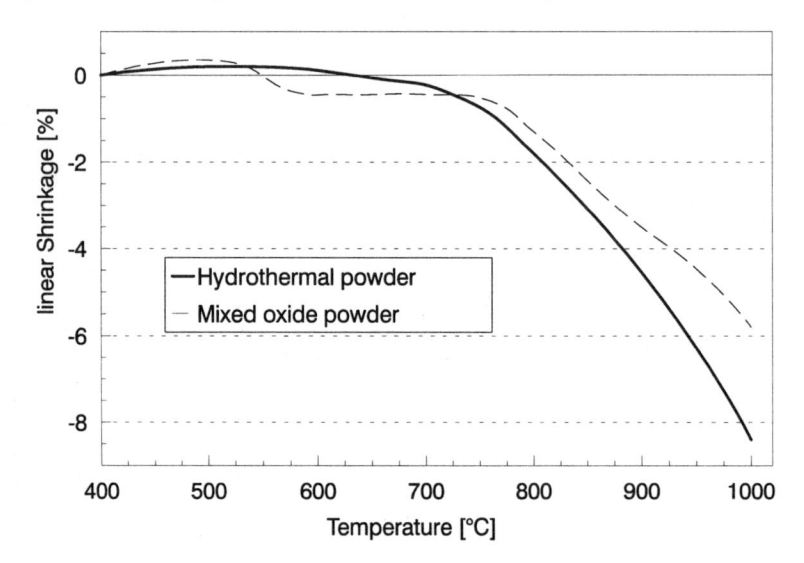

**Figure 3:** Shrinkage behavior of hydrothermal and conventional powders

**Table 1:** XRD phase composition depending on the hydrothermal reaction conditions

| Isothermal residence time [h] | 200 °C | 212 °C | 225 °C | 250 °C | 300 °C |
|---|---|---|---|---|---|
| 7 | | | | | Z 1<br>F 39<br>M 8<br>1.000 |
| 10 | | Z 11<br>F 69<br>M 9<br>0.071 | | Z 0<br>F 45<br>M 5<br>0.889 | Z 0<br>F 43<br>M 6<br>0.897 |
| 20 | Z 11<br>F 69<br>M 14<br>undetect. | Z 10<br>F 68<br>M 14<br>0.163 | Z 6<br>F 59<br>M 10<br>0.260 | Z 0<br>F 44<br>M 6<br>0.933 | Z 0<br>F 41<br>M 5<br>0.851 |
| 50 | Z 10<br>F 68<br>M 11<br>0.153 | Z 6<br>F 60<br>M 10<br>0.443 | Z <1<br>F 44<br>M 5<br>0.883 | Z 0<br>F 44<br>M 5<br>0.892 | |
| 72 | Z 0<br>F 68<br>M 12<br>0.198 | | Z 0<br>F 30<br>M 5<br>0.589 | | |

Conventional powder: Z 0    (Z = ZnO in Mol%)
(Reference)         F 40    (F = $Fe_2O_3$ in Mol%)
                        M 4      (M = $Mn_xO_y$ in Mol%)
                        0.779   (relative franklinite content RF)

$$RF = \frac{measured.\,franklinite}{franklinite.\,obtained.\,hydroth.\,at.\,300°C\,/\,7h}$$

The franklinite crystals have a grain size distribution between 20 and 200 nm (TEM). The appearance of the crystals is formed with respect to the cubic crystal system. The ferrite particles are almost perfectly grown because of low reaction temperatures and steadily available dissolved initial substances during the hydrothermal synthesis. However, the grain size distribution of the hydrothermal ferrite powders shows a clear shift to larger grain sizes with respect to the franklinite agglomerates and superposed residual oxides (Fig. 4). Therefore the subsequent processing of the hydrothermal ferrite powders requires well-defined conditions for desagglomeration and pressing powder preparation.

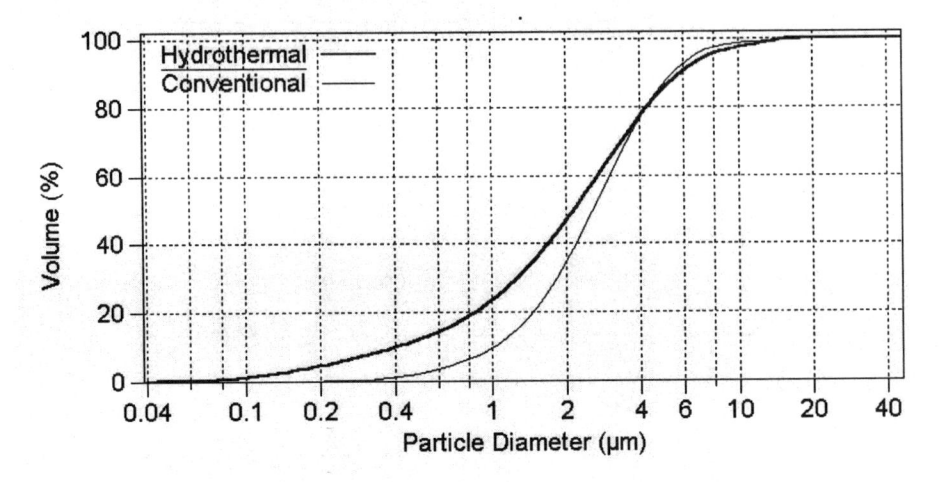

**Figure 4:** Grain size distribution of a hydrothermal compared to a conventional ferrite powder

The processing of the hydrothermal ferrite powders is considerably influenced by the fractions < 1 µm. In case of hydrothermal prepared powder they represent 75 % of the surface area and are consequently decisive for the agglomeration-, pressing- and sintering behavior.

After hydrothermal treatment the ferrite powder has the tendency to settle out and to form agglomerates. It can be prevent by slight stirring.

The adaptation to the available processes for shaping requires a pressable powder that can be achieved by different ways. The common process is spray drying after adding pressing aids to the slurry. It has to be carried out very carefully because hard agglomerates are formed otherwise. Nevertheless the pressing behavior is different to that of conventional powders. Fine-grained powders < 1 µm require adjusted organic pressing aids in order to keep the pressing forces low.

The sintering behavior is strongly dependent on the grain size distribution of the powder. Hydrothermal powders could be sintered at least 50 K lower temperature. This positively affects the Zn volatilization as well as the energy consumption of the sintering process.

The fine grained hydrothermal prepared powders have the tendency to form a very homogeneous microstructure of the sintered ferrite related to the porosity.

The grain size distribution is shifted to larger grain sizes under the same sintering conditions compared to a conventional processed powder (Fig. 5).

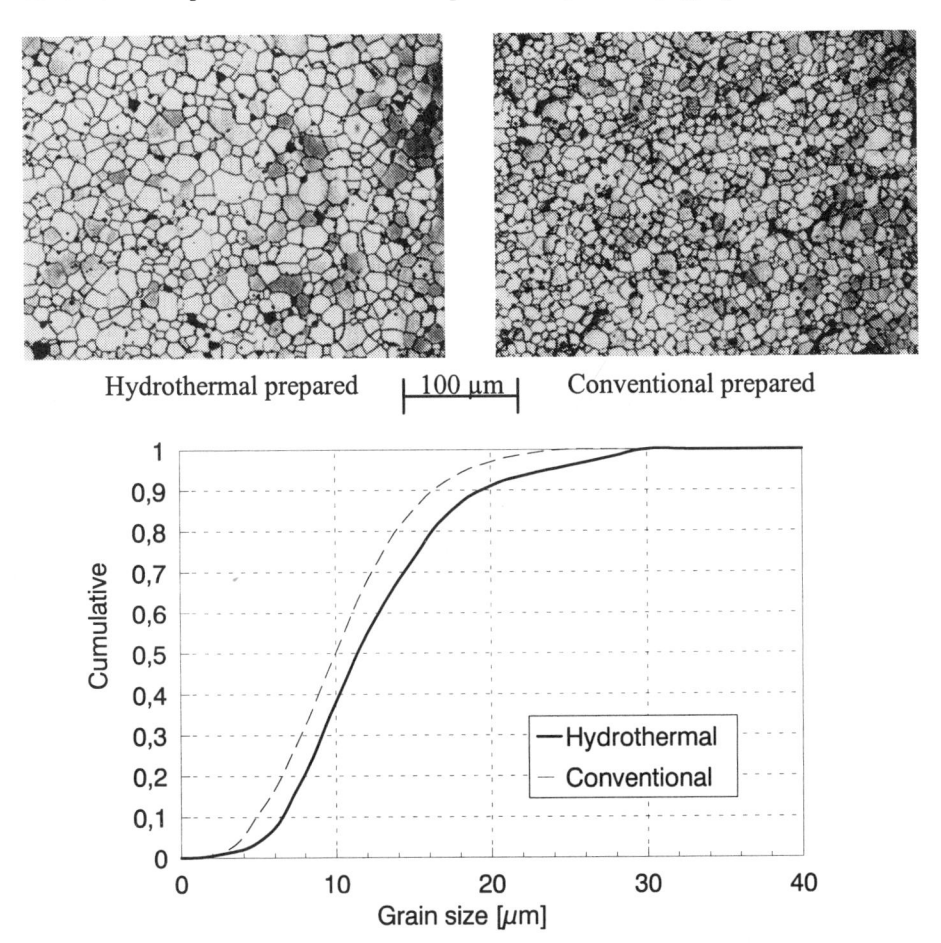

**Figure 5**: Microstructure and grain size distribution of ferrite cores with the same composition and different preparation (high permeability composition, 10 mm toroid)

Magnetic properties

In the case of the hydrothermal ferrites with high permeability material behavior a higher shrinkage at the same sintering temperature could be detected. Moreover the microstructure shows a coarser grain size distribution (Fig. 5). This is an interesting feature to obtain higher permeabilities. The differences in the microstructure correspond with different magnetic properties (Fig. 6). There is a 12 % higher initial permeability for the hydrothermal based toroids.

**Table 2:** Magnetic properties of the hydrothermal prepared high permeability material in comparison to the conventional produced material

|  | Symbol | Unit | Hydrothermal material | Conventional material |
|---|---|---|---|---|
| Initial permeability[a] (f = 10 kHz) | $\mu_i$ |  | 12,100 | 10,800 |
| Flux density (near saturation) (H = 1200 A/m)  25 °C     100 °C | $B_S$ | mT | 415  250 | 375  235 |
| Remanent flux density [a] | $B_r$ | mT | 255 | 220 |
| Coercive field strength [a] | $H_C$ | A/m | 12.5 | 13.0 |
| Relative loss factor [a] 10 kHz  1 MHz | $\tan \delta / \mu_i$ |  | $4.2 \cdot 10^{-6}$  $2.1 \cdot 10^{-3}$ | $4.7 \cdot 10^{-6}$  $2.0 \cdot 10^{-3}$ |
| Hysteresis material constant [a] | $\eta_B$ | $10^{-3}/mT$ | 0.24 | 0.28 |
| Density |  | kg/m³ | 4,950 | 4,840 |

a) T = 25°C

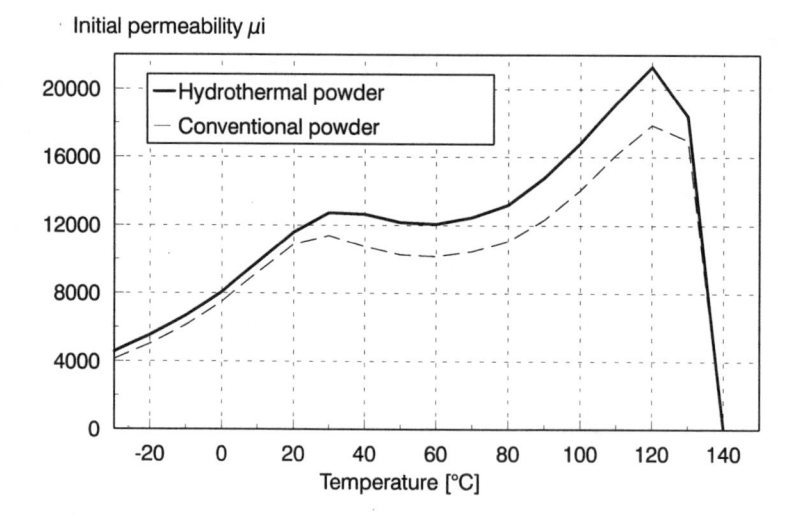

**Figure 6:** Temperature dependent permeability of toroids (∅ = 10 mm) made of hydrothermal and conventional powder

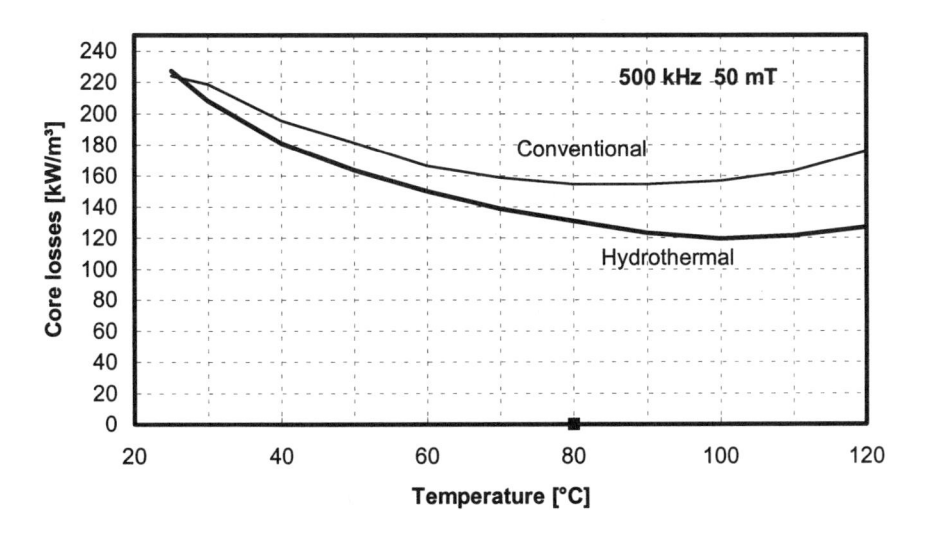

**Figure 7:** Power losses measured at 500 kHz for hydrothermal and conventional prepared ferrite powders (toroids ∅ = 29 mm)

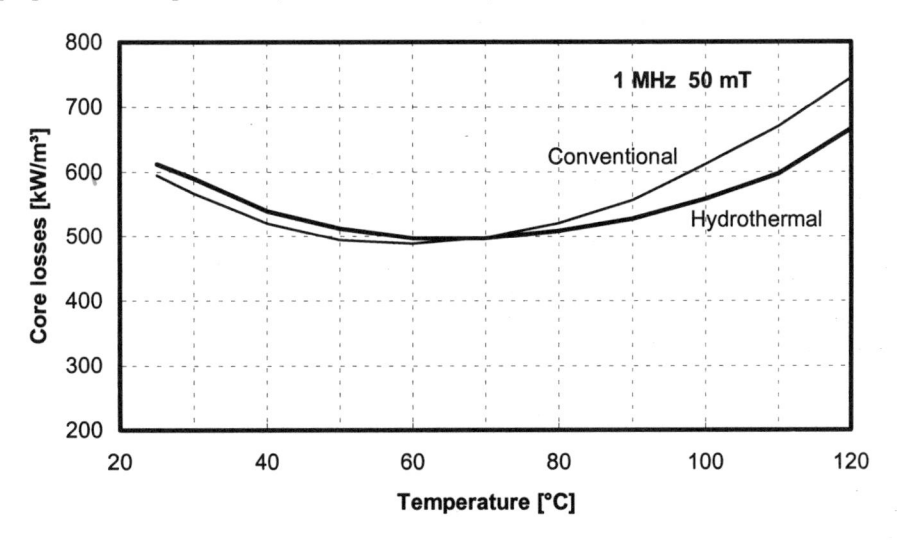

**Figure 8:** Power losses measured at 1 MHz for hydrothermal and conventional prepared ferrite powders (toroids ∅ = 29 mm)

Latest results of high frequency power transformer materials show lower losses with hydrothermal ferrites at frequencies ≥ 500 kHz. The relevance of decreased eddy current losses is responsible for 10 - 20 % lower power losses at 100 °C (see Fig. 7 and 8).

CONCLUSIONS

Manganese zinc ferrite powders were obtained from hydrothermal preparation of commercial oxidic raw materials. The subsequent handling of the nanoscaled powders has to be adjusted to the changed powder properties. Nevertheless the magnetic properties partially exceeded that of conventional produced ferrite cores. The microstructure is the key for improved magnetic properties. It can be optimized by using reactive powders with well-defined forming and sintering processes.

Considering the whole process, hydrothermal powder synthesis can provide the opportunity to overcome limits of the conventional processing.

REFERENCES

[1] G. Economos, "Solid Reactions in Ferrites"; pp. 243-254 in *Kinetics of High-Temperature Processes,* Edited by W. D. Kingery. J. Wiley & Sons, Inc., New York, 1959

[2] E. C. Snelling, "Soft ferrites" - 2nd ed.Butterworths, London, 1988

[3] L. Michalowsky, H. Baumgartner and W. Ernst, "Microstructure of Manganese-Zinc Ferrites", *Ceram. Int.* **19** 77-85 (1993).

[4] D. Stoppels, "Development in soft magnetic power ferrites", *J. Magn. & Magn.Mat,* **160** 323-328 (1996).

[5] R. Akiyama, "Low-Loss Ferrite Materials Hold Key to Future Power Supplies", *JEE* **7** 74-76 (1993).

[6] M. Bogs, M. Esguerra and W. Holubarsch, "New Ferrite Material for High Frequency",*Power Transformers. Proc. 26th Power Conversion Conf.*, pp. 361-370, Nuernberg, 1991.

[7] M. Bogs and W. Holubarsch, "Design Principles for High Frequency Power Transformer Materials", *J. Phys. IV France* **7** C1-117/118 (1997).

[8] H. Meuche, "Untersuchung des Zusammenhangs zwischen den Verlusten und dem Permeabilitätsspektrum von Leistungsferriten sowie deren Beeinflussung durch galvanische Ströme", Dissertation, University of Leipzig, 1998.

[9] E. Schlegel, "Grundlagen technischer hydrothermaler Prozesse", *Freiberger Forschungsheft* A655, Deutscher Verlag für Grundstoffindustrie, Leipzig, 1982

[10] D. L. Segal, "Chemical preparation of powders", pp. 69 - 98 in *Materials Science and Technology,* Vol. 17A, Processing of Ceramics, Edited by R. W. Cahn et. al

[11] R. Lucke, E. Schlegel and R. Strienitz, "Hydrothermal Preparation of Manganese Zinc Ferrites", *J. Phys. IV France* **7** C1-63/64 (1997).

[12] S. Komarneni, E. Fregeau, E. Breval etc., "Hydrothermal Preparation of Ultrafine Ferrites and Their Sintering", *J. Am. Ceram. Soc.* **71** [1] C26-28 (1988)

[13] M. Rozman and M. Drofenik, "Hydrothermal Synthesis of Manganese Zinc Ferrites", *J. Am. Ceram. Soc.* **78** [9] 2449-55 (1955)

# HYDROTHERMAL SYNTHESIS OF ZINC OXIDE PARTICLES

Chung-Hsin Lu, Yuan Cheng Lai, and Chi-Hsien Yeh
Department of Chemical Engineering,
National Taiwan University, Taipei, Taiwan, R.O.C.

## ABSTRACT

Zinc oxide particles have been prepared by a hydrothermal process using monoethanolamine (MEA) as the alkaline source. Reaction between zinc nitrate solutions and MEA at as low as 100°C results in crystallized zinc oxide powders. The concentration of MEA influences the precipitation state of zinc hydroxide as well as the subsequent crystallization processes of zinc oxide. At low concentrations of MEA, the formed particles exhibit an intertwined morphology; while, the high concentrations of MEA lead the shape of particles to become spheroidal. Adding ammonia in MEA during hydrothermal reaction causes ZnO particles to have an a-axis preferential orientation with a prismatic shape. The microstructure and crystal structure of ZnO particles are substantially influenced by the types and concentration of alkaline solutions used.

## INTRODUCTION

The properties of electronic and structural ceramics are significantly affected by the characteristics of the powders used in the fabrication processes. The ideal powders should possess the characteristics of well-controlled chemical composition, uniform morphology, fine grain size, and narrow size distribution. Using the conventional solid-state process hardly achieves the above ideal powders, especially for the oxides containing multiple cations. Various kinds of solution processes have been investigated to circumvent the problems encountered in solid-state reaction. Among the solution processing routes, recently the hydrothermal process has been proposed to be an effective method for synthesizing fine ceramic powders with a narrow size distribution and uniform particle morphology [1-7]. In addition, the hydrothermal process in general can sufficiently reduce the required temperature for preparing ceramic powders in light of enhanced reactivity in a closed solution system with a high autogeneous pressure. With the benefit of low-temperature processes, the required energy during heat-treatment is reduced, and the grain size of synthesized particles can be also decreased.

This reserach is aimed at the hydrothermal preparation of zinc oxide particles. ZnO is one type of important ceramic materials, and has been found to have diversified applications in electronic devices such as gas sensors, varistors, and transducers. Different routes such as precipitation [8-10], spray pyrolysis [11], and thermal decomposition [12] have been utilized for preparing ZnO powders; however, only a few studies have focused on the hydrothermal synthesis [13, 14]. In this study, ZnO particles were prepared by a hydrothermal process using monoethanolamine (MEA) as the alkaline source. The effects of the concentration of MEA on the variation of crystal structure and the morphology of ZnO particles were investigated. In addition, the influence of the mixed alkaline sources (MEA and ammonia) on the microstructure of the obtained powders was also examined.

## EXPERIMENTAL

Reagent grade zinc nitrate was adopted as the source material for zinc species. In the first part of the experiment, zinc nitrate was dissolved in distilled water and mixed with various concentrations of monoethanolamine (MEA). The mixed solutions were introduced into a teflon-lined autoclave apparatus, and heated at 100°C for 1 h. During the hydrothermal reaction, a mechanical stirrer was used with a rotation speed of 300 rpm. After heating, the hydrothermal products were cooled to room temperature, and the powders collected by filtration were washed with distilled water and dried. In the second part of the experiment, ammonia was added into the solution prepared in the above experiment to investigate the effect of mixed alkaline sources on the obtained powders. The following hydrothermal treatment was the same as described above. X-ray powder diffraction (XRD) was used to identify the compounds present in the obtained powders. Infrared spectroscopy (IR) was performed to examine the residual organic species remaining in powders. The morphology and the particle size of the powders were examined via scanning electron microscopy (SEM).

## RESULTS AND DISCUSSION

### Hydrothermal synthesis using MEA

Zinc nitrate solution was mixed with various amounts of MEA. The concentration of zinc nitrate was adjusted to be 0.1 M, and that of MEA was varied from 0.5 M to 2.0 M. When [MEA] was 0.5 M and 1.0 M, a large amount of precipitates were formed immediately during the mixing of two solutions. In the above two mixed solutions, the pH values after mixing were 9.9 and 10.4, respectively. As [MEA] increased to 1.5 M, the amount of precipitates markedly decreased. At [MEA] = 2 M, a clear solution without any precipitates was obtained, meanwhile the pH value was increased to 10.8. The precipitation state of the above solutions predominantly depended on the pH value. At [MEA] = 0.5 M and 1.0 M, the hydroxyl groups in MEA reacted with zinc cations to produce zinc hydroxide as the precipitates. Since zinc hydroxide is an amphoteric species, further increasing [MEA] results in an increase in pH value, thereby dissolving the formed precipitates. At [MEA] equal to 2 M, no precipitates but only clear solution was obtained.

The mixed solutions with or without precipitates were hydrothermally heated at 100°C for 1 h, and the XRD results of as-prepared powders are shown in Fig. 1. As shown in this figure, all specimens exhibited well-developed crystallinity, indicating that zinc oxide powders were directly prepared by the hydrothermal process. All diffraction peaks were assigned to ZnO as reported in JCPDS file: 36-1451 [15], revealing that the produced powder was monophasic zincite with a hexagonal structure. It was also noted that the intensity of the diffraction patterns for the powders prepared at [MEA] equal to 0.5 M, 1.0 M, and 1.5 M were similar, but different from that at [MEA] = 2 M. The relative intensity of (100) peak to that of (002) peak became increased in the specimen prepared at [MEA] = 2 M.

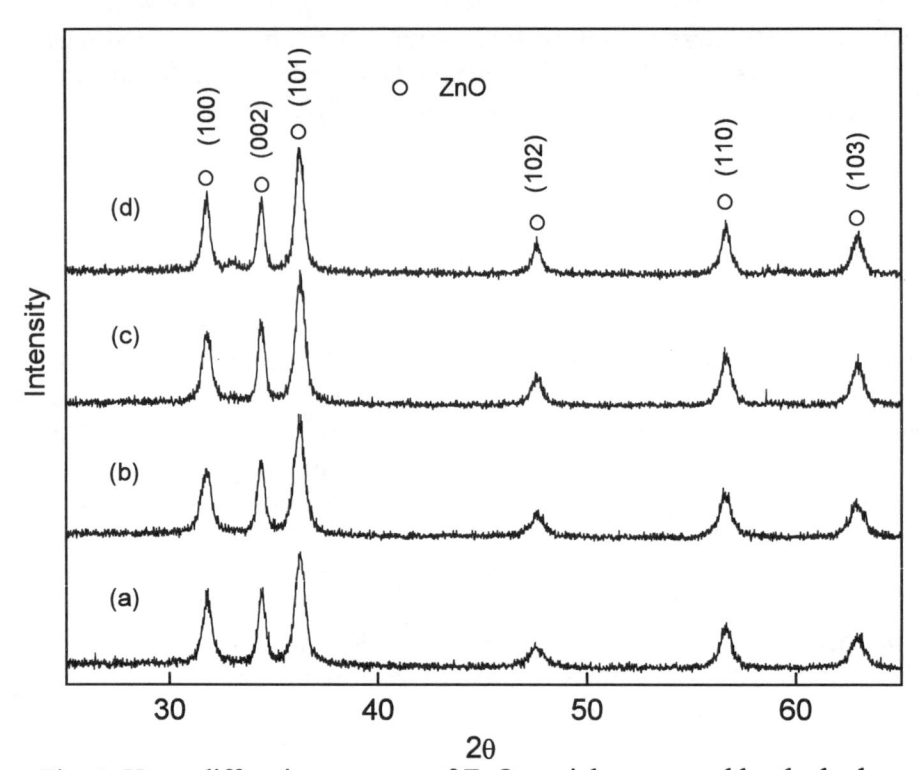

Fig. 1. X-ray diffraction patterns of ZnO particles prepared by the hydrothermal process at [$Zn^{2+}$] = 0.1 M. [MEA] = (a) 0.5 M, (b) 1.0 M, (c) 1.5 M, and (d) 2 M.

The microstructure of the above prepared ZnO powders was examined by SEM. The powders prepared at [MEA] = 0.5 M exhibited an intertwined morphology as shown in Fig. 2 (a). At higher magnification (Fig. 2 (b)), it was found that these powders were constituted by a large amount of fine primary particles of around 0.1 μm. As shown in Fig. 3 (a), at [MEA] = 1 M, the particles maintained the similar shape, but the particle size slightly increased. At [MEA] = 1.5 M, the morphology of ZnO particles was greatly varied (see Fig. 3 (b)). Two types of particles were observed: one looked like a hollow donut consisting of plate-like particles, and the other was spheroidal. When [MEA] increased to 2 M, only spheroidal particles of around 1.8 μm were produced (see Fig. 3 (c)). The various morphology of the obtained ZnO powders were strongly related to the precipitation state prior to the hydrothermal treatment. Under the conditions of [MEA] equal to 0.5M and 1.0 M, zinc hydroxide precipitated in the solutions and acted as nuclei for the subsequent formation of zinc oxide during the hydrothermal treatment. On the other hand, at [MEA] = 2 M, no nuclei were present in the initial stage of reactions. Therefore, the mechanisms of nucleation in the above two situations appeared to be different. The homogeneous nucleation in the latter case tended to facilitate the formation of spheroidal particles which had a preferential growth in a-axis as indicated in Fig. 1. In addition, it was found that the morphology of ZnO powders prepared by using MEA in the hydrothermal process was significantly different from that by using triethanolamine [14].

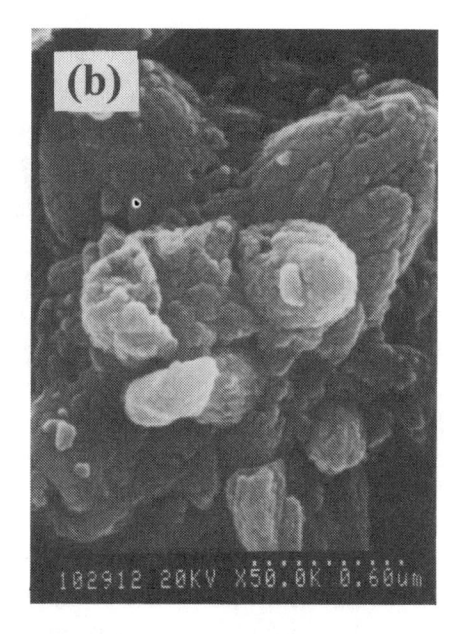

Fig. 2. Scanning electron micrographs of ZnO particles prepared by the hydrothermal process at $[Zn^{2+}] = 0.1$ M and [MEA] = 0.5 M. (b) is the magnification of (a).

Fig. 3. Scanning electron micrographs of ZnO particles prepared by the hydrothermal process at [$Zn^{2+}$] = 0.1 M. [MEA] = (a) 1.0 M, (b) 1.5 M, and (c) 2 M.

The IR spectra of the obtained ZnO powders are shown in Fig. 4. The characteristic adsorption peaks of ZnO were detected at 400 cm$^{-1}$ in all specimens. The hydroxyl groups were also found at around 3500 cm$^{-1}$. In addition, with a rise in [MEA], the amounts of residual amine groups from MEA appearing at around 1400 and 1500 cm$^{-1}$ were increased. These residues could be completely removed after 400°C-calcination.

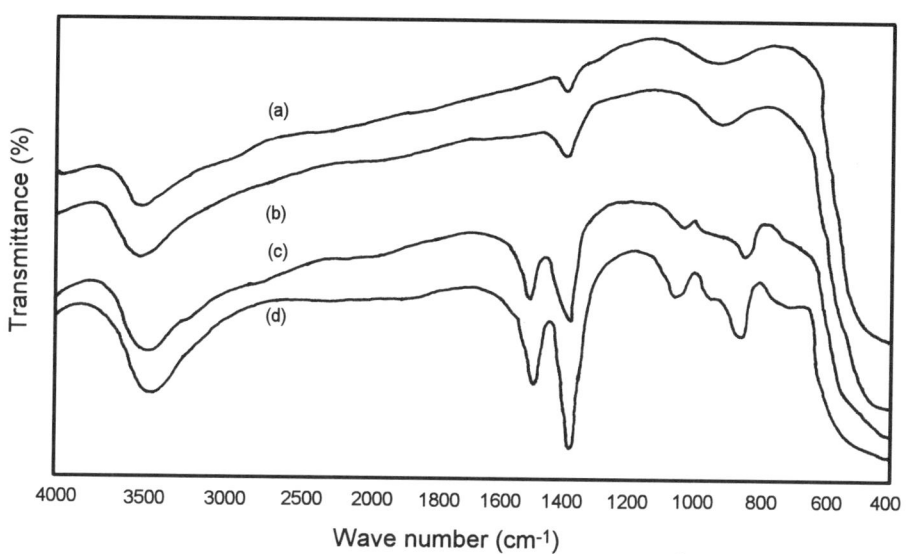

Fig. 4. Infrared spectra of ZnO particles prepared at [$Zn^{2+}$] = 0.1 M. [MEA] = (a) 0.5 M, (b) 1.0 M, (c) 1.5 M, and (d) 2 M.

**Hydrothermal synthesis using mixed alkaline sources**

Zinc hydroxide precipitates were formed during mixing at [MEA] = 1 M. For obtaining a clear solution prior to the hydrothermal reaction, ammonia of 0.18 M was added into the above solution to increase the pH value. After hydrothermal treatment, ZnO powders were produced. The XRD pattern and the SEM photograph of yielded ZnO powders are shown in Fig. 5 (a) and 6 (a), respectively. The XRD data indicate that the formed particles exhibited a preferential growth in a-axis with a high ratio of the intensity of (100) peak to that of (200) peak. Furthermore, the morphology of the formed powders was substantially transformed into a prismatic shape, which was owing to the fact that the formed particles have a preferential orientation in a-axis. The addition of ammonia was found to have a profound effect on altering the orientation of crystal structure as well as the morphology of ZnO particles. A parallel experiment using only ammonia (1 M) as alkaline sources in the hydrothermal reaction was also carried out. The obtained powders were found to have the similar characteristics as those of the above powders (shown in Fig. 5 (b) and 6 (b)). These results revealed that the presence of ammonia in the hydrothermal reaction significantly influenced the crystal orientation and morphology of ZnO particles formed.

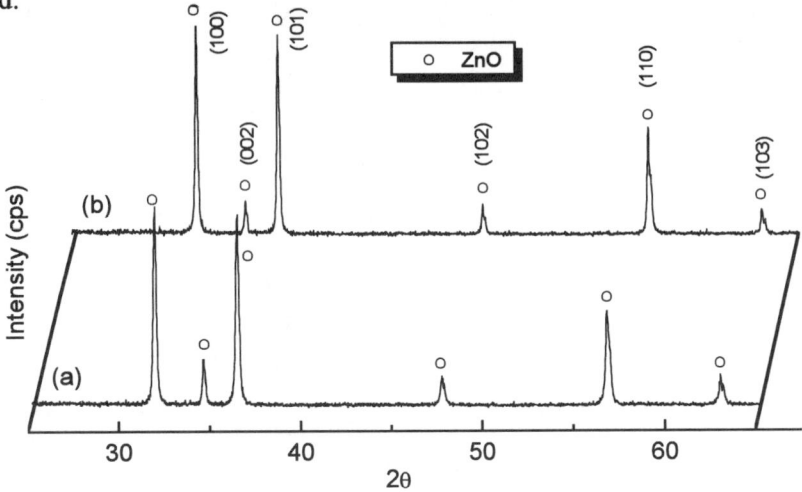

Fig. 5. X-ray diffraction patterns of ZnO particles prepared by the hydrothermal process at $[Zn^{2+}] = 0.1$ M. (a) [MEA] = 1 M and [NH$_4$OH] = 0.18 M, and (b) [NH$_4$OH] = 1 M.

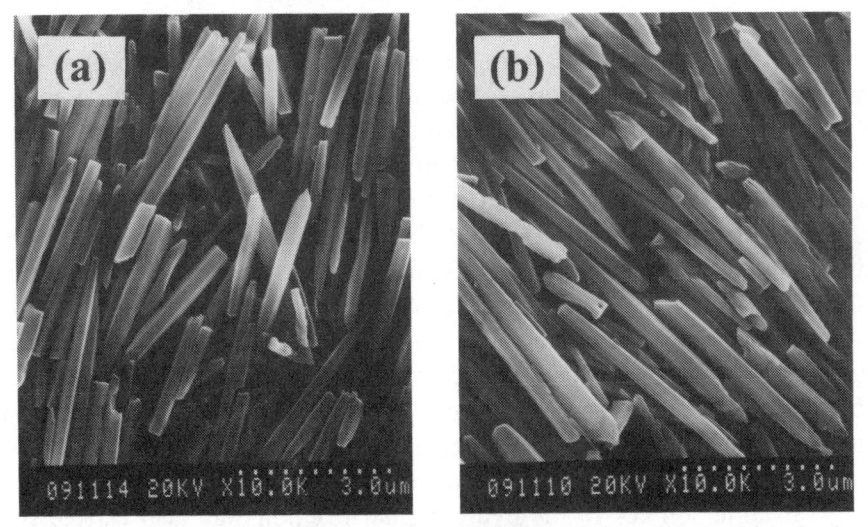

Fig. 6. Scanning electron micrographs of ZnO particles prepared by the hydrothermal process at $[Zn^{2+}] = 0.1$ M. (a) [MEA] = 1 M and [NH$_4$OH] = 0.18 M, and (b) [NH$_4$OH] = 1 M.

## CONCLUSIONS

The synthesis processes of ZnO particles have been investigated in this study by adopting the hydrothermal treatment. Crystallized zinc oxide particles have been successfully prepared at 100°C in hydrothermal reaction using monoethanolamine (MEA) as the base source. At low concentrations of MEA, the hydrothermally-prepared powders have an intertwined morphology; whereas, the morphology of ZnO particles becomes spheroidal at high concentrations of MEA. When ammonia is present in the hydrothermal reaction, the shape of ZnO particles is varied to be prismatic, and the formed particles exhibit a preferential orientation in a-axis. The types and concentrations of the alkaline sources used in the hydrothermal process have been found to play an important role in influencing the microstructure and crystal structure of ZnO particles.

## ACKNOWLEDGMENT

The authors would like to thank the National Science Council, Taiwan, the Republic of China, for financial support of this study under Contract No. NSC 86-2745-E002-002R.

## REFERENCES

1. T. R. N. Kutty, R. Vivekanandan, and S. Philp, "Precipitation of Ultrafine Powders of Zirconia Polymorphs and Their Conversion of $MZrO_3$ (M = Ba, Sr, Ca) by the Hydrothermal Method," *J. Mater. Sci.*, 25 [8] 3649-58 (1990).
2. G. A. Rossetti, Jr., D. J. Watson, R. E. Newnham, and J. H. Adair, "Kinetics of the Hydrothermal Crystallization of the Perovskite Lead Titanate," *J. Cryst. Growth*, 116 [3/4] 251-259 (1992).
3. Prabir K. Dutta and J. R. Gregg, "Hydrothermal Synthesis of Tetragonal Barium Titanate," *Chem. Mater.*, 4 [4] 843-846 (1992).
4. Y. C. Zhou and M. N. Rahaman, "Hydrothermal Synthesis and Sintering of Ultrafine $CeO_2$ Powders," *J. Mater. Res.*, 8 [7] 1680-86 (1993).
5. C. H. Lu and N. Chyi, "Fabrication of Fine Lead Metaniobate Powder Using Hydrothermal Processes," *Mater. Lett.*, 29 [1-3] 101-105 (1996).

6. C. H. Lu and S. Y. Lo, "Lead Pyronibate Pyrochlore Nanoparticles Synthesized via Hydrothermal Processing," *Mater. Res. Bull.*, 32 [3] 371-378 (1997).
7. C. H. Lu and W. J. Hwang, "Preparation of Pb(Zr, Ti)$O_3$-Pb(Ni$_{1/3}$-Nb$_{2/3}$)$O_3$ Solid Solution Powder from Hydrotthermally-treated Precursors," *Mater. Lett.*, 27 [4] 229-232 (1996).
8. S. M. Haile, D. W. Johnson, G. H. Wiseman, and H. K. Bowen, "Aqueous Precipitation of Spherical Zinc Oxide Powders for Varistor Applications," *J. Am. Ceram. Soc.*, 72 [10] 2004-2008 (1989).
9. M. E. V. Costa and J. L. Baptista, "Characterization of Zinc Oxide Powder Precipitated in the Presence of Alcohols and Amines," *J. Eur. Ceram. Soc.*, 11 [4] 275-281 (1993).
10. T. Trindade, J. D. Pedrosa de Jesus, and P. O'Brien, *J. Mater. Chem.*, "Preparation of Zinc Oxide and Zinc Sulfide Powders by Controlled Precipitationn from Aqueous Solution," 4 [10] 1611-1617 (1994).
11. T. Q. Liu, O. Sakurai, N. Mizutani, M. Kato, "Preparation of Spherical Fine ZnO Particles by the Spray Pyrolysis Method Using Ultrasonic Atomization Techniques," *J. Mater. Sci.*, 21 [10] 3698-3702 (1986).
12. M. Andres-Verges, M. Martinez-Gallego, "Spherical and Rod-like Zinc Oxide Microcrystals: Morphological Characterization and Microstructural Evolution with Temperature," *J. Mater. Sci.*, 27 [7] 3756-3762 (1992).
13. H. Nishizawa, T. Tani, and K. Matsuoka, "Crystal Growth of ZnO by Hydrothermal Decomposition of Zn-EDTA," *J. Am. Ceram. Soc.*, 67 [6] (1984) C-98-C-100.
14. A. Chittofrati and E. Matijevic, "Uniform Particles of Zinc Oxide of Different Morphologies," *Coll. Surf.*, 48 [1-3] 65-78 (1990).
15. Powder Diffraction File, Card No. 36-1451. Joint Committee on Powder Diffraction Standards, Swarthmore, PA.

# SYNTHESIS OF POTASSIUM NIOBATE CERAMIC POWDER UNDER HYDROTHERMAL CONDITIONS

Chung-Hsin Lu and Shih-Yen Lo
Department of Chemical Engineering,
National Taiwan University, Taipei, Taiwan, R.O.C.

## ABSTRACT

Crystallized potassium niobate (KNbO$_3$) powders have been directly synthesized using a hydrothermal process at temperature as low as 200°C. This temperature is 450°C lower than that required in the conventional solid-state reaction. For synthesizing KNbO$_3$, the concentration of KOH is found to be a crucial parameter, which has to be at least 8 M. The morphology of the as-hydrothermally prepared and calcined KNbO$_3$ powders significantly depends on the concentration of niobium cations used in the hydrothermal process. Decreasing the concentration of Nb$^{5+}$ solution tends to produce elongated particles and results in nonstoichiometry in composition.

## INTRODUCTION

Potassium niobate (KNbO$_3$), an important electro-optic material, exhibits a large bandgap, and its optical properties can be sensitively altered by the application of electric field. Because KNbO$_3$ possesses high nonlinear optical coefficients and excellent photorefractive properties, it has been widely applied in optical waveguides and holographic storage systems [1-3]. Although a large number of researchers have focused on the crystal growth and optical properties of KNbO$_3$, the study concerning the preparation of KNbO$_3$ powders is few. In the conventional solid-state reaction, thorough mixing of reactants and high-temperature heating for a prolonged period are necessary for preparing pure KNbO$_3$ powders [4]. For ameliorating the mixing state of reactants in solid-state reaction, solution synthesis such as precipitation process and alkoxide process has been investigated [5, 6]. However, these processes could not directly produce the crystalline phase of $KNbO_3$, and the subsequent calcination at elevated temperatures is still required.

For improving the drawbacks in the above processes, a hydrothermal process has been developed in this study for synthesizing $KNbO_3$ powders. The hydrothermal process has been recognized as an efficient route to yield ceramic powders at rather low temperatures [7-14]. In addition, the formed powders exhibit a well-developed crystal structure; therefore, the processes of high-temperature calcination for the prepared powders are not necessary. This work describes the hydrothermal preparation of $KNbO_3$ powders using various concentrations of constituent species. The conditions for obtaining $KNbO_3$ in the hydrothermal reaction are first determined. Then, the effects of the concentration of niobium cations on the morphology and particle size of $KNbO_3$ powders are investigated.

**EXPERIMENTAL**

Reagent grade potassium hydroxide and niobium hydrogenoxalate were used as raw materials which were first individually dissolved in distilled water. Then the aqueous solutions containing various cation concentrations were mixed together, and were heated in a teflon-lined autoclave apparatus at 200°C for 2 h. A mechanical stirrer with a rotation speed of 200 rpm was used during the hydrothermal reaction. After washing and drying, the hydrothermally prepared powders were obtained. On the other hand, in solid-state reaction, potassium hydroxide and niobium oxide ($Nb_2O_5$) were weighed in a molar ratio equal to 2: 1 and ball-milled for 48h in ethyl alcohol, using zirconia balls in a polyethylene jar. The mixed reactants were calcined at temperatures ranging from 400°C to 650°C. X-ray powder diffraction (XRD) was used to identify the compounds present in the powders prepared by both processes. Infrared spectroscopy (IR) and thermal analysis (DTA and TGA) were performed for detecting the residual organic species present in powders. The microstructural evolution and the particle size of $KNbO_3$ powders were examined via scanning electron microscopy (SEM). The composition of the particles were analyzed by energy dispersive X-ray spectroscopy (EDS).

## RESULTS AND DISCUSSION
### Hydrothermal Preparation

Potassium hydroxide and niobium cation solutions with various concentrations were hydrothermally heated at 200°C for 2h. The XRD results of the as-obtained products are summarized in Fig. 1. As seen in this figure, three different regions are classified. In region A, no precipitates or powders are formed, indicating that no occurrence of any reactions. In region B, a small amount of crystallized $KNbO_3$ is formed. It is found that raising [KOH] to 6 M at $[Nb^{5+}]$ = 0.01 M results in the initial formation of $KNbO_3$. When $[Nb^{5+}]$ reaches 0.1 M, the required concentration of KOH for forming $KNbO_3$ is reduced to 2 M. On the other hand, in region C when [KOH] is as high as 8 M, a large amount of $KNbO_3$ particles is generated. The above results reveal that the concentration of KOH is a critical controlling parameter for preparing $KNbO_3$. Since sufficient amounts of $KNbO_3$ powders for further analysis can only be obtained at [KOH] = 8 M, the following study focuses on the powders prepared under this condition.

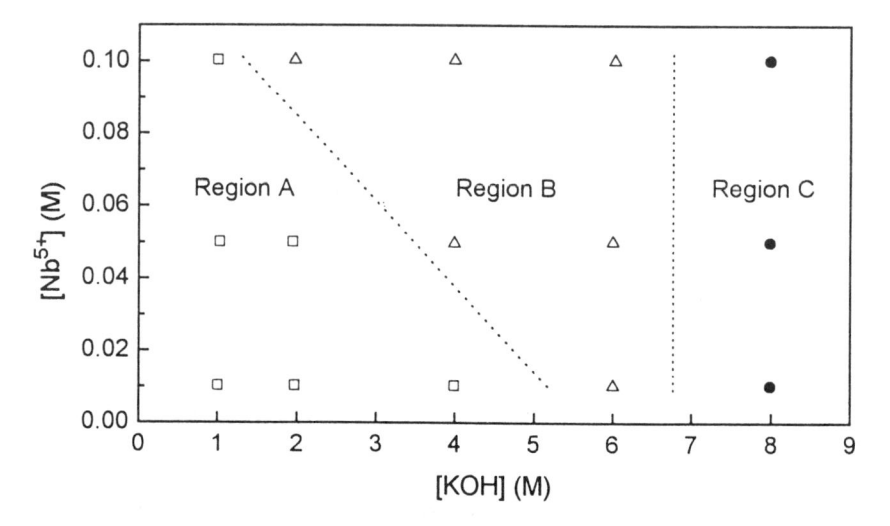

**Fig. 1.** Formation diagram of $KNbO_3$ in the hydrothermal process.

The XRD diffractograms of the specimens prepared under [KOH] = 8 M with various [Nb$^{5+}$] are illustrated in Fig. 2. As shown in this figure, monophasic KNbO$_3$ without any other residual compounds is obtained. All diffraction peaks are attributed to KNbO$_3$, revealing that KNbO$_3$ powders are successfully synthesized by the hydrothermal process at temperature as low as 200°C. It is noted that increasing the concentration of niobium cation slightly enhances the crystallinity of KNbO$_3$ powders, and the obtained powders exhibit an orthorhombic structure. The obtained powders were further analyzed by DTA and TGA. After heating up to 1000°C, no weight loss and thermal anomalies were found during thermal analysis. IR analysis also reveals that only the characteristics adsorption band of KNbO$_3$ appeared at around 650 cm$^{-1}$. These results indicate that no other organic species or water are present in the KNbO$_3$ powders; in other words, pure crystallized KNbO$_3$ powders were obtained through the developed hydrothermal process.

In a parallel experiment, KNbO$_3$ was also synthesized by the conventional solid state reaction. After Nb$_2$O$_5$ and KOH were mixed by ball-milling, the mixtures were calcined at various temperatures for 2 h. As shown in Fig. 3,

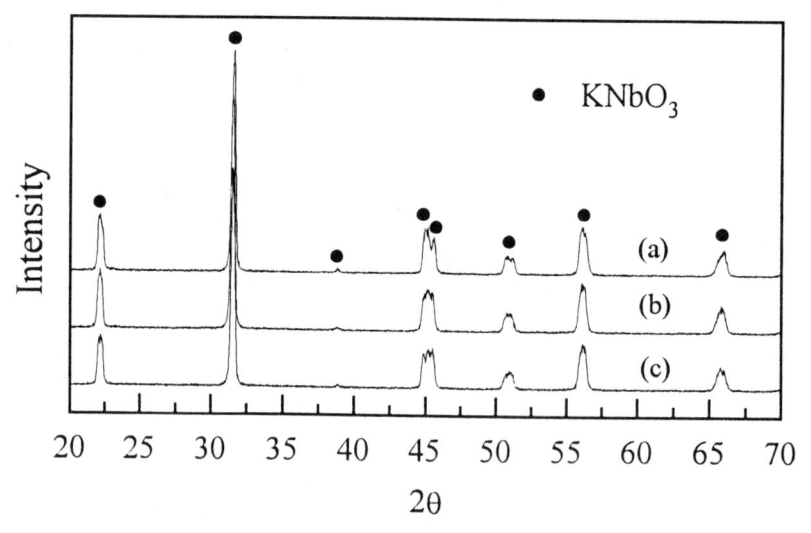

Fig. 2. X-ray diffraction patterns of the hydrothermally prepared KNbO$_3$ powders at [KOH] = 8 M. [Nb$^{5+}$] = (a) 0.01 M, (b) 0.05 M, and (c) 0.1 M.

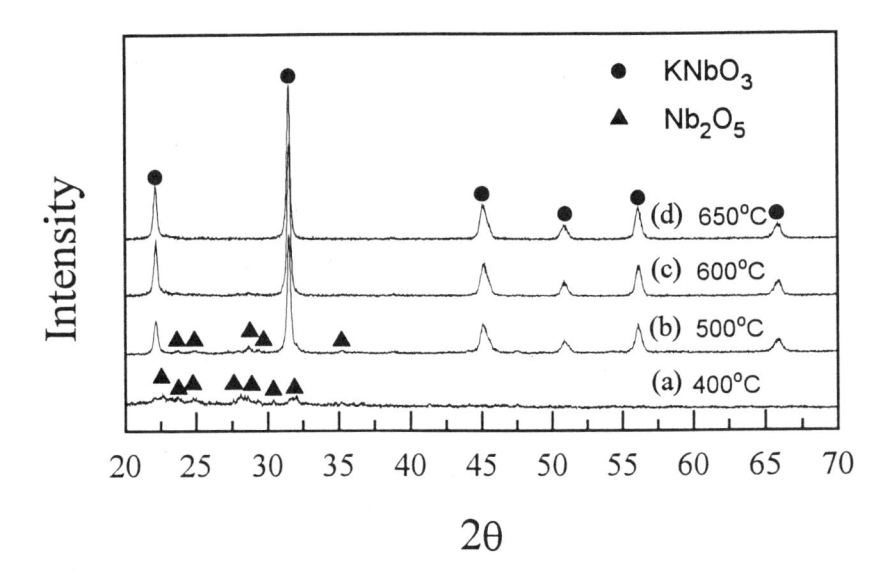

**Fig. 3.** X-ray diffraction patterns of the mixed reactants heated at various temperatures in solid-state reaction for synthesizing KNbO$_3$.

no formation of KNbO$_3$ is found at 400°C. After raising the temperature to 500°C, a small amount of KNbO$_3$ begins to appear. On 650°C-calcination, the formation of KNbO$_3$ is complete. Compared with solid-state reaction and other solution processes [5, 6], the hydrothermal process is clearly confirmed to significantly facilitate the formation process of KNbO$_3$, by reducing the synthesis temperature for obtaining KNbO$_3$ to be as low as 200°C. The low-temperature synthesis of KNbO$_3$ is attributed to the enhanced reactivity between the constituent species under hydrothermal conditions.

**Microstructure**

The microstructures of the hydrothermally prepared KNbO$_3$ powders are shown in Fig. 4. This figure indicates that the morphology of the obtained

KNbO$_3$ particles greatly depends on the concentration of niobium cations. At [Nb$^{5+}$] equals to 0.01 M, the KNbO$_3$ particles exhibit two types of morphology: a cubic shape and an elongated shape with a high aspect ratio. As [Nb$^{5+}$] increases, the number of the particles with an elongated shape is significantly reduced, and the particles become highly aggregated. In addition, the size of the primary particles is found to increase from 0.5-0.8 µm at [Nb$^{5+}$] = 0.01 M to 0.9-1.8 µm at [Nb$^{5+}$] = 0.1 M. From the EDS analysis, the particles with an elongated shape have a potassium-rich composition (K: Nb = 1.22: 1). On the contrary, the composition of other particles is near stoichiometric. The excess amount of potassium is probably ascribable to the low concentration of niobium cations used in the hydrothermal process.

Fig. 4. Scanning electron micrographs of the as-hydrothermally prepared KNbO$_3$ powders at [KOH] = 8 M. [Nb$^{5+}$] = (a) 0.01 M, (b) 0.05 M, and (c) 0.1 M.

## Heat Treatment

The as-hydrothermally prepared powders were further calcined at elevated temperatures for 2 h. Regardless of the concentration of niobium solution used, with a rise in heat treatment temperature, the orthorhombic structure of all three specimens becomes well-developed and the split of the diffraction peaks (220) and (002) becomes more obvious at high temperatures. The representative XRD patterns showing the evolution of crystal structure of the specimen with $[Nb^{5+}] = 0.1$ M are illustrated in Fig. 5. The microstructures of the 1000°C-calcined specimens are shown in Fig. 6. This figure indicates that these three specimens have different types of morphology. The elongated shape of as-prepared $KNbO_3$ particles under $[Nb^{5+}] = 0.01$ M becomes near equal-axial shape, and the particle size increases to 1-2 μm. The size of $KNbO_3$ particles prepared at $[Nb^{5+}] = 0.1$ M shows a large increase to 2-4 μm, demonstrating the most significant coarsening behavior. On the other hand, the size of $KNbO_3$ prepared by solid-state reaction at 1000°C is 3-6 μm which is larger than that prepared by hydrothermal processing. Consequently, the developed hydrothermal process developed is confirmed to successfully synthesize $KNbO_3$ powders with small particle size at low temperature. In addition, the morphology of the as-prepared and calcined $KNbO_3$ powders is significantly dependent on the concentration of $Nb^{5+}$ solution used in the hydrothermal process.

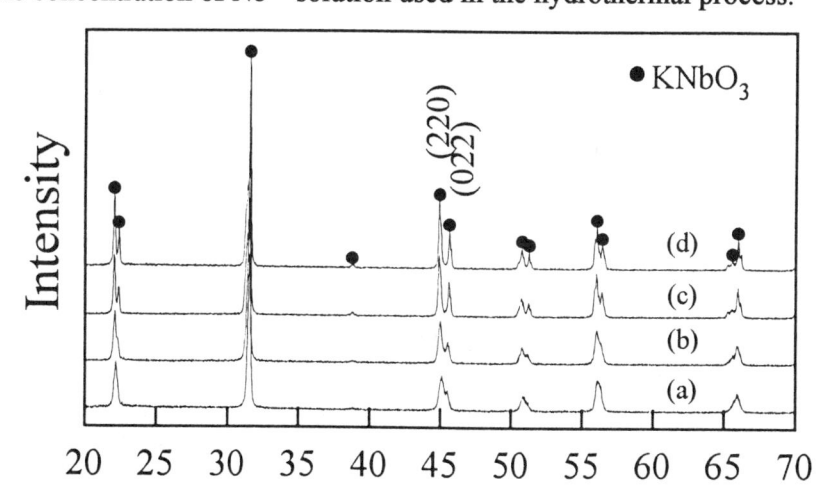

**Fig. 5.** XRD patterns of the hydrothermally prepared $KNbO_3$ ($[Nb^{5+}] = 0.1M$) heated at (a) 700°C, (b) 800°C, (c) 900°C, and (d) 1000 °C for 2 h.

**Fig. 6.** Scanning electron micrographs of hydrothermally prepared KNbO$_3$ powders heated at 1000°C. [Nb$^{5+}$] = (a) 0.01 M, (b) 0.05 M, and (c) 0.1 M.

## CONCLUSIONS

The preparation processes of KNbO$_3$ powders have been investigated in this study by using the hydrothermal process and the conventional solid-state reaction. The relation between the formation of KNbO$_3$ and the concentrations of KOH and niobium cations during the hydrothermal reaction has been studied. When [KOH] equals to 8 M, large amounts of crystallized KNbO$_3$ powders are successfully synthesized at temperature as low as 200°C. This temperature is remarkably lower than that required in solid-state reaction. This hydrothermal process is confirmed to have a profound effect on enhancing the formation of KNbO$_3$. The morphology and particle size of the as-prepared and calcined KNbO$_3$ powders are greatly influenced by the concentration of niobium cations used. Using low concentration of niobium cations results in the formation of KNbO$_3$ particles with an elongated shape and causes nonstoichiometry in composition.

## ACKNOWLEDGMENT

The authors would like to thank the National Science Council, Taiwan, the Republic of China, for financial support of this study under Contract No. NSC 85-2214-E002-014.

## REFERENCES

1. P. Gunter, "Holographic, Coherent Light Amplification and Optical Phase Conjugation with Photorefractive Materials," *Phys. Rep.*, 93 [4] 199-299 (1982).
2. M. K. Chun, L. Goldberg, and J. F. Weller, "Second-Harmonic Generation at 421 nm Using Injection-Locked GaAlAs Array and $KNbO_3$," *Appl. Phys. Lett.*, 53 [13] 1170-71 (1988).
3. B. A. Tuttle, "Electronic Ceramic Thin Films: Trends in Research and Development," Mater. Res. Soc. Bull., 12 [7] 40-45 (1987).
4. U. Fluckiger, H. Arend, and H. R. Oswald, "Synthesis of $KNbO_3$ Powder," *J. Am. Ceram. Soc.*, 56 [6] 575-577 (1977).
5. M. M. Amni and M. D. Sacks, "Synthesis of Potassium Niobate from Metal Alkoxides," *J. Am. Ceram. Soc.*, 74 [1] 53-59 (1991).
6. K. J. Kim and E. Matijevic, "Preparation and Characterization of Uniform Submicrometer Metal Niobate Particles: Part II. Magnesium Niobate and Potassium Niobate," *J. Mater. Res.*, 7 [4] 912-918 (1992).
7. S. I. Hirano, "Hydrothermal Processing of Ceramics," Am. Ceram. Soc. Bull., 66 [9] 1342-44 (1987).
8. T. R. N. Kutty, R. Vivekanandan, and S. Philp, "Precipitation of Ultrafine Powders of Zirconia Polymorphs and Their Conversion of $MZrO_3$ (M = Ba, Sr, Ca) by the Hydrothermal Method," *J. Mater. Sci.*, 25 [8] 3649-58 (1990).
9. G. A. Rossetti, Jr., D. J. Watson, R. E. Newnham, and J. H. Adair, "Kinetics of the Hydrothermal Crystallization of the Perovskite Lead Titanate," *J. Cryst. Growth*, 116 [3/4] 251-259 (1992).
10. Prabir K. Dutta and J. R. Gregg, "Hydrothermal Synthesis of Tetragonal Barium Titanate," *Chem. Mater.*, 4 [4] 843- 846 (1992).
11. Y. C. Zhou and M. N. Rahaman, "Hydrothermal Synthesis and Sintering of Ultrafine CeO2 Powders," J. Mater. Res., 8 [7] 1680-86 (1993).

12. H. Kumazawa, T. Kagimoto, and A. Kawabata, "Preparation of Barium Titanate Ultrafine Particles from Amorphous Titania by a Hydrothermal Method and Specific Dielectric Constants of Sintered Discs of the Prepared Particles," J. Mater. Sci., 31 [10] 2599-2602 (1996).
13. C. H. Lu and N. Chyi, "Fabrication of Fine Lead Metaniobate Powder Using Hydrothermal Processes," *Mater. Lett.*, 29 [1-3] 101-105 (1996).
14. C. H. Lu and S. Y. Lo, "Lead Pyronibate Pyrochlore Nanoparticles Synthesized via Hydrothermal Processing," *Mater. Res. Bull.*, 32 [3] 371-378 (1997).

# PROCESSING OF PURE AND Mn, Ni AND Zn DOPED FERRITE PARTICLES IN W/O MICROEMULSIONS

Doruk O. Yener and Herbert Giesche
NYS College of Ceramics, Alfred University
Alfred, NY 14802

## ABSTRACT

In recent years, the materials research focused towards finer and finer microstructural features. The unique properties of nanosized particles outweigh their higher production costs. Precipitation in microemulsion is one technique, which promises to produce small particles of controlled size and morphology at reasonable cost. In this study, the processing of nanosized pure and Mn, Ni or Zn doped ferrite particles was studied. The particles were synthesized in W/O microemulsions and the influences of the dopants on the morphological properties of the particles were examined. Ultra centrifugation was used to consolidate the particles.

## INTRODUCTION

The main goal of this experiment is to develop a new microemulsion synthesis technique and to demonstrate the consolidation of nanometer sized particles into dense and uniform green structures by centrifugation. Finally the designed particle microstructure should lead to an extremely fine grained sample.

The present study focuses on processing of Mn, Ni, or Zn doped ferrites, which have numerous practical applications in transformers, high frequency applications or data storage. Because of their ceramic nature, ferrites exhibit a much higher electrical resistivity than metals. They also use less expensive raw materials, which give them an economic advantage. In addition, electrons are scattered at the grain boundaries of nanostructured materials, which reduces eddy current losses in electrical applications. Therefore, these types of materials are well suited for cores in inductive components at high frequencies.

A limited but sufficient literature exits, describing the powder synthesis in microemulsions. With respect to the iron oxide nano-powders the work by

Kitahara et al.[1,2] or Ziolo et al.[3,4] should be mentioned. They produced $Fe_3O_4$ or $\gamma$-$Fe_2O_3$ powders approximately 5 to 10 nm in size. Kitahara et al.[1,2] used a microemulsion synthesis method. They synthesized monodispersed magnetite particles in the following systems: ($FeCl_3$ + $NH_3$) sol - AOT – isooctane, ($FeCl_3$ + $NH_3$) sol - AOT – isooctane, or ($FeCl_3$ + $FeCl_2$ + $NH_3$) sol - NP-6 - cyclohexane. Ziolo et al.[3,4] used a different precipitation technique, applying an ion exchange resin. Moreover several hydrothermal synthesis methods have been described in the literature[5] and frequently crystalline phases develop at higher temperatures whereas the room temperature products are amorphous.

A similar study to the one by Kitahara was done by Inouye et al.[6] and Chhabra et al.[7] Inouye studied the oxygenation of ferrous ions in microemulsions using ($Fe^{+2}$ + $NH_3$) sol - AOT - heptane and ($Fe^{+2}$ +$NH_3$) sol - AOT - cyclohexane systems. Chhabra used ferrous ammonium sulfate - Triton X-100 - n-hexanol - cyclohexane.

Wang et al.[8] used a microemulsion synthesis technique for the manganese zinc ferrite processing in which ferric nitrate, zinc nitrate and manganese nitrate sols. - NP5 & NP9 - cyclohexane was used as ingredients.

The present study uses a water - isooctane ($C_8H_{18}$) - Aerosol OT (sodium di-2-ethylhexylsulfosuccinate) system. Since only three components are present (no co-surfactant), this leads to a somewhat simpler phase diagram.

From a literature phase diagram of the given chemical system a composition was selected, which should form reverse micelles. AOT, isooctane, water mixtures as indicated in the figure below (fig. 1) were used for the experiments. The schematic illustration of the transition from microemulsion to lamellar liquid crystals due to a variation of the chemical composition can be seen in figure 1. The purpose of using AOT as the anionic surfactant is to obtain better steric stabilization due to its two non-polar tails and the polar head group (see fig. 2).

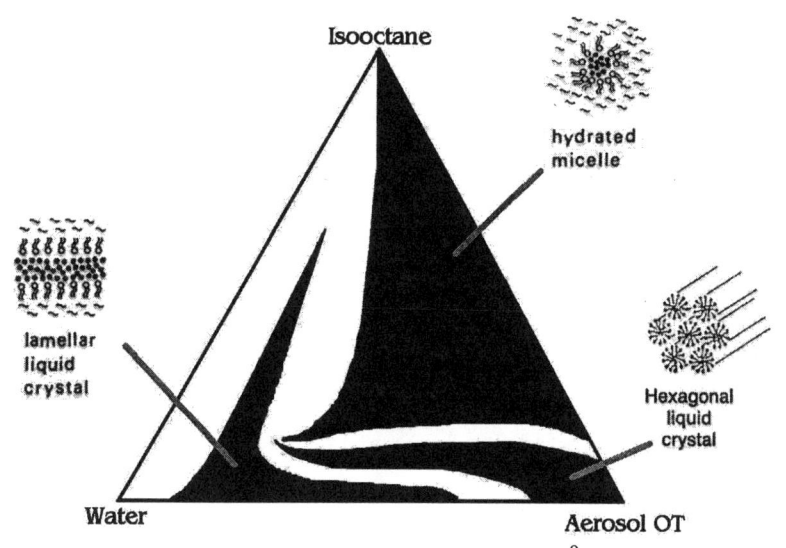

Fig. 1. Water-Isooctane-AOT phase diagram at 15°C[9]

Fig. 2. Structure of AOT

EXPERIMENTAL PROCEDURE:

Two sets of experiments were performed.

1) $Fe_2O_3$ synthesis and characterization
2) $Ni_{1-\delta}Zn_{\delta}Fe_2O_4$ and $Mn_{1-\delta}Zn_{\delta}Fe_2O_4$ synthesis and characterization

The chemicals used in the present study were: $Fe(NO_3)_3 \cdot 9H_2O$ (99.99%), $Mn(NO_3)_2 \cdot xH_2O$ (99.99%), $Ni(NO_3)_2 \cdot 6H_2O$ (99.999%), $Zn(NO_3)_2 \cdot xH_2O$ (99.999%), $NH_4OH$ (reagent grade), 2,2,4-Trimethylpentane (99.7%) [all chemicals from Aldrich, USA] and Aerosol OT (solid) [Fisher Scientific, USA]

## 1) $Fe_2O_3$ synthesis and characterization:

Two separate microemulsions were prepared in order to obtain submicron ferrite particles, as seen on the flow diagram (fig. 3).

During the synthesis, effects of different concentrations of cation and anion sources (Table 1), on the particle size and morphology were examined. The isooctane-AOT ratio was kept constant and the experiments were carried out at 15°C and 25°C, respectively.

Table 1. Anion and cation concentrations, as used in the different experiments

| $Fe_3(NO_3)_3$ | $NH_3$ |
|---|---|
| 0.7 M | 6.75 M |
| 1.4 M | 13.5 M |
| 1.76 M | 17 M |
| | |
| 0.7 M | 7.5 M |
| 1.4 M | 15 M |
| 1.6 M | 17 M |

## 2) $Ni_{1-\delta}Zn_\delta Fe_2O_4$ and $Mn_{1-\delta}Zn_\delta Fe_2O_4$ synthesis and characterization:

The $Ni_{1-\delta}Zn_\delta Fe_2O_4$ and $Mn_{1-\delta}Zn_\delta Fe_2O_4$, were synthesized by following the flow chart shown in figure 4.

During the experiments, 0.35 M $Ni(NO_3)_2$, 0.35 M $Zn(NO_3)_2$, and 1.4 M $Fe(NO_3)_3$ solutions in a microemulsion were mixed with 15 M $NH_3$ solution in another microemulsion for the $Ni_{0.5}Zn_{0.5}Fe_2O_4$ synthesis.

The same molar ratios of $Mn(NO_3)_2$, $Zn(NO_3)_2$, $Fe(NO_3)_3$ and $NH_3$ solutions were used during the synthesis of $Mn_{0.5}Zn_{0.5}Fe_2O_4$.

All samples, pure $Fe_2O_3$ as well as the doped $Ni_{0.5}Zn_{0.5}Fe_2O_4$ and $Mn_{0.5}Zn_{0.5}Fe_2O_4$, were calcined at 300°C, 450°C and 600°C and examined with XRD (Philips Norelco Type 42273/0 X-Ray Diffractometer, USA), TEM (JEOL 2000 FX, Japan) and submicron particle size analyzer (Coulter N4-MD, USA). Some of the powders were consolidated by centrifugation (Intertest E50 model 2001-E; equipped with a HB-6 rotor, USA) before calcination. Nitrogen adsorption provided information of the specific surface area and the porosity (<100 nm) in the samples (Coulter SA 3100$^+$, USA).

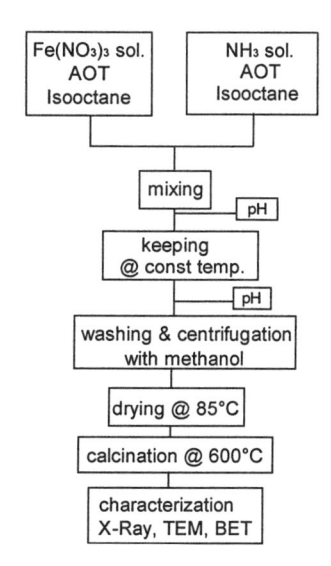

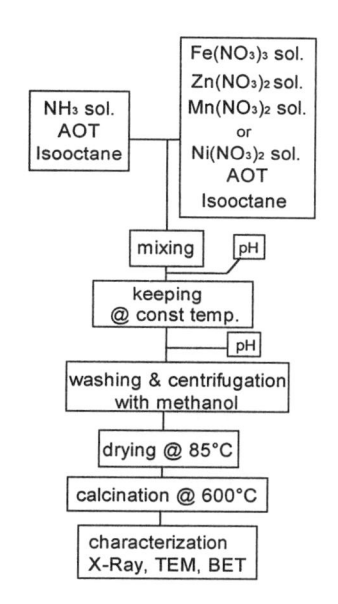

Fig. 3. Fe$_2$O$_3$ synthesis flow diagram

Fig 4. Ni$_{1-\delta}$Zn$_\delta$Fe$_2$O$_4$ and Mn$_{1-\delta}$Zn$_\delta$Fe$_2$O$_4$ synthesis flow diagram

RESULTS AND DISCUSSIONS:

*1) Fe$_2$O$_3$ synthesis and characterization:*

As explained in the experimental procedure, the synthesis was carried out at two different temperatures and with different chemical compositions of the microemulsion system.

The pH values were taken during these experiments, in order to follow the progress of the reactions inside the water droplets.

Since the solutions contained ~10% water, the pH data were not absolutely correct, but provided sufficient information about the trend of the reactions. The accuracy of these pH data were tested by measuring the pH values of different HCl and NH$_3$ solutions and then comparing these values with the ones that were obtained form the corresponding microemulsions. The observed difference for the basic range was about ± 0.6 and much larger for the acidic solutions. The present microemulsions showed pH values around 10 and thus, the measured pH could be taken as a rough guideline to monitor the progress of the reaction.

As seen from the pH-time plots (fig. 5), the solutions were equilibrated after about 6 days.

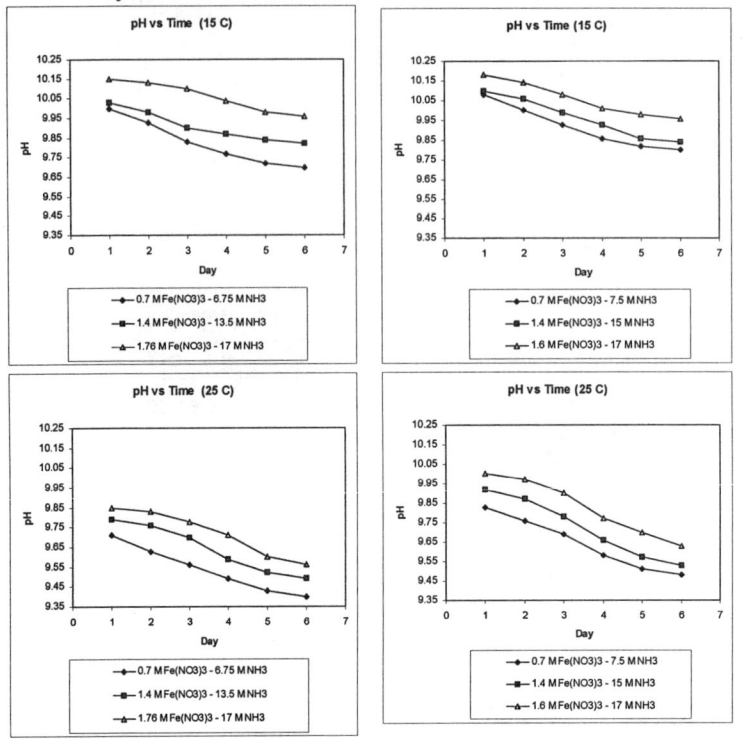

Fig. 5. pH as a function of time for different compositions.

The particle sizes of the samples were analyzed with a submicron particle size analyzer, which uses the photon correlation spectroscopy (see table 2).

Table 2. The average particle size at different reaction conditions

| Composition | T = 15°C | T = 25°C |
|---|---|---|
| 0.7 M Fe(NO$_3$)$_3$ - 6.75 M NH$_3$ | 135 nm | 140 nm |
| 1.4 M Fe(NO$_3$)$_3$ - 13.5 M NH$_3$ | 22 nm | 23 nm |
| 1.76 M Fe(NO$_3$)$_3$ - 17 M NH$_3$ | 59 nm | 62 nm |
| 0.7 M Fe(NO$_3$)$_3$ - 7.5 M NH$_3$ | 140 nm | 165 nm |
| 1.4 M Fe(NO$_3$)$_3$ - 15 M NH$_3$ | 34 nm | 50 nm |
| 1.587 M Fe(NO$_3$)$_3$ - 17 M NH$_3$ | 70 nm | 82 nm |

The sample, which was prepared at 25°C in the following microemulsion system (54 wt% AOT, 38 wt% isooctane, 8 wt% {1.4 M Fe(NO$_3$)$_3$, 13.5 M NH$_3$}) was selected for a more detailed characterization and calcination study.

The synthesized particles were characterized by XRD and TEM. The diffraction pattern of the crystallized phase matched perfect with the JCPDS PDF 33-0664 for hematite (syn) (fig. 6). The TEM pictures (fig.7) were obtained from samples centrifuged at room temperature, or subsequently calcined at 450°C and 600°C. From the peak width in the XRD spectra the crystallite size was calculated, and showed an average crystallite size of 67 nm for the same sample that was calcined at 600 °C, which compares reasonably well with the particle size of 40 to 85 nm as obtained from the TEM micrographs.

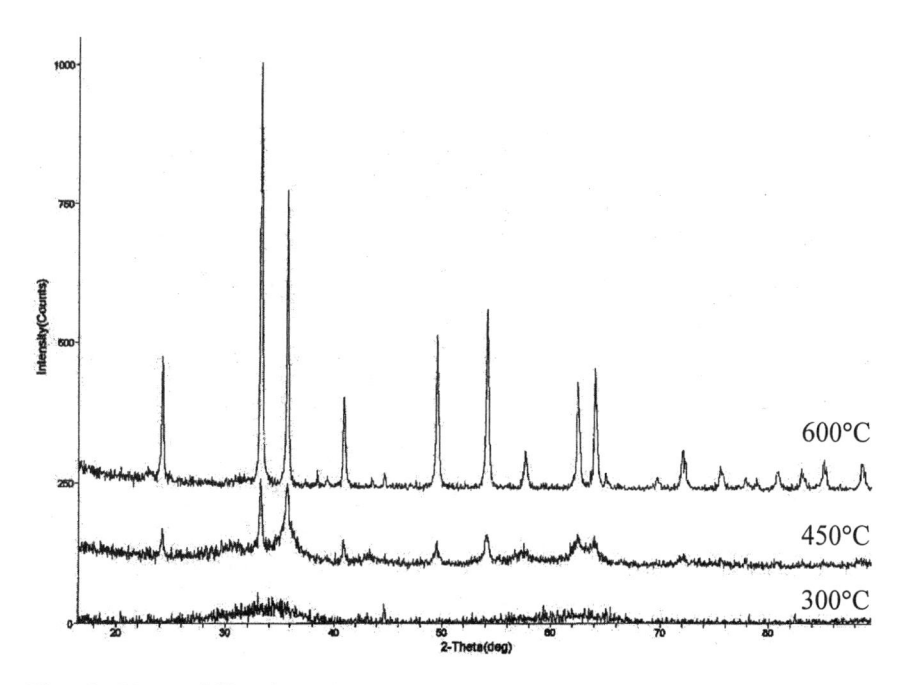

Fig. 6  X-ray diffraction plot showing the formation of Fe$_2$O$_3$ (Hematite) above 450°C [1.4 M Fe(NO$_3$)$_3$, 13.5 M NH$_3$, @ 25 °C]

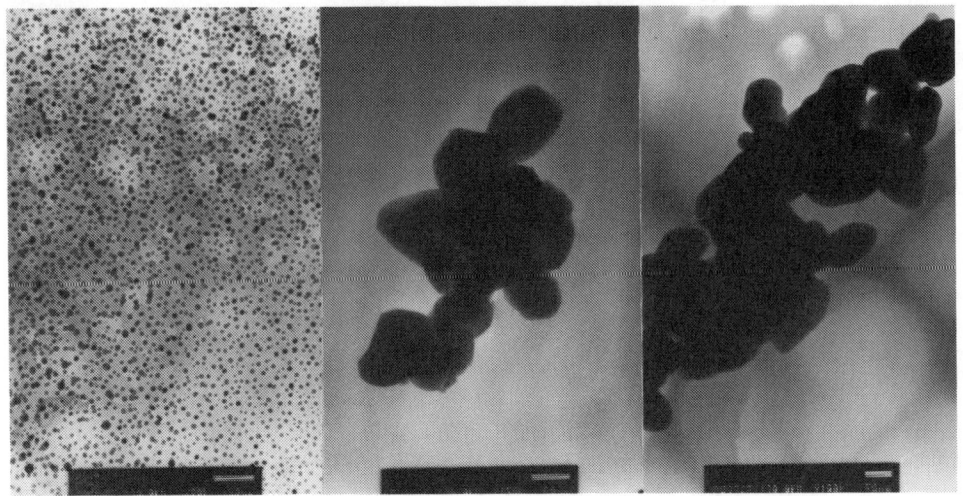

Fig. 7. TEM pictures of $Fe_2O_3$ particles. [1.4 M $Fe(NO_3)_3$, 13.5 M $NH_3$, @ 25°C] From left to right: without any calcination (bar = 100 nm), calcined @ 450°C for 6 hr. (bar = 50 nm), calcined @ 600°C for 6 hr. (bar = 50 nm)

Table 3. The particle size that is obtained form TEM pictures [1.4 M $Fe(NO_3)_3$, 13.5 M $NH_3$, @ 25°C]

| Calcination Temp. | Average particle size |
| --- | --- |
| Without calcination | 10 - 30 nm |
| 450°C, 6 hr | 35 - 70 nm |
| 600°C, 6 hr | 40 - 85 nm |

Moreover, several samples were compacted by centrifugation (6000 rpm, 5 min.) and then calcined at 600°C for 6 hr. The pore volume and pore size was analyzed by nitrogen adsorption. It showed a pore volume of 0.0104 $cm^3$/g and a pore size of 30 – 140 nm. Krypton adsorption at liquid nitrogen temperature indicated a specific surface area of 6.0 $m^2$/g.

## 2) $Ni_{1-\delta}Zn_{\delta}Fe_2O_4$ and $Mn_{1-\delta}Zn_{\delta}Fe_2O_4$ synthesis and characterization:

The synthesized $Ni_{0.5}Zn_{0.5}Fe_2O_4$ or $Mn_{0.5}Zn_{0.5}Fe_2O_4$, particles were characterized by x-ray, TEM and gas adsorption.

TEM images (fig. 8) showed that the particle size for $Ni_{0.5}Zn_{0.5}Fe_2O_4$ was between 5 – 30 nm and between 5 – 37 nm for $Mn_{0.5}Zn_{0.5}Fe_2O_4$.

Both samples were calcined at 300°C for 6 hr, 450°C for 6 hr and 600°C for 25 min. A longer calcination time of 6 hours at 600°C converted the sample completely to hematite. In the Ni$_{0.5}$Zn$_{0.5}$Fe$_2$O$_4$ sample, the diffraction pattern of the crystallized phase matched exactly with the JCPDS PDF 08-0234 (fig. 9). The diffraction pattern of the Mn$_{0.5}$Zn$_{0.5}$Fe$_2$O$_4$ also indicated the spinel structure (fig. 10).

The particles were compacted by centrifugation (6000 rpm, 5 min.) and calcined at 300°C, 450°C, and 600°C. Pore volume and pore size were analyzed by nitrogen gas adsorption. (table 4 and table 5)

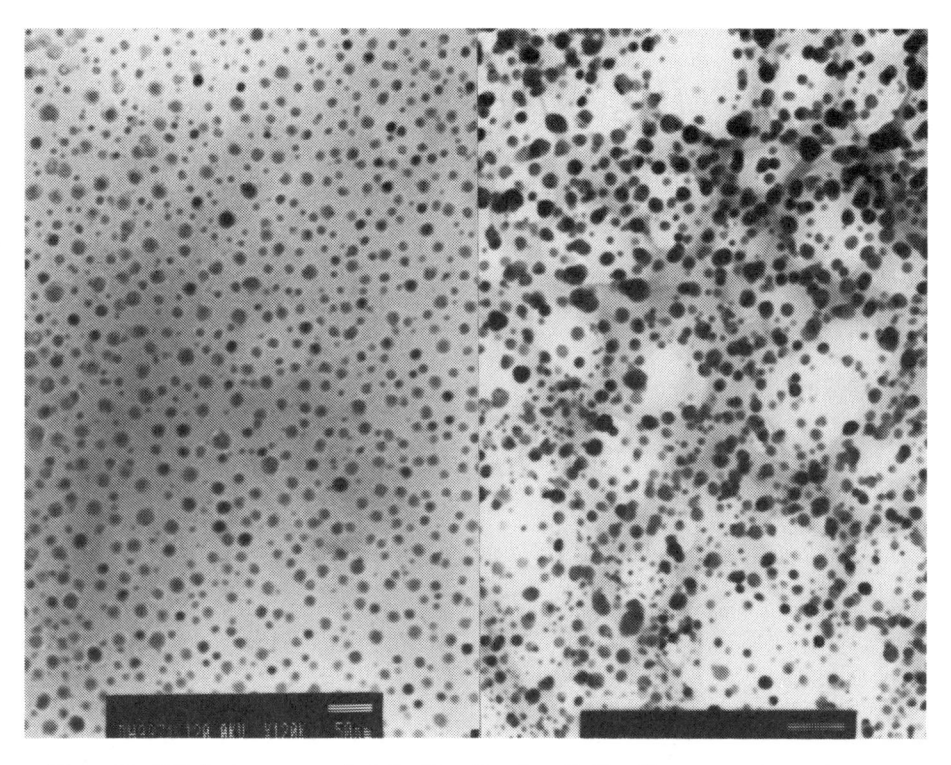

Fig. 8 TEM images of Ni$_{0.5}$Zn$_{0.5}$Fe$_2$O$_4$ (left) [bar = 50 nm] and Mn$_{0.5}$Zn$_{0.5}$Fe$_2$O$_4$ (right) [bar = 50nm]; original sample after washing with methanol at room temperature.

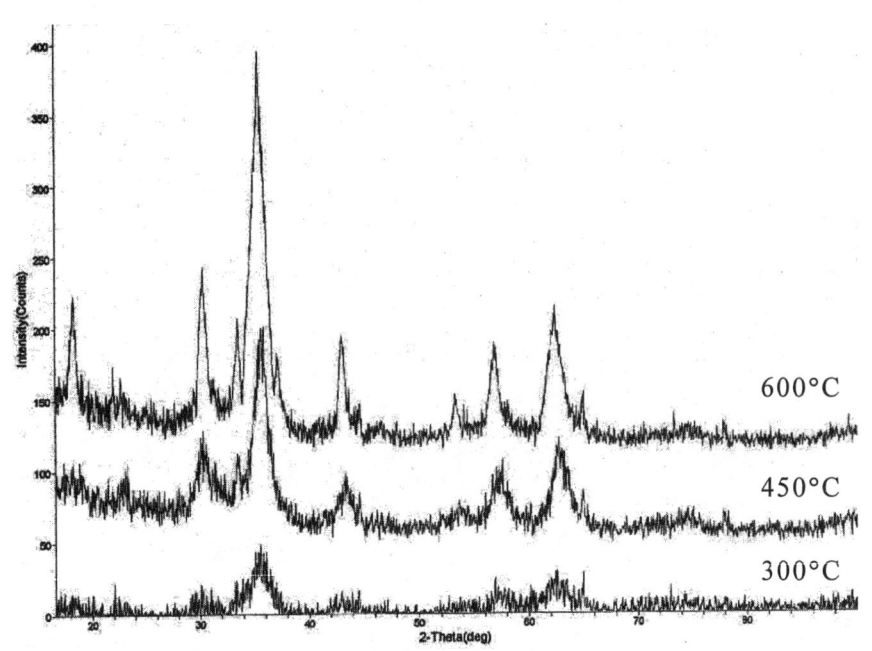

Fig. 9 X-ray diffraction pattern of $Ni_{0.5}Zn_{0.5}Fe_2O_4$

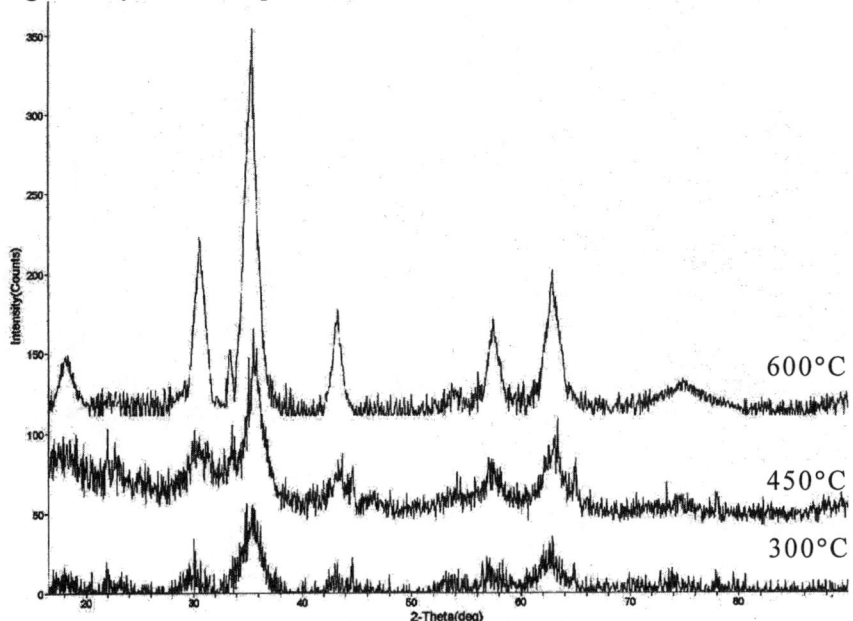

Fig. 10 X-ray diffraction pattern of $Mn_{0.5}Zn_{0.5}Fe_2O_4$

Table 4. Pore size and volume of the $Ni_{0.5}Zn_{0.5}Fe_2O_4$ sample

| Temperature (°C) | Pore Volume (cm³/g) | Average Pore Size (nm) |
|---|---|---|
| 300 (6 h) | 0.034 | 9 |
| 450 (6 h) | 0.123 | 14 |
| 600 (25 min.) | 0.061 | 10 |

Table 5. Pore size and volume of the $Mn_{0.5}Zn_{0.5}Fe_2O_4$ sample

| Temperature (°C) | Pore Volume (cm³/g) | Average Pore Size (nm) |
|---|---|---|
| 300 (6 h) | 0.045 | 9 |
| 450 (6 h) | 0.133 | 15 |
| 600 (25 min.) | 0.07 | 10 |

In both samples an increase in the pore volume was observed at 450°C. This could be a result of volatilization of remaining organics (AOT) in the sample even so it was washed several times with methanol. This can also be seen from the TGA plots. (fig. 11)

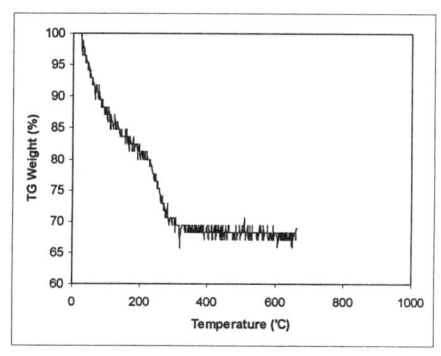

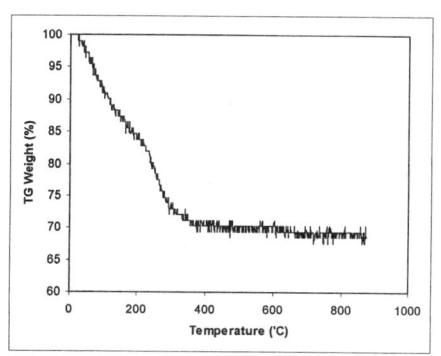

Fig. 11 TGA plots of $Ni_{0.5}Zn_{0.5}Fe_2O_4$ (left) and $Mn_{0.5}Zn_{0.5}Fe_2O_4$ (right)

CONCLUSIONS:

The present study demonstrated the synthesis of nanocrystalline $Fe_2O_3$, $Ni_{0.5}Zn_{0.5}Fe_2O_4$ and $Mn_{0.5}Zn_{0.5}Fe_2O_4$ particles in microemulsions. Nanosized particles of a relative uniform particle size distribution were obtained and extremely well dispersed and stable suspensions were observed.

The nanosized powders could be consolidated by centrifugation and dense green bodies were obtained.

Consolidation of these powders with the electrophoretic deposition technique and the magnetic characterization of these bodies will be studied in the near future.

ACKNOWLEDGMENTS

The support of Dr. Egon Matijevic and Dr. Andrei Zelenev during the particle size analysis at Clarkson University is acknowledged.

REFERENCES:

1. Masao Gobe, Kijiro Kon-no, Kazuhiko Kandori, Ayao Kitahara; "Preparation and characterization of monodisperse magnetite sols in W/0 microemulsion"; J. Colloid Interface Sci; **93** [1], 293-5 (1983)
2. Shunji Bandow, Keisaku Kkimura, Kijiro Konno, Ayao Kitahara; "Magnetic properties of magnetite ultrafine particles"; Japanese J. Apply. Phys.; **26** [5], 713-7 (1987)
3. Ronald F. Ziolo, Emmanuel P. Giannelis, Bernard A. Weinstein, Michael P. O'Horo, Bishwanath N. Ganguly, Vivek Mehrotra, Michael W. Russell, Donald R. Huffman; "Matrix-mediated synthesis of nanocrystalline $\gamma$-$Fe_2O_3$: a new optically transparent magnetic material"; Science, **257,** 219-23 (1992)
4. J. K. Vassiliou, Vivek Mehrotra, M. W. Russell, E. P. Giannelis, R. D. McMichael, R. D. Shull, R. F. Ziolo; "Magnetic and optical properties of $\gamma$-$Fe_2O_3$ nanocrystals"; J. Appl. Phys.; **73** [10], 5109-16 (1993)
5. Sridhar Komarneni, Elizabeth Fregau, Else Breval, Rustum Roy; "Hydrothermal preparation of ultrafine ferrites and their sintering"; J. Am. Ceram. Soc.; 71[1], C26-8 (1988)
6. K. Inouye, R. Endo, Y. Otsuka, K. Miyashiro, K. Kaneko, T. Ishakawa; "Oxygenation of ferrous ions in reversed microemulsion"; J. Phys. Chem.; **86** [8], 1465-69 (1982)
7. V. Chhabra, P. Ayyub, S. Chattopadhyay, A. N. Maitra; "Preparation of acicular $\gamma$-$Fe_2O_3$ particles from a microemulsion-mediated reaction"; Materials Letters; **26,** 21-26 (1996)
8. J. Wang, P. F. Chong, S. C. Ng, L. M. Gan; "Microemulsion processing of manganese zinc ferrites"; Materials Letters; **30,** 217-21 (1997)
9. H. Kunieda, K. Shinoda; "Solution behavior of aerosol OT/water/oil system"; J. Colloid Interface Sci.; **70** [3], 577-83 (1979)

# GRAIN GROWTH INHIBITION OF HARD FERRITES THROUGH SOL-GEL PARTICULATE COATING

Jung W. Lee, Yong S. Cho and Vasantha R. W. Amarakoon
New York State College of Ceramics at Alfred University, Alfred, NY 14802

## ABSTRACT

This preliminary work presents a particulate coating procedure to inhibit grain growth of a nonstoichiometric Sr ferrite, $SrO \cdot 5.9Fe_2O_3$, by incorporating small amounts of additives, Si and Ca. Tetraethyl orthosilicate and calcium acetate corresponding to 0.6 wt% $SiO_2$ and 0.7 wt% CaO were added to a calcined submicron ferrite powder using a sol-gel reaction. Abnormal grain growth observed in a sample without the additives was found to disappear by adopting this chemical addition method and after sintering at 1200°C for 4 hrs in ambient atmosphere. A uniform distribution of small grains having an average grain size of ≈ 0.7 μm was obtained with a high bulk density of 4.90 g/cm³. These results were attributed to a homogeneous distribution of the chemical additives, and compared with samples processed by batch-mixing of the same additives.

## INTRODUCTION

Hard ferrites having a hexagonal magnetoplumbite structure have been widely used for applications including permanent magnets, particularly as an anisotropic form where the crystallites are aligned in a direction to improve magnetic hysteresis properties.[1,2] Magnetic properties of the hard ferrites are known to depend on their microstructural characteristics such as grain size and porosity.[3-5] High remanence requires a high sintered density, whereas high coercivity demands a small grain size. Usually, two approaches, i.e., tailoring compositions or controlling microstructures, have been tried to achieve these goals.[6,7]

On the other hand, several particulate coating techniques have been studied to improve final microstructures via a homogeneous distribution of small amounts of additives.[8-10] Various electronic ceramics such as LiZn and MnZn ferrites, yttrium iron garnets, $BaTiO_3$, $SnO_2$ and superconductors processed by the chemical coating method have been investigated.[8-11] In particular, Cho and Amarakoon[10,11] showed in the study of yttrium iron garnets that the coating method utilizing small amounts of Si and Mn improved significantly microwave magnetic properties due to the resultant desirable microstructural characteristics.

In this work, a particulate coating process utilizing a sol-gel reaction was investigated to distribute two additives, Si and Ca, uniformly onto the surface of a nonstoichiometric $SrO \cdot 5.9Fe_2O_3$ particles. The addition of Si is known to be effective in suppressing grain growth by exerting a drag force against grain boundary movement.[12,13] A small amount of Ca has been recognized to be beneficial in inducing anisotropy during growth and also in impeding grain growth.[14,15] The main concern of this study is to examine the effects of the chemical additives processed by the novel sol-gel particulate coating method on the microstructure and grain growth of the Sr ferrite samples.

EXPERIMENTAL

Raw materials, $Fe_2O_3$ and $SrCO_3$ corresponding to a nonstoichiometric Fe-deficient composition, $SrO \cdot 5.9Fe_2O_3$, were mixed using steel ball media in ethanol for 20 hrs. The mixed powder was dried at ≈ 120°C and then calcined at 1000°C for 4 hrs. After milling the calcined powder using yttria-stabilized zirconia balls for 15 hrs, the additives, Si and Ca, were added by two different methods, sol-gel coating vs. batch-mixing.

In the case of sol-gel coating, tetraethyl orthosilicate (TEOS, $Si(OC_2H_5)_4$) and calcium acetate monohydrate (($CH_3CO_2)_2Ca \cdot H_2O$) corresponding to given quantities of 0.6 wt% $SiO_2$ and 0.7 wt% CaO were used. Tetraethyl orthosilicate was first dissolved in ethanol, and partially hydrolyzed by adding a tiny amount of HCl as a catalyst. An aqueous solution of calcium acetate separately prepared was slowly added to the hydrolyzed TEOS solution. The ball-milled Sr-ferrite powder was inserted into the Si and Ca solution and then dispersed by stirring with a magnetic bar, resulting in a slurry. The stirring process continued until the complete evaporation of the solvent took place at room temperature. A gel reaction is believed to occur during this evaporation. The final rigid powder containing a gel of the additives was crushed using a mortar/pestle. In the other case of batch-mixing, reagent-grade $SiO_2$ and $CaCO_3$ corresponding to the same amounts were added to the calcined powder by wet-mixing using zirconia ball media.

Pressing of the powders (derived by the two different addition methods) was conducted at 80 MPa after mixing with a 2 wt% PVA aqueous solution. The pressed pellets were sintered at 1200°C for 4 hrs after burn-out of the binder and organics at 600°C for 2 hrs. For the microstructure observation of the surface of the sintered specimens, a scanning electron microscope (1810 SEM, Amray Co.) was used after polishing/thermal etching (at 1100°C for 30 min). Bulk densities of the samples were measured by the Archimedes' technique. Phase analysis for the powder and sintered pellets were performed by an x-ray diffractometer (XRD)

RESULTS AND DISCUSSION

Fig. 1 shows the powder characteristics of $SrO \cdot 5.9Fe_2O_3$ which was calcined at 1000°C for 4 hrs and then ball-milled for 15 hrs. A relatively uniform particle size less than 1 μm can be seen with some agglomerates. An average particle size was estimated to be around 0.6 μm. The phase of the powder was identified as shown in the XRD pattern of Fig. 2 (A). A single hexagonal $SrFe_{12}O_{19}$ phase was observed without any detectable second phase.

Using the single-phase submicron Sr ferrite powder, the sol-gel particulate coating technique was applied to uniformly distribute additives, Si and Ca. A separate experiment was conducted to confirm that the additive composition became a gel after the sol-gel procedure described in the experimental section. Fig. 2 (B) shows a XRD pattern of the sol-gel derived additives, Si and Ca. It is clear that the

Fig. 1. SEM micrograph of SrO·5.9Fe$_2$O$_3$ powder calcined at 1000°C for 4 hrs before sol-gel coating.

procedure formed a gel. As reported previously,[10,11] the gel additive is believed to wet and thus coat the surface of the Sr-ferrite particles on the nanoscale as the sol-gel reaction occurs in a slurry of the Sr-ferrite. The gel additive has been thought to contribute further to the uniform distribution as temperature increases because the gel becomes a viscous liquid at a sufficiently high temperature. Penetration of the gel into agglomerates can be another merit of the coating process.[10,16]

The effect of the sol-gel coating procedure on the final microstructure of the Sr ferrite was significant. Fig. 3 shows the microstructures of the SrO·5.9Fe$_2$O$_3$ samples sintered at 1200°C for 4 hrs in ambient atmosphere. The sample without additives exhibited abnormal grain growth with enlarged acicular grains (Fig. 3 (A)). On the other hand, the sol - gel coating process remarkably improved the

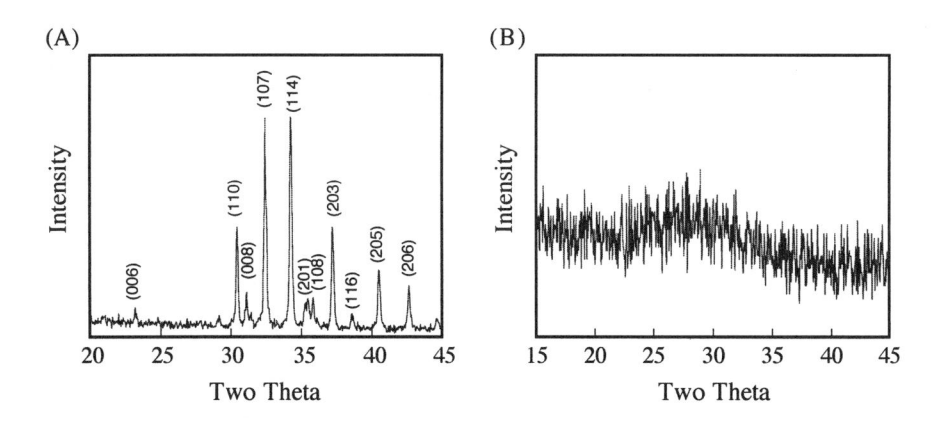

Fig. 2. XRD patterns of (A) the calcined SrO·5.9Fe$_2$O$_3$ powder (showing a single SrFe$_{12}$O$_{19}$ phase) and (B) the gel additive of Si and Ca.

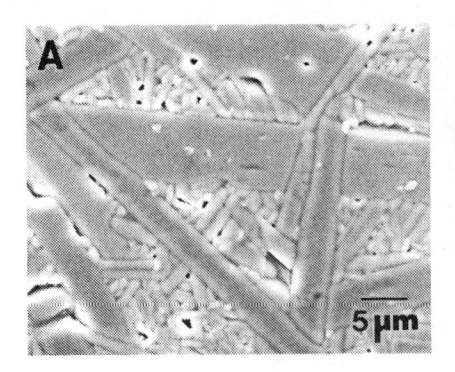

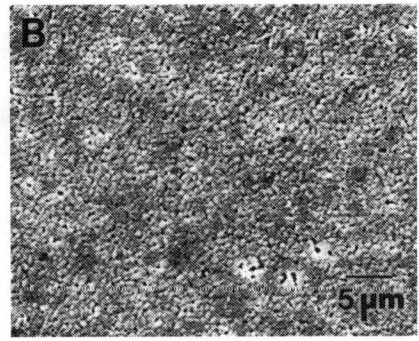

Fig. 3. Surface microstructures of SrO·5.9Fe$_2$O$_3$ samples containing (A) no additives and (B) sol-gel derived 0.6 wt% SiO$_2$ and 0.7 wt% CaO, sintered at 1200°C for 4 hrs in ambient atmosphere.

microstructural characteristics as seen in Fig. 3 (B). The sample containing the sol-gel derived 0.6 wt% SiO$_2$ and 0.7 wt% CaO showed an apparent suppression of abnormal grain growth. A uniform distribution of grain size (less than 1 μm) with less porosity can be clearly seen. It is believed that the intimate mixing of the additives via a sol-gel reaction led to homogeneous grain growth. The additives (having a low solubility) segregated along the grain boundaries is likely to act as a grain growth inhibitor during sintering.[13] Table I represents the bulk density values of the fired samples. The comparison in the density values support the observed tendency of microstructures. A higher bulk density of the sol-gel modified sample was 4.90 g/cm$^3$ (96% of theoretical).

Fig. 4 shows XRD patterns of the two samples corresponding to the microstructures of Fig. 3. Both samples showed a phase-pure Sr ferrite phase after sintering at 1200°C for 4 hrs. It is noticeable that the intensities of some peaks, particularly corresponding to (006) and (008) planes, are considerably different between the two patterns of Fig. 4. The prominent peaks of (006) and (008) planes in Fig. 4(A) are likely to be associated with observed anisotropic growth of the enlarged grains observed in Fig. 3(A).[17] This implies that the anisotropic growth occurred predominantly along a direction perpendicular to the c-axis. Note that the pressing direction was along the c-axis.

Fig. 5 demonstrates the microstructure of SrO·5.9Fe$_2$O$_3$ containing batch-

Table I. Density and microstructural characteristics of SrO·5.9Fe$_2$O$_3$ sintered at 1200°C for 4 hrs in ambient atmosphere.

| Additive Compositions | Addition Method | Linear Shrinkage (%) | Bulk Density (g/cm$^3$) |
|---|---|---|---|
| no additives | - | 16.3 | 4.81 |
| 0.6 wt% SiO$_2$ & 0.7 wt% CaO | Sol-gel coating | 18.6 | 4.90 |
| 0.6 wt% SiO$_2$ & 0.7 wt% CaO | Batch-mixing | 17.7 | 4.83 |

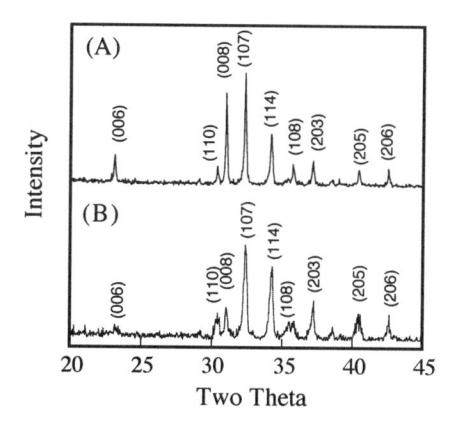

Fig. 4. XRD patterns of SrO·5.9Fe$_2$O$_3$ samples containing (A) no additives and (B) sol-gel derived 0.6 wt% SiO$_2$ and 0.7 wt% CaO, sintered at 1200°C for 4 hrs in ambient atmosphere.

mixed 0.6 wt% SiO$_2$ and 0.7 wt% CaO, sintered under the same condition of 1200°C and 4 hrs. Inferior microstructural characteristics (such as exaggerated grains and large pores) can be easily seen. The enlarged grains can be attibuted to inhomogeneous distribution of the additives processed by the conventional batch-mixing. A lower density value of 4.83 g/cm$^3$ was obtained in this case (Table I). From Fig. 3 (A) and Fig. 5, we can see the advantages of the sol-gel coating method in terms of microstructure. This is critical in determining the final magnetic

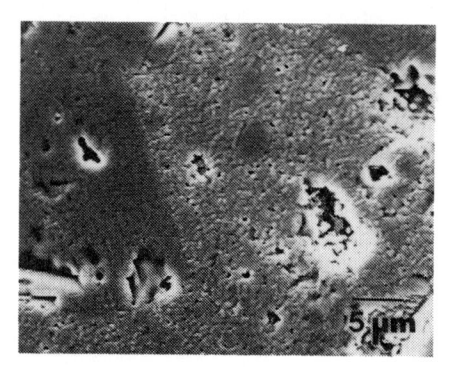

Fig. 5. Surface microstructures of SrO·5.9Fe$_2$O$_3$ containing batch-mixed 0.6 wt% SiO$_2$ and 0.7 wt% CaO, sintered at 1200°C for 4 hrs in ambient atmosphere.

properties and performance. The influences of the sol-gel derived additives on the magnetic properties are under investigation.

CONCLUSION

A partculate coating method utilizing a sol-gel reaction was investigated to uniformly distribute known additives, Si and Ca, on the particle surface of a Fe-deficient Sr hexaferrite (SrO·5.9Fe$_2$O$_3$). A phase-pure submicron Sr ferrite powder was used for the sol-gel coating. After sintering at 1200°C for 4 hrs, a Sr-ferrite sample without additives showed abnormal grain growth, resulting in acicular-shaped enlarged grains. On the other hand, the sol-gel derived additives corresponding to 0.6 wt% SiO$_2$ and 0.7 wt% CaO led to the inhibition of grain growth under the same sintering condition. A unifrom distribution of small grains having an average grain size of ≈ 0.7 μm was obtained with a high density of 4.90 g/cm$^3$. These characteristics were believed to originate from distingushable features of the sol-gel coating process, i.e., homogeneous distribution of the gel additive and penetration of the gel into agglomerates. The hypothesis was supported by a comparison between microstructures containing the same additives, but processed by different addition methods, sol-gel coating vs. batch-mixing.

ACKNOWLEDGMENTS

The support by NYS Center for Advanced Ceramic Technology (CACT) at Alfred University and Hoosier Magnetics Inc., and Army Research Office (ARO) under Grant No. DAAH04-95-1-0278) is acknowledged.

REFERENCES

[1] F. E. Luborsky, "The Application of Ultrafine Particles to the Fabrication of Permanent Magnets," pp. 488-513 in *Ultrafine Particles*, John Wiely and Sons, New York, 1963.

[2] R. A. McCurrie and S. Jackson, "Rotational Hysteresis in Anisotropic Barium and Strontium Ferrite Permanent Magnets," *J. Appl. Phys.*, **62** [2] 627-31 (1987).

[3] T. Fujiwara, "Magnetic Properties and Recording Characteristics of Barium Ferrite Media," *IEEE Trans on Magnetics*, **23** [5] 3125-30 (1987).

[4] M. H. Hodge, W. R. Bitler and R. C. Bradt, "Deformation Texture and Magnetic properties of BaO·6Fe$_2$O$_3$," *J. Am. Ceram. Soc.*, **56** [10] 497-501 (1973).

[5] J. S. Reed and R. M. Fulrath, "Characterization and Sintering Behavior of Ba and Sr Ferrites," *J. Am. Ceram. Soc.*, **56** [4] 207-9 (1973).

[6] K. Haneda, C. Miyakawa and H. Kojima, "Preparation of High Coercivity BaFe$_{12}$O$_{19}$," *J. Am. Ceram. Soc.*, **57** [8] 354-57 (1974).

[7] S. Besenicar, T. Kosmac and M. Drofenik, "The influence of ZrO$_2$ Additions on Magnetic and Mechanical Properties of Sr Ferrite," *Br. Ceram. Trans. J.*, **86**, 44-6 (1987).

[8] F. A. Selmi and V. R. W. Amarakoon, "Sol-Gel Coating of Powders for Processing Electronic Ceramics," *J. Am. Ceram. Soc.*, **71** [11] 934-37 (1988).

[9] K. G. Brooks and V. R. W. Amarakoon, "Sol-Gel Coating of Lithium Zinc Ferrite Powders," *J. Am. Ceram. Soc.*, **74** [4] 851-53 (1991).

[10] Y. S. Cho and V. R. W. Amarakoon, "Nanoscale Coating of Silicon and Manganese on Ferrimagnetic Yttrium Iron Garnets," *J. Am. Ceram. Soc.*, **79**(10) 2755-58 (1996).

[11] Y. S. Cho, V. L. Burdick and V. R. W. Amarakoon, "Enhanced Microwave Magnetic Properties in Nonstoichiometric Yttrium Iron Garnets for High Power Applications," *IEEE Trans. on Magnetics* (in press).

[12] F. Cools, "The Mechanism of Grain-Growth Impediment in Strontium Hexaferrite with Silica Addition," *Adv. in Ceramics*, **15**, 177-85 (1984).

[13] F. J. A. Den Broeder and P. E. C. Franken, "The Mechanism of Sintered Hexaferrite with Silica Addition, Investigated by ESCA and TEM," *Adv. in Ceramics*, **15**, 494-501 (1984).

[14] H. Yamamoto and T. Mitsuoka, "Effect of CaO and $SiO_2$ Additives on Magnetic Properties of $SrZn_2$-W Type Hexagonal Ferrite," *IEEE Trans. on Magnetics*, **30** [6] 5001-07 (1994).

[15] P. E. C. Franken, H. V. Doveren and J. A. T. Verhoeven, "The Grain Boundary Composition of MnZn Ferrites with CaO, $SiO_2$ and $TiO_2$ Additions," *Ceramurgia International*, **3**, 122-23 (1977).

[16] Y. S. Cho, V. L. Burdick, and V. R. W. Amarakoon, "Microstructural Aspects of Nanocrystalline LiZn Ferrites Densified with Chemically Derived Additives," *Mater. Res. Soc. Symp. Proc.*, **494**, 27-32 (1998).

[17] F. K. Lotgering, "Topotactical Reactions with Ferrimagnetic Oxides Having Hexagonal Crystal Structures," *J. Inorg. Nucl. Chem.*, **9**, 113-123 (1959).

# PREPARATION AND CHARACTERIZATION OF LITHIUM-TITANIUM OXIDE BY ALCOHOL BURNING OUT PROCESS

## CHEN-FENG KAO AND CHARNG-LIH JENG
Department of Chemical Engineering, National Cheng Kung University,
Tainan, 70101, TAIWAN

## ABSTRACT
$Li_{1+x}Ti_{2-x}O_4$ ($0 \leq x \leq 1/3$) is a spinel compound. However for x=0 the compound $LiTi_2O_4$ exhibits metallic characteristics and for x=1/3 the compound $Li_{4/3}Ti_{5/3}O_4$ is an insulator. In this study the lithium-titanium oxide was synthesized both by coprecipitation combined with high temperature melting and the alcohol burn out process. The powder with anatase or rutile type $TiO_2$ was produced at 1073K to be used as an anode for lithium ion battery. The resistivity of the anode was $8 \times 10^7$ $\Omega$-m at room temperature.

## INTRODUCTION
The transition metal compounds are used as a cathode in most secondary lithium ion battery for wherein they can reversibly vary the Li content as the battery charged and discharged. $Li_{1+x}Ti_{2-x}O_4$ is a spinel compound, but when x=0 the compound $LiTi_2O_4$ exhibits metallic characteristics and for x=1/3 the compound $Li_{4/3}Ti_{5/3}O_4$ is an insulator[1-7]. The various changes in the properties are caused by

---
To the extent authorized under the laws of the United States of America, all copyright interests in this publication are the property of The American Ceramic Society. Any duplication, reproduction, or republication of this publication or any part thereof, without the express written consent of The American Ceramic Society or fee paid to the Copyright Clearance Center, is prohibited.

the distribution of the lithium ion in the spinel structure. We prepare $LiTi_2O_4$ by coprecipitation with different alkaline solution and molten salt reaction, then describe the local structure of $LiTi_2O_4$ determined by the XRD patterns. Anatase type and rutile type $TiO_2$ appear in the $LiTi_2O_4$ with different alkaline solution process[8-10].

In this study, we investigate the homogeneity, particle size distribution, calcination temperature and phase evolution of the precursors derived by chemical coprecipitation.

Coprecipitation combined with high temperature melting[11] was used to synthesize the lithium-titanium oxides. The advantage of this method is described in previous study of $LaNiO_3$. The melting salt, NaCl or $NH_4Cl$, coprecipitated product not in need of washing out, is considered as a solvent of coprecipitates in high temperature, and dissolves the coprecipitates to react to give the oxides[12-18]. After reaction, the temperature was decreased directly to give the product and the salt which remained on the surface of the oxides. The salt was washed out with distilled water and the particle sizes of pure oxides obtained were very fine and were easily sinterable. The control of particle sizes is much easier in this process than the other, and the dissolved salt is used as an auxiliary agent in the reaction. The oxides from this method are more uniform than those obtained by the solid-state reaction or the general coprecipitation methods. Moreover, the process is easy and the mechanisms are as follows:

(1) $LiCl + 2\ TiCl_4 + 9\ NaOH \longrightarrow LiOH + 2\ Ti(OH)_4 + 9\ NaCl$
$\xrightarrow{\Delta} LiTi_2O_4 + 9\ NaCl + 9/2\ H_2O \uparrow + 1/4\ O_2 \uparrow$

(2) $LiCl + 2\ TiCl_4 + 9\ NH_4OH \longrightarrow LiOH + 2\ Ti(OH)_4 + 9\ NH_4Cl$
$\xrightarrow{\Delta} LiTi_2O_4 + 9\ NH_4Cl \uparrow + 9/2\ H_2O \uparrow + 1/4\ O_2 \uparrow$

Owing to the existence and the function of liquid melting salt, one can effectively give much more uniformity to enchance the reaction rate over those obtained by the conventional coprecipitation and the solid-state reaction. Therefore one can calcine at lower temperatures in this process to save energy.

**EXPERIMENTAL**

A mixture of lithium chloride and titanium chloride with the ratio of Li : Ti being

1:2 was dissolved in alcohol and diluted to 0.2 M solution. An alkali solution, such as NaOH or NH₄OH, was used to mix with the above solution to form a lithium titanium hydroxide precursor. The oven drying or alcohol burn out directly were succeeded to obtain dry precursors with different kind of salts that corresponds to the coprecipitated reagent.

Solid-phase precursors were calcined at 973K, 1073K and 1173K, respectively. The salt was acted as a flux to improve the reactivity of precursor. The calcined samples were washed with distilled water then filtered to obtain fine oxide particles. The silver nitrate solution was used to prove the presence or absent of chloride ions in the oxides or not. X-ray diffractometer (XRD) was used to identify the oxides. Fig. 1 shows the flow chart for coprecipitation combined with high temperature melting. The shapes and particle sizes of the powders were examined by scanning electron microscope (SEM). The conductivities of the bulk samples of the oxides were measured with a digital electrometer.

## RESULTS AND DISCUSSION

Thermogravimetry and calcination

The precursors were calcined to form the desired oxides. From the results of thermogravimetry and differential thermal analyses, as shown in Table 1, the calcination temperature was set to be at 973K, 1073K or 1173K by using a 10 $C^0$/min heating rate for 8 h. These calcination temperatures also contain the range of melting point of the NaCl and sublimation point of NH₄Cl.

Table 1 Thermogravimetric analysis of the precursor

|  | 973K°C | 1023K | 1073K | 1173K |
|---|---|---|---|---|
| NH₄Cl, not burning out | 68.69% | 71.30% | 74.21% | 77.78% |
| NH₄Cl, burning out | 64.70% | 66.00% | 70.92% | 72.28% |
| NaCl, not burning out | 69.73% | 73.00% | 74.00% | 78.13% |

It was shown by the SEM micrograph of the precursor that the particle size distribution of the coprecipitated powders was very wide. In addition, the larger

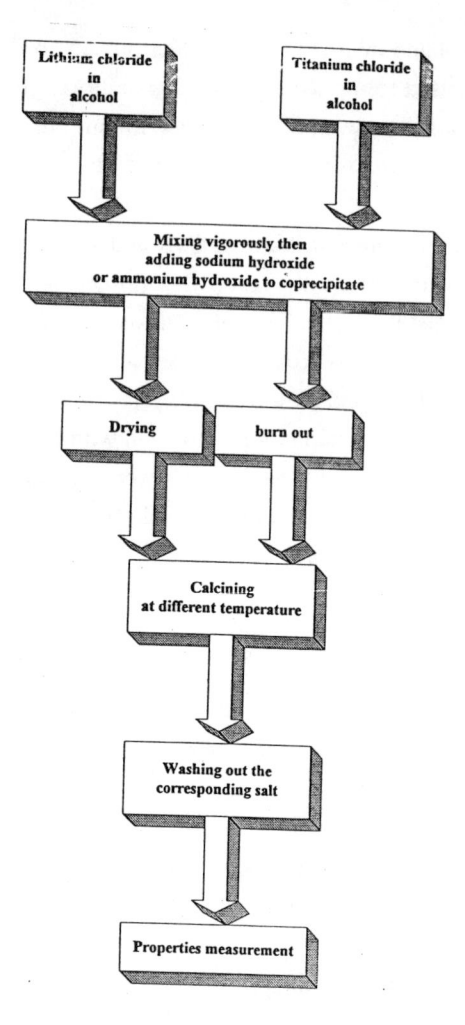

Fig. 1 Flow chart for this process

particles are mostly agglomeration of small particles. There are many visible crystals of NaCl observed by polarized microscope that the particles cause agglomeration of the precursors. Appropriate grinding, improvement of the drying mode or treating with other solvents will decrease the agglomeration. The agglomeration was not attended to because this did not affect the particle size distribution after calcination.

## SEM analysis of the calcined powder

The SEM images of the calcined powders dried by oven drying and by burning out directly were shown in Fig. 2 and Fig. 3, respectively. The uniformity of these powders is better than that of the coprecipitated powders, and the degree of agglomeration of the calcined particle is less. The particle size distribution is also much more uniform, and the shapes of grain is polygonal at 1073K. The particle size is about 0.1 µm.

(a).

(d).

(b).

(e).

(c).

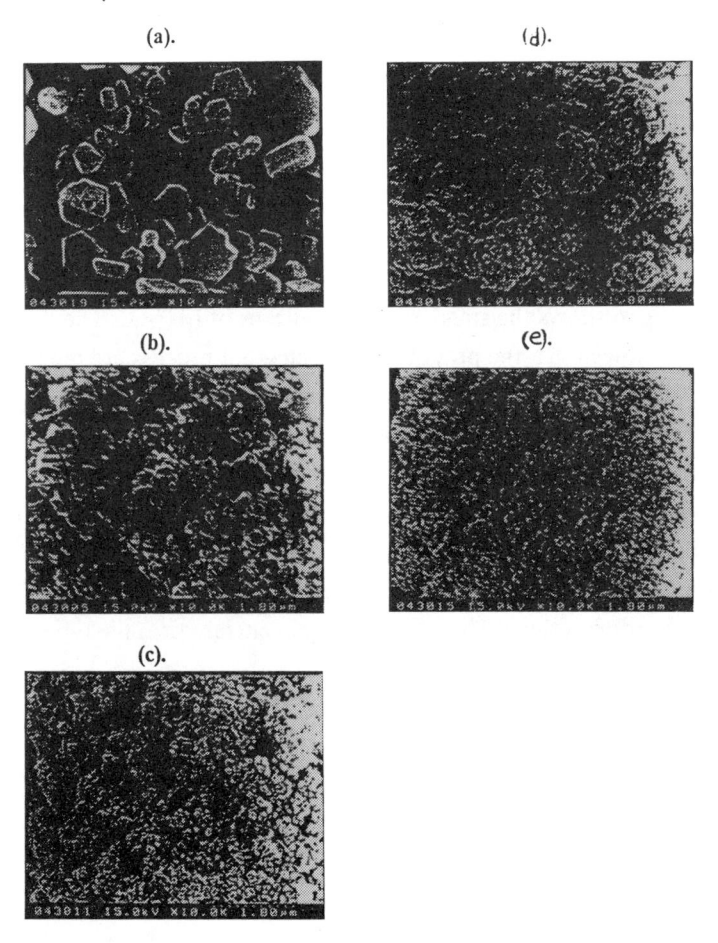

Fig. 2 SEM images for the samples dried by oven drying were calcined for 8 h at (a)1173K (b)1073K (c)973K (d)873K (e)773K.

Innovative Processing/Synthesis: Ceramics, Glasses, Composites II

(a). (b).

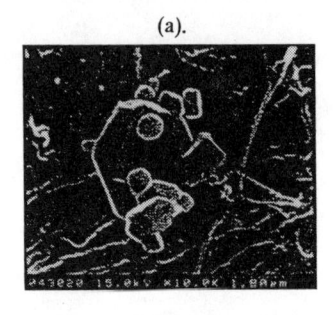

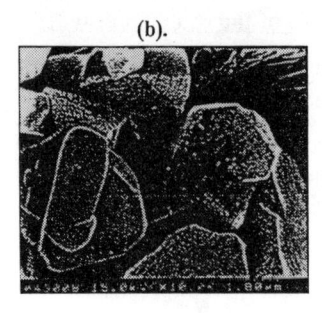

Fig. 3 SEM images for the samples dried by burning out directly were calcined for 8 hr. at (a)1173K and (b)1073K.

X-ray diffraction(XRD) analysis

The XRD analyses of the samples coprecipitated with NaOH and NH$_4$OH after drying by oven and burn out of the ethanol process are shown in Figs. 4 and 5, respectively. Fig. 6 shows the standard XRD patterns of LiTi$_2$O$_4$. The differences between them showed that the products by burning out process are much complicate than those used by conventional drying process. Fig. 7 and Fig. 8 showed that the XRD patterns for the samples coprecipitated with NH$_4$OH then drying by conventional and by burning out ethanol process, respectively.

Fig. 9 and Fig. 10 showed that the XRD patterns for the samples coprecipitated with NaOH and NH$_4$OH calcined at different temperatures, respectively. Using NaCl or NH$_4$Cl as the molten salt, the formation of LiTi$_2$O$_4$ is best at 1073K for both. There are some residual phases appearing in anatase or rutile type titanium oxide.

Compared with those above patterns, the anatase type TiO$_2$ would be obtained if NaCl was used as auxiliary melting salt; otherwise the rutile type TiO$_2$ would be taken.

Measurement of resistivity

The resistivity of the material at various ratios of Li/Ti has the relationship with its apparent density, or the denser the sintered body is, the better its conductivity is. This is very helpful to the preparation and research of the electrode.

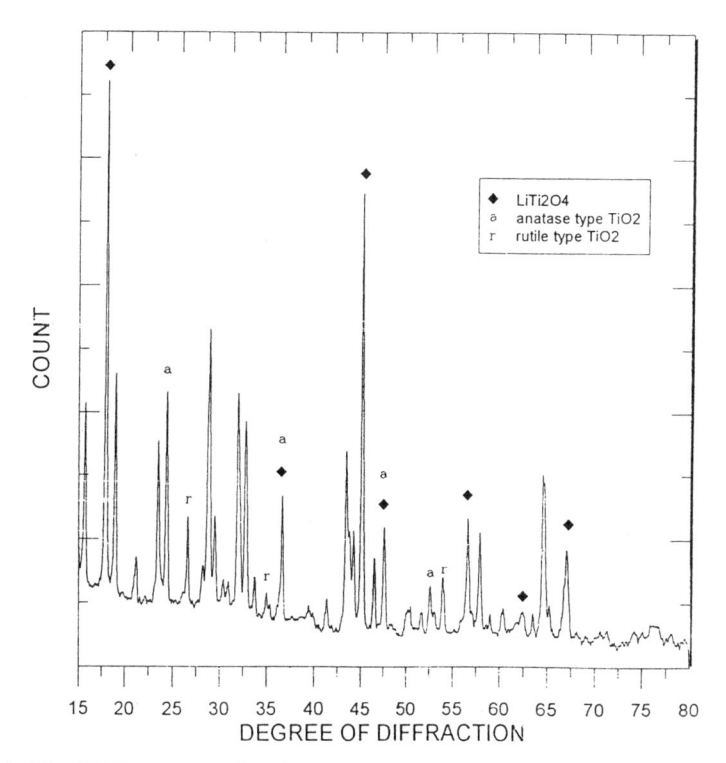

Fig. 4 The XRD patterns for the sample that coprecipited with sodium hydroxide and dried by conventional process was calcined at 1073K.

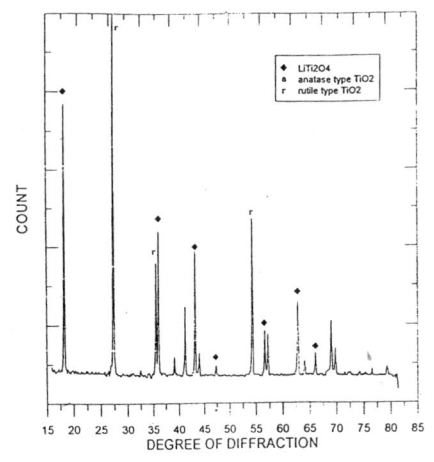

Fig. 5 The XRD patterns for the sample that coprecipited with ammonia water and dried by burning out ethanol process was calcined at 1073K.

Innovative Processing/Synthesis: Ceramics, Glasses, Composites II

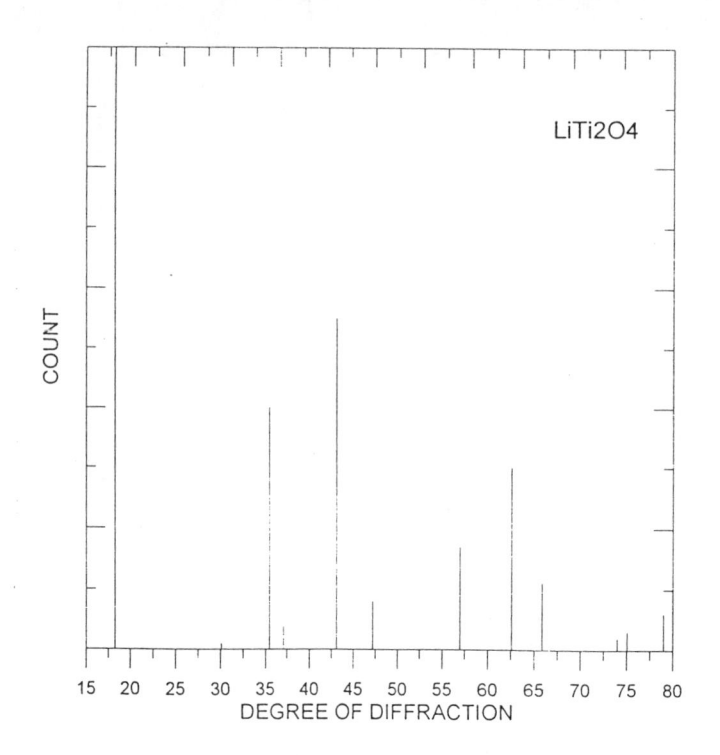

Fig. 6 The standard XRD patterns for the LiTi$_2$O$_4$.

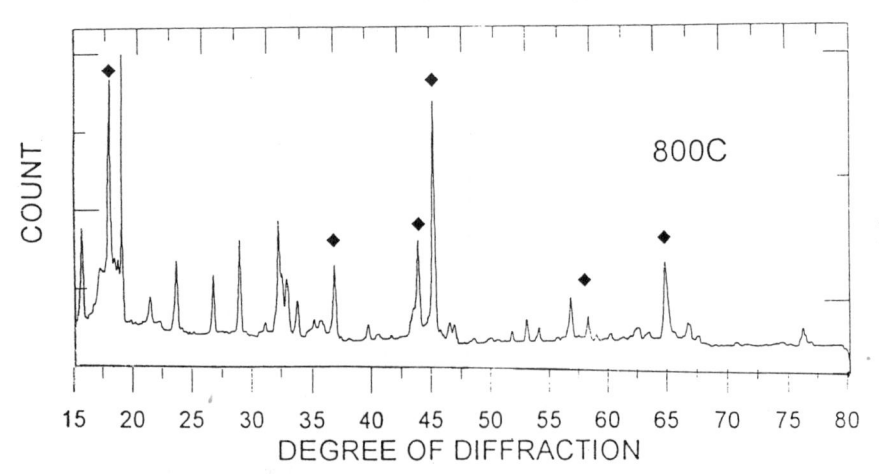

Fig. 7 The XRD pattern for the sample coprecipited with ammonia water and dried by conventional process was calcined at 1073K.

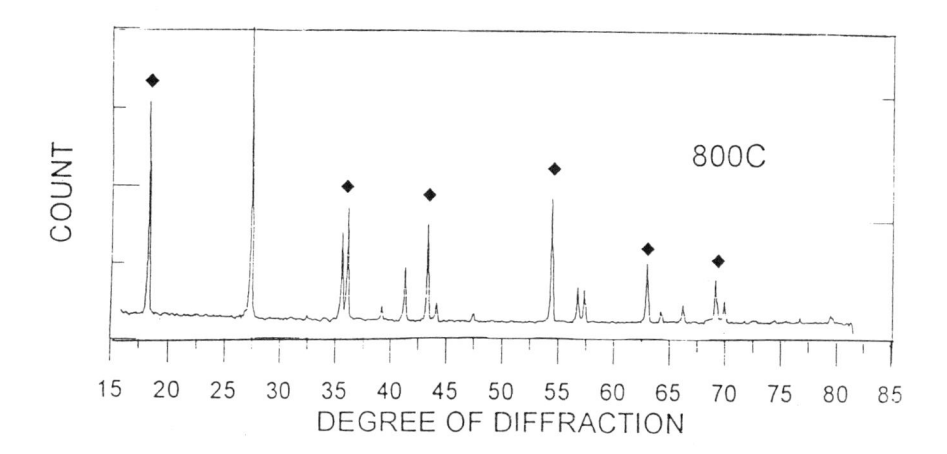

Fig. 8 The XRD pattern for the sample coprecipited with ammonia hydroxide and dried by burning out of the ethanol process was calcined at 1073K.

The percentages of the shrinkages of the sintered sample are 10% for drying by burning out of the ethanol process and 9% for drying by conventional process. The resistivity of the sintered sample is $8 \times 10^7$ Ω-m, showing that it is a semiconductor.

**CONCLUSION**
In this study the lithium-titanium oxide was synthesized by coprecipitation combined with high temperature melting. The powder with anatase $TiO_2$ was produced at 1073K for use as an anode for lithium ion battery. The resistivity of the anode was $8 \times 10^7$ Ω-m at room temperature.
Evaporating lithium oxide in high temperature would cause a sufficient titanium oxide. From this research it is indicated that the anatase type $TiO_2$ found by coprecipitating with NaOH and the rutile type $TiO_2$ appeared by coprecipitating with $NH_4OH$.
Better electrode energy potential would be expected when the anatase type $TiO_2$ appearing in these samples.

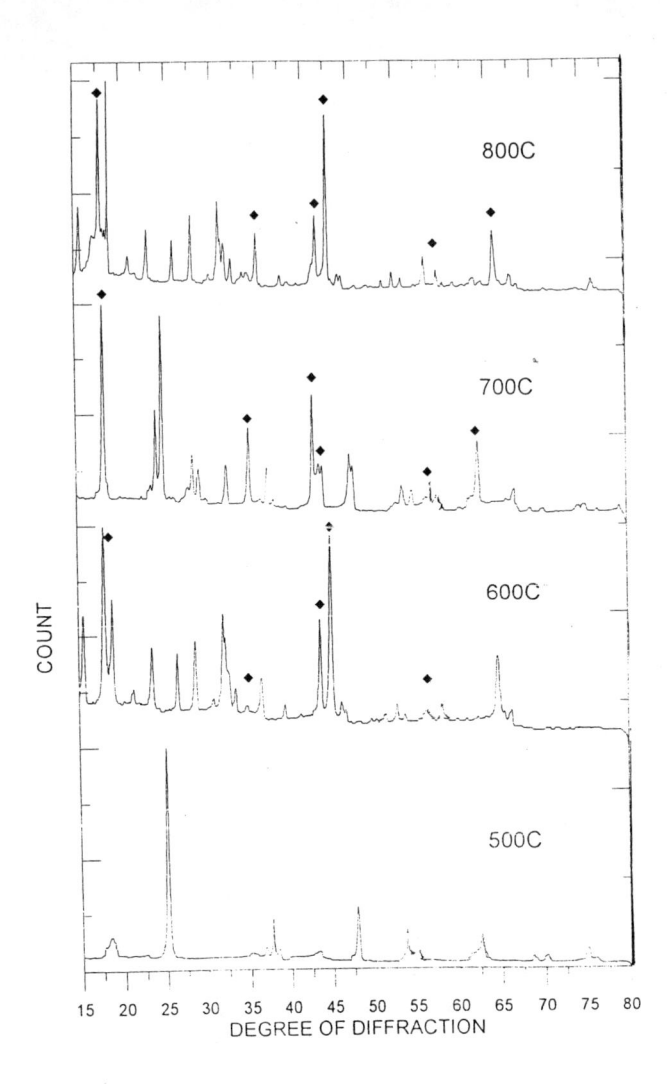

Fig. 9 The XRD pattern for the samples coprecipited with sodium hydroxide solution were calcined at (a)1073K(b)973K(c)873K(d)773K.

REFERENCES
[1] E. Moshopoulou, P. Bordet, A. Sulpice and J. J. Capponi,"Evolution of

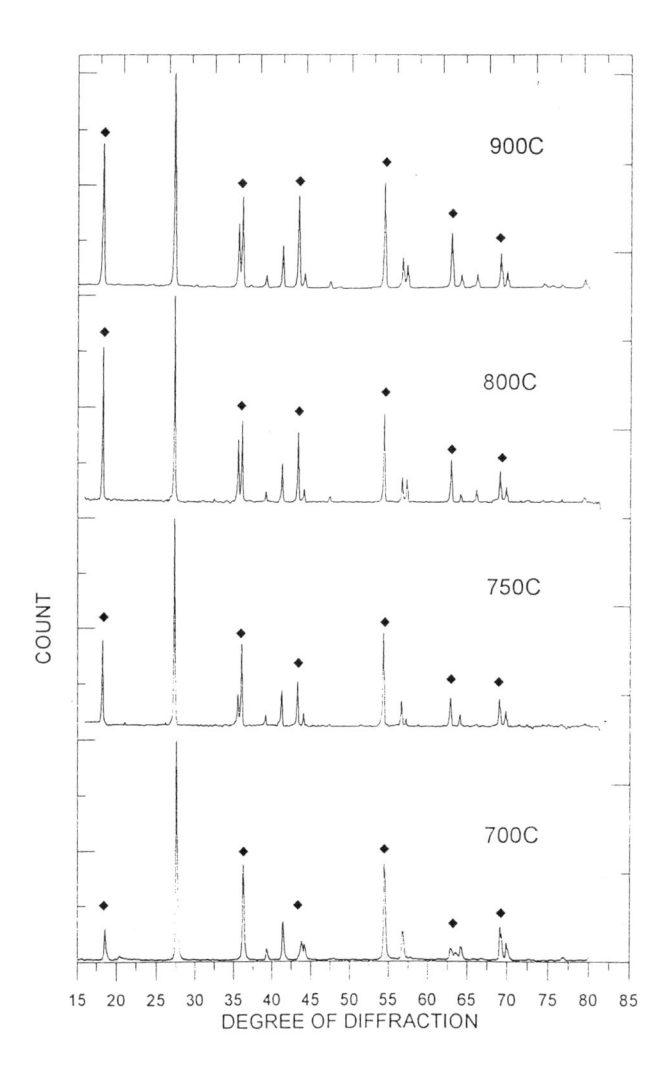

Fig. 10 The XRD pattern for the samples coprecipited with ammonia water were calcined at (a)1173K(b)1073K(c)1023K(d)973K.

Structure and Superconductivity of $Li_{1-x}Ti_2O_4$ Single Crystals without Ti Cation Disorder", Physica C235-240, 747-748 (1994).

[2] M. Dalton, DP. Tunstall, J. Todd, S. Arumugam and PP. Edwards,"NMR Studies in the Superconducting Spinel System $Li_{1-x}Ti_2O_4$", Physica C235-240, 1729-1730 (1994).

[3] M. Dalton, I. Gameson, A. R. Armstrong and P.P. Edwards,"Structure of the $Li_{1-x}Ti_2O_4$ Superconducting System : A Neutron Diffraction Study", Physica C221, 149-156 (1994).

[4] K. Sauv, J. Conard and M. Nicolas,"Modification of the Cuprate Superconducting Properties by $Li^+$ Doping : Resistivity and NMR", Physica C235-240, 1731-1732 (1994).

[5] Y. Ueda, T. Tanaka, K. Kosuge, M. Ishikawa and H. Yasuoka,"Superconducting Properties in the $Li_{1-x}Ti_2O_4$ System with the Spinel Structure", Journal of Solid State Chemistry, 77, 401-406 (1988).

[6] D. Z. Liu, W. Hayes, M. Kurmoo, M. Dalton and C. Chen,"Raman Scattering of the $Li_{1-x}Ti_2O_4$ Superconducting System", Physica C235-240, 1203-1204 (1994).

[7] M. Itoh, Y. Hasegawa, H. Yasuoka, Y. Ueda and K. Kosuge,"Li NMR in the Superconducting $Li_{1-x}Ti_2O_4$ Spinel Compound", Physica C157, 65-71 (1989).

[8] P. M. Lambert, M. R. Harrison and P. P. Edward,"Magnetism and Superconductivity in the Spinel System $Li_{1-x}M_xTi_2O_4$ (M=$Mn^+$, $Mg^{2+}$)", Journal of Solid State Chemistry, 75, 332-346 (1988).

[9] A. Campos, P. Quintana and A. R. West,"Order-Disorder in Rock Salt-Like Phases and Solid Solutions, $Li_2(Ti_{1-x}Zr_x)O_3$", Journal of Solid State Chemistry, 86, 129-130 (1988).

[10] I. Abrahams, P. G. Bruce, W. I. F. David and A. R. West,"Refinement of the Lithium Distribution in $Li_2Ti_3O_7$ Using High-Resolution Powder Neutron Diffraction", Journal of Solid State Chemistry, 78, 170-177 (1989).

[11] Zong-Yu Zheng, Bi-Jun Guo and Xue-Ming Mei, "A New Technology of Coprecipitation Combined with High Temperature Melting for Preparing Single Crystal Ferrite Powder", J. Magnetism and Magnetic Materials, 78, 73-76 (1989).

[12] J. A. Dean, "Solubility products", Lange Handbooks of Chemistry, Ch. 5, 11th ed., McGraw-Hill book Company, New York (1970).

[13] Burtron H. Davis,"Efeect of pH on Crystal Phase of $ZrO_2$ Precipitated from Solution and Calcined at $600^0C$", Communications of the American Ceramic

Society, C-168(1984).

[14] H. Nishizzawa, N. Yamasaki and K. Matsuoka, "Crystallization and Transformation of Zirconia Under Hydrothermal Conditions", J. Amer. Ceramic Soc., 65, No.7, 343-346(1982).

[15] S. Al. Dallal, M. N. Khan and Ashfaq Ahmed,"Structural,Transport and Infrared Studies of Oxidic Spinels $Zn_{1-x}Ni_xFeCrO_4$", J. Mat. Sci. 25, 407-410 (1990).

[16] A. Benedetti, G. Fagherazzi, F. Pinna and S. Polizzi,"Structural Properties of Ultra-Fine Zirconia Powders Obtained by Precipitation Methods", J. Mat. Sci., 25, 1473-1478 (1990).

[17] Boro Durici,"Synthesis and Characteristics of Zirconia Fine Powders from Organic Zirconium Complexes", J. Mat. Sci.,1132-1136 (1990).

[18] L. Lerot, F. Legrand and P. De Bruycker,"Chemical Control in Precipitation of Spherical Zirconia Particles", J. Mat. Sci., 26, 2353-2358 (1991).

# Nanotechnology

# STRONG MACHINABLE NANO-COMPOSITE CERAMICS

T.Kusunose, Y. H. Choa, T. Sekino and K. Niihara
ISIR, Osaka University, Ibaraki, Osaka 567, Japan

## ABSTRACT

$Si_3N_4$/BN nano-composites with dispersed, nano-sized BN up to 40 vol% were successfully fabricated by hot-pressing $\alpha-Si_3N_4$ powders with turbostratic BN (t-BN) coating, which was prepared through chemical processes using urea and boric acid. For comparison, the microcomposites were also prepared from mixed powders using the conventional powder mixing method. TEM observations revealed that the nano-sized hexagonal BN particulates were homogeneously dispersed within the $Si_3N_4$ grains as well as at the grain boundaries. Young's modulus of both micro- and nano-composites decreased with h-BN content. However, the strength of the nano-composites were significantly improved, compared to conventional micro-composites. Furthermore, the $Si_3N_4$/BN nano-composites exhibited excellent machinability like metals.

## INTRODUCTION

Ceramics are increasingly used in such engineering components as rotors, nozzles, valves and sliding gates because of their high strength, high hardness and good chemical inertness. Among ceramics, silicon nitride ($Si_3N_4$) is well-known due to its high fracture toughness, strength, excellent wear resistance and oxidation resistance. However, such properties as corrosion resistance to molten metals, thermal shock fracture resistance and machinability are insufficient for their practical use as various machine parts. Especially, the machinability is thought to be important, because most of engineering components have complex shapes and hence require machining by diamond tools. In order to improve these disadvantageous properties, recently, the h-BN has been studied as second phase dispersions into the $Si_3N_4$ matrix [1,2]. In the $Si_3N_4$/BN micro-composites, machinability similar to metals was found due to the laminar structure of h-BN dispersoids, although the siginificant decrease in fracture strength was observed with an increase in BN content[3]. This decrease in fracture strength of conventional $Si_3N_4$/BN composites could be attributed to the aggregation of h-BN grains caused during mixing of commercial powders. Therefore, the homogeneous dispersions of h-BN particles into the matrix is probably effective for enhancing the fracture strength of machinable $Si_3N_4$/BN composites.

Niihara and his colleagues[4-8] have investigated ceramic based nanocomposites in which the nano-sized particles are dispersed within the matrix grains and/or at the grain boundaries, and revealed that the dispersion of nano-sized particulates remarkably improve the mechanical properties of oxide and non-oxide ceramic materials. In early nanocomposites, hard and strong dispersoids were mainly incorporated into the matrix to improve the mechanical properties. But in later years soft and weak materials like a metal have also been used as dispersoids, and the enhancement of the fracture strength was also found by addition of even soft and weak dispersoids[9,10]. These results imply that the nano-sized h-BN dispersion could be expected to improve the mechanical properties of $Si_3N_4$ ceramics, if the homogeneous dispersion of nano-sized BN is realized in the composites. However, it has not been reported that the mechanical and machinable properties of $Si_3N_4$ ceramics are improved by nano-sized soft and weak ceramic dispersoid such as h-BN.

The purpose of this study is to find a new process to homogeneously disperse fine h-BN particulates up to 40 vol% into the $Si_3N_4$ matrix, and to clarify the microstructure and mechanical property relationship for fully sintered $Si_3N_4$/BN nanocomposites using the new processes. Special emphasis was placed to understanding the effect of micro and nanostructure on the machinability.

## EXPERIMENTAL PROCEDURE

*Preparation*

To have homogeneous dispersion of h-BN into $Si_3N_4$ ceramics, a novel chemical process to precipitate the BN precursor on $\alpha$-$Si_3N_4$ powders was developed. Figure 1 shows a schematic picture of the formation of $\alpha$-$Si_3N_4$/t-BN precursor mixed powders and $Si_3N_4$/BN nanocomposites.

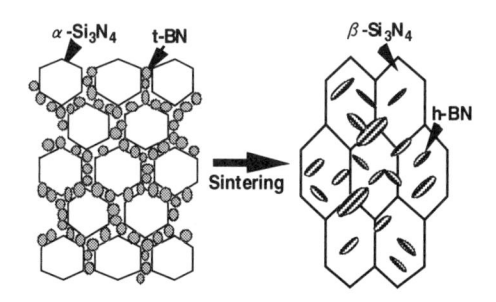

Fig.1. Fabrication image of $Si_3N_4$/BN nanocomposites.

In this study, the boric acid[11-13] and urea[11] were selected for producing the t-BN coating films on $\alpha$-$Si_3N_4$ powders. The experimental procedure is presented schematically in Figure 2. The BN content was adjusted to be 10, 20, 30 and 40 vol%. In this process, the $\alpha$-$Si_3N_4$ powder was ball milled with boric acid and urea in a plastic bottle using $Si_3N_4$ balls and ethanol. Before drying, 300 cm$^3$ of ion exchanged water was added to the slurry to prevent boric acid from evaporating. The dried mixtures were reduced at 1100°C in hydrogen gas, and then heated at 1500°C in nitrogen gas to produce the $Si_3N_4$-BN composite powder. The composite powders were hot-pressed at 1800°C in nitrogen gas after the second step of ball-milling with the sintering aid (2wt%Al$_2$O$_3$+6wt%Y$_2$O$_3$) and drying. For comparison, commercial BN powder with an average grain size of 9 μm was also used to fabricate Si$_3$N$_4$/BN microcomposites.

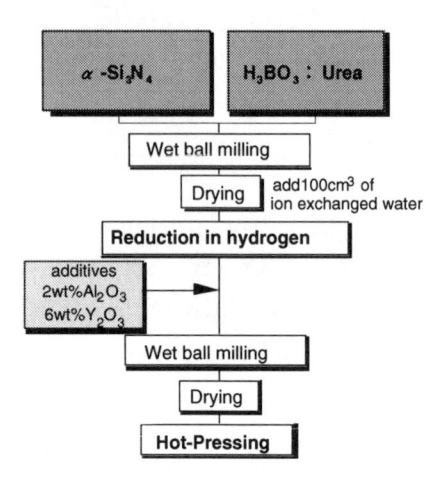

Fig.2. Flow chart of fabrication processes of the Si$_3$N$_4$/BN composites in this present work.

*Characterization*

The crystalline phases of the powder mixtures and hot-pressed bodies were identified by X-ray diffraction analysis (XRD). The microstructure was observed by a scanning electron microscope (SEM) and a transmission electron microscope (TEM). The microchemical analysis was done using an energy dispersive X-ray (EDX) analyzer attached to TEM.

Bulk density was measured by the Archimedes immersion technique in toluene. Young's modulus was determined by the resonance vibration method with first-mode resonance. Fracture strength was evaluated by a three-point bending test using rectangular bars (3x4x37 mm). The tensile surfaces of specimens were perpendicular to the hot-pressing direction. The span length and crosshead speed were 30 mm and 0.5 mm/min, respectively. Machinability was tested at 660 rpm using a WC/Co cermet drill 1 mm in diameter. Examination of sub-surface contact damage by ball-indentation was made using a bonded-interface technique[14-16], consisting of two polished half blocks joined together by adhesive. Indentations were made symmetrically across the traces of the interface at loads of 980 and 1960N with a tungsten carbide (WC) sphere of radius 1 mm. After testing, the bonded materials were separated and the surfaces and sections cleaned with acetone. The polished specimens were viewed in an optical microscope using Nomarski interface contrast to reveal the macroscopic damage patterns[17].

## RESULT AND DISCUSSIONS

*Fabrication of Nano-sized BN Dispersed $Si_3N_4$ Composites*

Figure 3 illustrates the transformation of crystalline phases in the processes for preparing the $Si_3N_4$/BN composite through the chemical route using boric acid and urea as a BN source. As seen in this figure, the starting powder before $H_2$ reduction consisted of boric acid and urea, and then the precipitatied boric acid and urea do not react on $\alpha-Si_3N_4$ powders during slurry drying process. The reduced product showed almost same peaks as starting powder except for the disappearance of

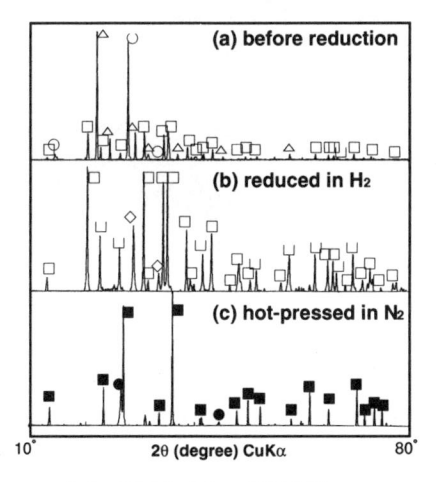

Fig.3. XRD patterns of the $Si_3N_4$/30 vol% BN precursor powder and sintered composite synthesized by the chemical route. □: $\alpha-Si_3N_4$, ■: $\beta-Si_3N_4$, ○:boric acid, ●: h-BN, △: urea, ◇: additive.

those of boric acid and urea and the appearance of a broad peak arising from turbostratic-BN(t-BN). This is in good agreement with the XRD data for the mixture consisting of only boric acid and urea reduced under the same condition.

Figure 4 shows representative TEM pictures and EDX analysis for the reduced powder corresponding to (b) in Fig. 3. It can be observed that the $\alpha-Si_3N_4$ powders are surrounded with a low contrast phase. Based on the observation of a disordered layer using higher magnification TEM and identification of boron using EDX, it was found that the reduced powder was $\alpha-Si_3N_4$ particles were partly coated with t-BN. It is well known that t-BN is transformed into h-BN[18], analogous to the structure change of graphite at 2000°C in nitrogen. Actually, the broad peak of t-BN at around 25°(2θ) developed into the peak of (002) reflection of h-BN at 26° (2θ) during hot-pressing of the reduced powder at 1800°C for 1h in nitrogen gas. The observed sintered body could be identified as a composite consisting of $\beta-Si_3N_4$ and h-BN.

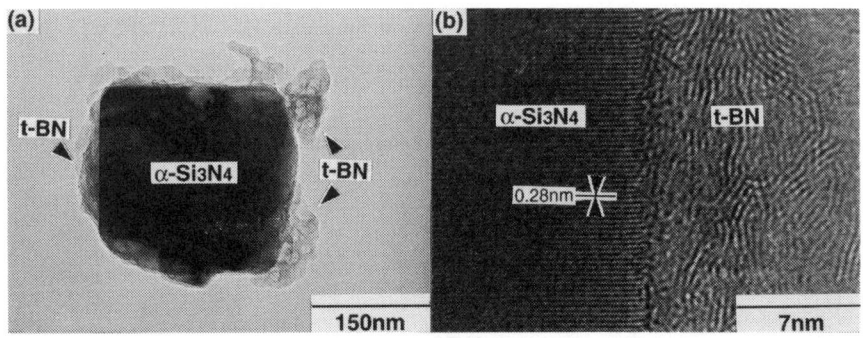

Fig.4. TEM micrographs of t-BN formed on the surface of $Si_3N_4$ powder heated at 700°C in $H_2$. (a) Low magnification and (b) higher magnification.

Figure 5(a) shows TEM observations processed of the hot-pressed $Si_3N_4$/BN composite processed through the chemical route. Nano-sized BN particles were homogeneously dispersed within $Si_3N_4$ grains as well as at grain boundaries. Impurity phases were not observed at the interphase between the $Si_3N_4$ matrix and intergranular BN dispersions. Instead, the microstructure evolved a graphite-like structure, as shown in Fig.5(b). Delaminations parallel to basal plane was also observed due to thermal expansion mismatch between β-Si$_3$N$_4$ and h-BN, which is characteristic of h-BN composites[19]. These observation means that the Si$_3$N$_4$/BN nanocomposites with homogeneous BN dispersions were successfully fabricated through a chemical process, followed by reduction and hot-pressing.

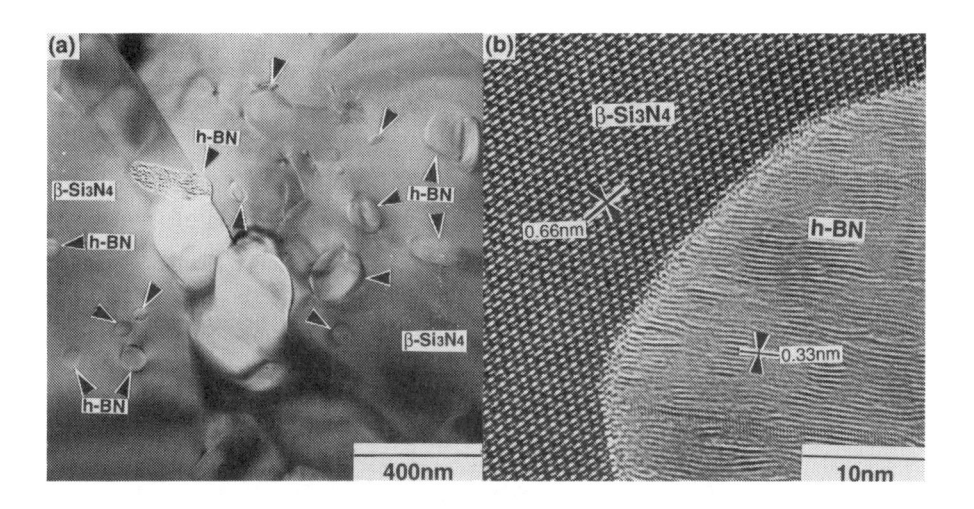

Fig.5. TEM micrographs of Si$_3$N$_4$/BN nanocomposites. (a) Low magnification, (b) interface structure between Si$_3$N$_4$ matrix and intragranular h-BN particulate.

*Mechanical Properties*

Figure 6 showed the variation in Young's modulus and fracture strength with BN content for the Si$_3$N$_4$/BN composites. In comparison to conventionally fabricated microcomposites, the nanocomposites fabricated in this work were successful in retaining relatively high strength in spite of Young's modulus decrease due to the soft h-BN addition (Young's modulus of h-BN and Si$_3$N$_4$ are about 65 and 350 Gpa, respectively). In special, the fracture strength of the nanocomposites was approximately two times higher than that of the microcomposite with BN content up to 40 vol%.

Based on the linear fracture mechanics concept, the strength of brittle ceramics is controlled by the fracture toughness and the critical flaw size, which is strongly related to processing defects (such as agglomerates of second phases, large pores, and abnormally grown grains). The fracture toughness was not remarkably influenced by the nano-sized BN dispersions. From SEM observations of fracture surfaces, it was found that, in the microcomposites, the large or aggolomerated h-BN particles initiated fracture. However, such large agglomerations due to BN dispersions were not observed. In addition, the matrix grain size of the nanocomposites was considerably refined by the nano-sized BN dispersions in comparison with the $Si_3N_4$ monolith and $Si_3N_4$/BN microcomposites. This should be attributed to the inhibition of the grain growth of $Si_3N_4$ caused by grain boundary pinning due to the nano-sized h-BN dispersions. Thus, it is concluded that the observed high strength of $Si_3N_4$/BN nanocomposites must be chiefly attributed to the inhibition of grain growth by the nano-sized BN dispersion, and also due to the reduction of flaw size.

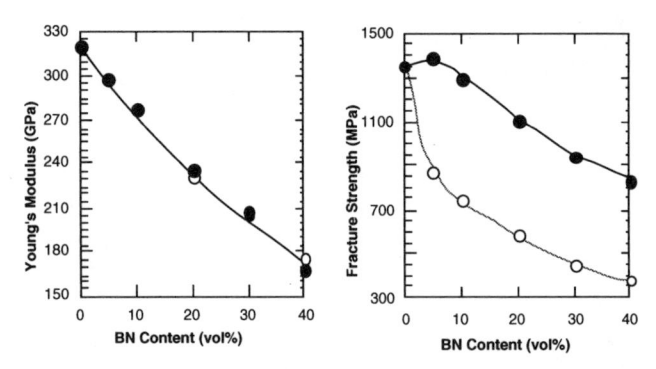

Fig. 6. Effects of h-BN content on fracture strength and Young's modulus of $Si_3N_4$/BN composites. ●: $Si_3N_4$/BN nanocomposites and ○: $Si_3N_4$/9 μm BN microcomposites.

*Machinability*

The $Si_3N_4$/BN composites containing more than 20 vol% BN in the present work showed good machinability like metals, which is an important feature in practical uses as engineering ceramics. Figure 7 shows section views of drilled surfaces in the micro and nanocomposite. In the microcomposite, the damage caused by drilling was severe. By contrast, the damage in the nanocomposites was relatively minor. The difference of the damage between microcomposites and nanocomposite is presumably attributed to the homogeneous dispersion of finer h-BN particles, which is expected to absorb and to disperse the damage force during the machining operations.

An especially simple experiment for the potential absorptivity of damage is the Hertzian contact test, in which a hard spherical indenter is loaded onto a polished specimen surface. Figure 8 presents the half-surface (upper) and side section (lower) views of Hertzian contact damage for $Si_3N_4$ monolith and $Si_3N_4$/30 vol% BN nanocomposite, respectively. The damage in the monolithic $Si_3N_4$ is a typical Hertzian cone fracture, whereas the damage observed in the zone beneath the contact in the nanocomposites appears to be quasi-plastic, reminiscent of the plastic zone in metals. It was reported that quasi-plasticity is driven by a strong shear component[14-16]. In the case of the nanocomposites, one of the reasons for good machinability should be quasi-plasticity caused by the weak interfaces between the $Si_3N_4$ matrix and the BN dispersion. In fact, significant interface cracking between the matrix and the BN dispersion were observed in the damaged zone by SEM.

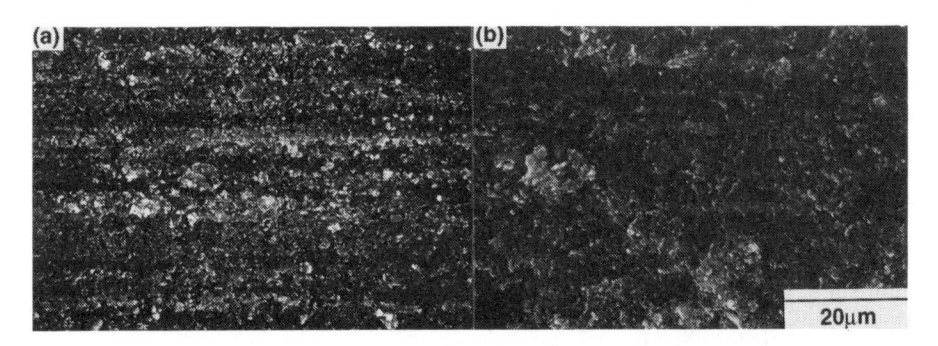

Fig.7. SEM observation of the surface of a hole machined using a WC cermet drill of 1 mm in diameter. (a) : $Si_3N_4$/30 vol% BN nanocomposite and (b) : $Si_3N_4$/30 vol% BN microcomposite.

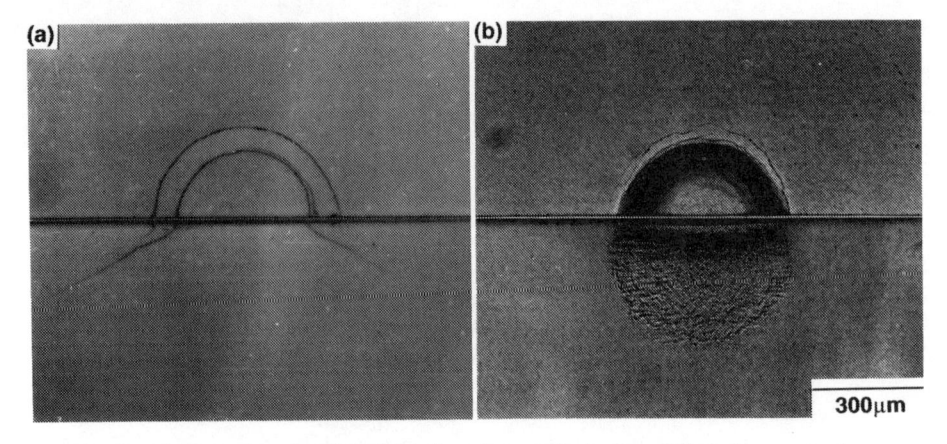

Fig. 8. Half-surface (top) and section (bottom) views of Hertzian contact damage in monolithic $Si_3N_4$ (a) and $Si_3N_4$/30 vol% BN nanocomposite (b) using a WC sphere of radius 1 mm at load 1960 and 980 N, respectively.

## CONCLUSION

(1) The $Si_3N_4$/BN nanocomposites, including the nano-sized BN up to 40 vol%, were fabricated by hot-pressing the $\alpha-Si_3N_4$ powders covered partly with t-BN. TEM observations revealed that the nano-sized h-BN particles were homogeneously dispersed within the $Si_3N_4$ grains as well as at the grain boundaries.

(2) Young's modulus of both microcomposites and nanocomposites decreased with increasing h-BN content according to the rule of mixtures, but the strength of the nanocomposites was significantly improved, compared with the conventional microcomposites. The enhancement of the fracture strength was attributed to the inhibition of grain growth and also due to a decrease in fracture origin size by nano-sized h-BN dispersions.

(3) The observed excellent machinability was attributed to quasi-plasticity, which originated at the weak interfaces between $Si_3N_4$ matrix and the h-BN dispersion.

## REFERENCES

[1]K. S. Mazdiyasni and Robert Ruh, "High/Low Modulus $Si_3N_4$-BN

Composite for Improved Electrical and Thermal Shock Behavior," J. Am. Ceram. Soc., 64 [7] 415-418 (1981)

[2]T. Funabashi, K. Isomura, A. Harita, and R. Uchimura, "Mechanical Properties and Microstructures of $Si_3N_4$-BN Composite Ceramics," Ceramic Materials & Components for Engines., 968-976(1986)

[3]D. Goeuriot-Launay, G. Brayet, F. Thevenot, "Boron Nitride Effect on the Thermal Shock Resistance of an Alumina-Based Ceramic Composite," J. Mater. Sci. Lett.., 5 (1986) 940-942

[4]K. Niihara, "New Design Concept of Structural Ceramics -Ceramic Nanocomposites-," J. Ceram. Soc. Japan., 99[10]974-982(1991)

[5]F. Wakai, Y. kodama, S. Sakaguchi, N. Murayama, K. Izaki & K. Niihara, "A Superprastic Covalent Crystal Composite." Nature, 344, 421-423(1990)

[6]K. Niihara, "Nanostructure Design and Mechanical Properties of Ceramic Composites," J. Jpan. Powd. and Powd Metal., 37[2]348-351(1990)

[7]K. Niihara, K. Izaki and A. Nakahira, "The $Si_3N_4$-SiC Nanocomposites with High Strength at Elevated Temperature," J. Jpan. Powd. and Powd Metal., 37[2]352-356(1990)

[8]K. Niihara and T. Hirai, "Surper-Fine Microstructure and Toughness of Ceramics," Ceramics, 21[7]598-604(1986)

[9]T.Sekino, T. Nakajima, S. Ueda and K. Niihara, "Reduction and Sintering of a Nickel-Dispersed-Alumina Composite and Its Properties," J. Am. Chem. Soc., 80[5]1139-1148(1997)

[10]T.Sekino and K. Niihara, "Fabrication and Mechanical Properties of Fine-Tungsten-Dispersed Alumina-Based Composites," J. Mater. Sci.., (1997) in print.

[11]T. E. O'Connor, "Synthesis of Boron Nitride," J. Am. Chem. Soc., 84, 415-418 (1962)

[12]R. T. Paine and C. K. Narula, "Synthetic Routes to Boron Nitride," Chem. Rev., 90, 73-91(1990)

[13]T. Hagio, K. Kobayashi, and T. Sato, "Formation of Hexagonal BN by Thermal Decomposition of Melamine Diborate," J. Ceram. Soc. Japan., 102[11] 1051-1054(1994)

[14] A.C. Fischer-Cripps and B.R. Lawn, "Indentation Stress-Strain Curves for "Quasi-Ductile" Ceramics," Acta mater., 44[2]519-527(1996)

[15] F. Guiberteau, N. P. Padture, H. Cai and B. R. Lawn, "Indentation Fatigue a Simple Cyclic Hertzian Test for Measuring Damage Accumulation in Polycrystalline Ceramics," Phil. Mag., A68[5] 1003-1016(1993)

[16] F. Guiberteau, N. P. Padture and B. R. Lawn, "Effect of Grain Size on Hertzian Contact Damage in Alumina," J. Amer. Ceram. Soc., 77[7] 1825-183`1(1994)

[17] H. Cai, M. A. Stevens Kalceff, and B. R. Lawn, J. Mater. Res., 9[3]762-770(1994)

[18] V. Brozek and M. Hubacek, "A Contribution to the Crystallochemistry of Boron Nitride," J. Solid State Chem., 100, 120-129(1992)

[19] W. Sinclar and H. Simmons, "Microstructure and Thermal Shock Behavior of BN Composites," J. Mater. Sci. Lett., 6, 627-629(1987)

# NEW MECHANISM AND KINETICS OF NANOAMORPHOUS METALS SYNTHESIS PROCESS.

## R.T.MALKHASYAN*, S.L.GRIGORYAN**.

*Scientific Production Enterprise "ATOM" Ministry of Industry and Trade, Republic of Armenia 67 Arshakouniats st.,376061, Yerevan. Armenia. "Dvin" Joint - stock company, E-mail: dvininfo@Tpa.am
**Chemical Physics Institute, National Academy of Science, Republic of Armenia. 5/2 P. Sevaki St. 375044 Yerevan, Armenia.. E-mail: rmalkhas@aua.am

## ABSTRACT

A new quantum - chemical technology for preparation of amorphous nanosize single-component metals by reducing their oxides with nonequilibrium hydrogen molecules highly vibrationally excited with an excitation energy up to $3,52 \times 10^{-19}$ Joule (2,2 eV) is suggested. It is shown, that such non-equilibric technology in which the equilibric heat influence practically completely is excluding, creates the new unknown possibilities for creation of a number of new nanoamorphous metals.

## INTRODUCTION

One of the first indication of the nonequilibrium influence was obtained in the plasma-chemical processes, which are realized in high heat emission conditions (by high temperature) [1].
Later it was shown by us [2-3]and in the other works [4-5] that in the elementary endothermic acts of chemical transformation the internal energy of excitation of the reagents is much more effective than the analogous quantity energy of heat movement of the same molecules, and that is in complete accordance with the conclusions of atomic collisions physics. The observing growth of reaction rate constant reaches 4-5 orders.
Quantum-chemical technology (QCT), which is developed by us in contrast to plasmachemistry, for the first time allows to carry out the especially

nonequilibrium processes with participation of vibrationally highly excited molecules practically at room heat energy of their translation movement.

This new field of nonequilibrium thermodynamics requires further studies yet, but showing essentially non Arrenius behavior already it allows to realize a row of new processes with the corresponding unknown products obtaining.

The creation of comparatively simple method of highly excited nondipole (homonuclear) molecules obtaining, with the energy of excitation reaching half of links energy of those molecules is the basis of QCT. Thus, for example, the vibrationally excited till sixth quantum level molecules of hydrogen with $\leq 3,52 \pm 0,2 \times 10^{-19}$ J (2,2eV) energy were obtained by us. Time of life of these excited molecules measured by us [6] make up more than $10^{-5}$ second magnitude, which is enough for their chemical transformation.

## EXPERIMENTAL PROCEDURE

The range of applicability of the quantum-chemical technology is defined by the range of conditions under which excitation of the vibrational degrees of freedom of interacting molecules remains in the nonequilibrium state. This signifies that the processes underlying this technology are gas-phase ( homogenous and heterogeneous), because in liquids, and more so in solids, it is impractical to produce nonequlibrium vibrationally excited states of individual molecules.

Generally, the problems of QCT can be divided into two parts: (1) QCT of treating surfaces with the aim of their modification [7-8] and (2) QCT of preparing chemical products [9-10].

The solution of a problem of full interaction of gas phase with the whole volume of this solid phase is also necessary for obtaining of the new substances.

For all this it is necessary to take into account that in QCT the depth of penetration of excited molecules comparatively is not great through their fast relaxation on the surface of metals.

For this reason it should be used as a row material the small dispersions powders with the grain size no more than $2 \pm 1 \times 10^{-6}$ m [9]. The depth of treatment in general case exponentially depends on the treatment time and on the relaxation constant of vibrational excitement in the every concrete case.

It is clear that the use of nanosize powders oxide as an initial material for processes conducting by QCT is the most expedient and will be accomplished in future.

As is known, procedure used to prepare amorphous metal alloys is based on fast cooling melts of these metals at rates of $10^6$ Kelvin/s [11]. It is also known that single-component amorphous metals can not be prepared by the method of fast cooling, because it requires a very high cooling rate (of about $10^{10}$-$10^{12}$ Kelvin/s) of their melts having low viscosity, which is technically unattainable [13].

The quantum-chemical technology of preparing amorphous materials developed by our group is based on direct reduction of the desired product, e.g., from its oxide or sulfide by highly vibrationally excited hydrogen molecules rather than on hardening of melts. The reduction reaction is conducted as follows. The initial oxide, with grain size < 50 mm to be reduced ( e.g., $MoO_3$. $WO_3$, CuO NiO, etc.) is placed in a tilted evacuated quartz reactor, into which after evacuation vibrationally excited hydrogen molecules $H_2(v)$ are admitted. The excited molecules are generated in immediate vicinity of the zone where they react with oxide. Hydrogen molecules in our experiments had a vibrational excitation energy equivalent to a temperature of 11650 K (1,5 eV). The reacted gas together with the reaction products (i.e., with water) is pumped out through a trap. In reducing sulfides the reaction products are hydrogen sulfide and free sulfur. By shaking the reactor or by other procedure a new portion of oxide to be treated is introduced, while the treated one reduced to the amorphous state is displaced to the opposite lower reactor end. Treated is normally only the surface oxide layer, therefore the material should be treated repeatedly with selection of treatment time optimal for a given oxide.

The quantum-chemical (QC) reduction of oxides with different value of reduction heat from $+1,7 \times 10^{-19}$ J ( for $WO_3$) , $\approx 0$ J (for NiO) and $- 1,3 \times 10^{-19}$ J (for CuO). So the study of as strong endothermic ( in the case of $WO_3$ reduction) as strong exothermic processes (as in the case of CuO reduction) is curried out. The practically thermoneutral reaction of NiO reduction also was studying. The dependents of amorphous metal yield on the treatment time and on the temperature of the QC processes was investigated too.

## RESULTS AND DISCUSSION

Till today QCT of obtaining of amorphous Mo the reaction heat of which is $0,32 \times 10^{-19}$ J (0,2 eV) is the most studied [9-10]. In this work we at first time obtained the nano size amorphous (or nanoamorphous [13]) tungsten , copper and nickel according the reaction (1-3).

$$WO_3 + H_2(v) \rightarrow W^* + H_2O \quad -1,1 \text{ eV} \qquad (1)$$

This kind of high value of endothermity of this reaction right away indicates, that this reaction which is realizing at room temperature proceeds, just at the expense of internal energy of excited hydrogen, i.e., is realized by QC means.

On the x-ray diffraction spectrum we do not observe the full disappearance of spectrum (as in the case of reduction of $MoO_3$ oxide).

The residual spectrum $WO_3$ on the level $\approx 10$ % from initial state remains. But the appreciable widen of spectral lines till 2-3 times and some peaks are shifted to the side of large angles on 0,3-0,5 degree was observed. And it indicates that the registered oxide has enough destroyed lattice [13] and it is possible to suppose that the QCT process was insufficiently full. However the increasing of the treatment time and the other methods of influence for the time being do not bring to the full reduction of initial oxide, which apparently indicates the subsequent oxidation of nanoamorphous tungsten by atmospheric oxygen when it is extracted from the reactor. The grain size of nanoamorphous W powders is smaller than 50 nm.

In the contrary of previous process the CuO reduction is accompanied by the extraction of almost the same large quantity of energy $1,3 \times 10^{-19}$ J (or 0,82 eV):

$$CuO + H_2(v) \rightarrow Cu^* + H_2O + 0,82 \text{ eV} \quad (2)$$

However also in this case when reducing by QCT, practically full disappearing of its spectrum without appearing of the spectrums of the other crystalline compositions, as before [10], is observed.

The large exothermity of the copper oxide reduction process was leading to the fact, that obtaining free metal in some cases not only crystallize, but also sintering in surface layer. The crust by thickness $\approx 10^{-4}$ meter, is easily come off from the bulk of the powder.

For the purpose of preventing from the crystallization and sintering it was necessary to decrease the energy deposit to the QCT by the way of vibration quantum level of molecule lowering from third ( or from $2,4 \times 10^{-19}$ J energy) till the first level (or $0,72 \times 10^{-19}$ J energy).

It was necessary to conduct the QC reduction in the conditions of additional cooling by the help of heat conduction lining.

The influence of heat factor in QCT become apparent most distinctly in the case of NiO reduction :

$$NiO + H_2(v) \rightarrow Ni^* + H_2O + 0,02 \text{ eV} \quad (3)$$

The crystalline lattice of the main modification $\alpha$-Ni has similar to NiO cubic lattice with the close values of lattice constants.

In this case one atom of metal will get one atom of oxygen and not three ones each, as in the case of $WO_3$ reduction and it is not required considerable deformations of initial oxide lattice for the formation of the new crystalline metallic lattice.

The x-ray diffraction spectrum of the initial crystalline NiO (curve 1) is offered on the Fig. 1. The spectrum of the same powder after QCT treatment is offered on the same figure (curve 2 ), which, as is seen, diminishes practically for an order,

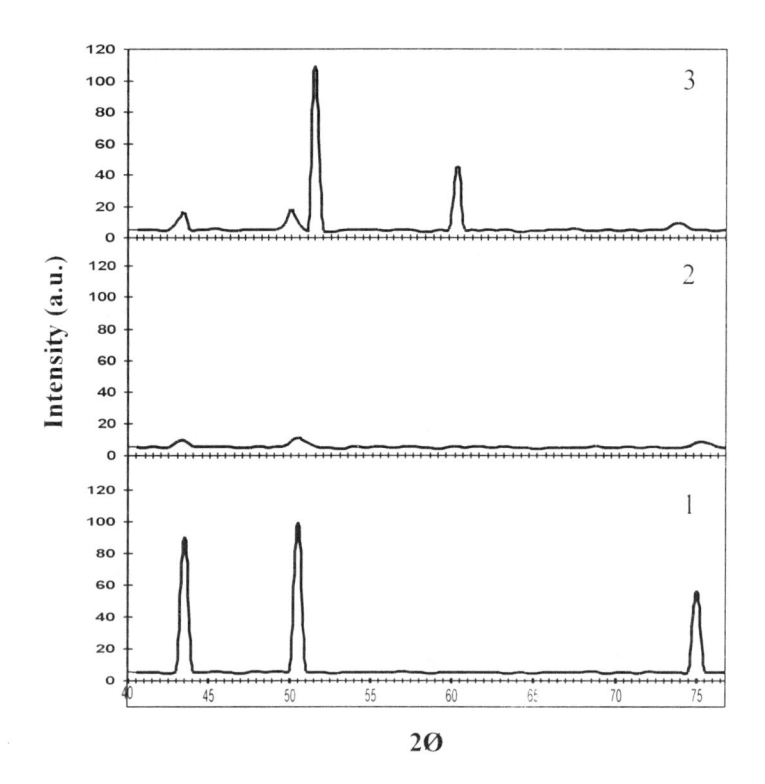

Fig.1 X-ray diffraction spectrum NiO:
1-initial crystalline pattern, 2-pattern with amorphous Ni,
3-pattern with crystalline Ni

without formation of the other crystalline combinations i.e., only nanoamorphous Nickel with the grain size 25÷10 nanometer is formed.

The amorphity of the obtained metallic nickel is determined according to the observing halo (Fig.2) [14], which is obtained on the same our transmitting electron microscope BS-500. Here the considerable widening of the spectral lines of oxide by 2-3 times with their shift to the side of large angles is observed, which is additional indication of the nanosizity of these amorphous products and its structural disorder. As a result we have extremely heightened Grain Boundary energy [13].

In the case when the process of QCT reduction is conducted at small heating <373K under the general diminishing of peaks by 5-6 times the appearing of the considerable quantity of crystalline Ni is observed as it is seen on the same picture (Fig.1 curve 3).

Fig 2. Electron-diffraction image of amorphous Ni.

As it is seen from the comparison of 2 and 3 curves, even in case of more deep transformations, (curve 2) without heating, the new crystalline phase of metal is not formed (as on curve 3). But at the presence of the small heating at once a crystalline phase appears, even at the less transformations (curve 3).

Thus, these experiments show undoubtedly that exactly equilibrium heat influences are responsible for the appearing of crystalline structures. The possibility of the synthesizing of new class of nanosize amorphous single-component metals appears only during the realization of purely non-equilibrium quantum-chemical technological processes.

**Conclusions**

The purely non-equilibrium new field of chemical technology is created, in which the equilibrium heat influence practically completely is excluded.

It is shown, that such non-equilibrium technology creates the new unknown possibilities, which are inaccessible till now under usual equilibrium technology.

Thus, at the first time it is occurred to be possible to synthesize nanoamorphous single component metal powders. They are not enough investigated yet, but undoubtedly are very interesting and very promising materials with the grain which size is smaller than 50 nanometer. It is enough to note their high (quadratic) energy saturation as well as all nanosize materials and as all amorphous metals.

We can also indicate the discovered by us some semiconductors properties of these metals. Detailed and full study of these new materials would allow to enlarge considerably the class of well-known self-semiconductors.

**REFERENCES**
1. L.S.Polak and D.I.Slovecky Plazmochimicheskie processy.(Plazmachemical processes) Edited by L.S.Polak p.336 "Nauka" Moscow (1979)
2. N.N.Tunitski,R.T. Malkhasyan, E.S.Jurkin "Excited ions affect on cross-section ion-molecular reaction $H_3^+$ + Ar → $ArH^+$ +$H_2$."Abstract of Papers,VI USSR Conference on Phys of Electron and Atomic Collisions Tbilisi, USSR, .176.(1975).
3. R.T.Malkhasyan,E.S.Jurkin,Tunitski."The ions $H_3$+ excitation and impact this excitation energy on cross-section ion-molecular reaction $H_3^+$ + Ar→ $ArH^+$ + $H_2$."//J Khim.Vysok. Energy 11, No.6.USSR pp.400-402 (1977)
4. Z.Karny, R.Estler, R.Zare "The reaction rate of the Ca + HF (v=1) → CaF + H and Sr + HF → SrF + H reaction"//J Chem.Phys.V.69,p 5199-5122 (1978).
5. G.C.Light "Production OH (v=1) in the reaction $O(^3P)$ +$H_2$(v=1)" //J Chem. Physic V68 2831-2836.(1978)
6. R.T.Malkhasyan ,G.L.Movsesyan , V.K.Potapov, "New method of determine vibrationally excited energy homonuclear molecules" Khim Vysok. Energy 26, No.1, 63 (1992) USSR.
7. R.T.Malkhasyan,E.B.Agababyan "Increasing of anticorrosion properties of technical steels by their treatment of vibrationally excited molecules."//J. Phys.and Chem.Treatment of Materials.No10,p 20-24,(1996)
8. E.B.Agababyan, R.T.Malkhasyan, "Investigating of process modification of condenser foil surfaces by treatment theirs vibrationally excited molecules (quantum-chemical methods)" //J. Physic. and Chem. Treatment of Materials. No1, p.73-77. (1996)
9. R.T.Malkhasyan,E.B.Agababyan,R.K,Karachanyan " Production of a new class of amorphous metals" // Chem.Phys.Reports V15.No10, p 1409-1417,(1996).
10. R.T.Malkhasyan Mat. Res. Soc. Symp.Proc. Vol.400, p. 77-82.(1996)
11. F.E.Luborsky "Amorphous Metallic Alloys". Edited by F.E.Luborsky. Butterworths and Co,USA, pp.41-50. (1983)
12. K.Sudzuki. Kh.Fudzimory, K.Khasimoto "Amorphous metals" pp.37-57 Metallurgia, Moscow,(1987)
13. R. Birringer, "Structure of nanostructured materials" in the book edited by G.G.Hadjipanayis and R.W.Siegel, "Nanophase materials", Kluwer Academic Publishers Dodrecht/Boston/ London/ ,165 (1993).
14. D .J.Sordelet and J.M.Dubois. "Quascrystals: Perspectives and Potential Applications"// MRS Bulletin p.40,V 22 No.11 (1997)

# Fibers/Whiskers

# POLYCRYSTALLINE CERAMIC FIBERS BY WAY OF CONVENTIONAL POWDER PROCESSING

Markus Wegmann, Beat Gut, and Karl Berroth

The Swiss Federal Laboratories for Materials Testing and Research (EMPA)
Ueberlandstrasse 129
CH-8600 Duebendorf
Switzerland

## ABSTRACT
Submicron powders of several advanced ceramics have been successfully extruded by conventional means as fibers and fiber tubes. Alumina ($Al_2O_3$), partially stabilized zirconia (PSZ), silicon nitride ($Si_3N_4$), silicon carbide (SiC), lanthanum strontium manganate (LSM), and zirconium diboride ($ZrB_2$) have each been blended with polyethylene-base thermoplastic binders or water soluble cellulose or polyethylene glycol-base binders and extruded with a laboratory single screw extruder. The green extrudates have diameters ranging between 50 and 150 µm, and the polyethylene-base 150 µm fibers can be drawn down to approximately 40 µm diameter. Tubes with 150 µm outer and 90 µm inner diameter can also be produced. The green LSM and PSZ fibers have been further processed into chopped fiber felts for use as gas distributors/current collectors in an experimental SOFC, and the first attempts at producing simple weaves have been successful. The fibers, tubes, and felts have been successfully debound and sintered, and characterization of the sintered PSZ fibers has revealed densities in excess of 99% and tensile strengths up to 1.0 GPa.

## INTRODUCTION
The demand for high performance ceramic fibers is steadily on the increase for use in a diverse range of applications such as fiber reinforcement, thermal insulation, power transmission, and fluid filtration. This has resulted in a tremendous volume of processes being developed with which to produce high quality material [1]. The majority of these processes are extremely complex, involving the formation of fibers from their constituent elements by a diversity of chemical reactions in the gaseous, liquid and/or solid phase. The goal of this

---

To the extent authorized under the laws of the United States of America, all copyright interests in this publication are the property of The American Ceramic Society. Any duplication, reproduction, or republication of this publication or any part thereof, without the express written consent of The American Ceramic Society or fee paid to the Copyright Clearance Center, is prohibited.

work was to develop a simple method to fabricate continuous polycrystalline fibers of several technologically interesting ceramics, fibers being fabricated using conventional ceramic powder extrusion and ready-made submicron powder grades.

EXPERIMENTAL PROCEDURE

The powders used in this work are listed in Table I. The synthesized powders were conditioned via wet milling, sieving, air-classifying, and spray drying while the commercial powders were used directly without conditioning. Each powder was characterized for particle size (light scattering; Mastersizer X, Malvern Instruments Ltd, UK), density (pycnometry; AccuPyc 1330, Micrometrics Instrument Corp., USA), morphology (scanning electron microscopy; JSM-6300F, Jeol Corporation Ltd., Japan,), and specific surface area (BET; Automated BET Sorptometer, Porous Materials Inc., USA).

Table II lists the polymeric binders selected for the study which were analyzed for characteristic temperatures (TGA/DTA) and density (pycnometry).

Table I. Powders conditioned for fiber extrusion

| Ceramic | $Si_3N_4$ | SiC | $Al_2O_3$ | PSZ | $(La_{0.8}Sr_{0.2})_{0.98}MnO_3$ | $ZrB_2$ |
|---|---|---|---|---|---|---|
| Source | commercial | commercial | commercial | commercial | synthesized | synthesized |
| additives (wt%) | 1.4 MgO  3.6 $Al_2O_3$ | 3.0 C  0.5 B | - | 5.4 $Y_2O_3$ | - | - |
| $\rho_{th}$ (g/cm$^3$) | 3.18 | 3.21 | 3.98 | 6.08 | 6.50 | 4.80 |
| $d_{50}$ (μm) | 0.78 | 0.67 | 0.30 | 0.25 | 0.88 | 0.59 |
| SSA (m$^2$/g) | 12.3 | 20.5 | 11.1 | 11.1 | 7.0 | 9.6 |

Table II. Polymers selected as potential binder systems

| Abbrev. | PEW | HOS | PEG4K PEG10K | MEC |
|---|---|---|---|---|
| Type | Thermoplastic | Thermoplastic | Water-based | Water-based |
| Source | in-house composition | commercial | commercial | commercial |
| Composition | PE/wax blend | polyolefin/wax blend | polyethylene glycol PEG4K: $M_n \approx 4000$ + 70wt% DI $H_2O$ PEG10K: $M_n \approx 10,000$ + 85wt% DI $H_2O$ | methylcellulose $M_n \approx 86,000$ + 85wt% DI $H_2O$ |

The extrusion pastes were prepared in an electrically heated/water cooled mixing head equipped with roller blades attached to an instrumented torque rheometer (Rheocord 9000, HAAKE Mess-Technik GmbH, Germany). The thermoplastic pastes were mixed at a temperature of 150°C while the water-based pastes were mixed at room temperature. A lightly loaded paste was first prepared and then ceramic powder was incrementally added to increase the powder loading up to either the highest possible loading or a maximum run time of 2 hours. Temperature and torque were monitored throughout the procedure and a given paste was considered sufficiently homogenized when the measured torque remained constant with time.

A laboratory increasing-core single-screw extruder attached to the torque rheometer used previously was used to extrude the pastes. The water-based feedstocks were forced onto the screw by a hydraulic ram and extruded at room temperature. The cooled thermoplastic feedstocks were first granulated and then fed into the extruder via a vibratory funnel and extruded between 140 and 160°C. Sapphire dies of 150, 100, and 50 μm diameter were used either singly or in 3-, 5-, or 7-die clusters. Incorporated into the system between the screw end and the die land was a sieve which served to remove any large agglomerates not removed during powder conditioning or destroyed during compounding. Extrusion pressures with the 150 μm dies were generally below 7.5 MPa for the water-based pastes and between 15 and 25 MPa for the thermoplastic pastes.

Single fibers were wound directly onto a variable speed rotating drum while the fiber bundles were first drawn away from the dies by a compressed air venturi before being picked up on the drum. By increasing the rotating speed of the drum the fibers could, depending on the binder system, be drawn down in diameter before solidifying. A relatively simple die modification permitted the 150 μm fibers to be extruded hollow with a 90 μm inner diameter. A schematic of the fiber extrusion setup and a micrograph of the sapphire dies are shown in Figure 1.

The resulting fibers were prepared for debinding and sintering as single strands and in the case of LSM and PSZ as disk-shaped felts. The felts were fabricated by chopping the green fibers to a length of about 10 mm and dispersing them in an aqueous cellulose solution at room temperature. The resulting fiber slurries were then filter pressed and the filter cakes completely dried. Continuous LSM single fibers were also rolled back off the pick-up drums and onto a special spindle which produced a simple woven structure.

Debinding the fibers was achieved by thermal decomposition in air, or in the case of SiC and $ZrB_2$ under nitrogen. Pressureless sintering was performed according to schedules optimized previously for green components made of the same powders used in this study, but formed by processes other than extrusion (slip casting, pressing, etc.). The $Al_2O_3$, LSM, and PSZ fibers were sintered under

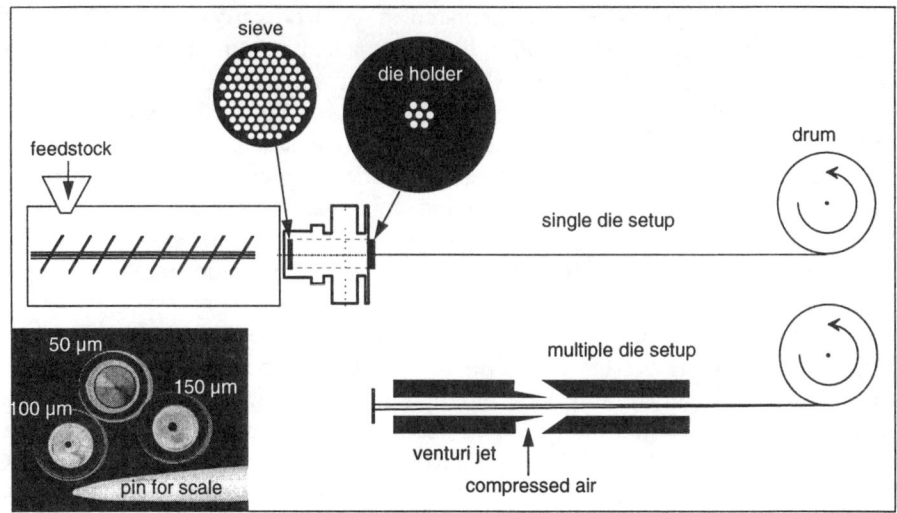

Figure 1. Schematic of the fiber extrusion and drawing process. (Inset: SEM-micrograph of the sapphire dies.)

air, while SiC and $ZrB_2$ material was sintered under argon and $Si_3N_4$ under nitrogen.

The sintered fibers were characterized for density (Archimedes method or pycnometry), microstructure (SEM), and in the case of PSZ for tensile strength (Zwick 1478, Zwick GmbH, Switzerland).

RESULTS/DISCUSSION

All the powders considered were successfully extruded as 150 μm fibers with at least one of the chosen binder systems. Two examples are shown in Figures 2 and 3. The mixtures attempted and their respective powder loadings are summarized in Table III. Obstruction of the dies with agglomerates and binder separation from the powder under high pressures became a serious problem with the 50 μm dies. The latter problem was very pronounced with the water-based pastes and pointed to an unoptimized interaction chemistry between the binders and the powder surfaces. The relatively low powder loadings achieved for these systems adds weight to this argument. While the sieve incorporated into the extruder greatly reduced the problem of agglomerates blocking the dies, the key to solving the problem completely is improved powder conditioning.

Drawing the fibers down in diameter was only possible with the PEW binder system. Drawing of the LSM/PEW fibers was extensively studied and at the optimum drawing temperature 40 μm was determined to be the minimum achievable fiber diameter, regardless of the starting diameter of the fiber.

Table III. Summary of the powder/binder combinations attempted.

| | $Si_3N_4$ (vol%) | SiC (vol%) | $Al_2O_3$ (vol%) | PSZ (vol%) | LSM (vol%) | $ZrB_2$ (vol%) |
|---|---|---|---|---|---|---|
| PEW | 55.1 | 55.1 | x | 53.4 | 58 | 55 |
| HOS | 59.7 | 54.1 | x | x | x | - |
| PEG4K | 47.5 | - | 50.0 | - | x | - |
| PEG10K | - | 52.5 | - | 42.0 | x | - |
| MEC | 49.0 | 52.0 | 51.5 | 42.3 | x | - |

x: not successful     -: not attempted

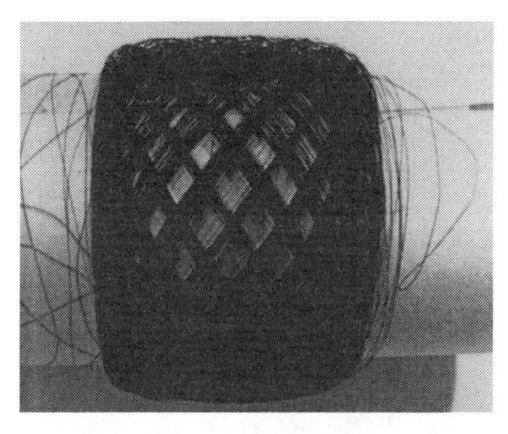

Figure 2. A simple weave of green 150 μm diameter LSM/PEW fibers.

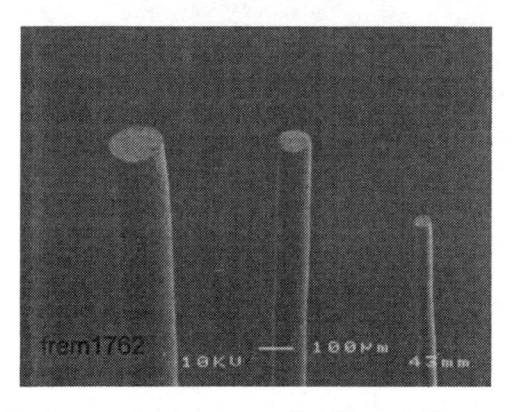

Figure 3. An SEM micrograph of green 150, 100, and 50 μm diameter SiC/HOS fibers.

The best average results of the physical characterization of the solid cross-section sintered fibers are presented in Table IV. A sintered hollow fiber is shown in Figure 4. While the densities achieved for SiC and Al$_2$O$_3$ are not optimal, the excellent sinterability of the utilized submicron powders is apparent considering that between 40 and 50 vol% of the fibers in the green state was occupied by binder. As expected, pressureless sintering of the Si$_3$N$_4$ fibers was not successful, the material decomposing at elevated temperatures, and sintering under elevated nitrogen pressure has not yet been attempted.

Table IV. Maximum achieved properties for sintered solid fibers.

| Fiber | Si$_3$N$_4$ | SiC | Al$_2$O$_3$ | PSZ | LSM | ZrB$_2$ |
|---|---|---|---|---|---|---|
| Binder | - | PEW | PEG4K | PEW | PEW | PEW |
| $\rho$ (g/cm$^3$) | - | 3.06 | 3.93 | 6.08 | 6.40 | 4.80 |
| % of $\rho_{th}$ | - | 96.1 | 98.7 | 100 | 99.0 | 100 |
| shrinkage (%) | - | -15 | -15 | -22 | -25 | - |
| tensile strength (GPa) | - | - | - | 0.8 (⌀118 µm) 1.0 (⌀78 µm) | - | - |

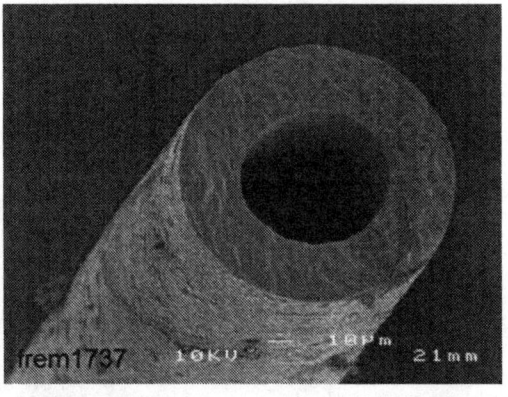

Figure 4. A SEM-micrograph of a sintered hollow LSM/PEW fiber.

First attempts at producing woven structures from the LSM fibers extruded with the PEW binder were successful, as shown in Fig. 2. However, while the green tensile strength of the PEW-based fibers was the highest of the four binder systems considered, it is still too low to permit continuous mechanical weaving and fiber breakage was extremely frequent. The woven material has not been sintered.

The LSM and PSZ felts, however, have been sintered and the LSM material was tested in an experimental SOFC application [2]. A representative disk is shown in Figure 5. While these initial felts prove that the method works, the fibers are still too thick to exhibit any appreciable elasticity in bending and as a result the felts are very stiff and brittle. To counter this problem the fiber diameter must be reduced below the current minimum of 40 µm into the range of 10 µm or less.

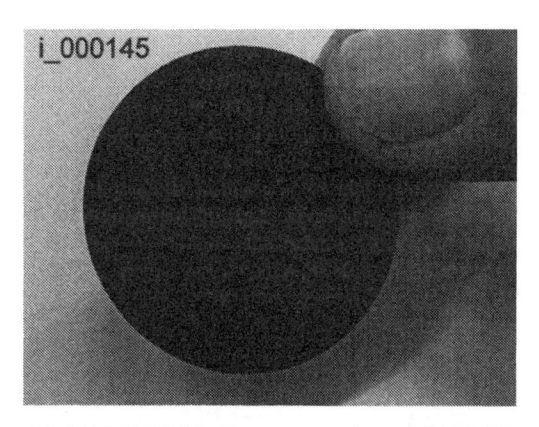

Figure 5. A sintered LSM/PEW felt for an experimental SOFC.

SUMMARY

Thus, fibers down to 50 µm diameter of submicron grades of $Si_3N_4$, SiC, $Al_2O_3$, PSZ, LSM, and $ZrB_2$ have been successfully extruded with various water-based and thermoplastic binder systems. With an appropriate binder system, the fiber diameter can be further reduced by drawing down the fiber before it solidifies. Hollow fibers with 150 µm outer diameter and 90 µm inner diameter have also been extruded. The green fibers have been processed further into simple woven structures and felts. While the general procedure from powder conditioning through fiber extrusion has been proven, several problems remain to be solved. Powder conditioning must be improved to prevent periodic obstruction of the dies with oversized material and therefore enable true continuous fiber extrusion. The binder/powder compatibility must be improved to prevent separation of the binder from the powder under pressure. The binders must also be improved with respect to green strength and drawability if continuous green fibers are to be mechanically woven and if the sintered material is to have the desired elasticity.

ACKNOWLEDGMENTS
The authors would like to thank G. van de Goor, R. Bächtold, H. Schindler, M. Kampus, S. Fuso and D. Heusser for their respective contributions to this work, EMPA Dept. 174 for their metalworking services, and the Swiss Federal Department of Energy (BFE) and the Commission for Technology and Innovation (KTI) for their contributions to the project funding.

REFERENCES

[1] T. F. Cooke, "Inorganic Fibers - A Literature Review," *J. Am. Ceram. Soc.*, **74** (12) 1991, 2959-2978.

[2] B. Gut and M. Wegmann, "La-Perovskite Felts as Air Distributors/Current Collectors in SOFCs," submitted to *the Proceedings of the 3rd European SOFC Forum*, Nantes, France, March 1998.

# SYNTHESIS OF $Ta_{0.5}Ti_{0.5}C$ AND $Ta_{0.33}Ti_{0.33}Nb_{0.33}C$ WHISKERS

M. Johnsson, M. Carlsson, N. Ahlén, and M. Nygren
Department of Inorganic Chemistry, Stockholm University, S-106 91 Stockholm, Sweden.

## ABSTRACT

A route for synthesis of solid solution (Ta,Ti,Nb)C whiskers and fibers in a yield of 70-80 vol% has been established. The whiskers are 0.1-1µm in diameter and 10-30µm in length. They are straight and have smooth surfaces.
The (Ta,Ti,Nb)C whiskers have been synthesized carbothermally by a Vapour-Liquid-Solid (VLS) growth mechanism in the temperature region 1150-1400°C in argon atmosphere. The starting materials consisted of $Ta_2O_5$, $TiO_2$, $Nb_2O_5$, C, Ni, and NaCl.
The VLS-mechanism is complex and involves formation of gaseous $MO_yCl_x$ species (M = Ta, Ti, or Nb), which are transported to the nickel catalyst. However, the overall reaction is a straightforward carbothermal reduction. The main impurity in the whiskers is oxygen, the product also contains carbide particles, minor amounts of unreacted carbon and remnants of the Ni catalyst.

---

To the extent authorized under the laws of the United States of America, all copyright interests in this publication are the property of The American Ceramic Society. Any duplication, reproduction, or republication of this publication or any part thereof, without the express written consent of The American Ceramic Society or fee paid to the Copyright Clearance Center, is prohibited.

# INTRODUCTION

Whiskers can be used as reinforcing material in ceramics. The thermal expansion coefficient of the reinforcer and the matrix should be matched and they should be chemically compatible. In this article we present a carbothermal preparation route for whiskers that are solid solutions of TaC, NbC, and TiC synthesized by a carbothermal vapor-liquid-solid (VLS) growth mechanism.

The basic VLS growth mechanism involves a vapor phase transport of one or more of the reacting species to a melted catalyst droplet where the desired whisker is grown, see figure 1. The metal catalyst must have a melting point or a eutectic point that is lower than the synthesis temperature – otherwise the VLS-mechanism would not be operative. It must also be able to dissolve the whisker components. Nickel has proved to be a good catalyst for the carbide whiskers we present in this article.

The whiskers were grown using $Ta_2O_5$ as Ta-source, $Nb_2O_5$ as Nb-source and $TiO_2$ as Ti-source. Nickel was used as catalyst, NaCl was added as a volatilizing agent for transportation of Ta, Nb, and Ti in the gas phase. The reaction took place in an inert Ar atmosphere. Equilibrium calculations made with the program HSC [1] indicate that Ta is mainly transported in gas phase as $TaOCl_3(g)$, Nb as $NbOCl_3(g)$ while Ti is transported mainly as $TiCl_3(g)$.

The carbothermal synthesis of TaC and $TiC_xN_{1-x}$ whiskers has recently been discussed [2-4]. The overall chemical reaction is a straightforward carbothermal reduction of the transition metal oxides, but the reactions that actually takes place are much more difficult to predict and to study. The following reactions or very similar ones are thought to represent the actual mechanisms in the different systems.

Formation of Cl(g):

$$NaCl(l) \longrightarrow Na(g) + Cl(g) \quad (1)$$

Formation of transition metal oxochlorides and chlorides:

$$Ta_2O_5(s) + 6Cl(g) + 3C(s) \longrightarrow 2TaOCl_3(g) + 3CO(g) \quad (2a)$$
$$Nb_2O_5(s) + 6Cl(g) + 3C(s) \longrightarrow 2NbOCl_3(g) + 3CO(g) \quad (2b)$$
$$TiO_2(s) + 3Cl(g) + 2C(s) \longrightarrow TiCl_3(g) + 2CO(g) \quad (2c)$$

Reactions at the Ni catalyst (not balanced):

$$TaOCl_3(g) + C(s) + Ni(l) \longrightarrow Ni(Ta,C)(l) + CO(g) + 3Cl(g) \quad (3a)$$
$$NbOCl_3(g) + C(s) + Ni(Ta,C)(l) \longrightarrow Ni(Ta,Nb,C)(l) + CO(g) + 3Cl(g) \quad (3b)$$
$$TiCl_3(g) + C(s) + Ni(Ta,Nb,C)(l) \longrightarrow Ni(Ta,Nb,Ti,C)(l) + 3Cl(g) \quad (3c)$$

Whisker growth (not balanced):

$$Ni(Ta,Nb,Ti,C)(l) \longrightarrow (Ta,Nb,Ti)C(s) + Ni(l) \quad (4)$$

Overall carbothermal reaction:

$$Ta_2O_5(s) + Nb_2O_5(s) + 2TiO_2(s) + 20C(s) \longrightarrow 6(Ta,Ti,Nb)C(s) + 14CO(g) \quad (5)$$

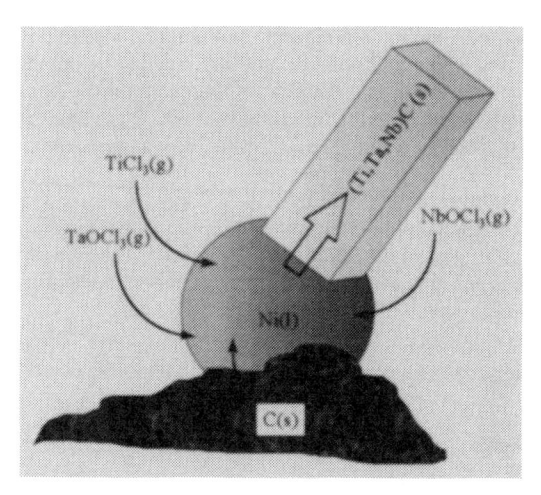

Figure 1. The carbothermal vapor-liquid-solid (VLS) growth mechanism involves transportation of the transition metals as chlorides and oxochlorides to the nickel catalyst that is in contact with carbon. The transition metal and carbon dissolves into the catalyst droplet that becomes supersaturated and precipitate the carbide whisker (Ta,Ti,Nb)C.

EXPERIMENTAL

The starting materials giving the best whisker yield are given in Table I.

Table I. Raw material used in the synthesis.

| Substance | Purity (wt%) | Manufacturer | Particle size | Comment |
|---|---|---|---|---|
| $Ta_2O_5$ | 99.9 | Cerac | <40µm | Ceramic grade |
| $Nb_2O_5$ | 99.9 | Cerac | <40µm | Ceramic grade |
| $TiO_2$ | 99.9 | Aldrich | 0.2-0.6µm | Anatase |
| C | - | Degussa (FW200) | 13nm forming flocculent agglomerates | Carbon black (containing 21wt%volatiles) |
| NaCl | 99.5 | Akzo | <5µm | |
| Ni | 99.9 | Cerac | <40µm | |

The aim was to synthesize $Ta_{0.5}Ti_{0.5}C$ and $Ta_{0.33}Ti_{0.33}Nb_{0.33}C$ whiskers. The molar ratios giving the highest whisker yield were:

$Ta_{0.5}Ti_{0.5}C$;        $Ta_2O_5 : TiO_2 : C : Ni : NaCl = 1 : 2 : 13 : 0.15 : 1.5$

$Ta_{0.33}Ti_{0.33}Nb_{0.33}C$;    $Ta_2O_5 : TiO_2 : Nb_2O_5 : C : Ni : NaCl = 1 : 2 : 1 : 20 : 0.20 : 2$

The starting materials were mixed in a blender, and portions of about 15 g were placed into a graphite crucible covered with a lid having a number of small holes in order to allow a controlled gas exchange between the reactor and the surrounding atmosphere. The reaction mixtures were heated to the reaction temperature; 1250°C and held at this temperature for 4h. The experiments were carried out in a graphite furnace (Thermal Technology) at atmospheric pressure in flowing Ar atmosphere.

The obtained whiskers were characterized by their X-ray powder diffraction patterns (XRD), obtained in a Guinier-Hägg focusing camera. The morphology of the whiskers was investigated in light and scanning electron microscopes (SEM, JEOL 880). The whisker yields were estimated from SEM micrographs and are expressed in vol%. The whisker composition was analyzed by use of EDS.

RESULTS AND DISCUSSION

The synthesis route described above gave $Ta_{0.5}Ti_{0.5}C$ and $Ta_{0.33}Ti_{0.33}Nb_{0.33}C$ whiskers in a yield of 70 - 80 vol%. The whiskers produced had typically diameters in the range 0.5 - 1 µm. All whiskers had a length of 10 - 30 µm (see figure 2).

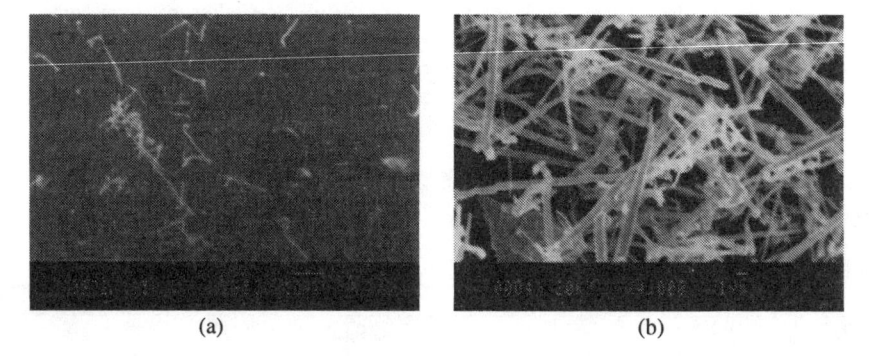

Figure 2. SEM micrograph of (a) the $Ta_{0.5}Ti_{0.5}C$ whisker product and (b) the $Ta_{0.33}Ti_{0.33}Nb_{0.33}C$ whisker product. Typical whisker diameters were 0.1 - 0.5 µm having a length in the range 10 - 30 µm. The whisker yield is estimated to be 70 - 80 vol%.

There are two competing reactions taking place at the synthesis temperature, the VLS growth of whiskers and the direct reaction between carbon and the transition metal oxides resulting in carbide particles. The VLS growth mechanism can be favored by optimizing the molar ratios of the starting materials and the reaction temperature. The reaction temperature should be kept as high as possible to reduce the residual amounts of oxygen as much as possible. However, at a too high synthesis temperature the whiskers formed start to sinter together and the oxochloride and chloride gaseous species show a tendency to escape from the reaction chamber before reacting at the catalyst. The optimum synthesis temperature was found to be close to 1250 °C. Overall chemical analysis show that the whisker product contain some oxygen and remnants of Ni (Table II). The Ni residuals could easily be removed by washing in diluted HCl(aq) indicating that Ni is not incorporated in the whisker lattice. The average compositions of whiskers and particles according to chemical analysis data are $Ta_{0.49}Ti_{0.51}C_{0.91}O_{0.09}$ and $Ta_{0.32}Ti_{0.34}Nb_{0.34}C_{0.90}O_{0.10}$ respectively assuming that oxygen replaces carbon and that some excess carbon remains after synthesis.

Table II. Overall chemical analysis of the whisker products.

| Content (at.%) | (Ta,Ti)C | (Ta,Ti,Nb)C |
|---|---|---|
| Ta | 21.48 | 15.38 |
| Ti | 22.22 | 16.03 |
| Nb | - | 15.97 |
| Ni | - | 1.77 |
| C | 52.27 | 46.15 |
| O | 4.04 | 4.69 |

EDS analyses of the metal content in individual whiskers gives for hand that there is a spread in the whisker composition (see Figure 3 and Table III). It is obvious that the whisker composition is not the same as the overall composition indicating a composition difference between whiskers and particles.

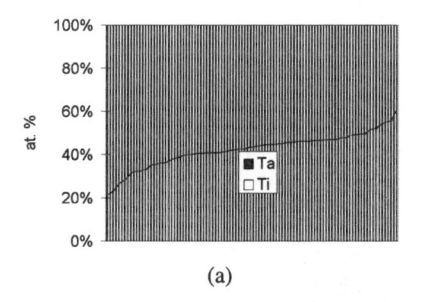

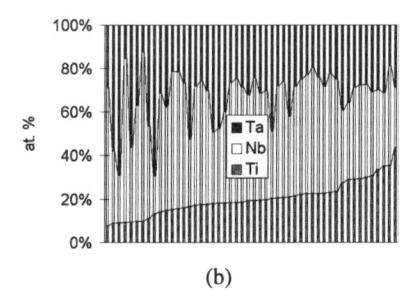

Figure 3. The content of Ta, Ti, and Nb in (a) $Ta_{0.5}Ti_{0.5}C$ whiskers based on 100 EDS analysis on different whiskers, and (b) $Ta_{0.33}Ti_{0.33}Nb_{0.33}C$ whiskers based on 50 EDS analysis on different whiskers.

Table III. Transition metal content in the whisker products according to EDS analysis.

| Content (at.%) | $Ta_{0.5}Ti_{0.5}C$ | $Ta_{0.33}Ti_{0.33}Nb_{0.33}C$ |
|---|---|---|
| Ta | 58 ± 8 | 33 ± 13 |
| Ti | 42 ± 8 | 20 ± 8 |
| Nb | - | 47 ± 13 |

The synthesis result is very much depending on the carbon source used. The best carbon source found is a type of amorphous carbon black. Thermodynamic calculations with the program HSC [1] gives for hand that amorphous carbon react with the transition metal oxides at a much lower temperature compared to graphite (see Figure 4). The carbon black powder contain 21wt.% volatile matter that were slowly burnt away during the heating procedure. We believe that the fluffy consistence of this carbon source promotes the formation of whiskers by providing space for the growing whiskers and by improving the possibility for the gaseous species to reach the Ni-droplets located at the surface of the flocculent agglomerates.

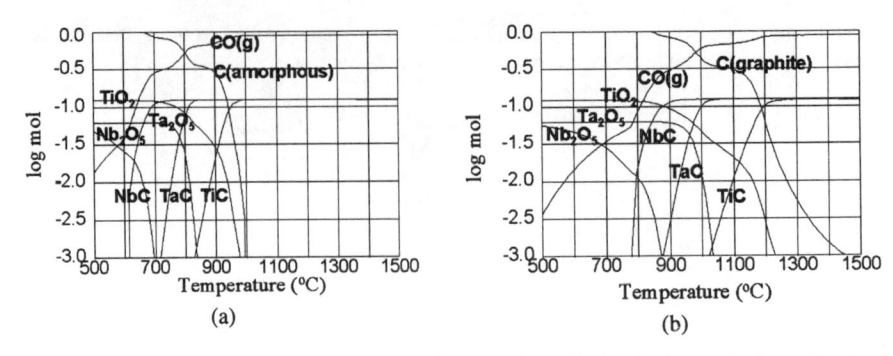

Figure 4. Calculated formation temperatures of TaC, TiC and NbC during carbothermal reduction of $Ta_2O_5$, $TiO_2$ and $Nb_2O_5$, respectively, with (a) an amorphous carbon source and (b) with graphite.

A requirement for the VLS-growth mechanism to work is that the liquidus temperature of the catalyst droplet is lower than the reaction temperature. A Ni-droplet terminating a $Ta_{0.33}Ti_{0.33}Nb_{0.33}C$ whisker proving that the VLS-mechanism is operative at 1250°C is shown in Figure 5.

The eutectic temperatures for the systems Ni-Ta-C, Ni-Ti-C, and Ni-Nb-C where a melt is in contact with TaC, TiC and NbC are found to be higher than 1250°C (see Table IV). However, the eutectic temperature is be expected to be lower when Ta and Ti and/or Nb are dissolved in the Ni-droplet. In addition also oxygen may help to lower the eutectic temperatures.

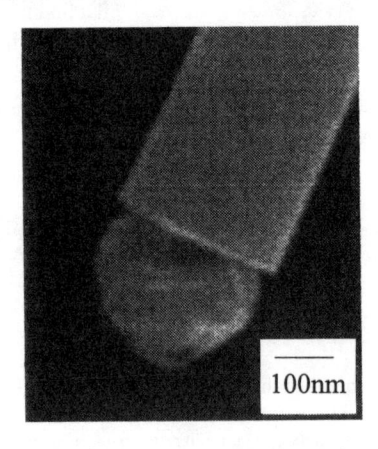

Figure 5. Ni-droplet terminating a $Ta_{0.33}Ti_{0.33}Nb_{0.33}C$ whisker synthesized at 1250°C.

Table IV. The lowest temperature where liquid is in contact with the carbide phase MC (M = Ta, Ti or Nb).

| System | Eutectic temperature [5] | Calculated eutectic temperature | Calculated composition at the eutectic temperature |
|---|---|---|---|
| Ni-Ti-C | 1280°C | 1318°C | Ni-2.7at%Ti-10at%C |
| Ni-Nb-C | 1115°C / 1320°C* | 1262°C | Ni-8.8at%Nb-6.8at%C |
| Ni-Ta-C | 1370°C | - | - |

The program ChemSage [6] was used to calculate eutectic temperatures and eutectic compositions using data from the SGTE binary alloy database; no data were available for Ta.
* Two values are given.

CONCLUSIONS

A carbothermal reduction route has been used to synthesize $Ta_{0.5}Ti_{0.5}C$ and $Ta_{0.33}Ti_{0.33}Nb_{0.33}C$ whiskers. The synthesis temperature was 1250 °C.
The whiskers are 0.5 - 1 µm in diameter and 10-30 µm in length. The as-synthesized products contained 70 - 80 vol% whiskers and 20 - 30 vol% carbide particles. The main impurities are oxygen and nickel.
The whiskers were grown via a vapor-liquid-solid (VLS) growth mechanism, using $Ta_2O_5$, $TiO_2$, and $Nb_2O_5$ respectively, as sources for Ta, Ti and Nb. Nickel was used as catalyst, NaCl was used as a volatilizing agent for transportation of Ta, Ti and Nb in gas phase. Equilibrium calculations indicate that Ta is transported mainly as $TaOCl_3(g)$, Nb as $NbOCl_3(g)$ and Ti as $TiCl_3(g)$. Carbon black was used as carbon source.

ACKNOWLEDGMENT

This work has been financially supported by the Swedish Foundation for Strategic Research and the Swedish Research Council for Engineering Sciences.

REFERENCES

[1] HSC 3.0, Computer program developed by A. Roine *et al.*, Outokumpo research Oy, Pori, Finland (1997).
[2] N. Ahlén, M. Johnsson, and M. Nygren, *J. Am. Ceram. Soc. 79 (1996) pp. 2803.*
[3] M. Johnsson and M. Nygren: "Carbothermal synthesis of TaC whiskers *via* a vapour-liquid-solid growth mechanism", *J. Materials Research, 12 (1997) pp. 2419-2427.*
[4] N. Ahlén, M. Johnsson, and M. Nygren, In *"Processing and Handling of Powders and Dusts"*, Eds. T.P. Battle and H. Henein, The Minerals, Metals & Materials Society, (1997), pp. 57.
[5] P. Villars, A. Prince, and H. Okamoto: "Handbook of ternary alloy phase diagrams", ASM International (1995).
[6] Chemsage 4.0, Computer program developed by G. Eriksson *et. al*, GTT-Technologies, Herzogenrath, Germany (1997).

# Crystal Growth

# A NEW APPROACH TO GROWTH OF BULK ZnO CRYSTALS BY MELT SOLIDIFICATION

J. E. NAUSE, G. AGARWAL
Cermet, Inc., 1019 Collier Road, Suite C1 Atlanta GA 30318

D.N. HILL
Georgia Institute of Technology, Atlanta, GA 30332

## ABSTRACT

Zinc oxide crystals were obtained by solidification of a ZnO melt. The ZnO melt achieved using a novel, patented technique involving pressurized radio frequency induction heating of ZnO powder, contained in a water-cooled crucible. The ability to obtain a pool of molten ZnO would enable large diameter ZnO crystals, using conventional melt growth processes. Centimeter-sized crystals were obtained in the preliminary experiments by cooling the ZnO melt. These crystals were analyzed for crystalline perfection and stoichiometry using x-ray diffraction and photoluminescence. The semiconducting properties of these crystals were also measured. This technology can potentially provide large, low cost ZnO single crystal wafers for use in the fabrication of GaN blue diodes and blue lasers, high temperature / high power FETs, as well as homoepitaxy of ZnO wide band-gap devices. Single crystal zinc oxide also has potential application in the piezoelectric device market.

## INTRODUCTION

High quality single crystal ZnO has a large number of potential commercial applications. These include substrates for group III-V based optoelectronic devices, active blue diodes and acoustic wave devices. The group III-V based materials (primarily GaN) have attracted much attention in recent years for their emerging applications such as detectors and sensors[1, 2], blue light emitting diodes (LED) [3] and laser diodes [4] in the ultraviolet regions. The emergence of group III nitride based devices promises to substantially open new markets with their applications. Full color large area displays using red, green and blue LEDs will become a large market for use in video billboards, scanners, optical storage, traffic lights and flat panel TVs. These short wavelength devices could be used in producing white light, which has an even larger potential market [5]. Besides the applications listed above, these nitride devices have potential applications in high temperature electronics [6] and radiation resistant devices used in nuclear reactors.

---

To the extent authorized under the laws of the United States of America, all copyright interests in this publication are the property of The American Ceramic Society. Any duplication, reproduction, or republication of this publication or any part thereof, without the express written consent of The American Ceramic Society or fee paid to the Copyright Clearance Center, is prohibited.

Most of the commercially manufactured GaN LED's and laser device are grown on sapphire or silicon carbide substrates. In GaN growth on (0001) sapphire substrates, one has to cope with a 16% lattice mismatch [7], yielding dislocation densities as high as $10^7$ - $10^8$ / $cm^2$. ZnO has a wurtzite crystal structure (isomorphous with GaN) with a lattice constant of 3.2498 Å in the $a$ direction. This results in a lattice mismatch of only 2.2% between ZnO and pure GaN and a perfect match between ZnO and $In_{0.22}Ga_{0.78}N$ [8, 9]. These properties make ZnO an attractive alternative for sapphire and silicon carbide. Several research groups have recommended the use of ZnO crystalline substrates for GaN growth. [10, 11]. However the high cost of these crystals and limited availability of substrate quality single crystals have restricted research and investigation in this area. Studies of GaN growth on ZnO have been limited by the instability of ZnO in reducing atmospheres and at high temperatures. In a recent article [12], researchers at Bell Labs and Lucent Technologies reported growing GaN films using ZnO substrates. They reported that ZnO had a good lattice match and ideal structure for GaN epitaxy. However, the major drawback in using ZnO as a substrate material was the lack of availability of high quality crystals, cost and instability of the crystal at high temperature in a reducing atmosphere.

It is known that zinc oxide has a band gap of 3.5 eV, which is very similar to that of GaN. Recently it has been demonstrated that ZnO can be used as a source for short wavelength lasing [13]. The intensity of optical excitation in ZnO epitaxial films grown on sapphire substrates was comparable to GaN. This is perhaps the most tantalizing use for ZnO. By combining established ZnO MOCVD technology with readily available, low cost, high quality ZnO substrates for homoepitaxy of ZnO films, the major problem of lattice mismatch currently facing the GaN technology is avoided.

Single crystal ZnO is unavailable because the material can not be grown using standard melt growth processes. In standard melt growth atmospheres, the ZnO decomposes upon heating into a very defective structure. Currently, the only techniques for growing bulk single crystal ZnO are sublimation and hydrothermal growth [14, 15]. The hydrothermal technique is relatively less expensive, but the crystal quality is poor, and the ultimate size is limited. The sublimation technique is a slow process, which increases the cost of crystals produced.

Standard crystal growth techniques rely on growth from a melt or liquid. The melting point of stoichiometric ZnO has been difficult to obtain, since zinc oxide decomposes above 1800°C. At atmospheric pressure and temperatures near the melting point, ZnO rapidly loses lattice oxygen and reduces to a highly defective, non-stoichiometric $ZnO_{(1-x)}$. Estimates for the melting point are approximately 1880°C. At this temperature, only noble metal crucibles are candidates for the containment of molten ZnO. However, the melt atmosphere required to protect the crucible from oxidation reduces the ZnO lattice. Conversely, the melt atmosphere required to maintain ZnO stoichiometry quickly oxides even iridium at 1880°C. The oxidation from the crucible would readily react with the ZnO in such hostile growth environments, introducing impurities into the lattice.

The technique used by Cermet eliminates the problem of crucible reactivity, crystal contamination and decomposition by using a high pressure variation of the established skull melting technique (for which patents will issue in Fall of 1998). The system is capable of melting materials at temperatures in excess of 3600°C and pressures in excess of 100 atmospheres.

EXPERIMENTAL PROCEDURE

The schematic of the high pressure skull melting apparatus is shown in Figure 1. The vessel consists of steel chamber capable of withstanding over 20 atmospheres of internal gas pressure. The vessel was constructed in a bell-type arrangement, the top of which was raised and lowered between melts. A two inch diameter, high pressure window was installed on the top flange of the vessel. Through this window, the crucible could be sighted for *in-situ* temperature measurements using a disappearing filament pyrometer. Pressure-tight feed-throughs for the induction coils, cooling water and pressurizing gas were also built into the vessel. A 50 kW radio frequency generator capable of oscillating in the range of 0.5 to 5 MHz was used to energize the induction coils. A motorized variac was used to vary power level in a controlled and reproducible fashion. The temperature of the crucible cooling water and generator cooling water was closely monitored.

In general, radio frequency melting is accomplished by placing a suitably conductive material in an alternating magnetic field. Induced fields in the material produce eddy currents which produce joule heating in the charge. The "skull melting" technique utilized here consisted of a segmented, water-cooled, copper crucible, which cooled the outer surface of the material being melted. This resulted in a thin layer of material adjacent to the crucible wall, which enclosed the melt in a shell of material of the same composition. This technique eliminated the possibility of any crucible related impurities by eliminating any reaction of the crucible with the molten material.

The powder used for this study was a 99.9 % pure ZnO powder (KADOX-930, Zinc Corporation of America, Monaca PA). This powder was loaded in a 3 inch-diameter crucible, and placed inside the vessel. The generator coils were placed around this crucible and the vessel was closed and sealed. After evacuating for 30 minutes, the vessel was pressurized using oxygen gas. The over-pressure of oxygen was varied from 2 to 10 atmospheres in different melting experiments. The melting was carried out at power levels of 2-20 kW. The melt was held at the maximum power level for 30-60 minutes and then cooled at ~50°C/min.

After each melt experiment, the resulting crystals were visually inspected for appearance and color. A combination of optical and scanning electron microscopy (SEM) was used to evaluate the crystal texture and microstructure. X-ray diffraction (using a Philip PW 1800 Diffractometer) was used to determine crystal structure and the presence of any impurities. A Bede QC2a, double crystal x-ray diffractometer was used to generate rocking curves, indicating crystalline perfection. In addition to these techniques, photoluminescence spectrum were gathered by radiating the sample with 5 mW of power from a HeCd laser at 325 nm. The spectrometer was a Spex 1404, and was detected with a GaAs Cathode Photomultiplier Tube. Electrical properties such as resistivity, intrinsic charge carrier concentrations, and mobility were also measured. For these measurements, a 0.253 cm crystal was used and ultrasonic iron was used to apply the contacts.

RESULTS AND DISCUSSIONS

The melting of ZnO occurred in the range of 9 to 16 kW. Due to formation of a cold crust above the melt, the temperature of the melt could not be obtained accurately. The estimated temperature of the melt was >1600°C. Due to fast uncontrolled cooling in the present set-up, the molten pool of liquid ZnO solidified as elongated crystals ~ 3-4 mm in cross-section and 15-25 mm long. The ZnO crystalline phase was verified by X-ray diffraction. The color of the crystals varied from clear colorless to yellow and in some cases, dark brown crystals. These color variations were a result of variations in the melt atmosphere. It has been reported [16] that the dark color in the ZnO crystals resulted from presence of oxygen deficient ZnO lattice. Upon heating, ZnO has a tendency to reduce, which causes zinc interstitials. Figure 2 shows picture of some clear to light yellow colored crystals.

Rocking curve analysis confirmed that the colorless crystals were of a very high quality, with a full width at half maximum of approximately 125 arcseconds (Figure 3). A broader FWHM (300-500 arcsecond) was evident for the brown-colored crystals. To further evaluate the crystal properties and to verify the stoichiometry of the colorless crystals, photoluminescence spectroscopy was performed. Data collected at 78 K and room temperature yielded peaks at 373 nm (~3.23 eV) and 377 nm (~3.29 eV), respectively (Figure 4). These peaks were slightly lower in energy than the bandgap energy, but no deep state emission was evident.

Table I summarizes the electrical properties of a clear, colorless ZnO crystal. The crystal was n-type as expected. The charge carrier density was $5 \times 10^{17}$ cm$^{-3}$ as compared to $1 \times 10^{17}$ cm$^{-3}$ typically found in vapor grown ZnO crystals.

Table I. Summary of Electrical Properties for ZnO.

| Structure: ZnO | | Excitation Current: 0.1 A | |
|---|---|---|---|
| Sample Type: N | | Current Reversal: On | |
| Thickness: 0.253 cm | | Field Reversal: On | |

| Field (G) | Rh (cm$^3$/C) | Resistivity (ohm.cm) | Density (cm$^{-3}$) | Mobility (cm$^2$/V.s) | Temp (K) |
|---|---|---|---|---|---|
| 9999 | -1.2372E+01 | 9.430E-02 | -5.045E+17 | -1.312E+02 | 296 |
| 5000 | -1.2379E+01 | 9.429E-02 | -5.043E+17 | -1.312E+02 | 296 |
| 9999 | -1.7195E+02 | 5.770E-01 | -3.640E+16 | -2.980E+02 | 78 |

The resistivity and charge carrier density remained virtually constant at different applied fields. As expected, temperature had a significant effect on these properties. The crystals were conductive, and the crystal contacts were ohmic with respect to the field. The charge carrier density and hence the mobility and conductivity of these crystals can be easily manipulated by varying the oxygen over-pressure during melting. Melt processing also allows doping the crystals with an n-type or p-type element.

CONCLUSIONS

In summary, we have demonstrated that clear, colorless zinc oxide can be melted and crystallized using Cermet's melt growth apparatus. The technique will enable stoichiometric zinc oxide to be melted, and crystals to be formed from the melt. Large diameter crystals can be pulled from this melt using the Czochralski technique. The crystal formation costs will be significantly lower than currently available techniques, since growth rates in a molten liquid process are 10 to 100 times faster than vapor growth rates, and capital costs are relatively low. This technique would allow ZnO wafers to be produced at a reasonable price.

The surface and bulk acoustic wave device market will also get a boost from the availability of these crystals. Currently, the only way to use the piezoelectric properties of ZnO are by thin film epitaxy [17, 18]. With bulk crystals, the device manufacturers will be able to use the high coupling coefficient and high conductivity of zinc oxide for unique filters and acoustic wave devices in the frequency range of 10-50 MHz.

Availability of bulk single crystal ZnO could accelerate the research of ZnO as a source for blue to ultraviolet lasers. A high quality, single crystal substrate for homoepitaxy of ZnO eliminates one of the major problems currently facing the GaN technology; substrate / epitaxy lattice mismatch. Readily available substrates, combined with existing ZnO MOCVD technology, make ZnO an attractive lasing system.

**References:**

[1] M. A. Khan, J. N. Kuznia, D. T. Olson , J. M. Van Hove, M. Blasingame and L. F. Reitz, Applied Physics Letters, Vol 60, page 2917, 1992.
[2] K. S. Stevens, M. Kinniburge and R. Beresford in Applied Physics Letters, Vol 66, page 3518, 1992.
[3] S. Nakamura, M. Senoh, N. Iwasa, S. I. Nagahama, T. Yamasa and T. Mukai in Journal of Applied Physics, Vol 34, page L1332, 1995.
[4] S. Nakamura, M. Senoh, N. Iwasa, S. I. Nagahama, T. Yamada and T. Matsushita in Applied Physics Letters Vol 31, page 498, 1995.
[5]. Bill Imler, "High Volume Growth and Fabrication of AlInGaN LEDs", at Fifteenth Conference on Crystal Growth and Epitaxy, Fallen Leaf Lake, CA, June 1-4, 1997.
[6] M. W. Shin and R. J. Trew in Electron. Lett. Vol 31 page 489 (1995).
[7]. Markus Kenp, et. al., in Compound Semiconductor, Special Issue 1997, pp. 26.
[8] M. Matsuoka, Y. Yoshimoto, T. Sasaki and A. Katsui in Journal of Electronic Materials, Vol. 21, page 157, 1992.
[9] S. Strite and Morkoc, Journal of Vacuum Science and Technology A, Vol 14, page 1237, 1992.
[10] T. Detchprohm, K. Hiramatsu, H. Amano, and I .Akasaki in Applied Physics Letters, Vol 61, page 2668, 1992.
[11] T. Matsuoka in Journal of Crystal Growth, Vol 124, page 433, 1992.
[12] E. S. Hellman, D. N. E. Buchanan, D. Wiesmann and I. Brener, "Growth of Ga-Face and N-Face GaN films using ZnO Substrates, in Nitride Semiconductor Research, Volume 1, Article 16, 1996.
[13]. D. M. Bagnall et al., in Appl. Phy. Lett. Vol 70, No. 17, page 2230, 1997.
[14]. E. S. Hellman, C. D. Brandle, L. F. Schneemeyer, D. Weismann, I. Brener, T. Siegrist, G. W. Berkstresser, D. N. E. Buchanan, E. H. Hartford., MRS Internet

J. Nitride Semiconductor Res., Vol 1, Article 1, 1996.
[15]. E. Klob, R. Laudise, J. Am. Ceram Soc., Vol 48, pp. 342, 1965.
[16] M. A. I. Johnson, Journal of Electronic Materials, Vol 25, No 5, pp 855-62, 1995.
[17]. Gordon S. Kino, "Acoustic Waves: Devices, Imaging and Analog Signal Processing", Prentice Hall Signal Processing Series, A. V. Oppenheim, editor, 1987 page 557.
[18]. N. Tamaki, A. Onodera, T. Sawada, and H. Yamashita, J. Korean Phys. Soc. Vol 29, suppl. issue, Nov 1996.

**Acknowledgment:**

Cermet would like to acknowledge Dr. April Brown and Dr. A. Doolittle at Georgia Institute of Technology, Atlanta GA. for their assistance in providing the photoluminescence data. We would like to thank Dr. Dave Look at Wright State University and Dr. Cole Litton at Wright Patterson Air Force Base for the electrical property measurements on the ZnO crystals.

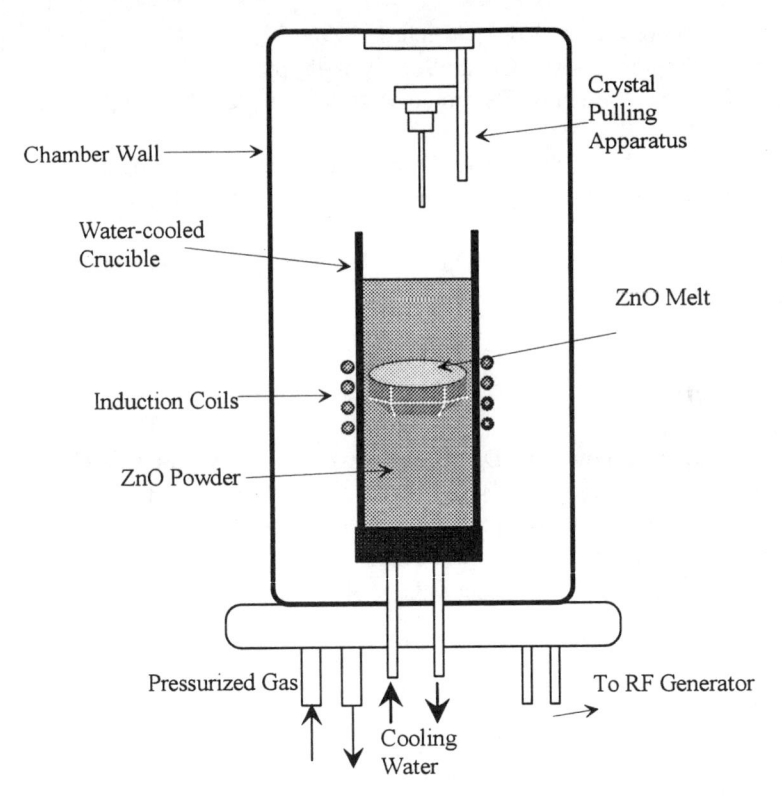

Figure 1. A Schematic Representation of the High Pressure Melting Apparatus Used in the Solidification of ZnO Single Crystals.

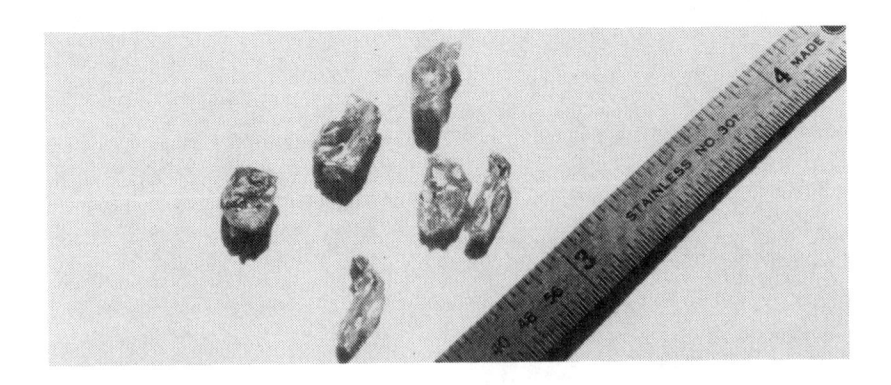

Figure 2. Bulk Crystals of ZnO Produced Using Cermet's Melt Growth Process (Scale in Inches).

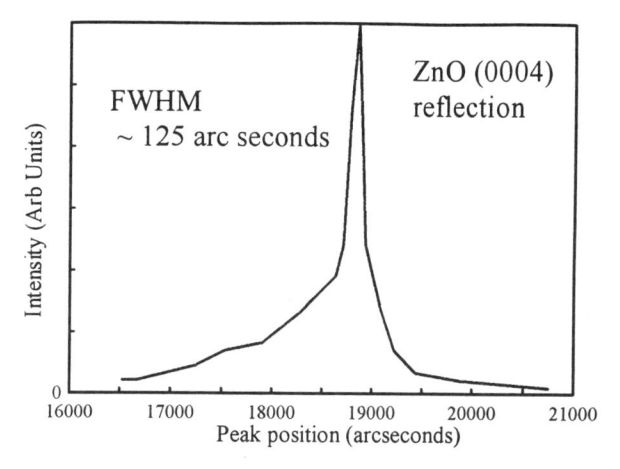

Figure 3. X-ray Rocking Curve for Melt Grown ZnO, Indicating Full Width at Half Maximum of ~125 Arcseconds.

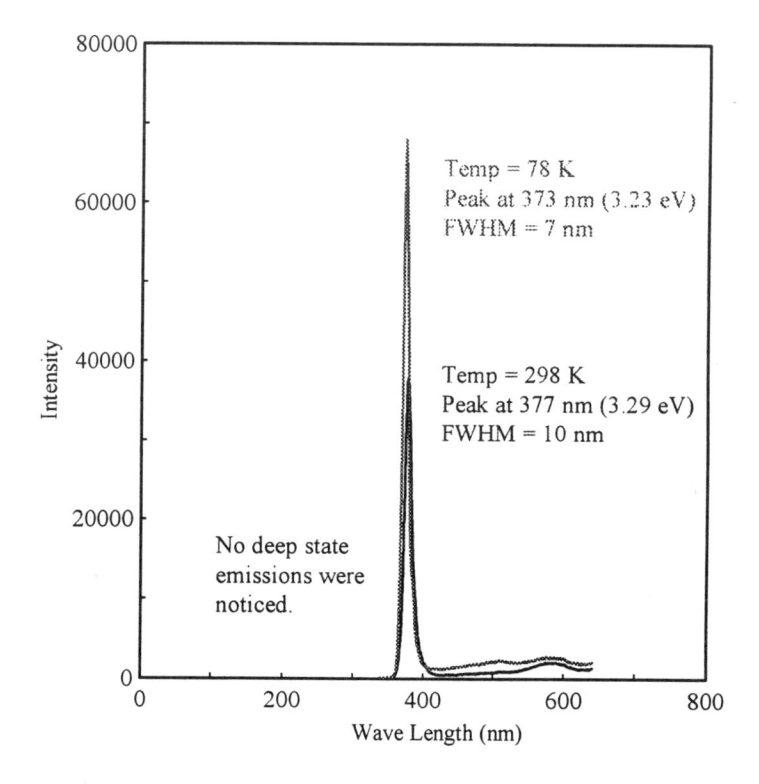

Figure 4. Photoluminescence Spectrum for Bulk ZnO, Obtained at 78K and 298K.

# Porous Metals

# PREPARATION OF POROUS METALS FROM CERAMIC PRECURSORS

Steven M. Landin and Dennis W. Readey
Colorado School of Mines
Golden, Colorado 80401

## ABSTRACT

Porous metals are used in filters, electrodes and as light weight structural materials. A method is presented for forming porous metals by reduction of porous oxides. In principle, by controlling the porosity of the initial oxide, the porosity of the resulting metal can be controlled. The porosity and the pore size of the oxide precursor can be controlled by vapor phase sintering in a reactive gas prior to reduction. The microstructures of porous Ni and Fe made by this process are presented and analyzed. It was found that the vapor phase sintered oxides shrunk far less during reduction. This permits the porosity and pore size of the resulting metal to be controlled in a way that cannot be accomplished by any other process.

## INTRODUCTION

Porous metals can be used in a broad spectrum of advanced applications. These include filters, electrodes, heat exchangers, and lightweight structural materials.[1,2,3] The advantages presented by these materials are determined by their porous networks, high surface areas, low resistances to flow and light weight.

Porous metals can be made by either liquid or solid state processing. A gas can be used to generate metallic foams but a limitation of this method is that a predominantly closed porous structure is formed.[1] Foams can also be produced through the addition of a pore-former that is removed, either thermally or chemically after fabrication. Also, metal foams can be fabricated by deposition or infiltration into open-cell polymer foam substrates.[3]

Sintering of loose fine metallic powders can be used to produce structures with densities in the range of 30 to 50% but control of the resulting microstructure can be difficult. Compacted metal powders that are partially sintered are commonly used for filtration applications. These processing methods lead to material with densities greater than 50% and usually poor permeabilities.

Reduction of porous oxides offers the opportunity of producing very low density, porous metallic parts. For example, assume that the starting material is a packed NiO powder having a green density of around 50 percent by volume. If this NiO can be reduced without shrinkage, then the resulting nickel will have a density of 2.88 g/cc or a porosity of 67 percent. To obtain an even lower starting density of the NiO, nickel sulfate hexahydrate, $NiSO_4 \cdot 6H_2O$ powder packed to 55 percent starting density could be decomposed to about a 90% porosity NiO. If this is subsequently reduced to Ni with no shrinkage, the final nickel density would be 0.25 g/cc or 97 percent porosity! Clearly, the formation of porous metals by the reduction of oxides offers the promise of very low densities.

Takahar and Fukuura[4] describe a method for sintering an oxide partially in air and then reducing it to the metal. They report porosities up to 64% with nickel. Their method relies on a very high organic content in the oxide prior to reduction since they report shrinkages in the range of 20%. A sizable portion of the pores formed with this process may actually be closed due to the shrinkages encountered during processing.

Vapor phase sintering offers the unique possibility of controlling both the grain size and the volume fraction of pores in the final compact[5] The presence of a reactive gas promotes volatile species formation and vapor phase transport. Vapor transport inhibits or even precludes any densification and causes significant grain growth.[5,6] The dominant reaction for $Fe_2O_3$ in the presence of HCl is:[7]

$$Fe_2O_{3\,(s)} + 6HCl_{(g)} \Leftrightarrow 2FeCl_{3\,(g)} + 3H_2O_{(g)}$$

Similarly, Figure 1[8] shows that for NiO in HCl the main vapor species are $NiCl_2$ and $H_2O$ generated by the reaction:

$$NiO_{(s)} + HCl_{(g)} \Leftrightarrow NiCl_{2\,(g)} + H_2O_{(g)}.$$

As has been shown,[5,6] when the partial pressures of the transporting gases exceeds about $10^2$ Pa ($10^{-3}$ atm) at normal sintering temperatures, then vapor transport dominates microstructure development. Vapor transport can be used to generate porous oxides with porosities and pore size controlled independently.[9] Vapor phase sintering of NiO and $Fe_2O_3$ powders was used in this study to control the initial oxide particle size independent of the starting density.

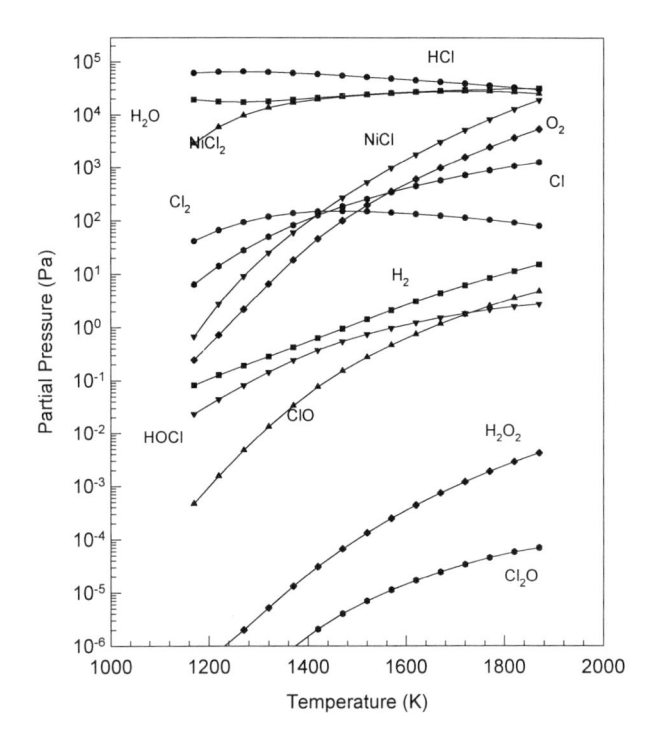

Figure 1. Calculated vapor pressures over NiO in HCl gas.

## EXPERIMENTAL PROCEDURE

Starting materials consisted of high purity $Fe_2O_3$[†] and NiO [‡]. The average particle size of these starting powders was measured to be 0.4 and 0.5µm respectively. Powders were pressed into pellets with no binders at pressures between 35 and 70 MPa. Densities of the pressed pellets were calculated from the sample weights and dimensions. Pellets were stored in an oven at 105 °C for a minimum of 24 hours prior to sintering.

Vapor phase sintering was performed in a "closed" system to minimize

---

[†]Alfa Aesar #14680, 99.99% purity

[‡]Baker #2796-01, 99.9% purity

weight loss by sublimation in HCl. Pellets were contained in overlapping crucibles. which were then placed into a larger "inner tube". The entire tube assembly was placed within the alumina furnace process tube, which was sealed by vacuum fittings. The system was flushed several times by first evacuating with a mechanical vacuum pump and then purging with argon. The system was evacuated a final time and sufficient HCl gas was introduced into the system such that atmospheric pressure was achieved at the sintering temperature. Sintering schedules consisted of 10 °C/min heating rates to maximum sintering temperatures of 1000, 1100 or 1200 °C for 30 minutes.

Reduction of the vapor phase sintered oxides was done in a flowing gas mixture of 10% hydrogen / 90% nitrogen at temperatures between 700 and 1000 °C. Time at temperature for reduction was either four or six hours.

## RESULTS AND DISCUSSION

### Densification

Table I summarizes the results for densification. Densities for the oxides were in the range of 50 to 60%, both before and after vapor phase sintering ($\rho[Fe_2O_3]$ = 5.24g/cc, $\rho[NiO]$ = 6.67g/cc).

**Table I. Densification results**

| Raw Material | Vapor Phase Sintering Temperature (°C) | Density after Vapor Phase Sintering (g/cc) | (%) | Reduction Temperature (°C) | Density after Reduction (g/cc) | (%) |
|---|---|---|---|---|---|---|
| $Fe_2O_3$ | 1200 | 3.2 | 61 | 1000 | 2.30 | 29 |
| NiO | 1200 | 3.5 | 52 | 700 | 2.43 | 27 |
| NiO | 1200 | 3.7 | 55 | 1000 | 2.91 | 33 |
| NiO | 1000 | 3.4 | 51 | 700 | 2.58 | 29 |
| NiO | None | --- | --- | 700 | 5.37 | 60 |

Vapor phase sintered Fe$_2$O$_3$ was reduced at 1000 °C for 6 hours. An original sample mass (Fe$_2$O$_3$) of 0.7341g (0.5135g Fe) resulted in a reduced sample mass of 0.5127g. This indicates complete reduction from iron oxide to iron metal. The sample density was calculated to be 2.3g/cc. Comparison of this density to the theoretical density of iron ($\rho$[Fe] = 7.86g/cc) indicates that the sample had 71 volume percent porosity.

NiO sintered at 1000 °C for 30 minutes had a geometric density of 3.4g/cc. After reduction at 700 °C for 4 hours, the bulk density was determined to be 2.58g/cc by the Archimedes' method. This gives a bulk density of 29% of the theoretical value for nickel ($\rho$[Ni] = 8.9g/cc). The weight loss of the compacts after reduction indicated that all of the NiO was reduced to nickel metal. Sintering at 1200 °C for thirty minutes, with the same reduction schedule, resulted in a density of 27% of theoretical.

It is interesting to note that the reduction step did not cause significant sample shrinkage. NiO was reduced at temperatures between 700 and 1000 °C with no further densification as long as vapor phase sintering was performed prior to reduction. In contrast, a pellet pressed from NiO and reduced at 700 °C for 4 hours without any prior vapor phase sintering resulted in a nickel pellet with a density of 60% of theoretical, Table I.

### Table II. Grain growth results

| Raw Material | Starting Average Particle Size (μm) | Vapor Phase Sintering Temperature (°C) | Grain Size after Vapor Phase Sintering (μm) | Reduction Temperature (°C) | Grain Size after Reduction (μm) |
|---|---|---|---|---|---|
| Fe$_2$O$_3$ | 0.4 | 1200 | 10-30 | 1000 | 10-30 |
| NiO | 0.5 | 1200 | 4-10 | 700 | 4-10 |
| NiO | 0.5 | 1200 | 4-10 | 1000 | 4-10 |
| NiO | 0.5 | 1000 | 0.5-1 | 700 | 0.5-1 |
| NiO | 0.5 | None | --- | 700 | 1-3 |

## Microstructure Analysis

Results for the grain size of pressed compacts after vapor phase sintering as well as after reduction are presented in Table II. Figure 2 shows the resulting microstructure of an $Fe_2O_3$ compact after vapor phase sintering and reduction. Large grains with sizes in the range of 10 to 30 μm are present. These large grains define a continuous pore network with pores in the size range of 0.5 to 3μm.

Figure 3 shows the Ni microstructure after vapor phase sintering at 1000 °C for 30 minutes and reduction to nickel metal. The resulting microstructure is a porous network of 0.5 to 1 μm nickel particles joined by extensive particle necking. No significant grain growth occurred during the reduction process. Porosity within the individual grains is not apparent for these processing conditions.

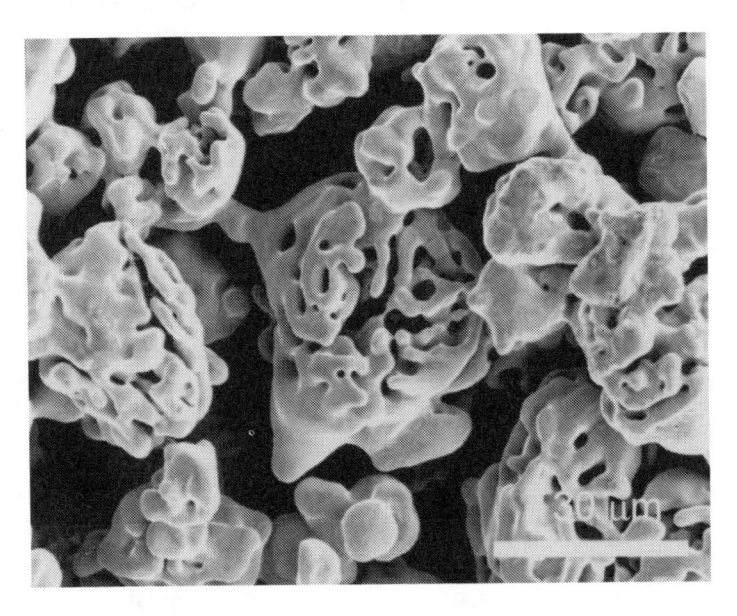

Figure 2. Microstructure of iron produced by vapor phase sintering $Fe_2O_3$ in HCl for 30 minutes at 1200 °C and then reducing at 1000 °C for 6 hours.

Figure 3. Microstructure of the porous nickel produced by vapor phase sintering NiO in HCl at 1000 °C for 30 minutes and reducing and reducing at 700 °C for 4 hours.

Figure 4. NiO sintered in HCl at 1200 °C for 30 minutes.

Figure 5. NiO sintered in HCl at 1200 °C for 30 minutes and reduced at 700 °C for 4 hours.

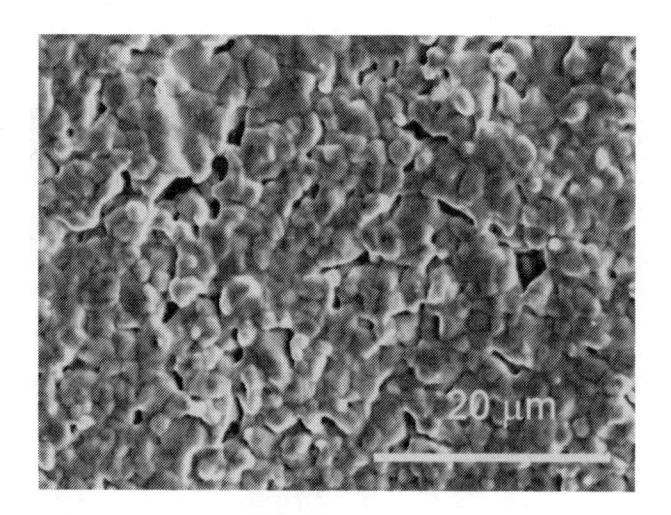

Figure 6. Surface microstructure of a NiO powder compact reduced at 700 °C for 4 hours without any sintering in HCl.

Figure 4 shows the NiO microstructure after vapor phase sintering at 1200 °C for 30 minutes. In contrast to sintering at 1000 °C, this sample is defined by large (4 to 10μm), well defined faceted grains. The vapor phase sintering temperature did not significantly affect the density, however it did have a dramatic effect on the grain size. Figure 5 shows the microstructure of a sample vapor phase sintered at 1200 °C and then reduced to nickel metal. The grain size and the porosity network that was present after vapor phase sintering has not changed, however as with the iron sample, the large grains contain a network of submicron pores as a result of the oxide reduction.

As a comparison to the effects of vapor phase sintering, Figure 6 shows the surface microstructure of a NiO pellet reduced at 700 °C. Note that most of the open porosity has been closed due to densification.

**Permeability Testing**

One application for these microporous metals is as filtration media. Permeability testing was performed on porous nickel samples, Table III. The samples were in the configuration of 2.54 cm diameter disks. Gas permeabilities were obtained by flowing nitrogen through the disk samples. From Darcy's Law, the independent permeability (k/L), where L is the thickness, was determined to be $1.04 \times 10^{-10}$m for a NiO sample vapor phase sintered at 1000°C for 30 minutes and reduced to nickel metal at 750 °C for 4 hours. This can be converted into a flux by dividing (k/L) by the fluid viscosity. For water, the resulting flux is 47.58 liters/(m$^2$ psi min). Bubble point testing for this sample, determined with n-octanol as the fluid, gave a maximum pore size of 5.8μm and a mean pore size of 1.02 μm. A NiO sample sintered at 1200 °C and reduced under the same conditions had a maximum pore size in the range of 20 to 50 μm. The exact pore size was difficult to determine by the bubble point method due to the very low pressures corresponding to this size range and the surface tension of n-octanol. The water flux was calculated to be 314 liters/(m$^2$ psi min). The grain growth associated with sintering the NiO at 1200 °C resulted in a much larger pore size which corresponded to a larger water flux.

**CONCLUSIONS**

The unique microstructure of these materials permits unique applications. These include lightweight structural materials, catalytic substrates, combustion

**Table III. Permeability results for vapor phase sintered NiO reduced to Ni**

| Vapor Phase Sintering Temperature (°C) | Reduction Temperature (°C) | Independent Permeability (m) | $H_2O$ Flux (Calculated) (l/m² psi min) | Max Pore Size (μm) | Mean Pore Size (μm) |
|---|---|---|---|---|---|
| 1000 | 750 | $1.04 \times 10^{-10}$ | 47.6 | 5.8 | 1.02 |
| 1200 | 750 | $6.83 \times 10^{-10}$ | 314 | 20-50 | --- |

substrates, electrode materials, filtration media, and heat exchangers. Concerning their potential as filter materials, these materials have the potential to provide state-of-the-art filtration. The key to this potential is the incredibly low bulk density, in combination with a relatively small pore size. Commercial metallic depth filters have densities in the range of 50% or greater. These porous metals are generally produced by simply performing a partial sintering on pressed powder shapes. The range in pore size for these materials is also large.

The measure permeabilities on pre-vapor phase sintered materials are better than any commercial metallic filter elements available with these corresponding pore sizes. Additionally, the process is very simple with limited equipment expense.

Another problem with current metallic filters is the uniformity of the filter microstructure. Ideally, a "graded" microstructure such that the pore size increases through the thickness of the filter would be ideal. In this way, one surface, with the small pore size, acts as the filtering surface while the other surface acts as a structural "support." A large pore size for the support (compared to the filtering surface) would result in higher permeabilities for the effective filtration size. This can be achieved with vapor phase sintering. Simply sintering in a temperature gradient would result in a graded microstructure. A gradient of 100 °C should be easy to achieve. Notice the significant difference between the permeabilities for nickel samples presintered at 1000 and 1200 °C.

After vapor phase sintering at 1000 °C and reduction, the structures presented here have virtually no closed porosity and densities of only 30%. Additionally, the uniform "net-like" structure as a result of the two-step process provides porous metals with superior permeability, pore uniformity, and strength.

## ACKNOWLEDGMENT

The authors wish to acknowledge the Center for Separations using Thin Films at the University of Colorado, Boulder for their measurements of independent permeabilities.

## REFERENCES

1. A. E. Simone and L. J. Gibson, "Efficient Structural Components using Porous Metals," Mat. Sci. Eng., A229 55-62 (1997).

2. S. Langlois and F. Coeuret, "Flow-Through and Flow-By Porous Electrodes of Nickel Foam, I. Material Characterization," J. Appl. Electrochem. 19 43-50 (1988).

3. L. A. Cohen, W. H. Power and D. A. Fabel, "New Metal Foams are Highly Porous and Uniform," Mat. Eng. [4] 44-46 (1968).

4. K. Takahar and K. Fukuura, "Metallic Porous Membrane and Method of Manufacture," U.S. Patent 5,417,917, Issued May 23, 1995.

5. D. W. Readey, D. J. Aldrich, and M. A. Ritland, "Vapor Transport and Sintering," pp. 53-60 in Sintering Technology, R. M. German, G. L. Messing, and R. G. Cornwall, eds., (Marcel Dekker, NY), 1996.

6. D. W. Readey, "Vapor Transport and Sintering," pp. 86-110 in Sintering of Advanced Ceramics, Ceramic Transactions, Vol. 7, C. A. Handwerker, J. E. Blendell, and W. A. Kaysser, eds., (The Am. Ceramic Society, Columbus), 1990.

7. J. Lee and D. W. Readey, "Microstructure Development of $Fe_2O_3$ in HCl Vapor," p. 145 in Sintering and Heterogeneous Catalysis, Vol. 16 of Materials Science Research, G. C. Kuczynski, A. E. Miller, and G. A. Sargent, eds. (Plenum, N.Y.), 1984.

8. A. Roine, HSC Chemistry for Windows, Ver. 3.0, Outokumpu Research, Finland. (thermodynamic software)

9. Marc A. Ritland and Dennis W. Readey, "Alumina-Copper Composites by Vapor Phase Sintering," Ceramic Engineering and Science Proceedings, 14 [9-10] 896-907 (1993).

**Joining**

# JOINING OF DIAMOND THIN FILM TO OPTICAL AND IR MATERIALS

J. G. Lee, K. Y. Lee and E. D. Case
Materials Science and Mechanics Department,

Michigan State University
East Lansing, MI 48824

## ABSTRACT

Many materials that are transparent in both the visible and infrared regions of the spectrum (such as ZnS and $MgF_2$) are easily scratched. Hard coatings may help to protect such materials, but direct deposition of diamond films onto such substrates is difficult, since the materials tend to degrade at the temperatures required for diamond deposition. We discuss attempts to make $MgF_2/MgF_2$ joins, as well as efforts to bond diamond films to $MgF_2$.

## 1. INTRODUCTION

Ceramic/ceramic joining has been done successfully for a variety of structural ceramic materials, including alumina, SiC, and zirconia. For example, Case et al. has joined alumina, Macor[R], and zirconia using spin-on layers [1-2]. In addition to structural ceramics, optically-transmitting ceramics have been joined, usually at high joining pressures. Yen et al. [3] joined $MgF_2$ specimens using a direct diffusion bonding technique at temperatures of 800°C and greater, using pressures of 17 MPa to 25 MPa. Yen et al. attributed a drop in the joined $MgF_2$ specimens' transmittance to grain growth in the polycrystalline $MgF_2$ specimens.

This paper discusses attempts to bond (1) $MgF_2$ to $MgF_2$ and (2) polycrystalline diamond thin films to $MgF_2$. For successfully joined specimens, the transmittance is measured as a function of wavelength and the optical absorption coefficient, $\alpha$, in order to compare the optical qualities of joined and unjoined specimens.

To the extent authorized under the laws of the United States of America, all copyright interests in this publication are the property of The American Ceramic Society. Any duplication, reproduction, or republication of this publication or any part thereof, without the express written consent of The American Ceramic Society or fee paid to the Copyright Clearance Center, is prohibited.

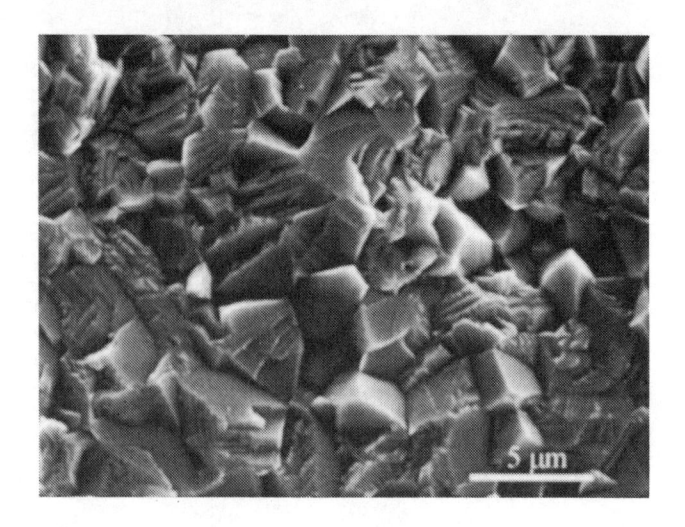

**Figure 1**. SEM image of the fracture surface of MgF$_2$.

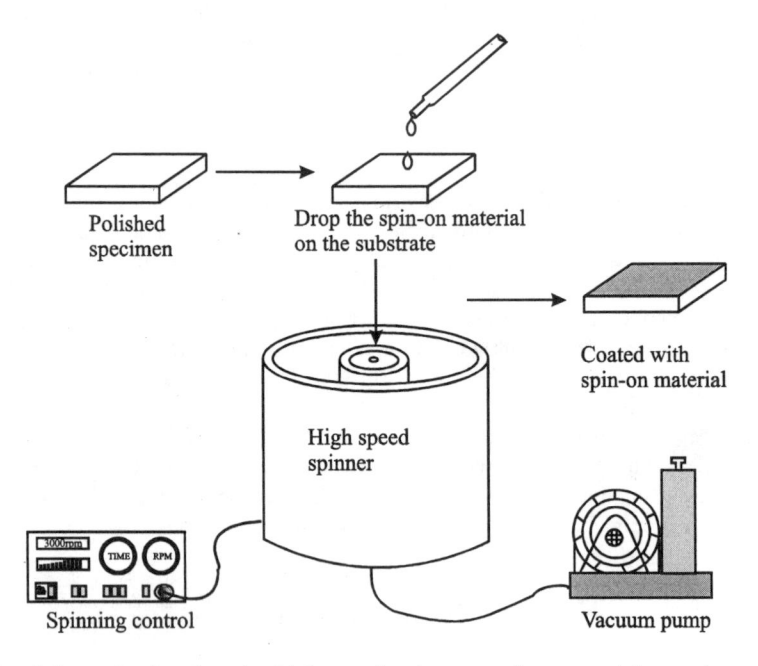

**Figure 2**. Schematic showing the high-speed substrate spinner used in coating to the MgF$_2$ substrates.

## 2. EXPERIMENTAL PROCEDURE
### 2. 1. Materials
As received commercial $MgF_2$ (Itran I, Eastman Kodak Company) is a hot-pressed polycrystalline material with a mean grain size of approximately 2.8µm (Figure 1). Specimens were sectioned into roughly 1cm X 1cm and 2mm thickness using a low speed diamond saw.

After cutting, the specimens were polished by an automatic polisher (LECO Corporation) using the following series of diamond grit sizes 17, 10, 6, and 1µm. The specimens were polished on both sides to facilitate IR transmittance measurements before and after joining.

The polished specimens were coated with a sodium silicate solution (Columbus Chemical Industries Inc.) using a high-speed spinner (Figure 2). After spinning at a rate of 2000 rpm to 5000 rpm, coating was uniform in thickness and smooth.

In this study, two silicate solutions were used as bonding agents. A commercial sodium silicate solution (Columbus Chemical Industries, Inc.) was used as one of the joining media. Although the sodium silicate was very viscous, after spinning on the high-speed substrate spinner, the coatings were quite smooth. A second bonding agent that was used was an organically-based silica solution, which pyrolyses to an amorphous silica film upon curing at 200°C. However, the silica film was not successful in either the $MgF_2/MgF_2$ bonding or the diamond film/$MgF_2$ bonding.

The diamond films used in this study were between 3 to 4 microns thick, with a mean grain size of approximately one micron. The films were microwave plasma-deposited at 400°C to 500°C under 28 torr pressure onto (100) oriented single crystal silicon wafers that were 0.048 to 0.053 cm thick and 5.08 cm in diameter. The diamond-coated silicon specimens were fractured into 1 cm X 1 cm sections using a razor blade. Following ultrasonic cleaning in deionized water, the diamond-coated silicon specimens were placed onto $MgF_2$ substrates that had been coated with sodium silicate solution. The specimens were placed such that the diamond film coating on the diamond-silicon specimens contacted the sodium silicate coating on the $MgF_2$ substrates.

Prior to joining, the bonding agents were spun onto the $MgF_2$ substrates. Using a pipette, a few drops of either sodium silicate or the silica film was placed onto a polished $MgF_2$ substrate and the film was spun between 2000 and 5000 rpm using a high-speed substrate spinner (Figure 2).

### 2.2. Joining Procedure
Coated specimens were heated in either a single mode microwave cavity (Sairem, Model MWPS 2000) or in a conventional electrical resistance furnace. For joining the $MgF_2$ to $MgF_2$ using the silica film, the annealing temperatures ranged from 500°C to 1200°C with applied dead weight loads of 20 gm to 85 gm. The $MgF_2/MgF_2$ joining using the sodium silicate solution utilized temperatures of 500°C to 800°C and dead weights of 20 gm to 85 gm.

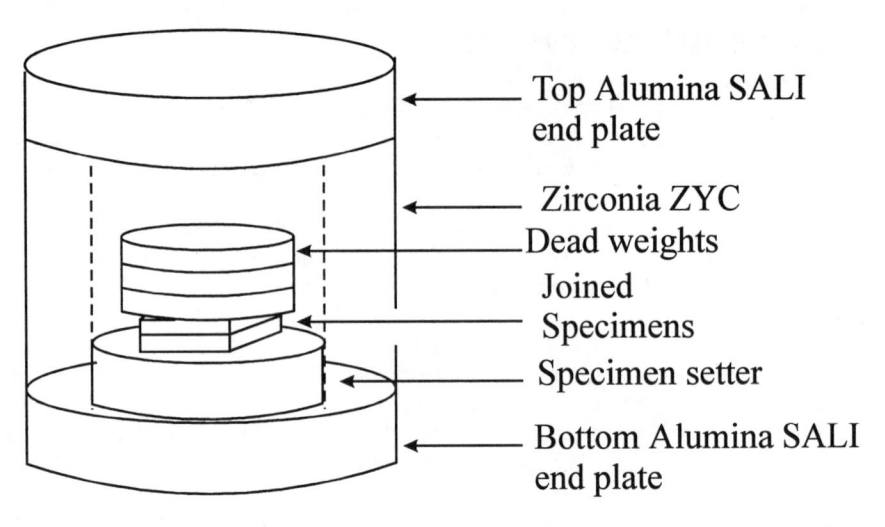

**Figure 3**. Schematic of the refractory casket used for microwave joining. The dead weight loading also is illustrated [2].

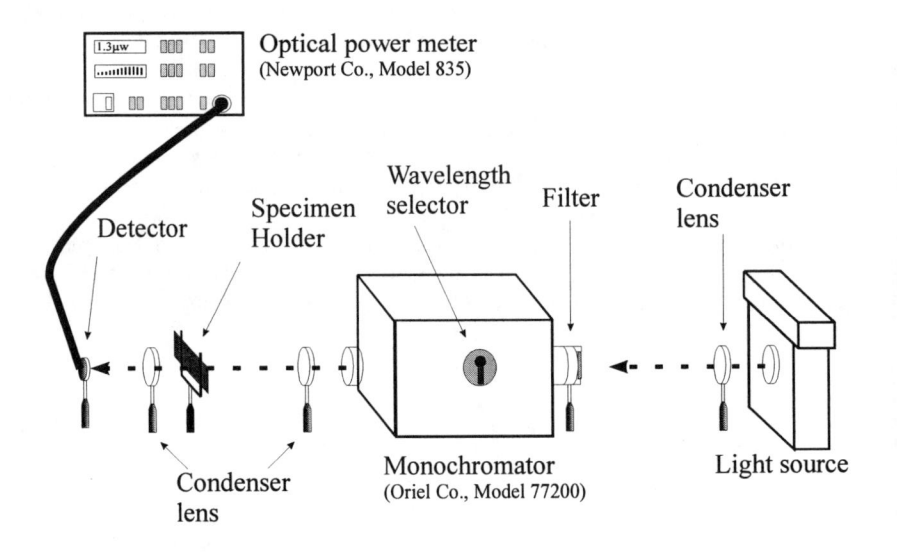

**Figure 4**. Schematic of the spectrophotometer used for the IR transmittance measurements.

## 2.3. IR Transmittance Measurements

All IR transmittance measurements were performed using a monochromator (Oriel Co, Model 77200) over wavelength range of 800 nm to 1600 nm (Table 1).

Table 1. Type of filters and detector for various wavelengths.

| Wavelength range | Types of filters | Types of detectors |
|---|---|---|
| 400-650nm | Transparent filter | 813-SL |
| 650-840nm | Red filter | 813-SL |
| 850-1650nm | Black filter | 800-IR |

Prior to the transmittance measurements on the joined specimens, a series of experiments was done to determine the effect of (1) heating the polished but uncoated $MgF_2$ at or near the temperature used for joining and (2) coating and heating the $MgF_2$ substrates (but not joining them). In order to examine these effects, transmittance measurements were done on the $MgF_2$ substrates having the following surface treatments:
(1) As-polished specimens,
(2) Polished specimens which were subsequently heated at 700°C in air,
(3) Polished specimens that were coated (without joining) and then heated at 700°C in air.

In addition to the three "comparison" surface treatment/heat treatment conditions listed above, polished specimens were coated, and then the coated surfaces of the two specimens were placed in contact. The specimens were then heated in an effort to join the specimens (either $MgF_2$/$MgF_2$ joining or diamond/$MgF_2$ joining).

Joined specimens were cross-sectioned using a low speed diamond saw. A portion of the sectioned specimen was mounted in thermosetting powder. Using an automatic polisher, the mounted specimens were polished using a series of diamond grit, including 17, 10, 6, and 1µm.

SEM examination determined the bond-layer thickness in the sectioned specimens. Since ceramic materials are not electrically conductive materials, the sectioned and mounted specimens were coated with a 0.7 nm thick layer of gold prior to SEM examination.

## 3. RESULTS AND DISCUSSION

### 3.1. Joining $MgF_2$ using sodium silicate solution

For the sodium silicate bonding agent, the $MgF_2$/$MgF_2$ joining was successful at 700°C using both microwave and conventional heating. For both the microwave and conventional heating, dead weight loading was used during joining (Table 2).

Table 2. The conditions of joining both microwave and conventional furnace.

| Heating method | $T_{max}$(°C) | Hold time(min) | Dead weight(gm) | Heating rate (°C/min) |
|---|---|---|---|---|
| Conventional | 700°C | 20min | 85gm | 10°C/min |
| Microwave | 700°C | 20min | 60gm | 45°C/min |

The mean bond layer thickness for the $MgF_2$/$MgF_2$ specimens was 10 microns for the microwave-joined specimens and 15 microns for the specimens joined in the conventional furnace. The microwave and the conventionally-joined MgF$_2$/MgF$_2$ specimens showed very different microstructures near the bond layer (Figure 5). Microstructural differences between the microwave-joined and the conventionally-joined MgF$_2$/MgF$_2$ specimens may be related to the nature of microwave heating itself. The coupling of a material with microwave energy is a function of both the material and the temperature of the material, but if the material does absorb microwave power, heat can be generated within the specimen [4,5]. In this case, the microstructural differences (Figure 5) may be a function of differing chemical reaction kinetics between the conventional and microwave heating (hence forming differing sodium silicate/MgF$_2$ reaction layers). Alternatively, the microstructural differences may be linked to differences in microwave and conventional heating at and near the interfacial region. Future studies should address these points.

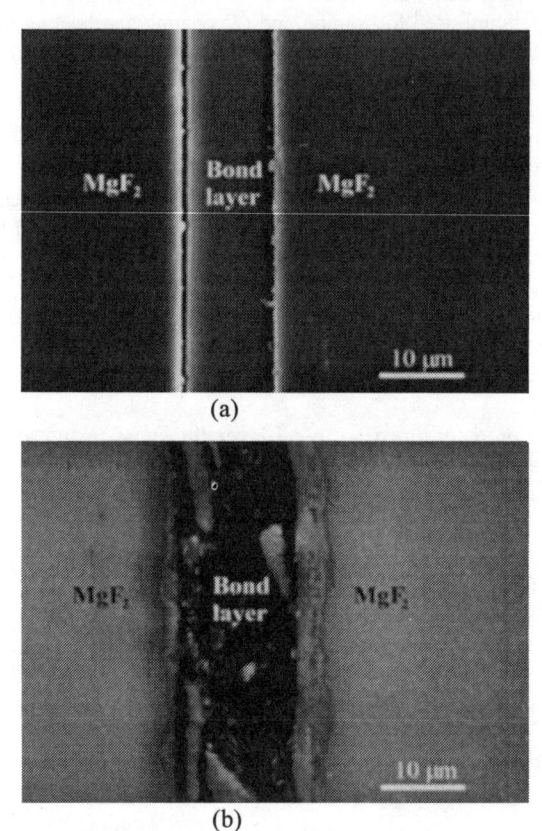

**Figure 5.** SEM micrographs of both the microwave and the conventionally joined MgF$_2$ showing the bond-layer. (a) Microwave joined MgF$_2$ (joining at 700°C for 20min with 60gm weight loading), (b) Conventional joined MgF$_2$ (joining at 700°C for 20min with 85gm weight loading).

A rough estimate of the bond-layer toughness was obtained by placing a 98N Vickers indentation crack near the bond-layer of a sodium silicate bonded $MgF_2/MgF_2$ specimen (Figure 6). For the microwave joined $MgF_2$, the Vickers indentation crack stopped at bond-layer, indicating the bond-layer material is not as strong as the matrix material. Similar results were obtained for Vickers indentation cracks placed near the bond layer for conventionally-joined specimens.

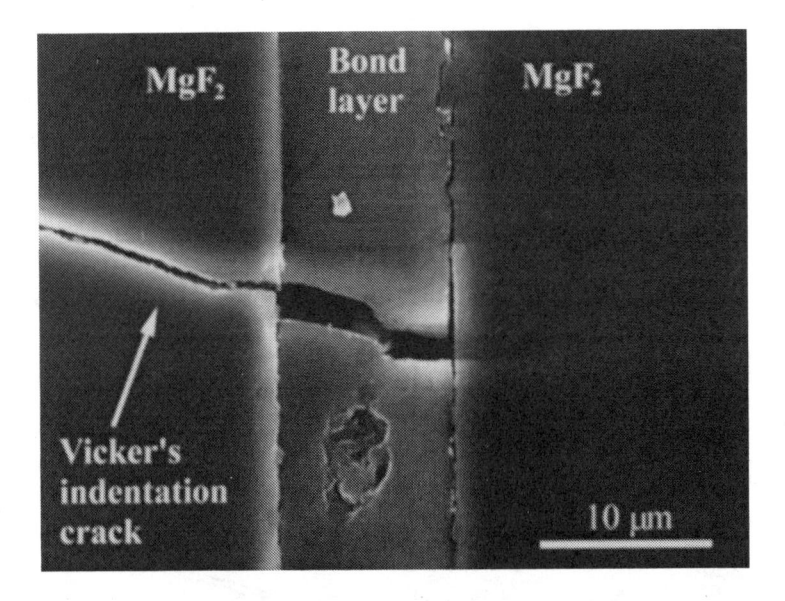

**Figure 6.** SEM image of a $MgF_2/MgF_2$ specimen joined using the sodium silicate solution. The specimen was joined by microwave heating at 700°C for 20min with a 60gm dead weight loading.

The IR transmittance of the polished and heated (but not coated) specimens and the IR transmittance of the polished coated, and heated (but not joined) were nearly identical to the IR transmittance of the polished specimens (Figure 7). Thus, (1) heating without coating, (2) coating, heating and not joining gave an IR transmittance that was very similar to the as-polished specimens, and therefore heating and coating alone do not seem to significantly affect the IR transmittance of the $MgF_2$ specimens.

The microwave joined $MgF_2/MgF_2$ shows a considerably higher IR transmittance than the conventionally joined $MgF_2/MgF_2$ specimens (Figure 8). The differences in transmittance lead to differences in the calculated optical absorption coefficients, as will be discussed in Section 3.2 of this paper.

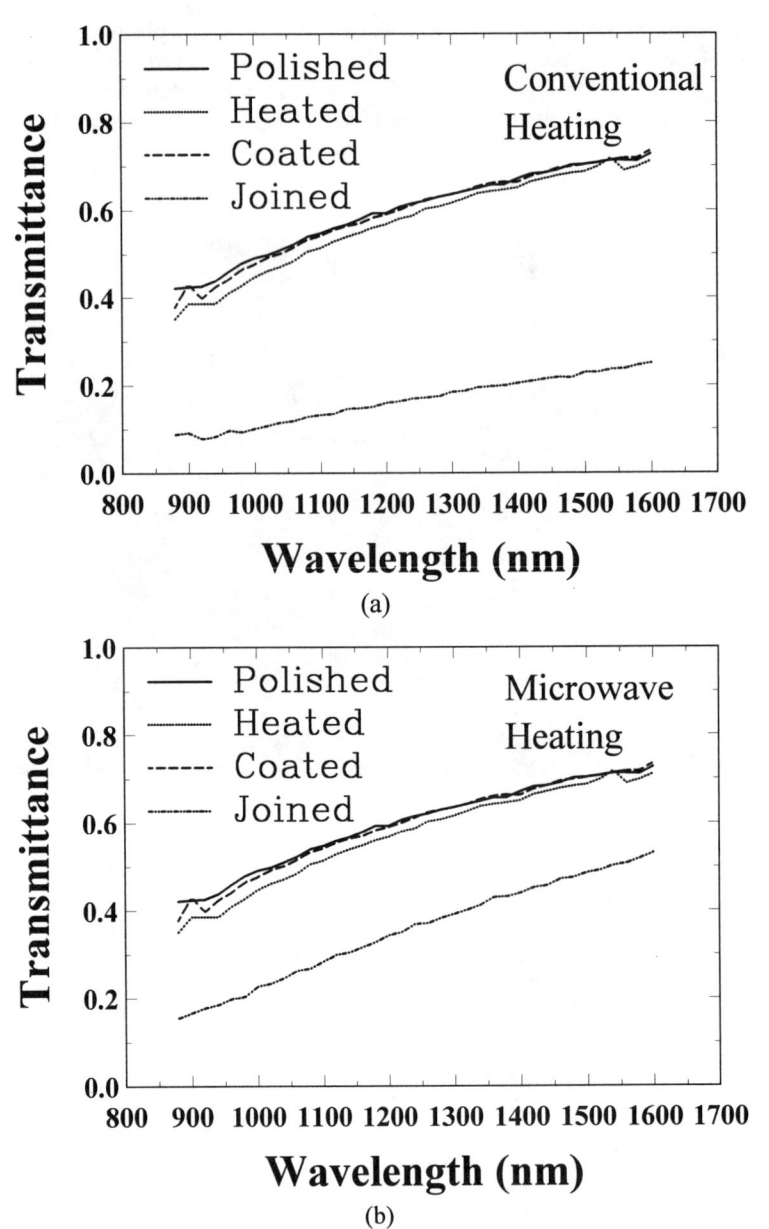

**Figure 7.** Transmittance for both microwave and conventionally heated $MgF_2$. (a) Conventionally-heated $MgF_2$ and $MgF_2/MgF_2$ joined and (b) Microwave-heated $MgF_2$ and $MgF_2/MgF_2$ joins.

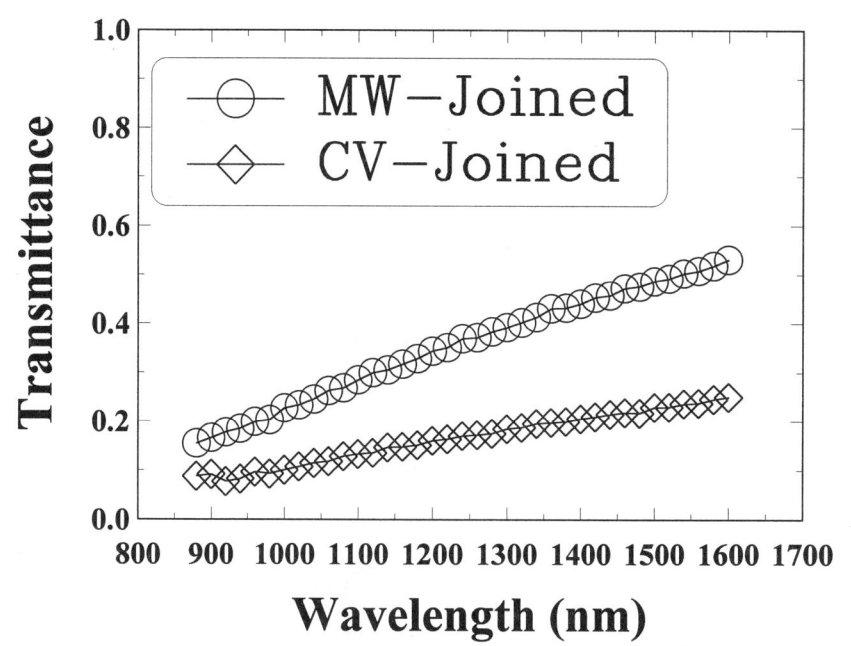

**Figure 8.** Comparison of transmittance for microwave and conventionally joined MgF$_2$(From Figure 7).

### 3.2. Calculation of Absorption Factor for microwave and conventionally joined MgF$_2$ using sodium silicate solution

To determine the optical absorption factor, $\alpha$, for both the conventionally and the microwave joined MgF$_2$/ MgF$_2$ specimens, the Lambert-Bouger law [6]

$$I_T = I_I\, e^{-\alpha x} \quad (1)$$

was used, where $I_T$ is intensity of transmitted radiation, $I_I$ is the intensity of the incident radiation, $\alpha$ is absorption factor, and $x$ is the specimen thickness.

If the optical reflection at each interface is taken into account, $I_T/I_I$ is given by

$$I_T/I_I = (1-R)^m e^{-\alpha x} = T \quad (2)$$

where R is the reflectivity for normal incidence angles. R is a function of the optical index of reflection, $n$, and $m$ is the number of interfaces for the specimen. The ratio $I_T/I_I$ is the transmittance, $T$. For a planar slab, $m = 2$.
At normal incidence, $R$, is given by

Innovative Processing/Synthesis: Ceramics, Glasses, Composites II

$$R = \frac{(n-1)^2}{(n+1)^2} \qquad (3)$$

Solving for optical absorption factor, $\alpha$, gives

$$\alpha = -\frac{1}{x}\ln\frac{T}{(1-R)^m} \qquad (4)$$

The optical properties of bond-layer are unknown, but we estimated $\alpha$ by assuming a zero thickness bond-layer in equation (2) for the joined $MgF_2$. For $MgF_2/MgF_2$ specimens, $m$ is set equal to 3 (Figure 9).

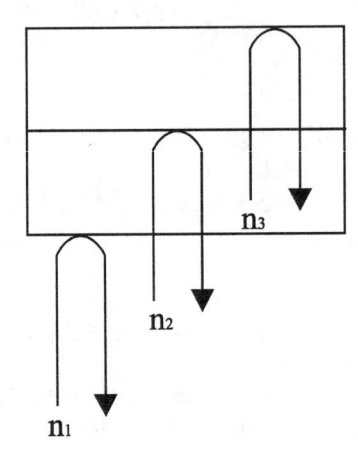

**Figure 9.** As schematic showing the reflections at the various interfaces of a two-layer specimen (which gives $m = 3$).

**Table 3.** Calculation of the optical absorption factor, $\alpha$.

| Processing | Measured $T$ | $R$ | $x$(mm) | $\alpha(mm^{-1})$ |
|---|---|---|---|---|
| Polished | 0.703 | 0.025 | 2.029±0.002 | 0.149 |
| MW-joined | 0.486 | 0.025 | 4.134±0.009 | 0.156 |
| CV-joined | 0.229 | 0.025 | 4.209±0.012 | 0.342 |

An optical absorption factor, $\alpha$, was calculated from equation (4) using an optical index of refraction, $n$, of 1.3749 at a wavelength 1500nm [7]. The calculated optical absorption coefficients, $\alpha$, were similar for the polished (unjoined) and for the microwave-joined specimens (0.149mm$^{-1}$ and 0.156mm$^{-1}$, respectively Table 3). The $\alpha$ values for the polished (unjoined) and the microwave-joined specimens implied that the polished MgF$_2$ and the microwave-joined MgF$_2$ had a similar optical quality. The higher $\alpha$ for conventionally-joined MgF$_2$ indicated optical losses resulting from IR scattering [8,9] induced by crystallographic phases formed in and near the bond region (Figure 5).

The scattering of electromagnetic radiation depends on a number of factors, including (1) the mismatch in optical indices of refraction between the scattering center and the matrix material, and the (2) the relative size of the scattering center, compared to the wavelength of the incident radiation [9, 10]. If we let a = characteristic size of the scattering center and $\lambda$ = the wavelength of the incident light, then for scattering is significant for $0.1 < \lambda/a < 10$, with a maximum in scattering at about $\lambda/a = 1$. As an example of how effectively scattering can reduce the transmitted light intensity in a ceramic, consider the scattering of visible light by pores in a ceramic [10]. At the wavelength of red light (700 nm or 0.7 micron), a 3 percent volume-fraction porosity consisting of pores with a diameter of 2 microns reduces the transmittance to about 0.01 % compared to the transmittance for a theoretically-dense material [10]. When the pores have a diameter of 0.7 micron, then 0.7 micron radiation will be reduced by a similar amount for a volume fraction porosity of only one-percent [10]. From Figure 5, we see that the characteristic dimension of the scattering centers (second phase particles) is roughly a few microns. Thus, second phases that are several micron diameter should scatter the incident radiation used in this study (a wavelength range from 880 nm to 1600 nm, or equivalently 0.88 micron to 1.6 micron), since $\lambda/a$ should be very roughly in the range from about 0.1 to 0.5. However, a detailed scattering study for non-spherical particles that are not uniform in size is extremely complex [8-10]. Therefore, attempts to extract further information from the transmittance curves (Figures 7 and 8) will be a topic of future study.

### 3.3. Diamond joining on MgF$_2$ using sodium silicate solution

Diamond/MgF2 joining was attempted for both the sodium silicate solution and for the silica film. For the sodium silicate bonding agent, the films were spun between 3000 rpm and 5000 rpm from for 20 seconds. Then, using a heating rate of $10^\circ$ C per minute and a dead weight loading of 85 gm, the diamond/MgF2 specimens were heated in a atmosphere of flowing nitrogen at maximum temperatures of 500°C to 800°C for hold times of 20 minutes. However, the joining was not successful at any of these conditions. In separate experiments involving single-slab specimens of magnesium fluoride, it was found that even in flowing nitrogen, the specimens were discolored at temperatures above 800°C, presumably due to point defects that evolve during heating. Due to the degradation of the magnesium fluoride's optical properties, no joining attempts were made at temperatures higher than 800°C.

In addition to the sodium silicate solution, the silica film was used to attempt both MgF$_2$/MgF$_2$ bonding and the diamond film/MgF$_2$ bonding. The specimens were heated by a conventional furnace at maximum temperatures between 500°C and 1200°C in an atmosphere of flowing nitrogen. Dead weight loads ranged from 20 gm to 85 gm. The MgF$_2$/MgF$_2$ bonding occurred only at for a maximum anneal temperature of 1200°C, but at that temperature (despite the flowing nitrogen atmosphere) the surface of the MgF$_2$ specimen was milky-colored (likely due to oxidation). The resulting transmittance of the MgF$_2$ specimen was severely degraded by the high-temperature reaction.

Had the joining been successful, the silicon substrate (on which the diamond film has been microwave-plasma deposited) would have been etched away using a nitric acid or a

KOH etching solution. The transmittance of the specimen, without the silicon substrate, could then have been measured.

## 4. CONCLUSIONS

The $MgF_2/MgF_2$ specimens were successfully joined using: (1) spun-on sodium silicate interlayers, (2) both conventional and microwave heating and (3) low externally applied pressures. The microstructure of the bonded region was quite different for the microwave and the conventionally bonded $MgF_2/MgF_2$ materials (Figure 5). In addition, the IR transmittance of the microwave-bonded specimens, as determined from the optical absorption factor, $\alpha$, was similar in optical quality to the polished but unjoined $MgF_2$ specimens (Table 3). In contrast, the $\alpha$ for the conventionally-joined $MgF_2/MgF_2$ indicates a considerably degraded optical transmittance (Table 3). The difference in optical quality between the microwave-joined specimens and the conventionally-joined specimens may be related to IR scattering induced by second phases [8,9] in the bond region (Figure 5).

## REFERENCES

1. K.Y. Lee and E.D. Case, "Microwave Joining and Repair of Ceramics And Ceramic Composites," *Ceramic Engineering and Science Proceedings, 21st Annual Cocoa Beach Conference and Exposition on Composites, Advanced Ceramics, Materials and Structure*, Cocoa Beach, Florida, **V18** 543-550, 1997.
2. K.N. Seiber, K.Y. Lee, and E.D. Case, "Microwave and Conventional Joining of Ceramic Composites Using Spin-On Materials," *Proceeding of the 12th Annual Technical Conference Dearborn, MI*, 941-949, 1997.
3. T.F. Yen, Y.H. Chang, D.L. Yu and F.S. Yen, "Diffusion Bonding of $MgF_2$ Optical Ceramics," *Material Science and Engineering*, **A147** 309-321, 1991.
4. M.L. Santella, "A review of Techniques for Joining Advanced Ceramics," *Journal of the American Ceramic Society Bulletin*, **71**[6] 947-953 (1992).
5. I. Ahmad and R. Silberglitt, "Joining Ceramics Using Microwave Energy," *Materials Research Society Proceedings*, **314**, 119-130, 1993.
6. I.W. Donald and P.W. Mcmillan, "Review Infrared Transmitting Materials," *Journal of Materials Science*, **13**, 1151-1176 (1978).
7. Moses "Refractive Index of Optical Materials in the Infrared Region"; pp7-16, Hughes Aircraft Company, Culver City, CA 1970.
8. P. Debye, H.R. Anderson, and H. Brumberger, *Journal of Apply Physics.*, **28**: 679, 1957.
9. C.F. Boren and D.R. Huffman, Chaper 3 in Absorption and Scattering Light by Small Particles, John Wiley and Sons, New York, 1983.
10. W. D. Kingery, H. K. Bowen, and D. R. Uhlmann, *Introduction to Ceramics*, 2nd Edition, pp 674-677, John Wiley & Sons, New York, NY (1976)

# Laminated Object Manufacturing

# LAMINATED OBJECT MANUFACTURING USING CERAMIC PAPER PRODUCTS

B.J.Kellett and Wei Guo
Department of Materials Science and Engineering
University of Cincinnati
Cincinnati, OH 45221-0012

## ABSTRACT

This article demonstrates the use of commercial inorganic paper in a Laminated Object Manufacturing (LOM) process. Inorganic papers have found widespread use as heat insulators and battery separators. This article will demonstrate the use of these commercial papers--without significant modification--as feed stock for direct LOM processing of inorganic fiber composites. Both polymer (polyester, polyamide, phenolic) and inorganic laminating agents (sodium silicate) have been tested with ceramic and glass fiber paper products. Preliminary results show that these materials and laminating agents can be cut with a $CO_2$ 25 watt laser. Polymer matrix composites have been fabricated with final tensile strengths of 25MPa. Silicate matrix composites have achieved bend strengths of 10 MPa after heat treatments to 700°C.

## 1. Introduction.

Laminated Object Manufacturing (LOM) is one of many rapid prototyping methods based on a finite element building strategy. For LOM, the finite element is a thin sheet (i.e. 2-dimensional element), others are based on thin, narrow ribbons, (i.e. 1 dimensional element) and others are finite cubes/points (i.e. a zero dimensional element).

The LOM process as developed by Helisys Inc.[1] is shown in **Figure 1**. The process is currently configured with regular Kraft paper coated with a hot melt adhesive (polyvinylacetate). The LOM 1015 machine can process parts with a foot print of 10x15 inches. The paper is fed through pinch rollers onto the work table, and a heated roller laminates the paper to the object. The laser cuts the paper as dictated by the computerized movement of the mirror, with a final cut to separate the roll of paper from the object. The paper roll is then advanced, the work table

---

[1] Helisys Inc. 24015 Garnier Street, Torrance, CA 90505

To the extent authorized under the laws of the United States of America, all copyright interests in this publication are the property of The American Ceramic Society. Any duplication, reproduction, or republication of this publication or any part thereof, without the express written consent of The American Ceramic Society or fee paid to the Copyright Clearance Center, is prohibited.

lowered by the thickness of the paper, and a new 10-15 sheet is laminated down to the top of the object. In this way a complete object is built. The thickness of the object, the thickness of the paper, the speed in which the laser can cut through the paper, and the total length of cut on each paper dictates the total time needed to manufacture an object. A typical part takes 5-6 hours.

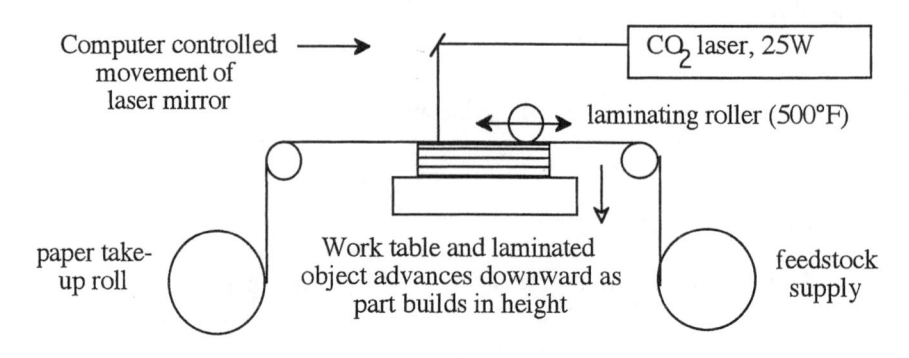

Figure 1 Schematic of the **L**aminated **O**bject **M**anufacturing Process (LOM)

Our interest is to develop the LOM process for new materials. Current technology uses regular paper, which in the end produces a laminated paper object with properties very similar to plywood. While laminated paper is sufficient for modeling purposes, it has limited use as a structural material. The direct LOM processing of real structural materials, i.e. ceramics, metals, cements, would have obviously commercial advantages. Significant activity has been directed at the direct manufacture of ceramic materials, as rapid design-manufacture of ceramic materials is more difficult than metals and polymers.[1,2,3]

Direct LOM processing of ceramic materials has been based on tape cast feed stock materials. Tape casting is an established industrial process for producing long and thin sheets of ceramic powder with a significant fraction of polymer added to produce needed rheology. LOM processing from tape cast sheet is in escence a near net shape process, i.e. a green state forming process requiring a final, post-LOM bonding process. This bonding process is usually a sintering process where mass transport occurs to particle-fiber contact areas creating direct, solid bonds. Other potential bonding methods are liquid metal infiltration--to produce a metal-ceramic fiber composite, vapor phase infiltration, or chemical precursor infiltration. Regardless of the bonding process, all require prior removal of the organic phase. While high quality tapes of ceramic powder are readily produced by tape casting, fairly large volume fractions of organics are required. The tape is nearly 50 volume

% polymer and has properties similar to video or cassette recording tape. Removal of this organic phase is a severe fundamental limitation of tape cast ceramic sheets. To its advantage, tape cast sheets are commercially produced.

Efforts on the LOM processing of ceramic has focused on tape casting. While highly uniform powder sheets are easily produced--as required for uniform densification--elevated loading of organic binders (50 volume %) are difficult to remove. Post-LOM binder burnout has proved to be extremely disruptive. Also, the polymer removal process is diffusional. The time needed for removal will increase with the square of the part thickness, and inversely with the powder size. Thus fine powder components, and large components can not be processed from this type of feed stock. Again, fundamental limitations may prove to be insurmountable.

This article discusses the process of using new feed stock materials for direct LOM processing of ceramic objects. The paper making process has already been used to produce ceramic paper. These commercial products are typically used for insulating purposes and have insufficient strengths for structural applications.

## 2. Building Ceramic Parts with Ceramic Paper and Sodium Silicate

Water glass is a highly soluble, polymerized silicate of sodium or potassium. Its adhesion properties are caused by both by its high viscosity and the reactivity of the Si-O-Na groups.[4] The adhesive silicates usually have composition from $Na_2O:2SiO_2$ to $Na_2O:4SiO_2$ and viscosity of 50 mPa-s and upward. They dry to hard vitreous films which do not really dissolve and which contain small amounts of residual water, depending on the drying conditions and on the alkali-silica ration of the silicates.[5]

The study of soluble silicates as adhesives starts from McBain's three reports.[6] A variety of adhesive materials were used including a series of sodium silicate solutions of typical soda-to-silica ratio with which strong bonds were obtained on glass, wood and paper.[7] These studies have discussed the behavior of using soluble silicates as adhesives. Sodium silicate from the PQ Corporation[2] was used in this study. Because the pure sodium silicate was too viscous to be adsorbed by the paper, de-ionized water was added until reasonable viscosity was reached.

A number of ceramic papers are commercially available, i.e. Zircar, Unifrax, Thermal Ceramics. The properties of two types of ceramic paper are presented in Table 1. These papers are available as cut sheet and continuous roll. An important property of the paper is the loss-on-ignition (LOI). Sodium silicates are viscous liquids which prevent outgassing during burn-out of the organic sizing agents. Processing of the refractory composites with sodium silicate laminating agents were

---

[2] Philadelphia Quartz Co, Philadelphia, PA.

based on the low LOI 970-H paper from Unifrax. A further requirement for LOM processing by the Helisys system is the ability to laser cut

**Table 1** Physical properties and chemical analysis of ceramic paper.

|  |  | Unifrax Corporation[3] |  | Thermal Ceramics[4] |  |
|---|---|---|---|---|---|
| Supplier / Cermic Paper |  | 970-H | 970 | 3000 | SF607 |
| Physical Properties | Color | white | white | white | white |
|  | Use limit (°F) | 2300 | 2300 | 2800 | 1832 |
|  | Melting point (°F) | 3260 | 3260 | 3600 | 2327 |
|  | Density (PCF) | 12 | 10 | 7~10 | 12 |
|  | Nominal thickness | 1/32" | 1/32" | 1/32" | 1/8" |
| Chemical Analysis | $Al_2O_3$ (wt%) | 49.2 | 49.2 | 95 | – |
|  | $SiO_2$ (wt%) | 50.5 | 50.5 | 5 | 65 |
|  | CaO (wt%) | – | – | – | 29 |
|  | MgO (wt%) | – | – | – | 5 |
|  | L.O.I * (%) | 0.1 | 5 | 6~10 | 8-10 |

* L.O.I: loss on ignition

The sodium silicate was directly applied to the top-most paper layer of the part with a brush and then imedieately covered with a new sheet of ceramic paper. The heated roller then bonds this new top-most sheet down onto the stack by pressing the sodium silicate into the pores of the paper. The laser is then used to cut through the top most paper layer. In this way the part is built, layer-by-layer. Adhesion is related to the viscosity of the sodium silicate, and the porosity of the paper. The heated roller is not, in-of-itself, sufficient to dry the soaked paper with a single pass of the roller. Thus the part required drying after building. There are some issues which need to be addressed. For example, the liquid adhesive will flow into the cut, and rebonds the cut. This can create difficulties in decubing, i.e. separating the

---

[3] Unifrax Corporation, 2351 Whirlpool Street, Niagara Falls, NY 14305
[4] Thermal Ceramics, P.O.Box 923, Augusta, Georgia 30903-0923

part from the excess paper. Two parts fabricated by this method are shown in **Figure 2**. After LOM processing, the two parts were decubed and heat treated for drying and strengthening. One part was heated at 400°C for 1 hour and the other was heated at 200°C for 1 hour.

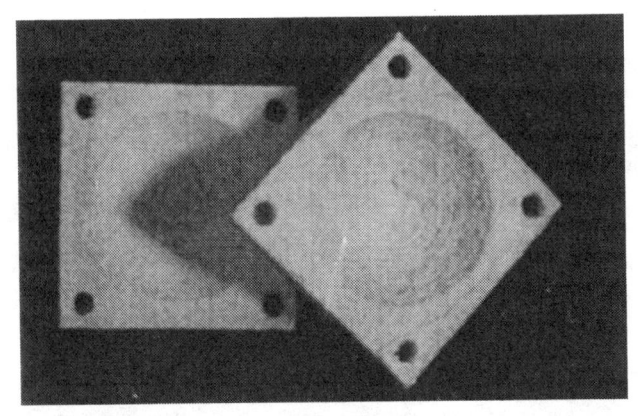

Figure 2  Two parts made using sodium silicate as adhesive with ceramic fiber paper.

**Figure 3** shows a microstructure of the part made using Unifrax 970-H paper and sodium silicate. The layer structure of the part is partially evident in this cross section. The layer thickness is about 200$\mu$m. **Figure 3** shows that the fibers of the ceramic paper have been partially disolved by the sodium silicate. For a strong microstructure it will be necessary to retain the fiberous structure of the paper while also having the sodium silicate completely infiltrate through the ceramic paper to create a uniform microstructure.

Further heat treatments showed that the remaining fibers have begun to sinter after heating to 1000°C. This sintering action is accompanied with gross shrinkage of the part. Heating further to 1500°C creates large mass loses due to evaporation of the water glass, leaving a glass-like residue.

In order to quantitatively characterize the strength of parts made of ceramic paper and sodium silicate, flexural strength test specimen were made using exactly the same processing, feedstock and adhesive. Specimens were heated to 400°C for 30 minutes before testing, which appears to be near the optimum post lamination processing temperature.

The test was done in accordance with ASTM standard C1161-94 with test bars measuring 50x6x6 mm. The test was performed with an Instron testing machine at a rate of 2 mm/min and the load versus displacement is recorded during the test.

Tests were performed with different laminate orientations, and surprissingly, no effect of orientation was seen. Figure 3 shows the flexural tests result. The average strength is about 6.5MPa. The Weibull analysis indicates a Weibull modulus of about 7. These low strengths are in part due to the reactivity of the water glass.

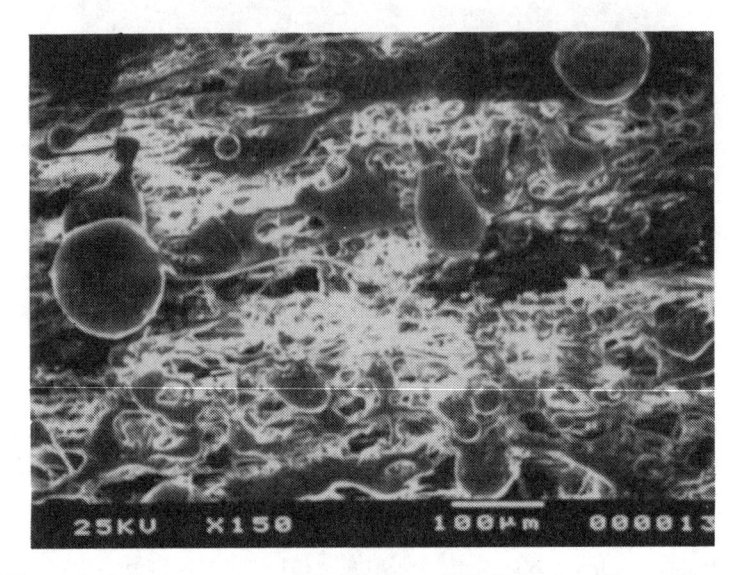

Figure 3   Microstructure of the Unifrax 970-H and sodium silicate after heating to 400°C for 30 minutes.

## 3. LOM Processing of Fiber Glass Reinforced Plastic

Fiberglass materials are widely used as reinforcements for plastics. The composite is usually called FRP (fiberglass-reinforced plastic). FRP has been used for various type of equipment in the chemical process industries since the early 1950s and its use continues to grow. Process vessels of various shapes and sizes, columns, scrubbers, hopper, ducts, fans, valve bodies, heat exchanger shells and tube sheets, floor topping, and tank lining systems are just a few examples of equipment made of fiberglass-reinforced plastics.[8,9] The chief reason for the popularity of these materials is their excellent resistance to corrosion, as well as their ability, when the appropriate additives have been incorporated, to resist heat and/or fire.

Fiberglass can also be used as the reinforcement of silicone.[10] The largest present user is the electrical industry. In this industry, flat sheet laminates are used to make slot wedges, spacers, mounting boards, insulating forms, and similar related parts for use wherever service requires the resistance to severe environmental conditions that is provided by silicone resins.

Alumina paper with sodium silicate adhesive

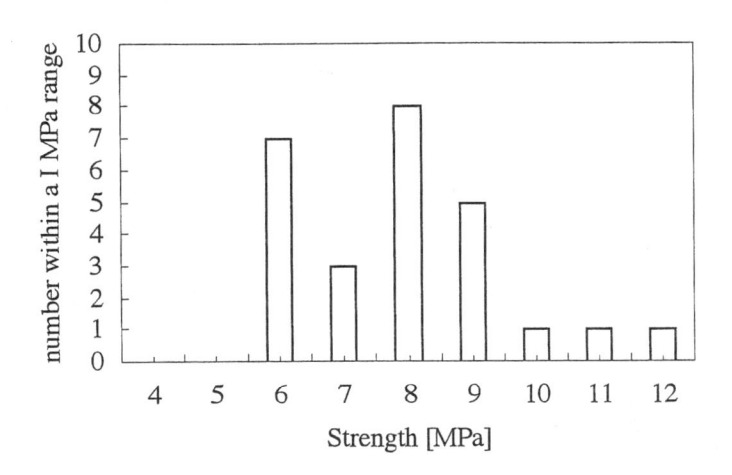

Figure 4  Histogram of strength data for alumina paper-sodium silicate composite heat treated at 400°C.

As mentioned before, the ability to process sheet material makes LOM uniquely capable of fabricating composite materials. Composites via rapid prototyping technique could be a truly direct and freeform process.[11] Compared to traditional processing methods for manufacturing FRP, such as contact molding, compression molding and resin transfer molding, rapid prototyping may provide much higher automation, lower cost and time.

Thin glass fiber sheets are commercially available as surface veil. Surface veil M524-C64 from Owens Corning Fiberglass Corporation[5] has been used in this study. The fibers are produced from C glass, a chemically resistant glass which is highly resistant to attack by both acid and alkaline environment. This type of fiberglass is chosen because it meets almost all the requirements of LOM processing. The nominal thickness is 0.010 inch, which is good for dimension control. It is supplied in roll which fits the LOM machine very well. This material is extremely inexpensive and is readily cut with a 25 Watt $CO_2$ laser. The ultimate

---

[5] Owens Corning Fiberglass Corporation, One Owens Corning Pkwy, Toledo, OH 43659.

tensile strength of as-received fiberglass veil reaches 2.5 MPa which is enough for the LOM machine. Although only hand lay-up processing has been used to make parts, fully automatic process can clearly be developed using this fiberglass product.

Thermogravimetric test result of the as-received fiberglass shows that the organic binders in the fiberglass starts to burn at about 200°C and the total weight loss is about 6%. The conclusion is that the binders will not burn out with many types of hot melt adhesives, which are used at temperatures below 180°C.

Composites have been fabricated with 3M[6] scotch-Weld™ Brand 583 bonding film. This bonding film forms a flexible, 100% solids, heat activated dry film adhesive composed of synthetic elastomer, thermoplastic and thermosetting resins (nitrile phenolic). The film softens and flows when heat is applied and provides a strong, permanent bond to the surface to which it is applied. It is a very thin film with nominal thickness of 0.002 inch.

3M 583 bonding film is one of the few heat activating adhesives that thermosets. It goes through a thermoplastic phase prior to thermosetting. The adhesive softens to a heavy liquid which flows to develop intimate surface contact and high adhesion. This occurs at 280-340°F (138-171°C). If the activating temperature is not increased it will remain a thermoplastic bond. When thermosetting the activating temperature should be in the 350-380°F (177-193°C) range. The adhesive goes through the thermoplastic phase and then chemically reacts to form a solvent and heat resistant thermoset bond.

The following LOM processing parameters are used: laser cutting speed is 7 inches per second; laser cutting power is 6 percent. The laser cut is very clean and fast. The hot roller temperature is set to 380°F which is the high limit for the bonding film to set.

**Figure** 5 is the picture of some specimen made using LOM processing with fiberglass and 3M bonding film. The microstructure is porous and it is very easy to notice the individual layers after LOM processing. Although the highest compression parameter has been used during the LOM processing, the pressure applied to the part is still not enough to squeeze the thermoplastic through the paper. The temperature of the hot roller had been set to the recommended high temperature limit of the machine. As the contact time of the hot roller and the top layer is short and the fiberglass is an effective insulating thermal barrier, higher laminating temperatures or lower viscosity resines would be desirable. Specimens were post cured in a heated press at 350°F and 20MPa for 2 minutes. After compaction, the

---

[6] 3M Aerospace Materials Department, St.Paul, MN 55144

thickness of specimens drops greatly and that means the void content in specimens decreased rapidly.

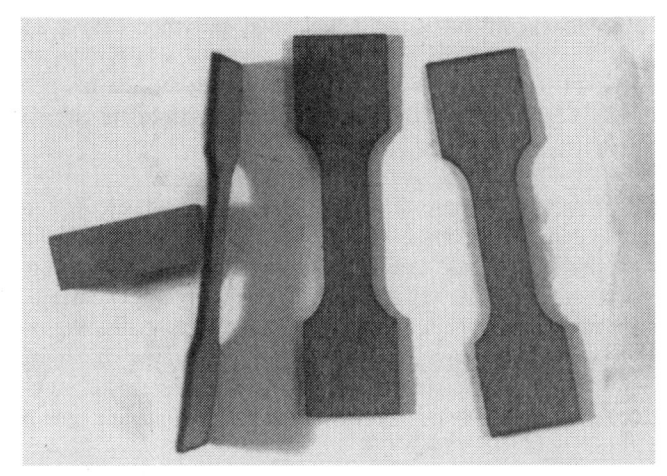

Figure 5 Tensile specimen made using Owens Corning fiberglass and 3M bonding film. The speciments are 70mm long, have a gauge width of 8.9mm, a thickness of 1.6mm.

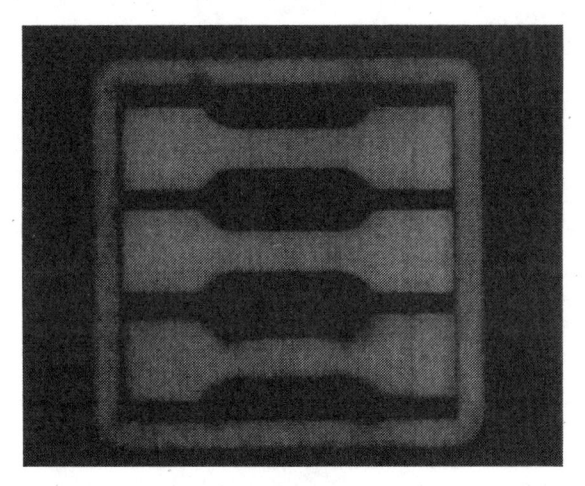

Figure 6 Tensile specimens made using fiberglass with PE105 web adhesive. These tensile specimens are 70mm long, have a gauge width of 8.9mm.

Besides 3M Brand 583 bonding film, two types of Bostik[7] hot melt adhesive are available in nonwoven form. Bostik PA115 web is a high performance polyamide hot melt polymer with a softening point of 125°C. PE105 is also a hot melt adhesive in nonwoven form and based on a polyester polymer with a softening point of 105°C. Both adhesives and glass veil were cut at a laser power of 9 percent and cutting speed of 7 inches per second. Hot roller surface temperature was set to 380°F for PA115 and 350°F for PE105. These specimens are shown in Figure 6 below.

The tensile properties of these three different composites are shown in **Figure** 7. The ultimate strengths and modulus of elasticity of these three different composites are less than that of the matrix polymer phase suggesting that complete consolidation and adhesion to the glass fibers has not yet occurred. The elastic modulus are much lower than the glass fiber suggesting slipping between the fibers and matrix phases. The glass fibers are not yet providing stiffening or strengthening to the composite. These are preliminary results which have not yet been optimized with improved interfacial bonding agents and more complete consolidation.

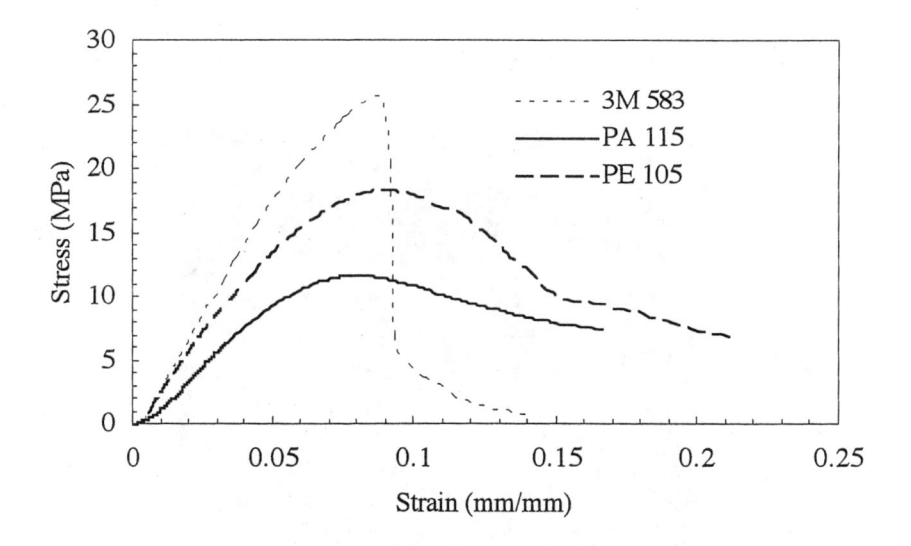

Figure 7  Tensile test of fiberglass with three different matrix resins: 3M phenolic, PE polyester, and PA nylon.

---

[7] Bostik, 211 Boston St., Middleton, MA 01949

## 6. FINAL COMMENTS

Rapid prototyping is a new approach for the manufacture of materials which combines computational design with direct manufacturing. A number of different composite materials have been processed by the LOM method. Preliminary results suggest that reasonable strengths of 25 MPa can be achieved with polymer matrix and 5 MPa strengths with inorganic matrix phase which can be used at temperature up to 700°C. Further research should be able to improve these strength values.

## REFERENCES

1. Klosterman, D., Chartoff, R., Osborne, N., Graves, G., Lightman, A., and Han, G., "Laminated Object Manufacturing (LOM) of Advanced Ceramics and Composites", *Proc. of the 7th International Conference on Rapid Prototyping*, University of Dayton and Stanford University, San Francisco, CA, April 1997.

2. Klosterman, D., Chartoff, R., Osborne, N., Graves, G., "Laminated Object Manufacturing, A New Process for the Direct Manufacture of Monolithic Ceramics and Continuous Fiber CMCs", *Proc. of the 21st Annual Cocoa Beach Conference & Exposition*, Cocoa Beach, Florida, January 12-16, 1997.

3. Sachs, E., Cima, M., Cornie, J., "Three Diminsional Printing: Ceramic Shells aand Cores for Casting and Other Applications", Proc. of the Second International Conference on Rapid Prototyping, Dayton, 1991.

4. W.A. Weyl, ASTM, *Proc., 46,* pp.1506, (1946).

5. W.F. Wegst, *unpublished records*, Philadelphia Quartz Co.

6. McBain, J.W., Great Britain, Dept. Scientific and Ind. Research, Adhesives Research Committee, *1st, 2nd, 3rd reports* (1922, 1926, 1932).

7. Vail J.G., *Soluble Silicate - Their Properties and Uses*, American Chemical Society, (1952).

8. Talbot, R.C., "Using fiberglass-reinforced plastics", *Chemical Engineering*, pp. 76-82, October 29, (1984).

9   Rolston, J.A., "Fiberglass composite materials and fabrication processes", Chemical Engineering, January 28, pp.96-110, (1980).

10  Elliott, E.C., *Reinforced Silicone Resins*, John Wiley & Sons, Inc., (1970).

11  Priore, B.E., *"Fabrication of Polymer Composites Using Laminated Object Manufacturing"*, M.S. Thesis, University of Dayton, August, 1996.

**Kinetics and Mechanism**

# NEUTRON DIFFRACTION STUDIES OF THE PARTIAL REDUCTION OF NiAl₂O₄: PHASE AND STRAIN EVOLUTION

E. Üstündag[*], R. H. Woodman and J. C. Hanan, Department of Materials Science, Keck Laboratory, California Institute of Technology, Pasadena, CA 91125

B. Clausen, T. Hartmann[†] and M. A. M. Bourke, Manuel Lujan Jr. Neutron Scattering Center and ([†]) Materials Science and Technology Division, Los Alamos National Laboratory, Los Alamos, NM 87545

## ABSTRACT

Metal-ceramic composites consisting of Ni particles embedded in an $Al_2O_3$ matrix were obtained by the partial reduction of $NiAl_2O_4$. The reaction was studied *in situ* with neutron diffraction using a special controlled-atmosphere furnace. The volume shrinkage that occurs during this reaction generates strain whose evolution was monitored. Neutron diffraction also yields kinetics data with very good time resolution. Experiments were performed to study the effect of processing parameters such as reduction temperature, reducing atmosphere and initial ceramic density on both kinetics and internal strains. The results indicate a complicated mechanism is operative that involves high temperature deformation, e.g., creep, and chemical effects, i.e., change in cationic disorder in the unreduced part of spinel.

## INTRODUCTION

The partial reduction of the spinel oxide $NiAl_2O_4$ to a metal ceramic composite of nickel and aluminum oxide has recently been studied [1-5]. The advantage of using this method to produce a composite derives from its inherent ability to control microstructure by manipulating processing variables such as reduction temperature, time, initial oxide density, dopants and oxygen activity in the atmosphere which acts as the driving force. A critical issue that profoundly affects the microstructure evolution during this reaction is the volume shrinkage that accompanies it. This volume change has been predicted to be about -18% theoretically (using the crystallographic densities of the phases involved) [4], but the measured values are between -2% and -14% [4]. The difference is compensated by porosity that forms inside the aluminum oxide matrix [4]. The value of porosity is a strong function of initial spinel density, reduction temperature and time [4].

---

[*] Corresponding author; e-mail: ersan@cco.caltech.edu.

Another consequence of the volume change during reduction is that it generates residual stresses due to the constraint imposed by the unreduced part of spinel. If not relaxed or controlled, these stresses can crack specimens and that was a recurring problem during the early stages of this study [1,2]. The solution to the cracking problem was the realization that it usually started at the original spinel grain boundaries; hence a finer initial spinel grain size required a higher strain energy to crack, therefore suppressing it [2]. The cracking does prove, however, that strain energy is generated due to the volume change during reduction.

The aim of this study was to follow the reduction *in situ* using neutron diffraction in a controlled-atmosphere furnace in order to better understand how phases and strains evolve. This would help identify the stress generation and relaxation mechanisms and hence provide opportunity to better manipulate the residual stresses. Neutron diffraction offers a unique advantage in this case since neutrons can penetrate much deeper into most materials than X-rays and hence allow *in situ* studies. Some preliminary results are presented and discussed in this article.

EXPERIMENTAL PROCEDURE

The neutron powder diffractometer (NPD) at Lujan Center, Los Alamos National Laboratory was utilized in this study. A special furnace (Fig. 1) was built for this project that had a retort tube made of amorphous silica (to reduce neutron absorption and avoid contribution to the diffraction pattern). The samples were placed inside the tube and were in contact with an alumina holder. The four specimens examined so far are listed in Table I and their geometry is shown in Fig. 2 together with the expected stress state during reduction. The spinel oxide was hot isostatically pressed (HIPed) into long cylinders in nickel cans.

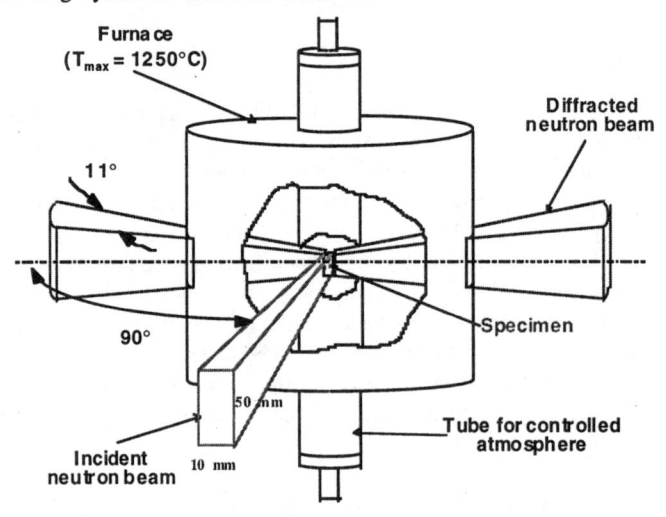

**Figure 1.** Schematic of the controlled-atmosphere furnace used in this study.

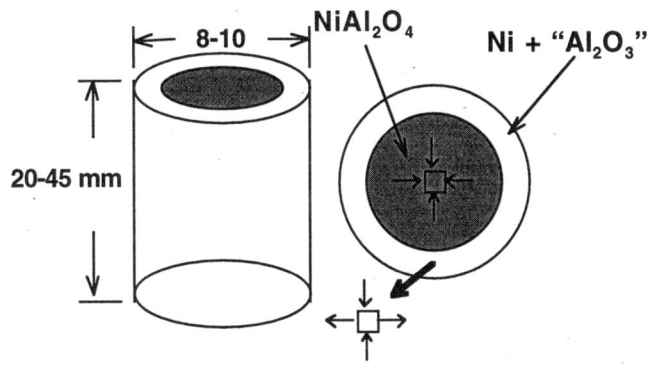

**Figure 2.** Geometry of NPD specimens. The dimensions indicate the volume sampled by neutrons (almost entire sample volume). The stress state is due to the volume change during reduction.

**Table I:** *In-Situ*-Reduced NPD Specimens and Experimental Details

| Specimens | T (±5°C) | Time (hr) | Atm. | $\rho_{initial}$ (%TD) | $\Delta V/V$ (%) | $f_{spinel}$ (%) |
|---|---|---|---|---|---|---|
| #1 | 1140 | 32 | 10%CO in "$N_2$" | 96.9 | -0.4 | 79 |
| #2 | 1210 | 100 | (95:5 $CO:CO_2$) in 50%"$N_2$" log $a_{O2} \approx$ -13 | 96.3 | +7.4† | 66 |
| #3 | 1220 | 17 | 10%CO in "$N_2$" | 87.6 | -5.1 | ~0 |
| #5 | 1220 | 42 | 10%CO in "$N_2$" | 97.5 | -2.9 | 27 |

† There was a surface layer buildup due to reaction with the BN sample holder.

The experiment started by heating a spinel sample in nitrogen up to the reduction temperature. In addition to several data acquisitions during heating a final data collection was performed at the reduction temperature. This allowed the determination of the lattice constant of spinel before the reaction started and was later used as the "strain-free" reference to calculate strain evolution.

The reduction reaction was initiated by introducing either a $CO/CO_2$ or a CO/"$N_2$" mixture into the furnace. "$N_2$" denotes commercial-purity nitrogen with about 100 ppm $O_2$ impurity. Except for #2, the gas mixtures were not buffered; therefore, the value of oxygen activity can not be accurately predicted. However, it is estimated to be on the order of $10^{-14}$ to $10^{-16}$ and well below the stability limit of spinel [1,6]. Neutron data acquisition runs of 0.5 to 1 hr each were performed to monitor phase and strain evolution.

The experimental details for specimens run so far are presented in Table I. In addition to the reduction temperature and time other data included are: the reduction atmosphere, the initial density of spinel ($\rho_{initial}$) as determined by the measurement of dimensions, the volume change during reduction (by comparing initial and final volumes measured by a micrometer) and the volume fraction of spinel left in the specimen before the cooldown ($f_{spinel}$) determined by Rietveld analysis using the GSAS package from Los Alamos [7].

## RESULTS

### Reaction Kinetics

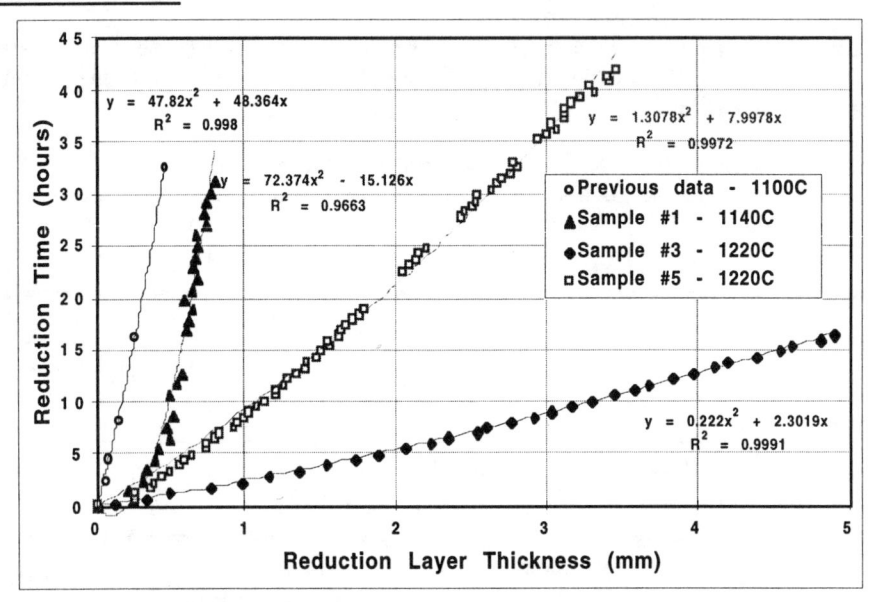

**Figure 3.** Comparison of reaction kinetics for different processing conditions. Previous data were obtained by cross section observations of reduced samples [3].

The phase evolution data revealed only Ni and $\alpha\text{-}Al_2O_3$ in addition to spinel while no $\theta\text{-}Al_2O_3$ was detected. This is surprising considering the fact that the latter phase would usually form during low temperature reductions [3]. However, it must be noted that the resolution of Rietveld refinement is about 2-3 wt.%, therefore, if $\theta\text{-}Al_2O_3$ is present, it is likely to be a thin layer at the reduction front [2,4]. Another important observation is that the alumina to nickel volume fraction is about 4 as predicted by the phase diagram [6]. Volume fraction information can be used to determine the thickness of the reaction layer as shown in Fig. 3. It is seen there that the reduction temperature and the initial density of spinel are very potent parameters that influence the reaction kinetics. The information presented in Fig. 3 is also useful in determining the rate controlling step of the reaction. A diffusion-controlled reaction will show a parabolic variation of reaction layer thickness vs. time, whereas interface-reaction-controlled kinetics are expected to be a linear function of time. As it is seen from the equations listed for each fitting curve, both mechanisms seem to control the reaction kinetics.

## **Strain Evolution During Reduction**

Figs. 4-7 illustrate the evolution of strain in spinel and the reduced composite after preliminary Rietveld analyses. The strain in spinel was calculated using its lattice constant at the reduction temperature before the reduction was started. There was no such reference for Ni and $\alpha$-$Al_2O_3$. The initial references for these phases were their lattice constants calculated from JCPDS files and coefficient of thermal expansion (CTE) data in the literature. After an almost complete reduction was obtained in sample #3, however, a more refined method was adapted. The lattice constants of the two phases obtained from the last data collection run before cooling down the sample were taken as the "strain-free" references. This was justified by the fact that no detectable spinel was left in the sample at that point leading to the full relaxation of all macrostrains due to the volume change during reduction. The difference between the strains calculated by this method and that by the CTE data was found for both phases and was later used in samples #2 and #5 to "correct" its strain values. In these figures, the strain in the reduced composite was found from the rule of mixtures assuming 20 vol.% Ni in an $\alpha$-$Al_2O_3$ matrix.

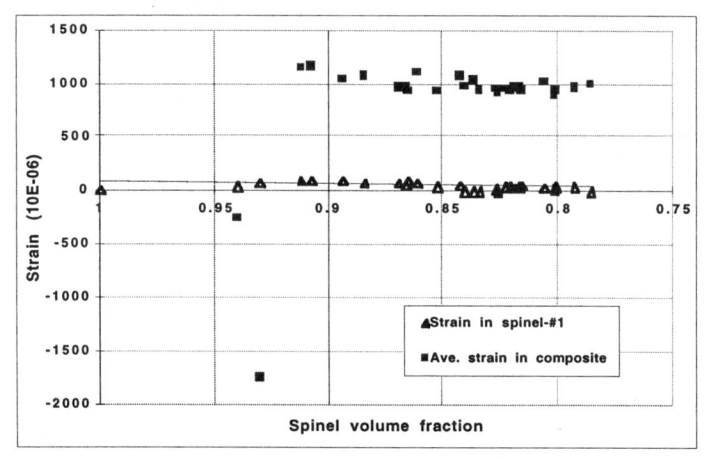

**Figure 4.** Strain evolution during the reduction of sample #1 (initial spinel density: 97%; reduction temperature: 1140°C).

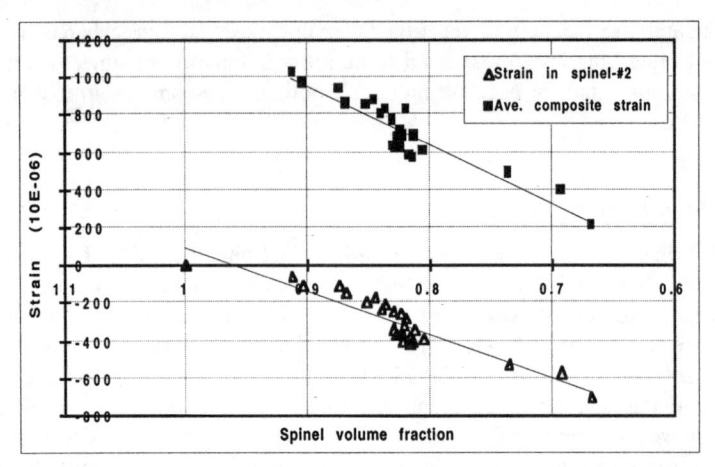

**Figure 5.** Strain evolution during the reduction of sample #2 (initial spinel density: 96%; reduction temperature: 1210°C).

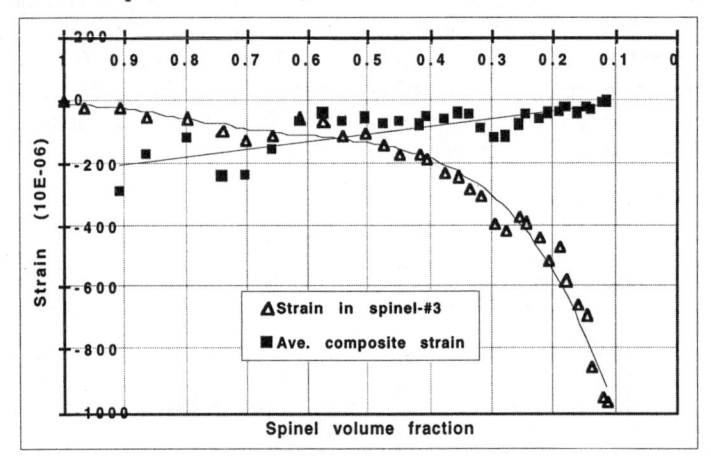

**Figure 6.** Strain evolution during the reduction of sample #3 (initial spinel density: 88%; reduction temperature: 1220°C).

Referring to Fig. 2, one can visualize a pseudo-hydrostatic compressive strain state in spinel during reduction. The value of this strain is expected to increase as the reaction proceeds (assuming no relaxation). The strain (and stress) state in the reduced composite is not so straightforward to interpret. Looking along a plane perpendicular to the axis of a sample, radial compression and hoop tension is predicted in this region (Fig. 2). However, since the whole specimen is sampled by the neutron beam, an appropriate averaging scheme is needed to see which one of these is more dominant. Finite element modeling and analytical calculations were used for this purpose and the details are presented in the next section.

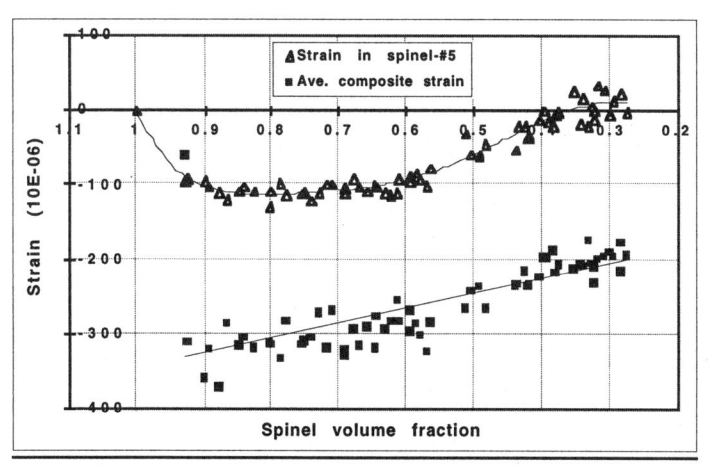

**Figure 7.** Strain evolution during the reduction of sample #5 (initial spinel density: 97%; reduction temperature: 1220°C).

## Finite Element Modeling

An axisymmetric model with elastic deformations only was assumed (Fig. 8a). Calculations were made using ABAQUS (version 5.6) in the standard non-linear implicit mode. The calculation of the average strain as measured by NPD and used in finite element calculations is shown in the Appendix (the details are explained elsewhere [5]).

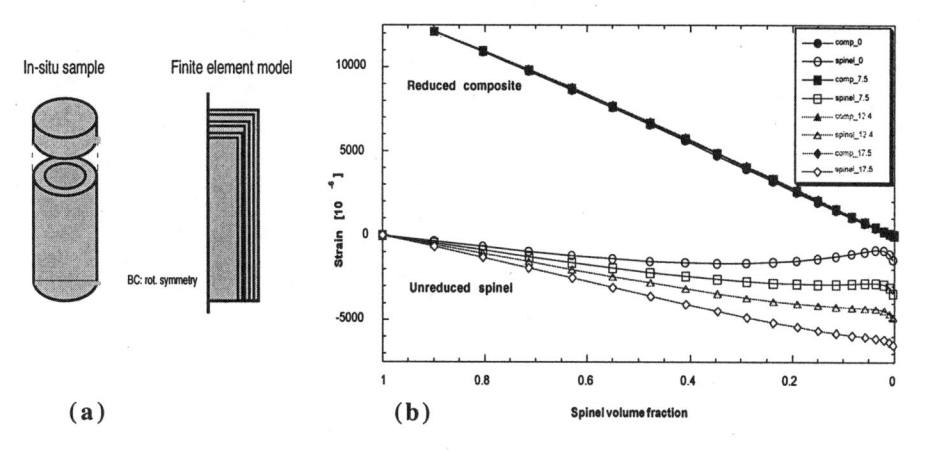

**Figure 8.** Finite element model: (a) geometry; (b) predictions for strain evolution during reduction as a function of initial spinel porosity. The calculations assume elastic deformations only.

The reduced composite was modeled as a single phase with experimentally-determined elastic constants at room temperature. The strain evolution for 10% shrinkage is shown in Fig. 8b. Here the initial porosity in spinel was varied to study its effect. The porosity values were chosen to approach experimental data in Table I. It is seen that the porosity in spinel tremendously affects the evolution and maximum value of compressive strain in it. Since the Young's modulus of spinel is a very strong function of porosity a highly porous spinel will have a low modulus. This in turn will reduce the constraint exerted by unreduced spinel on the reduction layer, hence affecting strain evolution during reduction. Comparing the finite element predictions with Figs. 4-7 the following observations can be obtained for each sample:

Sample #1: No detectable compressive strain was built in spinel (Fig. 4). The measured strain values are within the detection limit of the NPD suggesting that almost zero strain was generated due to the volume change during reduction. The variation of strain in the reduced composite is "flat" as well corroborating the behavior in spinel. (The high variability, esp. in alumina at the beginning, can be attributed to the difficulty of accurately fitting a diffraction profile to a phase of small amount.) The zero strain generation in spinel is also supported by the fact that the sample had cracked and broken into pieces during the reaction. The fracture surfaces of these pieces were also reduced and the reduction layer thickness there was similar to that around an uncracked region indicating that the cracking occurred early in the process. There were additional surface cracks on the sample that had not yet propagated fully.

Sample #2: This sample did show compressive strain buildup in unreduced spinel that was also corroborated by the behavior of the composite (Fig. 5). This observation proves, therefore, that the reduction of spinel can generate compressive stresses by avoiding their total relaxation. However, the role of the surface layer on this sample's strain behavior is not clear. Electron microscopy and X-ray diffraction investigations are underway to identify the structure and phases in this layer.

Sample #3: There was a large compressive strain generated in this sample (Fig. 6). The maximum compressive strain value corresponds to a residual stress of about -300 MPa. However, the behavior of the composite does not support that of spinel. The trend in the composite is opposite to what is expected.

Sample #5: Almost negligible compressive strain was built in spinel, though there is a definite trend in strain evolution (Fig. 7). Again, the behavior of the composite is not what is expected. No conclusions can be reached at this moment before further microscopy and data analyses.

**Disorder in Ni-Spinel**

Recently, careful Rietveld structure analyses were initiated to determine the evolution of spinel crystal structure during reduction [8]. Ni-spinel, although classified as a normal spinel, $(A^{2+})[B^{3+}_2]O_4$, can easily experience cationic disorder approaching an inverse spinel, $(B)[AB]O_4$, structure [9]. In this process, the larger $Ni^{2+}$ ions which occupy the tetrahedral (8a) sites in normal spinel move to the octahedral (16d) sites in the inverse spinel. At the same time, half of the smaller $Al^{3+}$ ions move in the opposite direction [9]. The extent of this move is measured by the disorder parameter $x$ which indicates the fraction of the tetrahedral sites occupied by the Al ions (x=0 for normal spinel and x=1 for inverse spinel). The most "random" distribution of the cations will lead to x=2/3 and this number is approached at high temperatures [9]. Recent studies [9,10] have used XRD and Rietveld analysis on samples quenched to room temperature after various heat treatments (assuming the high-temperature structure is "frozen"). They showed that the lattice constant of Ni-spinel decreases as the disorder parameter increases. These data are compared with results from sample #3 in Fig. 9. It must be noted that the disorder parameter in sample #3 steadily increases during reduction which in turn decreases the lattice constant of the spinel phase. This behavior is in contrast with the observations that during high temperature heat treatments of spinel (without reducing it) its disorder parameter should approach 2/3. In this case, the disorder is probably increasing to better accommodate the larger Ni ions in the inverse spinel lattice. This process could be driven by the residual stress generated due to the volume change of the reduction. If true, no appreciable stress evolution will be observed in the reduced composite whereas the unreduced spinel will exhibit a pseudo-compressive stress due to the disorder changes. Additional NPD experiments are planned to clarify this issue.

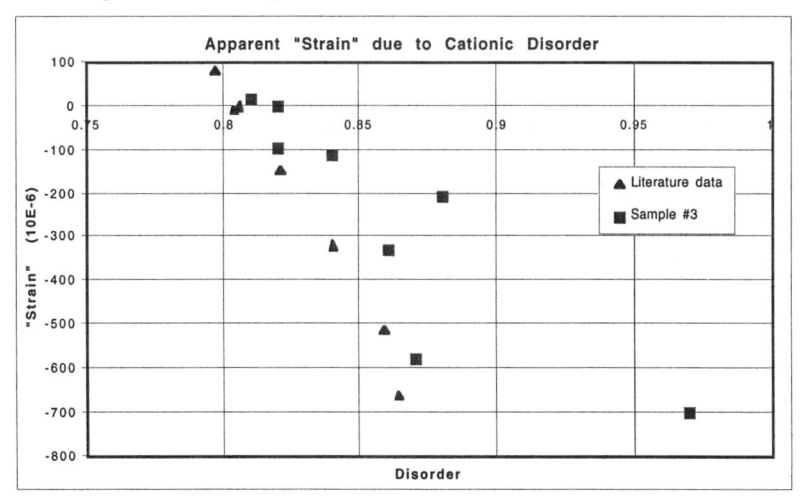

**Figure 9.** Evolution of cationic disorder in the unreduced spinel during reduction. The literature data is from ref. [10].

## CONCLUSIONS

Based on these preliminary results it can be concluded that compressive stresses of several hundred MPa can be generated during the reduction of Ni spinel depending on processing conditions. However, not all samples exhibited similar behavior. There are indications that structural changes in the unreduced spinel (in terms of disorder evolution towards inverse spinel structure) can also lead to a compressive "strain" buildup in this phase. It is proposed that the reduction involves several mechanisms that influence strain evolution. For example, high temperature mechanical deformation such as creep can also dramatically alter the strain state of the specimen. Finite element modeling of this process is underway.

Neutron diffraction is very a powerful method to investigate, *in situ*, the partial reduction of nickel spinel as well as other solid state reactions and phase transformations.

## ACKNOWLEDGMENTS

This study was supported by a Laboratory Directed Research and Development Project at Los Alamos and the Powell Foundation at Caltech. It also benefited from the national user facility at the Lujan Center supported by the Department of Energy under contract W-7405-ENG-36.

## REFERENCES

1. E. Üstündag, R. Subramanian, R. Dieckmann and S.L. Sass, *Acta Metall.*, **43**, 383 (1995).
2. E. Üstündag, P. Ret, R. Subramanian, R. Dieckmann and S.L. Sass, *Mat. Sci. Eng.*, **A195**, 39 (1995).
3. Z. Zhang, E. Üstündag and S.L. Sass, p. 489 in *Thermodynamics and Kinetics of Phase Transformations*, J.S. Im et al. (eds.), MRS Proceedings, vol. 398, 1996.
4. E. Üstündag et al., *Mat. Sci. Eng.*, **A238**, 50 (1997).
5. E. Üstündag, et al., to be submitted to *Appl. Phys. Lett.* (1998).
6. F.A. Elrefaie and W.W. Smeltzer, *J. Electrochem. Soc.*, **128**, 2237 (1981).
7. A.C. Lawson and R.B. Von Dreele, GSAS-General Structure Analysis System, LAUR 86-748, Los Alamos National Laboratory, 1986.
8. T. Hartmann, E. Üstündag, B. Clausen and M.A.M. Bourke, to be submitted to *J. Am. Ceram. Soc.* (1998).
9. K. Mocala and A. Navrotsky, *J. Am. Ceram. Soc.*, **72**[5], 826 (1989).
10. J. N. Roelofsen, R. C. Peterson and M. Raudsepp, *Amer. Mineralogist*, **77**, 522 (1992).

## APPENDIX: Calculation of Average Strain Measured by Neutron Diffraction [5]

The volume element in a cylindrical coordinate system is shown in figure A1.

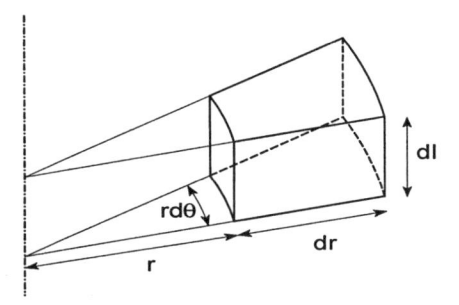

**Figure A1.** Volume element in cylindrical coordinate system.

The average strain in the direction of the Q vector in a cylindrical sample is found as

$$\bar{\varepsilon} = \frac{4\int_0^L \int_0^R \int_0^{\pi/2} (\varepsilon_r \cos\theta + \varepsilon_h \sin\theta) r d\theta dr dl}{4\int_0^L \int_0^R \int_0^{\pi/2} r d\theta dr dl}$$

If we look at the volume only, we find that

$$4\int_0^L \int_0^R \int_0^{\pi/2} r d\theta dr dl = Volume = \sum_{i=1}^{N} volE^i$$

$$4\int_0^L \int_0^R [\theta]_0^{\pi/2} r dr dl = \sum_{i=1}^{N} volE^i$$

$$2\pi \int_0^L \int_0^R r dr dl = \sum_{i=1}^{N} volE^i$$

where $volE^i$ is the volume of the $i$th element. Therefore the average strain can be found as

$$\bar{\varepsilon} = \frac{4\int_0^L \int_0^R [\varepsilon_r \sin\theta - \varepsilon_h \cos\theta]_0^{\pi/2} r dr dl}{2\pi \int_0^L \int_0^R r dr dl}$$

$$\bar{\varepsilon} = \frac{4\int_0^L\int_0^R (\varepsilon_r + \varepsilon_h) r\,dr\,dl}{2\pi\int_0^L\int_0^R r\,dr\,dl}$$

$$\bar{\varepsilon} = \frac{2\pi\int_0^L\int_0^R \frac{2}{\pi}(\varepsilon_r + \varepsilon_h) r\,dr\,dl}{2\pi\int_0^L\int_0^R r\,dr\,dl}$$

$$\bar{\varepsilon} = \frac{\sum_{i=1}^N \frac{2}{\pi}(\varepsilon_r^i + \varepsilon_h^i) volE^i}{\sum_{i=1}^N volE^i}$$

But as $\varepsilon_r$ and $\varepsilon_h$ are independent variables we can find the average strain as

$$\bar{\varepsilon} = \frac{\frac{2}{\pi}\left[\sum_{i=1}^N \varepsilon_r^i volE^i + \sum_{i=1}^N \varepsilon_h^i volE^i\right]}{\sum_{i=1}^N volE^i}$$

and if we define the average radial and hoop strains as

$$\bar{\varepsilon}_r = \frac{\sum_{i=1}^N \varepsilon_r^i volE^i}{\sum_{i=1}^N volE^i} \quad \text{and} \quad \bar{\varepsilon}_h = \frac{\sum_{i=1}^N \varepsilon_h^i volE^i}{\sum_{i=1}^N volE^i}$$

the overall average strain (measured by NPD) is

$$\bar{\varepsilon} = \frac{2}{\pi}\left(\bar{\varepsilon}_r + \bar{\varepsilon}_h\right).$$

# SURFACE AREA, AND OXIDATION EFFECTS ON NITRIDATION KINETICS OF SILICON POWDER COMPACTS

Ramakrishna.T. Bhatt
U.S. Army Research Laboratory
Lewis Research Center
21000 Brook Park Rd
Cleveland, OH 44135
and
A.R. Palczer
National Aeronautics and Space Administration
Lewis Research Center
21000 Brook Park Rd
Cleveland, OH 44135

## ABSTRACT

Commercially available silicon powders were wet-attrition-milled from 2 to 48 hr to achieve surface areas (SA's) ranging from 1.3 to 70 $m^2/g$. The surface area effects on the nitridation kinetics of silicon powder compacts were determined at 1250 or $1350^0C$ for 4 hr. In addition, the influence of nitridation environment, and preoxidation on nitridation kinetics of a silicon powder of high surface area ($\approx 63$ $m^2/g$) was investigated. As the surface area increased, so did the percentage nitridation after 4 hr in $N_2$ at 1250 or $1350^0C$. Silicon powders of high surface area (>40 $m^2/g$) can be nitrided to > 70% at $1250^0C$ in 4 hr. The nitridation kinetics of the high-surface-area powder compacts were significantly delayed by preoxidation treatment. Conversely, the nitridation environment had no significant influence on the nitridation kinetics of the same powder. Impurities present in the starting powder, and those accumulated during attrition milling, appeared to react with the silica layer on the surface of silicon particles to form a molten silicate layer, which provided a path for rapid diffusion of nitrogen and enhanced the nitridation kinetics of high surface area silicon powder.

---

To the extent authorized under the laws of the United States of America, all copyright interests in this publication are the property of The American Ceramic Society. Any duplication, reproduction, or republication of this publication or any part thereof, without the express written consent of The American Ceramic Society or fee paid to the Copyright Clearance Center, is prohibited.

## INTRODUCTION

Monolithic reaction-bonded silicon nitride (RBSN) material is typically fabricated by first consolidating 3 to 50-μm-sized silicon powders into the desired shape and then nitriding the object in $N_2(g)$ or in a $N_2$-$H_2(g)$ environment for 50 to 100 hr between 1350 and 1450°C [1]. This processing methodology, however, cannot be used for the fabrication of RBSN composites using small diameter ceramic fibers because of fiber degradation at high processing temperatures[2]. Also, it is difficulty to achieve high-volume-fraction composites using silicon powders ranging in diameter from 3 to 50 μm. Therefore for the development of RBSN composites, there is a need to develop a shorter time, and lower temperature nitridation cycle using sub-micron silicon powders.

Low temperature nitridation cycles have been developed by using laser synthesized high purity nano-sized silicon powders[3], or by adding transition metal additives to commercial purity silicon powders [4-5]. Both techniques cannot be adapted to composite fabrication for the following reasons. Haggerty et al [3] reported that laser synthesized silicon powders tend to agglomerate easily, and are prone to oxidation during air exposure or in contact with many of the polymer binders typically used in RBSN processing. They have also shown that oxidation delayed, or even completely stopped the nitridation reaction depending on the degree of oxidation. On the other hand, the transition metal nitride enhancing additives are known to react with SiC fibers, or with interface coatings such as boron nitride (BN) and SiC during RBSN processing causing fiber degradation [6].

The current study had two objectives. From a basic understanding point of view, the first objective was to determine the effects of powder lots, surface area, preoxidation, and nitridation environment on nitridation kinetics. From a practical point of view, the second objective was to develop a low temperature nitridation cycle without using nano-sized silicon or transition metal nitridation enhancing additives for future RBSN composite development with small diameter SiC fibers.

## EXPERIMENTAL PROCEDURE

Three lots of silicon powders were used. The powder lots were supplied by Union Carbide (Union Carbide, Linde Division, Tonawanda, NY), Kemanord-Sicomill grade-IV (Kemanord, Ljungaverk, Sweden), and Albemarle (Albemarle Corporation, Baton Rouge, LA). For brevity, these powders are henceforth referred to as Type-A, Type-B, and Type-C, respectively. All the as-received silicon powders were attrition-milled to reduce their particle size and, hence, to increase their surface area. A $Si_3N_4$ grinding medium was used for milling at room temperature for 2, 8, 32, and 48 hr, with Stoddard (kerosene-based fluid) as the grinding fluid. The weight ratio of silicon powder to grinding media was ≈40. The attrition milling was accomplished by the procedure detailed in reference 7. Afterwards, the excess grinding fluid was siphoned off from the grinding vessel. Then, the silicon slurry was poured into a rectangular pan and dried for 24 hr in a vacuum oven set at 600°C. The dried powder was transferred to a glass jar and stored in a glove box that was purged continuously with high purity nitrogen.

The impurities in, and the particle size range and specific surface area of the silicon powders were determined respectively, by wet chemistry, laser light scattering (Microtrac, Model 7991), and the three point Brunauer-Emmett-Teller (BET) adsorption (Micromeritics, Model ASAP 2010) techniques.

Silicon compacts, 12.7 mm in diameter and 2 to 3 mm thick, were prepared by uniaxially pressing attrition-milled silicon powder in a stainless steel die at 70 MPa. The compacts were ≈45% dense compared to the theoretical density of silicon (2.33 g/cc).

The silicon compacts were nitrided for 4 hr at 1250 or 1350°C in a thermogravimetric analysis unit (Model 429/409, Netzsch, Germany) equipped with a tungsten-element furnace. The nitriding atmospheres were flowing $N_2$(99.999%), or $N_2$+4%$H_2$ (99.99%), or $N_2$+5%$NH_3$(99.99%). These gases were percolated through several cartridges of gettering agents to reduce oxygen and water vapor content to less than 10 ppm. After gettering, these gases typically contained 3 to 4 ppm of water vapor and oxygen.

Following nitridation, the compacts were analyzed for impurities and phase composition by wet chemistry and X-ray diffraction (XRD), respectively. The XRD runs were made at a scanning speed of 1 deg/min using standard powder diffraction equipment with a Ni filter and Cu $K_\alpha$ radiation.

## RESULTS AND DISCUSSION

SEM examination indicates that Type A and B powders are flaky and faceted, and Type C powder is spherical. The particle size, surface area, and impurity analysis for the as-received powders are shown in Table I. According to this table, the as-received Type-A silicon powder contained significant amounts of Fe, Ni, Al, Cr, and oxygen impurities; whereas Type-B and Type-C silicon powders are relatively pure. All powders were attrition milled and then characterized similar to the as-received powders. The effect of attrition milling on particle size, surface area, and chemistry of Type-B silicon powder is shown in Table II as an example. In general, as the attrition-milling time increased, the average particle size decreased and the surface area increased, as expected, but the amount of Y, Al, and oxygen impurities also increased. The amount of Y and Al impurities varied with the

Table I. Particle size, surface area, and impurity analysis of the as-received silicon powders.

|  | Type A | Type B | Type C |
|---|---|---|---|
| Average particle size ($d_{50}$), µm | 8.9 | 23.12 | 0.88 |
| Specific surface area, $m^2/g$ | 4.9 | 1.3 | 8.17 |
| Impurities, wt% | | | |
| Carbon | 0.03 | 0.01 | 0.16 |
| Oxygen | 0.42 | 0.7 | 0.48 |
| Iron | 0.5 | 0.02 | 0.01 |
| Impurities, ppm | | | |
| Nickel | 460 | 0.006 | 6 |
| Aluminum | 1000 | 0.002 | 2 |
| Chromium | 730 | 0.02 | 20 |
| Yttrium | 100 | 0.002 | 2 |

Table II. Particle size, surface area, and impurity analysis of attrition-milled Type-B silicon powder.

| Milling time, hr | Average particle size, µm | Specific surface area, $m^2/g$ | Oxygen, wt% | Yttrium, wt% | Aluminum, ppm |
|---|---|---|---|---|---|
| 2 | 15.94 | 3.6 | 1.01 | 0.002 ppm | 30 |
| 8 | 1.44 | 9.6 | 1.45 | 0.002 ppm | 40 |
| 32 | 0.54 | 30 | 3.71 | 0.06 | 120 |
| 48 | 0.48 | 63 | 8.87 | 0.17 | 200 |

attrition-milling time and reached values of ≈700 and 200 ppm, respectively, for the 48-hr-attrition-milled Type-B powder. The source of these impurities was traced to the $Si_3N_4$ grinding medium which contained 6 wt% $Y_2O_3$ and 2 wt% $Al_2O_3$ as sintering additives.

**Effect of surface area on nitridation kinetics**

The nitridation kinetics were determined for the compacts prepared from the as-received silicon powder lots and the powder attrition-milled for 2, 8, 32, and 48 hr. The surface area of these powders, ranged from 1.3 to 70 m$^2$/g depending on the attrition milling time. Fig. 1 shows typical nitridation behavior of the as-received Type-B powder and the same powder attrition-milled for 8 or 48 hr, and then nitrided in N$_2$ at 1250°C for 4 hr. As the nitridation reaction progressed, the weight increased because large amounts of silicon were converted to Si$_3$N$_4$, but small amounts of silicon also evaporated as SiO(g) or Si(g), which resulted in weight loss. In the percentage nitridation plots shown in Fig. 1, the evaporation of these gaseous

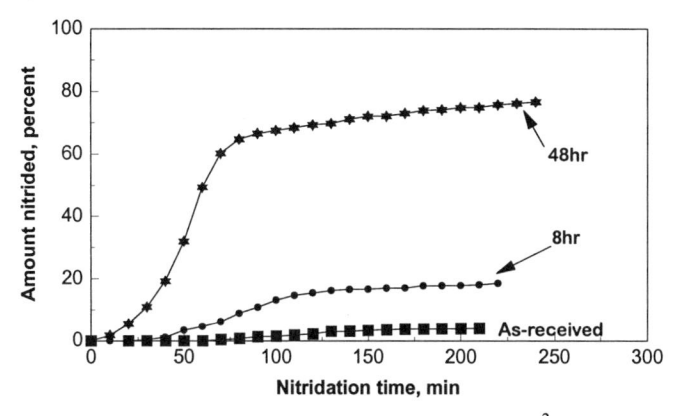

Fig. 1. Nitridation behavior of the as-received (SA≈1.3 m$^2$/g), and 8 hr (SA≈1.3 m$^2$/g) and 48 hr (SA≈1.3 m$^2$/g) attrition milled silicon powders in N$_2$ at 1250°C.

species during the nitridation reaction was not taken into account. It is clear from this figure that there are three nitridation regions: an induction period, a rapid nitriding period, and a nitridation saturation period. As the grinding time is increased - in other words, as the surface area of the silicon powder is increased - the slope of the second region and the percent silicon nitrided in 4 hr are increased. Also noticed in Fig. 1 is that the nitridation curves for the as-received and 8-hr-attrition-milled silicon powder compacts appear to reach a plateau after 120 to 150 min, but that for the 48-hr-attrition-milled silicon powder shows increased nitridation with time. If this trend continues, it appears that complete conversion of silicon to Si$_3$N$_4$ can be achieved in ≈ 8 hr.

Fig. 2 shows how percent nitridation varies with surface area for the three lots of attrition milled silicon compacts nitrided at 1250 or at 1350°C for 4 hr. Three main conclusions can be drawn from this figure: (1) as the surface area of the powder increases, the percentage of silicon converted to $Si_3N_4$ also increases; (2), at the higher nitridation temperature, the conversion rate is higher, especially for the Type-A silicon powder; (3) both the as-received and attrition milled Type-A silicon powders convert to silicon nitride to a greater extent at both nitridation temperatures than the other two lots of silicon powders. Extrapolation of the 1250°C nitridation curve with surface area suggests that the attrition-milled Type-B silicon powder with surface area >80 m²/gm can be completely nitrided in 4hr.

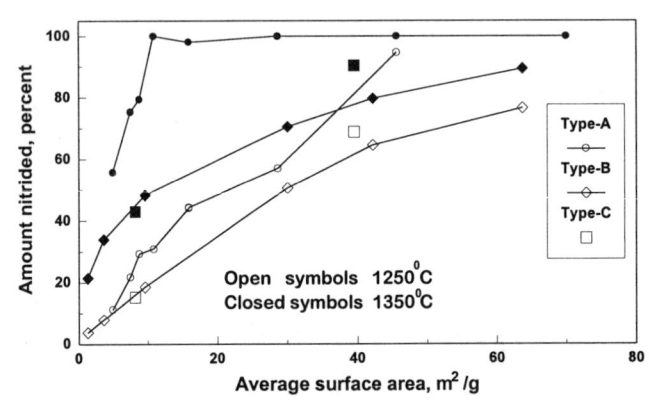

Fig. 2. Effect of surface area on amount of silicon in % converted to $Si_3N_4$ for the as-received and attrition milled Types A, B, and C silicon powder compacts nitrided at 1250°C and 1350°C in $N_2$ for 4hr.

The phase and oxygen analysis of the nitrided as-received and attrition-milled Type-B powder compacts are summarized in Tables III and IV; Nitridation of these compacts were performed at 1250°C and 1350°C in $N_2$ for 4hr. These tables demonstrate that as the surface area increases, the amounts of $\alpha$-$Si_3N_4$ and $\beta$-$Si_3N_4$ in the compacts increases continuously except in the case of the 63 m²/g silicon powder compact nitrided at 1350°C, but the ratio of $\alpha$-/$\beta$-$Si_3N_4$ (henceforth referred to as $\alpha/\beta$ ratio) decreases. At a given nitridation temperature, the ratio of oxygen after nitridation to that before nitridation (referred to in the tables as oxygen ratio) remains nearly the same with increasing surface area, but decreases with increasing temperature of nitridation. The compacts prepared from the highest surface area

(≈63 m²/g) silicon powder and nitrided at 1350°C for 4 hr contained an oxynitride phase, and lower amounts of α-Si₃N₄ than the compacts prepared from most of the

Table III. Surface area effects on phase composition and oxygen ratio of Type-B silicon compacts nitrided in N₂ at 1250°C for 4 hr.

| Surface area, m²/g | α–Si₃N₄, wt% | β-Si₃N₄, wt% | α/β ratio | Un-reacted silicon, wt% | Oxygen ratio@ | Wt. gain, % |
|---|---|---|---|---|---|---|
| 1.3 | 3.7 | 0 | - | 96.3 | 0.71 | 0.80 |
| 3.6 | 6.7 | 1 | 6.7 | 92.2 | 0.63 | 3.60 |
| 9.6 | 12.3 | 6.2 | 2.1 | 81.5 | 0.79 | 10.04 |
| 30 | 26.3 | 24.3 | 1.2 | 49.4 | 0.76 | 23.08 |
| 63 | 40.1 | 36.6 | 1.09 | 23.4 | 0.72 | 32.12 |

@ Ratio of oxygen after nitridation to oxygen before nitridation.

Table IV. Surface area effects on phase composition and oxygen ratio of Type-B silicon compacts nitrided in N₂ at 1350°C for 4 hr.

| Surface area, m²/g | α-Si₃N₄, wt% | β-Si₃N₄, wt% | α/β ratio | Un-reacted silicon, wt% | Si₂N₂O, wt% | Oxygen ratio@ | Wt. gain, % |
|---|---|---|---|---|---|---|---|
| 1.3 | 18.6 | 2.8 | 6.64 | 78.6 | 0 | 0.6 | 6.80 |
| 3.6 | 26.8 | 7.1 | 3.77 | 66.1 | 0 | 0.59 | 13.70 |
| 9.6 | 34.2 | 14 | 2.44 | 51.7 | 0 | 0.67 | 26.48 |
| 30 | 34 | 36.5 | 0.93 | 29.5 | 0 | 0.68 | 34.20 |
| 63 | 9.1 | 73.5 | 0.12 | 10.7 | 6.8 | 0.67 | 36.10 |

@ Ratio of oxygen after nitridation to oxygen before nitridation

lower surface area powders nitrided under similar conditions. Tables III and IV also show that the total weight gain after the 4 hr nitridation at each temperature also increased with increasing surface area. However, the maximum weight gain seen in the compacts nitrided to 80% or greater conversion is ≈36 wt%, which is much lower than the theoretical value of 66 wt%. The discrepancy in the weight gain is partially due to the loss of silicon as SiO (g) and Si (g) and partially due to the greater amounts of amorphous silica present in the finer powder, which did not nitride completely.

Increased amount of nitridation with surface area can be understood with current nitridation models[1]. It is generally accepted in RBSN literature that for initiation of the nitridation reaction, the silica layer on the surface of silicon must be disturbed. Sustaining the nitridation reaction depends on the rate of formation and growth of $Si_3N_4$ nuclei. Devitrification or formation of a low melting glassy phase can disrupt the silica layer. Impurities such as Na, Ca, Al, Fe, Cr, Ni and Y in the silicon powder are known to react with the silica layer to form a low melting glassy phase which promote rapid diffusion of nitrogen to the silicon substrate. Of these impurities, alkaline earth and transition metal impurities are known to be effective silicate formers [4,5]. On the other hand, for continuation of the nitridation reaction, surface area of the silicon particle is important because it controls rate of formation and growth of $Si_3N_4$ nuclei. Data shown in Tables I and II indicate that as-received Type-A silicon powder contains significant amounts of transition metal impurities. And all three powder lots show increasing amounts of Al and Y impurities with grinding time. Therefore, it appears that surface area and glassy phase formation are responsible for increased amount of nitridation in high surface area powders.

Influence of nitridation environment on the nitridation kinetics of a high surface area (≈63 $m^2/g$) attrition milled Type-B silicon powder was investigated at 1250° and 1350°C. The results shown in Fig. 3 indicate nitridation environment had no significant influence on the nitridation kinetics of this powder. In contrast, addition of small amounts (<5 vol%) of $H_2$ or of $NH_3$ to $N_2$ is proven to enhance nitridation kinetics of low surface area commercially available silicon powders [8-9]. Phase analysis and oxygen ratio of the compacts after 4 hr nitridation in $N_2$, $N_2$+4%$H_2$, and $N_2$+5%$NH_3$ at 1250°C or at 1350°C are summarized in Table V. This table shows that at 1250 or at 1350°C, $H_2$ and $NH_3$ additions to $N_2$ promote formation of $Si_2N_2O$ except for the compacts nitrided at 1350°C in $N_2$+4%$H_2$. At a fixed nitridation temperature, the nitridation environment appears to have had no effect on the $\alpha$-/$\beta$-$Si_3N_4$ ratio except for compacts nitrided for 4 hr at 1350°C in $N_2$+5%$NH_3$, but the ratio appears to decrease with increasing nitridation temperature. On the other hand, the oxygen ratio ranged from 0.66 to 0.93 and no discernible trend was noticed with nitridation environment or temperature.

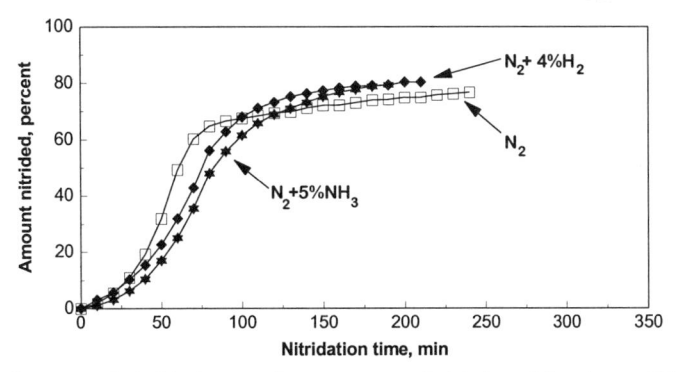

Fig. 3. Influence of nitridation environment on nitridation kinetics at 1250°C for 48hr attrition milled Type-B silicon powder compacts (SA ≈63 m²/g).

Table V. Effects of nitriding environment and temperature on phase compositions and oxygen ratio of nitrided 48-hr-attrition-milled Type-B silicon powder compacts (SA≈63 m²/g).

| Nitriding conditions | α-Si$_3$N$_4$, wt% | β-Si$_3$N$_4$, wt% | α/β ratio | Unreacted silicon, wt% | Si$_2$N$_2$O, wt% | Oxygen ratio@ |
|---|---|---|---|---|---|---|
| N$_2$, 1250°C, 4 hr | 15.4 | 48.5 | 0.32 | 33.1 | 3.4 | 0.77 |
| N$_2$+4%H$_2$, 1250°C, 4 hr | 15.5 | 58.0 | 0.27 | 19.7 | 6.8 | 0.83 |
| N$_2$+5%NH$_3$, 1250°C, 4 hr | 14.7 | 57.5 | 0.26 | 20.7 | 7.1 | 0.93 |
| N$_2$, 1350°C, 4 hr | 9.1 | 73.5 | 0.12 | 10.7 | 6.8 | 0.66 |
| N$_2$+4%H$_2$, 1350°C, 4 hr | 8.5 | 75.9 | 0.11 | 15.6 | 0 | 0.89 |
| N$_2$+5%NH$_3$, 1350°C, 4 hr | 3.1 | 77.3 | 0.04 | 7.6 | 11.9 | 0.87 |

@ Ratio of oxygen after nitridation to oxygen before nitridation

**Effect of preoxidation on nitridation kinetics**

To study the influence of oxidation on nitridation kinetics, the 48-hr-attrition-milled Type-B powder compacts (SA≈63 m$^2$/g) were first oxidized in air to 5 or 10 wt% gain to grow an additional layer of silica on the silicon. These compacts were then nitrided in N$_2$ or in a N$_2$+4%H$_2$ mixture at 1250 for 4 hr; see Fig. 4. For comparison purposes, the nitridation kinetics of the unoxidized silicon compacts are also included in the figure. Clearly, as the degree of preoxidation increases, the incubation period for the start of the nitridation reaction also increases and the nitridation rate decreases. The nitridation rate of the 5-percent oxidized compacts nitrided in N$_2$ was generally lower in the initial stages, but after 4 hr, the percentage converted to Si$_3$N$_4$ reached a level almost to that of the unoxidized powder. The 10-percent oxidized compacts, on the other hand, were only partially nitrided in N$_2$. In the N$_2$+4%H$_2$ environment, however both the 5- and 10-percent oxidized powder compacts were significantly nitrided, although the nitridation rate in the initial stages was slower.

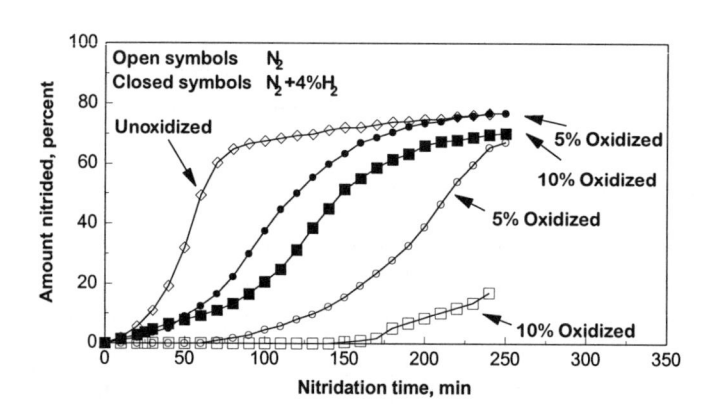

Fig. 4. Influence of nitridation environment on 48hr attrition milled and preoxidized silicon powder compacts (SA≈ 63 m$^2$/g) nitrided at 1250$^0$ C.

**SUMMARY OF RESULTS**

The surface area of commercially available silicon powders was increased by wet attrition milling to facilitate processing and conversion to Si$_3$N$_4$. The influence of surface area, the nitridation environment, and preoxidation on the nitridation kinetics of the silicon powder compacts were investigated. Key findings are as follows:

(1) With increasing attrition milling time, the surface area of silicon powder increases but impurities also accumulate in the powder due to wear of attrition mill components.
(2) At a given temperature, as the surface area of the silicon powder increased, so did the percentage nitridation after 4 hr, but the $\alpha/\beta$ ratio decreased. The silicon powders having surface area > 40 $m^2/g$ can be nitrided to 70-percent conversion or greater in 4 hr at $1250^0C$ or above.
(3) Variation in % nitridation of the as-received silicon powder lots is possibly due to different starting impurity contents and surface areas.
(4) The nitridation environment had no significant influence on the nitridation kinetics of high-surface-area silicon powder compacts.
(5) Preoxidation of high-surface-area silicon powder compacts retarded nitridation kinetics. However, the deleterious effects of oxidation could be overcome by nitriding the preoxidized compacts in an $N_2/H_2$ mixture.

## CONCLUSIONS

The high-surface-area silicon powders can be prepared by attrition milling commercially available silicon powders. However during attrition milling wear of the grinding medium invariably introduces impurities into the silicon powder. By controlling the particle size of, and limiting impurities in the attrition milled silicon powder, a low temperature processing cycle required for RBSN composite fabrication can be developed. However, additional studies are needed to determine stability of SiC fibers under these processing conditions.

## REFERENCES

[1] A.J. Moulson, "Review- Reaction-Bonded Silicon Nitride: Its Formation and Properties," *Journal of Material Science,* **14** 1017 (1979).

[2] J.W. Lucek, G.A. Rossetti, Jr., and S.D. Hartline, "Stability of Continuous Si-C(-O) Reinforcing Elements in Reaction Bonded Silicon Nitride Process environments," pp. 27 in *Metal Matrix, Carbon, and Ceramic Matrix Composites,* NASA CP-2406, Edited by J.D. Buckley (NASA Washington, D.C.,1985).

[3] B.W. Sheldon and J.S. Haggerty, "The Nitridation of High Purity, Laser-Synthesized Silicon Powder to Form Reaction Bonded Silicon Nitride," *Ceramic Engineering and Science Proceedings,* **9**[7-8] 1061(1988).

[4] W.R. Moser, D.S. Briere, R.C. Correria, and G.A. Rossetti, "Kinetics of Iron-Promoted Silicon Nitridation," *Journal of Materials Research*, **1** (6), 797-802 (1986).

[5] C.G. Cofer and J.A. Lewis, "Chromium Catalysed Silicon Nitridation," *Journal of Materials Science*, **29**, 5880-5886, (1994).

[6] R.T. Bhatt and D.R. Hull, "Effects of Fiber Coatings on Tensile Properties of Hi-Nicalon SiC/RBSN Tow Composites," NASA TM-113170 (1997)

[7] T.P. Herbell, T.K. Glasgow, and N.W. Orth, "Demonstration of a Silicon Nitride Mill for Production of Fine Si and $Si_3N_4$ Powders," *Bulletin of the American Ceramic Society*, **3**[9](1984)1176.

[8] M.N. Rahaman and A.J. Moulson, "The Removal of Surface Silica and its Effect on the Nitridation of High-Purity Silicon," *Journal of Materials Science*, **19**, 189-194 (1984).

[9] J.A. Mangels, "Effect of $H_2$-$N_2$ Nitriding Atmospheres on the Properties of reaction sintered $Si_3N_4$," *Journal of The American Ceramic Society*, **58** 354 (1975).

# SUPERCRITICAL DEBINDING OF CERAMICS

Thierry Chartier[1], Eric Delhomme[1], Jean-François Baumard[1], Philippe Marteau[2], Roland Tufeu[2]
[1] LMCTS-CNRS, ENSCI, 47 av. Albert Thomas, 87065 Limoges, France,
[2] LIMHP-CNRS, Institut Galilée, 93430 Villetaneuse, France

## ABSTRACT

The removal of organic additives from ceramic green parts remains one of the most critical stage in ceramic processing. An original method of extraction of hydrocarbon-type binders, based on the unique dissolving characteristics and transport properties of supercritical fluids, is presented. The binder extraction by supercritical fluid is controlled by two mechanisms, namely the solubilisation of the molecules and the diffusion of dissolved species within the porosity of the green sample. Solubility of the binder is experimentally determined and a model is developed that takes account of its composition. A second model allows the estimation of the kinetics of binder extraction by supercritical $CO_2$. This original method offers the great advantage of removing binders from ceramic green pieces without melting and without degradation of the organic compounds and then leads, in a shorter time, to better mechanical properties of ceramic pieces compared to the classical thermal pyrolysis.

## INTRODUCTION

Most high technology ceramic processing like dry-pressing, tape-casting, injection or extrusion-moulding requires the use of organic compounds as dispersants, binders, plasticizers and so on to confer such properties as cohesion, flexibility and workability in the green state. Amounts as large as 50 vol.% of organics are sometimes added to the ceramic powder during the forming step, and obviously have to be removed prior to sintering. Although the most widely used

---

To the extent authorized under the laws of the United States of America, all copyright interests in this publication are the property of The American Ceramic Society. Any duplication, reproduction, or republication of this publication or any part thereof, without the express written consent of The American Ceramic Society or fee paid to the Copyright Clearance Center, is prohibited.

method to remove the additives, thermal debinding only received attention in the last few years.[1-7] An effort was made to understand the many physico-chemical processes that occur during binder removal. Binder removal by thermal treatment involves (i) chemical mechanisms with the thermal degradation of organic compounds into volatile species and, (ii) physical mechanisms like the diffusion of these species to the surface as well as the changes in the binder distribution within the green body. Changes in the binder distribution, governed by diffusion and capillary migration, and the magnitude of the mass- and heat-transfer limitations are critical parameters for the design of efficient heating cycle to prevent stresses and the formation of defects in ceramic parts.[5] Thus, thermal debinding remains one of the most critical steps of ceramic processing, especially in the case of large size parts. Defects are generated which affect properties of the sintered pieces. It is very time-consuming and residues of pyrolytic degradation can be detrimental to the subsequent sintering stage.

This is why alternative techniques are needed to produce defect-free green bodies and to reduce debinding time. We have used an original method of extraction of organic compounds based on the unique dissolving characteristics and transport properties of supercritical fluids.[8-10] The binder extraction by supercritical fluid is controlled by the solubilisation of the molecules and the diffusion of dissolved species within the porosity of the sample. In a first part, the solubility of paraffins in supercritical $CO_2$, using an infrared spectroscopy method, is presented. In a second part, results concerning the kinetic aspects of binder removal are given. Finally, mechanical properties of sintered ceramics, previously debound by classical pyrolysis and by supercritical extraction, are compared.

EXPERIMENTAL
Samples
We have chosen an organic formulation, based on paraffin binders, both suitable for the ceramic forming technique and for $CO_2$ extraction. The solubility of two paraffins was studied, one paraffin melting at 42 °C (paraffin 42) and one melting at 52 °C (paraffin 52). The samples investigated are low pressure (0.6 MPa) injection-moulded bars ($20\times25\times40$ mm$^3$) made from a 0.6 μm mean particle size alumina powder (P172SB, Péchiney, France) and the paraffin 52.

Removal of paraffin
The extraction, using $CO_2$ in supercritical state (Tc=31°C, Pc=7.37 MPa), was operated from 12 to 30 MPa, at temperatures ranging from 35 to 120°C. The schematic of the equipment used is shown in Fig. 1. Green samples to be debound are introduced in the extractor. Carbon dioxide, taken from a storage tank, is cooled down to 0°C by means of a heat exchanger. The resultant liquid carbon

dioxide is pressurized up to the working pressure. Then, the pressurized liquid is heated up to the working temperature with a second heat exchanger, leading to supercritical carbon dioxide. The carbon dioxide, in the supercritical state, is flowing through the extractor, dissolving binders from the green sample. The dissolved binders are recovered in a separator cylinder and the carbon dioxide is allowed to flow again through the extraction vessel. The process temperature of the extractor and the temperatures of the two heat exchangers are regulated by thermostated baths. The $CO_2$ flow rate used during the experiments was set to 2.5 $l.h^{-1}$, which avoids any problem of confinement.

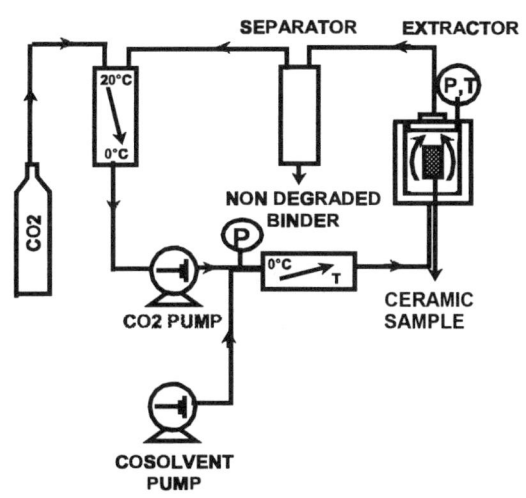

**Fig. 1.** Schematic diagram of the extraction device

In order to compare mechanical properties, paraffin removal was also performed by pyrolysis at 0.5 $°C.mn^{-1}$ up to 550°C, without dwell.

Solubility measurements

$CO_2$-paraffin mixtures and $CO_2$-alcane mixtures are compressed and heated in a high pressure cell fitted with two sapphire windows.[11] The temperature is stabilised within ± 0.2 °C. The pressure, adjusted by varying the internal volume of the cell with a piston, is measured with an accuracy of ± 0.05 MPa using a strain gauge pressure transducer calibrated with respect to temperature. The infrared light beam goes through the sapphire windows, the distance of which can be adjusted to any desired value from a few tenths of mm up to 16 mm in order to make the absorption always measurable. For the present measurements, the distance between the windows is set to 16 ± 0.05 mm. The heated cell can be rotated around the optical axis in order to observe either the gas phase or the liquid phase and also to mix the solute-solvent system to reach thermodynamic equilibrium. The infrared spectra are recorded with a spectrometer (Bohmem MB155) fitted with an InAs detector.

The molar densities d of each species in the $CO_2$ rich phase are easily deduced from the absorption spectra as the absorption bands of the alcanes are well separated from those of $CO_2$. All the n-alcanes and paraffin waxes have rigorously the same absorption spectra. Absorption intensity measurements are performed on the absorption bands located at 5000 cm$^{-1}$ for the carbon dioxide and at 4174 cm$^{-1}$ for the organic species. The molar densities, and subsequently the mass densities, are derived from the absorption intensities without any further mathematical treatment through the Beer-Lambert law :

$$d = \frac{1}{\alpha(v)l_a} \times \ln\frac{I_0(v)}{I(v)} \qquad (1)$$

where $l_a$ is the absorption pathlength, $\alpha(v)$ the absorption coefficient at the frequency $v$, $I_0(v)$ and $I(v)$ the transmitted intensities through the empty cell and through the sample respectively.

Characterisation

The compositions of the paraffins 42 and 52 were determined by gas phase chromatography.

The experimental extraction rate was estimated from the weight change of the sample during the supercritical treatment.

The room temperature strength was measured by the four-point bend test on both samples submitted to supercritical treatment or debound by pyrolysis, and sintered with the same heating cycle, i.e. 5°C.mn$^{-1}$ up to 1600°C with a dwell time of 3 hours. Weibull statistics were used to provide a means of comparison of the strength distribution and the flaw size distribution between the two series of samples.

RESULTS AND DISCUSSION

Solubility of paraffin in supercritical $CO_2$

Paraffins are mixtures of n-alcanes. In order to model the extraction of the paraffin, the knowledge of the total solubility of the paraffin in supercritical $CO_2$ is not sufficient. Indeed, it is necessary to know the composition of the paraffin in the $CO_2$ rich phase, which depends on the composition of the initial paraffin, and then to determine the partition coefficient $K_n$ of each n-alcane between the $CO_2$ rich phase and the paraffin rich phase. $K_n$ is defined as the ratio of the mass fraction ($W_n^s$) of the alcane n in the $CO_2$ rich phase to the mass fraction ($W_n^p$) of the same alcane n in the paraffin rich phase :

$$K_n = \frac{W_n^s}{W_n^p} \qquad (2)$$

In a first approximation, the paraffin rich phase is considered without solvent, then $W_n^p = 1$ and $K_n$ is equal to the mass fraction of the alcane n in the $CO_2$ rich phase in an equilibrium taking only into account the binary system alcane-$CO_2$.

Paraffins 42 and 52 are mixtures of n-alcanes with n ranging from 17 to 30 and from 20 to 34, respectively. The solubilities of five n-alcanes ($C_{19}$, $C_{21}$, $C_{24}$, $C_{26}$ and $C_{32}$) were measured at 70 °C under a pressure varying from 12 to 30 MPa. The $K_n$ values, experimentally determined, were expressed by the following equation :

$$\ln(K_n) = -\alpha n + \beta \qquad (3)$$

where $\alpha$ and $\beta$ are functions of the supercritical fluid density then, of the pressure and of the temperature (Fig. 2).
Paraffin being mixtures of linear hydrocarbons, a model was developed by analogy with the modelling of the solubility of a polydisperse polymer in a solvent. This model relates, at given pressure and temperature, the solubility of the n-alcanes to the solubility of the paraffin (W) in the supercritical $CO_2$ :

$$W = \sum_n K_n X_n \qquad (4)$$

where $X_n$ is the concentration of the alcane n in the paraffin determined by the gas phase chromatography.
This model of solubility is in good agreement with experimental values (Fig. 3).

Polymer diffusion through a porous medium containing pressurised $CO_2$

Once solubilised, dissolved molecules diffuse through the porous medium. As already shown,[10,12] diffusion is the limiting step of soluble paraffin extraction and a non-planar diffusing front is created in the sample during extraction. In the case of a parallelepipedic sample (L,l,h) with an initial organic concentration $\phi_0$, and under the assumption that the diffusivity D of the dissolved species remains constant, under constant temperature and pressure conditions, the problem is similar to that of heat conduction.[13] This diffusivity depends on the diffusion of dissolved species and on the microstructure of the green piece (porosity, tortuosity).

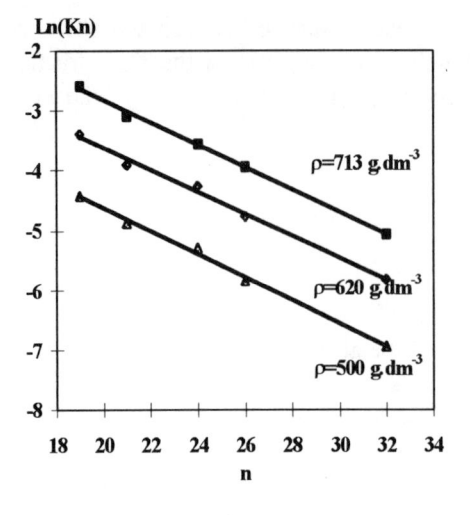

**Fig. 2.** Solubility of n-alcanes in supercritical $CO_2$ for three densities of $CO_2$.

**Fig. 3.** Experimental and theoretical solubility of paraffins 42 and 52 in supercritical $CO_2$ (70°C) under various pressures.

The concentration of polymer after a time t of extraction at coordinates (x,y,z) (reference at the centre of the sample) can then be expressed by:

$$\frac{\phi(x,y,z,t)}{\phi_0} = \frac{64}{\pi^3} \sum_{i=0}^{\infty} \sum_{j=0}^{\infty} \sum_{k=0}^{\infty} \frac{(-1)^{i+j+k}}{(2i+1)(2j+1)(2k+1)} \cos\frac{(2i+1)\pi x}{L} \cos\frac{(2j+1)\pi y}{l} \cos\frac{(2k+1)\pi z}{h} \exp(-\alpha_{i,j,k} t)$$

with $\alpha_{i,j,k} = D\pi^2 \left( \frac{(2i+1)^2}{L^2} + \frac{(2j+1)^2}{l^2} + \frac{(2k+1)^2}{h^2} \right)$ \hfill (5)

Integration of $\phi(t) = \dfrac{\int \phi(x,y,z,t)dxdydz}{\int dxdydz}$ \hfill (6) gives the mean concentration $\phi(t)$ in organic at a given time t, hence the extraction rate $\beta(t)$ ($\beta(t) = 1-\phi(t)/\phi_0$) :

$$\beta(t) = 1 - \frac{512}{\pi^6} \sum_{i=0}^{\infty} \sum_{j=0}^{\infty} \sum_{k=0}^{\infty} \frac{\exp(-\alpha_{i,j,k} t)}{((2i+1)(2j+1)(2k+1))^2} \qquad (7)$$

Samples prepared with paraffin 52 were treated under various experimental conditions. Theoretical curves (Fig. 4), plotted using mean calculated values of diffusivity (Eq. (7)) are in good agreement with experimental data, indicating that paraffin extraction is controlled by diffusion of dissolved species. The model was also used to plot the profiles of concentration of paraffin $\phi(x,y,z,t)$ from the centre to the surface of the sample as a function of treatment time (Fig. 5). The porosity becomes entirely opened and interconnected after extraction of 40% of the paraffin. Remaining paraffin can then be removed during sintering at a rather high heating rate without causing stresses.

Using mean calculated values of diffusivity, the diffusion model agrees satisfactorily with experimental results. However, the assumption that the diffusivity of solubilised species remains constant during extraction under constant pressure and temperature conditions does not represent the real extraction process.

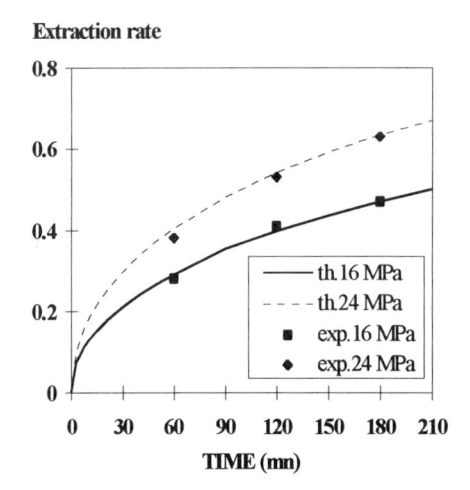

**Fig. 4.** Kinetics of extraction of paraffin 52 (50°C).

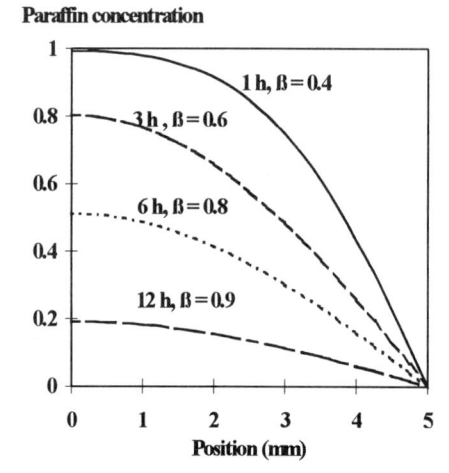

**Fig. 5.** Profiles of paraffin 52 concentration during supercritical treatment (50°C-24 MPa).

First, paraffin waxes are mixtures of n-alcanes that will not diffuse with the same velocity through the sample. Second, the composition of the paraffin varies during extraction, the lighter n-alcanes being preferentially removed in the early stage of debinding. Third, the pores of the sample, initially filled with the binder phase, become filled with the supercritical phase in which solubilised molecules diffuse more rapidly during the experiment. Thus, the global diffusivity coefficient varies during the experiment.

Solubility measurements discussed in the previous section allowed us to calculate the variations of this factor during the debinding process. With a similar approach, further experiments have to be performed to model the extraction kinetics.

Mechanical properties of sintered pieces

Weibull diagrams of sintered samples debound by pyrolysis and by supercritical treatment are given in Fig. 6. These samples subjected to thermal and supercritical treatments were prepared using the same forming technique. Hence, differences in strength-controlling defects can only be associated with the removal of organic compounds. The mean failure strengths, in the four-point bend test, were 219, 253 and 277 MPa for samples pyrolysed at rates of 1 and $0.5°C.min^{-1}$, and for samples submitted to supercritical treatment, respectively. The Weibull modulus was low, i.e. 5.4, for samples thermally debound at a rate of $1°C.min^{-1}$, whereas samples thermally debound at a rate of $0.5°C.min^{-1}$ and those supercritically treated exhibited a larger and similar value of the Weibull modulus, i.e. 12.4 and 13.7, respectively.

Supercritical debinding leads to a more homogeneous green state with smaller critical defects and lower scatter of defect sizes and/or shapes than classical thermal debinding. As far as the treatment duration is concerned, better mechanical properties were achieved after 3 h of supercritical treatment than after about 18 h of thermal treatment on identical samples. The time parameter then becomes critical for large ceramic parts.

Compared to thermal debinding which is governed by the capillary migration of the molten polymer and the diffusion of degradation products, supercritical debinding offers the great advantage to remove organics at a rather low temperature at which no redistribution of the binder in the green body takes place and at which no degradation of organic compounds occurs.

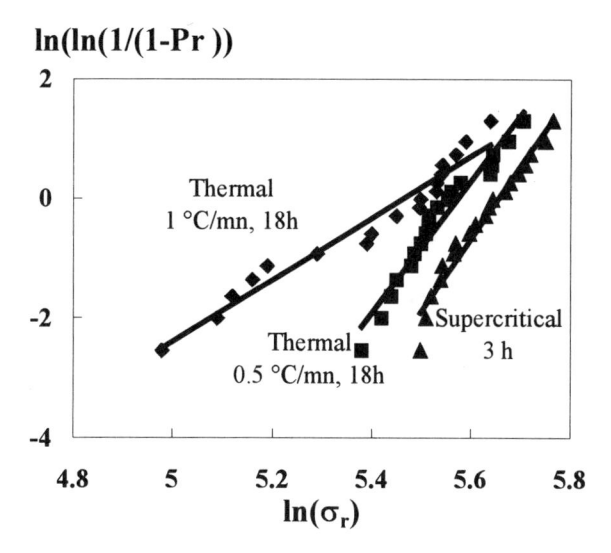

**Fig. 6.** Weibull diagrams of sintered alumina samples.

CONCLUSION

Organics removal from ceramic green parts is a critical stage in ceramic processing. An original method based on unique properties of supercritical fluids was developed. This low temperature technique (no melting and no degradation of organic binders) offers great advantages in comparison to the classical technique of pyrolysis :
- absence of deformation and of stresses in the green part,
- absence of residues of pyrolytic degradation,
- rapidity.

Two mechanisms control the extraction, namely the solubilisation of the binder and the diffusion of dissolved species. A first model allows the prediction of the solubility, in supercritical $CO_2$, of a paraffin used as ceramic forming additive. A second model allows the estimation of the diffusivity of dissolved species through the porous ceramic. These two models are in good agreement with experimental results.

# REFERENCES

[1] R.M. German, "Theory of Thermal Debinding," *Int. J. Powder Metall.*, **23** [4] 237-45 (1987).

[2] P.D. Calvert and M.J. Cima, "Theoretical Models for Binder Burnout," *J. Am. Ceram. Soc.*, **73** [3] 575-79 (1990).

[3] J.R.G. Evans, M.J. Edirisinghe, J.K. Wright and J. Crank, "On the Removal of Organic Vehicle from Moulded Ceramic Bodies," *Proc. Roy. Soc. (London)* **A432** 321-40 (1991).

[4] I.E. Pinwill, M.J. Edirisinghe and M.J. Bevis, "Development of Temperature-Heating Rate Diagrams for the Pyrolytic Removal of Binder Used for Powder Injection Moulding," *J. Mat. Sc.*, **27** 4381-88 (1992).

[5] M.J. Cima, J.A. Lewis and A.D. Devoe, "Binder Distribution in Ceramic Greenware During Thermolysis," *J. Am. Ceram. Soc.*, **72** [7] 1192-99 (1989).

[6] J.K. Wright and J.R.G. Evans, "Removal of Organic Vehicle from Moulded Ceramic Bodies by Capillary Action," *Ceram. Int.*, **17** 79-87 (1991).

[7] M.R. Barone and J.C. Ulicny, "Liquid-Phase Transport During Removal of Organic Binders in Injection-Molded Ceramics," *J. Am. Ceram. Soc.*, **73** [11] 3323-33 (1990).

[8] D.W. Matson and R.D. Smith, "Supercritical Fluid Technologies for Ceramic-Processing Applications," *J. Am. Ceram. Soc.*, **72** [4] 871-81 (1989).

[9] N. Nakashima, E. Nishikawa and N. Wakao, "Binder Removal from a Ceramic Green Body in the Environment of Supercritical Carbon Dioxide with/without Entrainers";pp.357-59 in *Proceedings of the 2nd International Symposium on Supercritical Fluids*, Edited by M.McHugh. Butterworth Publishers, Boston, 1991.

[10] T. Chartier, M. Ferrato and J.F. Baumard, "Supercritical Debinding of Injection Molded Ceramics," *J. Am. Ceram. Soc*, **78** [7] 1787-92 (1995).

[11] Ph. Marteau, J. Obriot and R. Tufeu, "Experimental Determination of Vapor-Liquid Equilibria of $CO_2$+Limonene and $CO_2$+Citral Mixtures," *J. of Supercritical Fluids*, **8** 20-24 (1995).

[12] T. Chartier, E Delhomme and J.F. Baumard, "Mechanisms of Binder Removal Involved in Supercritical Debinding of Injection Moulded Ceramics," *J. Phys. III*, 7 291-302 (1997).

[13] J. Crank, "The mathematics of diffusion";p. 49, 2nd ed., Clarendon Press, Oxford, 1975.

# KEYWORD AND AUTHOR INDEX

β-SiC coating layer, 319

Abnormal grain growth, 83
Additives, 419
Adhesion, 329
Agarwal, G., 483
Ahlén, N., 473
Aizawa, T., 273
Akinc, M., 347
Alcohol, 427
Aldinger, F., 307
Alloys, Mo-Si, 347
Alumina, 97
  sol-gel composites, 185
Aluminum nitride, 153
Aluminum tri-sec-butoxide, 163
Amarakoon, V.R.W., 419
Apblett, A.W., 205

Balandina, N., 2
Balmori-Ramírez, H., 83, 91
Bao, X., 241
Barbieri, L., 295
Barris, G.C., 107
Baumard, J.-F., 561
Bend strength, 523
Berroth, K., 465
Bhatt, R.T., 549
Bill, J., 307
Blackglas, 361
Boehmite, 83
Bondioli, F., 295
Borchert, R., 57
Boron nitride, 443
Bourke, M.A.M., 537
Brown, I.W.M., 107
Burning out, 427

Cake filtration, 337
Canañas-Moreno, J.G., 91
Cao, L., 37
Carbon dioxide separation, 337
Carbothermal reduction, 107, 119
Carbothermal synthesis, 119
Carlsson, M., 473

Case, E.D., 69, 509
Casket geometry, 69
Cationic lattice disorder, 537
Chartier, T., 561
Chemical vapor reaction, 319
Cho, Y.S., 419
Choa, Y.H., 443
Choi, S.-C., 319
Chwa, S.O., 329
Citak, R., 141
Clausen, B., 537
Clay, 107
Coatings, 227, 361
Colloidal processing, 227
Combustion synthesis, 2, 13, 23
Composites
  $MoSi_2$-SiCp/SiCw, 273
  oxide/metal, 141
  oxide/oxide, 215
  polymer-derived, 263
  $SiC$-$Si_3N_4$, 241
  Y-TZP, 97
Contreras, M.E., 83
Copper nitride, 329
Coprecipitation, 427
Crystal structure, 387
Crystal wafers, 483

De Guire, M.R., 307
De la Torre, S.D., 83, 91, 97, 287
Debinding, 561
Delhomme, E., 561
Densification, 45
Diamond thin films, 509
Diao, Y., 37
Diffusion, 561
Dire, S., 251
Displacement reaction, 141

Edirisinghe, M.J., 241
Ekström, T.C., 107
Eucryptite, 195

Fernando, G.F., 241
Ferrites, 407

Fiber coatings, 215
Fiber composites, 523
Fibers, 361
Finite element modeling, 537
Folkes, M.J., 241
Fracture toughness, 347
Freer, R., 45
Functionally gradient materials, 57

Gao, L., 97
Ghosh, N.N., 195
Giesche, H., 407
Grain growth, 419
Greil, P., 263
Grigoryan, S.L., 455
Guo, W., 523
Gut, B., 465

Halloysite, 107
Hanan, J.C., 537
Hard materials, 263
Hartmann, T., 537
Hellmann, J.R., 361
Heuer, A.H., 307
High-permeability material, 373
High-temperature ceramics, 2
Hill, D.N., 483
Hong, J.S., 97
Hydrazine, 163
Hydrothermal powders, 373
Hydrothermal process, 387, 397

Impurities, 549
Induction heating, 483
Inorganic pigments, 295
Intermetallics, 347
Ishihara, K.N., 287

Jeng, C.-L., 427
Johnsson, M., 473
Joining, 509

Kaindl, A., 263
Kakitsuji, A., 83
Kang, K.-T., 263
Kao, C.-F., 427
Kawai, C., 337
Kaza, S., 119
Kellett, B.J., 523

Kim, D.-J., 263
Kim, J.Y., 163
Kim, K.H., 329
Koc, R., 119
Koizumi, M., 2
Komeya, K., 153
Kramer, M.J., 347
Kumar, P., 129, 141
Kumta, P.N., 163
Kusunose, T., 443

Lai, Y.C., 387
Laminated object manufacturing, 523
Landin, S.M., 495
Lanthanumhexaluminates, 215
Lasers, 523
Lee, J.G., 509
Lee, J.W., 419
Lee, K.-Y., 69, 509
Lin, S.-C., 23
Liquid-metal carboxylates, 205
Lo, S.-Y., 397
Lu, C.-H., 387, 397
Lucke, R., 373
Luss, D., 23

Machinability, 443
Magnesium aluminate, 129
Magnesium infiltration, 129
Malkhasyan, R.T., 455
Manfredini, T., 295
Marteua, P., 561
Mayer, L., 215
McCauley, R.A., 227
Mechanical alloying, 273, 295
Mechanical properties, 141, 347
Mechanosynthesis, 295
Meguro, T., 153
Melting, 427
Metal-bearing precursors, 129
Metal-ceramics, 57
Metal oxidation, 129
Miao, S., 13
Microemulsions, 407
Microstructure, 153, 273, 319, 373, 387, 495
Microwave heating, 69
Microwave plasma heating, 37
Microwave sintering, 37, 45, 57

Mixing, 227
Miyamoto, H., 83, 91, 97, 287
Miyamoto, K., 83, 97
Morphology, 13, 387, 397, 407
Mullite, 91, 195, 215
Munir, Z.A., 2

Nakahata, S., 337
Nanoamorphous metals, 455
Nanocomposites, 443
Nause, J.E., 483
Near-net shape, 141
Nersesyan, M., 23
Neutron diffraction, 537
Nicalon, 361
Nickel-based alloys, 57
Niesen, T., 307
Niihara, K., 443
Nishioka, T., 337
Nitridation, 107, 119, 549
Nitrogen selectivity, 337
Nonoxide processing, 153
Nonoxides, 163
Nygren, M., 473

Ohyanagi, M., 2
Okumus, S.C., 347
Oliver, M., 251
Orlando, A.C., 227
Oxidation, 549
Oxide powders, 295
Oxides, Li-Ti, 427

Palczer, A.R., 549
Pantano, C.G., 361
Park, H.C., 329
Park, J.-J., 337
Particle size, 13
Particulate coatings, 419
Petervary, M.P., 361
Photoluminescence, 483
Piezoelectric devices, 483
Polycrystalline fibers, 465
Polymer architecture, 251
Polymer matrix composites, 523
Polymeric precursors, 241
Polysiloxanes, 251, 263
Pore size, 495

Porosity, 495
Porous metals, 495
Powder processing, 465
Powders
$\alpha$-$Si_3N_4$, 119
Cd-In-Ga-O, 23
$KNbO_3$, 397
ZnO, 387
Power transformer material, 373
Pramanik, P., 195
Preceramic polymers, 241, 251
Pressureless infiltration, 141
Pressureless sintering, 57
Priest, J.M., 361
Puszynski, J.A., 13
Pyrolysis, 251

Quantum-chemical technology, 455

Rathbone, S.M., 45
Reaction sintering, 91
Reactive fillers, 263
Readey, D.W., 495
Reduction of porous oxides, 495
Reduction reactions of $NiAl_2O_4$, 537
Residual stress, 537
Resistivity, 427
Rios-Jara, D., 287
Rocha-Rangel, E., 91
Rogers, K.A., 141
Romagnoli, M., 295

Sale, F.R., 45
Sampathkumaran, U., 307
Sandhage, K.H., 129, 141
Saphikon, 361
Sapphire, 307
Saruhan, B., 215
Schlegel, E., 373
Schneider, H., 215
Seeding, 83
Sekino, T., 443
Self-assembled monolayers, 307
Self-propagating high-temperature synthesis, 23
Shen, X., 37
Shin, D.W., 329
Shingu, P.H., 287
Shirai, K., 2

Shock synthesis, 273
SiAlON, 107
Silica, 195
 -based glass membranes, 337
Silicate matrix composites, 523
Silicon
 -germanium, 307
 high-surface-area, 549
Silicon carbide, 13, 307
Silicon nitride, 153, 227, 337, 443
Silicon oxycarbide, 361
 glasses, 251
Sintering aids, 227
Soft magnetic ferrites, 373
Sol clusters, 185
Sol-gel, 419
 -derived polymers, 251
 processing, 163, 195
Solid solutions, 2, 295, 473
 AlN-SiC, 2
Solidification, 483
Solubility, 561
Sorarù, G.D., 251
Spark plasma sintering, 83, 91, 97
Specific surface area, 13
Spinels, 129
Spodumene, 195
Sputtering, 329
Sriram, M.A., 163
Step sintering, 37
Strength, 347, 361
Strienitz, R., 373
Strontium ferrite, 419
Substrates, 23
 graphite, 319
 silicon oxide, 329
Supercritical fluids, 561
Synthesis, 153, 163, 205, 241, 387, 407, 455, 473

Takeuchi, H., 337
Thadhani, N.N., 273
Thermal barrier coatings, 57
Thin films, 307
 GaN, 23
Thio-sol-gel, 163
Toughening, 97
Transition alumina, 83
Transition metal carbides, 473

Traub, M.L., 69
Troczynski, T., 185
Tsuchiya, K., 91
Tufeu, R., 561
Tungsten carbide, 263

Umemoto, M., 91
Unal, O., 347
Üstündag, E., 537

Vapor phase sintering, 495
VLS growth, 473

Walker, E.H., Jr., 205
Wegmann, M., 465
Whiskers, 473
White, G.V., 107
Wilkins, R., 23
Willert-Porada, 57
Woodman, R.H., 537
Wroe, F.C.R., 45

X-ray diffraction, 295, 483

Yamagiwa, M., 337
Yamakawa, A., 337
Yang, Q., 185
Yang, Y., 37
Yeh, C.-H., 387
Yener, D.O., 407
Yun, Y.-H., 319

Zárate-M., J., 83
Zhang, J., 37
Zinc oxide crystals, 483
Zirconia, 57, 91, 97
 films, 307